Lecture Notes in Physics

Bisher erschienen/Already published

Vol. 1: J. C. Erdmann. Wärmeleitung in Kristallen, theoretische Grundlagen und fortgeschrittene experimentelle Methoden. II, 283 Seiten. 1969.

Vol. 2: K. Hepp, Théorie de la renormalisation. III, 215 pages. 1969.

Vol. 3: A. Martin, Scattering Theory: Unitarity, Analyticity and Crossing. IV, 125 pages. 1969.

Vol. 4: G. Ludwig, Deutung des Begriffs „physikalische Theorie" und axiomatische Grundlegung der Hilbertraumstruktur der Quantenmechanik durch Hauptsätze des Messens. 1970. Vergriffen.

Vol. 5: Schaaf, The Reduction of the Product of Two Irreducible Unitary Representations of the Proper Orthochronous Quantummechanical Poincare Group. IV, 120 pages. 1970.

Vol. 6: Group Representations in Mathematics and Physics. Edited by V. Bargmann. V, 340 pages. 1970.

Vol. 7: R. Balescu, J. L. Lebowitz, I. Prigogine, P. Résibois, Z. W. Salsburg, Lectures in Statistical Physics. V, 181 pages. 1971.

Vol. 8: Proceedings of the Second International Conference on Numerical Methods in Fluid Dynamics. Edited by M. Holt. 1971. Out of print.

Vol. 9: D. W. Robinson, The Thermodynamic Pressure in Quantum Statistical Mechanics. V, 115 pages. 1971.

Vol. 10: J. M. Stewart, Non-Equilibrium-Relativistic Kinetic Theory. III, 113 pages. 1971.

Vol. 11: O. Steinmann, Pertubation Expansions in Axiomatic Field Theory. III, 126 pages. 1976.

Vol. 12: Statistical Models and Turbulence. Edited by C. Van Atta and M. Rosenblatt. Reprint of the First Edition. VIII, 492 pages. 1975.

Vol. 13: M. Ryan, Hamiltonian Cosmology. VII, 169 pages. 1972.

Vol. 14: Methods of Local and Global Differential Geometry in General Relativity. Edited by D. Farnsworth, J. Fink, J. Porter, and A. Thompson. V, 188 pages.

Vol. 15: M. Fierz, Vorlesungen zur Entwicklungsgeschichte der Mechanik. V, 97 Seiten. 1972.

Vol. 16: H.-O. Georgii, Phasenübergang 1. Art bei Gittergasmodellen. IX, 167 Seiten. 1972.

Vol. 17: Strong Interaction Physics. Edited by W. Rühl and A. Vancura. V, 405 pages. 1973.

Vol. 18: Proceedings of the Third International Conference on Numerical Methods in Fluid Mechanics, Vol. I. Edited by H. Cabannes and R. Temam. VII, 186 pages. 1973.

Vol. 19: Proceedings of the Third International Conference on Numerical Methods in Fluid Mechanics, Vol. II. Edited by H. Cabannes and R. Temam. VII, 275 pages. 1973.

Vol. 20: Statistical Mechanics and Mathematical Problems. Edited by A. Lenard. VIII, 247 pages. 1973.

Vol. 21: Optimization and Stability Problems in Continuum Mechanics. Edited by P. K. C. Wang. V, 94 pages. 1973.

Vol. 22: Proceedings of the Europhysics Study Conference on Intermediate Processes in Nuclear Reactions. Edited by N. Cindro, P. Kulišic and Th. Mayer-Kuckuk. XIV, 329 pages. 1973.

Vol. 23: Nuclear Structure Physics. Proceedings 1973. Edited by U. Smilansky, I. Talmi, and H. A. Weidenmüller. XII, 296 pages. 1973.

Vol. 24: R. F. Snipes, Statistical Mechanical Theory of the Electrolytic Transport of Nonelectrolytes. V, 210 pages. 1973.

Vol. 25: Constructive Quantum Field Theory. The 1973 "Ettore Majorana" International School of Mathematical Physics. Edited by G. Velo and A. Wightman. III, 331 pages. 1973.

Vol. 26: A. Hubert, Theorie der Domänenwände in geordneten Medien. XII, 377 Seiten. 1974.

Vol. 27: R. K. Zeytounian, Notes sur les Ecoulements Rotationnels de Fluides Parfaits. XIII, 407 pages. 1974.

Vol. 28: Lectures in Statistical Physics. Edited by W. C. Schieve and J. S. Turner. V, 342 pages. 1974.

Vol. 29: Foundations of Quantum Mechanics and Ordered Linear Spaces. Advanced Study Institute, Marburg 1973. Edited by A. Hartkämper and H. Neumann. VI, 355 pages. 1974.

Vol. 30: Polarization Nuclear Physics. Proceedings 1973. Edited by D. Fick. IX, 292 pages. 1974.

Vol. 31: Transport Phenomena. Sitges International Schools of Statistical Mechanics, June 1974. Edited by G. Kirczenow and J. Marro. XIV, 517 pages. 1974.

Vol. 32: Particles, Quantum Fields and Statistical Mechanics. Proceedings 1973. Edited by M. Alexanian and A. Zepeda. V, 132 pages. 1975.

Vol. 33: Classical and Quantum Mechanical Aspects of Heavy Ion Collisions. Proceedings 1974. Edited by H. L. Harney, P. Braun-Munzinger, and C. K. Gelbke. VII, 311 pages. 1975.

Vol. 34: One-Dimensional Conductors GPS Summer School Proceedings, 1974. Edited by H. G. Schuster. VII, 371 pages. 1975.

Vol. 35: Proceedings of the Fourth International Conference on Numerical Methods in Fluid Dynamics, 1974. Edited by R. D. Richtmyer. V, 457 pages. 1975.

Vol. 36: R. Gatignol, Théorie Cinétique des Gaz à Répartition Discrète de Vitesses. II, 219 pages. 1975.

Vol. 37: Trends in Elementary Particle Theory. Proceedings 1974. Edited by H. Rollnik and K. Dietz. V, 472 pages. 1975.

Vol. 38: Dynamical Systems, Theory and Applications. Proceedings 1974. Edited by J. Moser. VI, 624 pages. 1975.

Vol. 39: International Symposium on Mathematical Problems in Theoretical Physics. Proceedings 1975. Edited by H. Araki. XII, 562 pages. 1975.

Vol. 40: Effective Interactions and Operators in Nuclei. Proceedings 1975. Edited by B. R. Barrett. XII, 339 pages. 1975.

Vol. 41: Progress in Numerical Fluid Dynamics. Proceedings 1974. Edited by H. J. Wirz. V, 471 pages. 1975.

Vol. 42: H II Regions and Related Topics. Proceedings 1975. Edited by D. Downes and T. L. Wilson. XII, 488 pages. 1975.

Vol. 43: Laser Spectroscopy. Proceedings 1975. Edited by S. Haroche, J. C. Pebay-Peyroula, T. W. Hänsch, and S. E. Harris. X, 466 pages. 1975.

Lecture Notes in Physics

Edited by J. Ehlers, München, K. Hepp, Zürich
R. Kippenhahn, München, H. A. Weidenmüller, Heidelberg
and J. Zittartz, Köln
Managing Editor: W. Beiglböck, Heidelberg

82

Few Body Systems and Nuclear Forces I

8. International Conference
Held in Graz, August 24–30, 1978

Edited by
H. Zingl, M. Haftel and H. Zankel

Springer-Verlag
Berlin Heidelberg GmbH 1978

Editors
H. Zingl
M. Haftel
H. Zankel
Institut für Theoretische Physik
der Universität Graz
Universitätsplatz 5
A-8010 Graz

ISBN 978-3-540-08917-9 ISBN 978-3-540-35760-5 (eBook)
DOI 10.1007/978-3-540-35760-5

Originally published by Springer-Verlag Berlin Heidelberg New York in 1978

2153/3140-543210

EDITORIAL PREFACE

The series of "International Conferences on Few-Body Systems and Nuclear Forces" started 1959 in London and is now regularly scheduled every second year. The main topics covered by these conferences are reactions of two and few hadrons in the low and intermediate energy region. New experimental information is now becoming available from polarization experiments in meson factories and other laboratories. It provides an increased challenge to nuclear and particle theory. Also, relativistic generalizations and the problem of the four-body system are at present under intensive investigation. Beside these topics the off-shell behavior of the strong interaction, the electromagnetic form-factors of two- and few-nucleon systems and the N-body problem, just to name a few, continue to be of current interest.

This first volume contains the contributed papers to the conference, about 160 in number, which were selected by a subcommittee of the International Advisory Board. The invited review papers will be published in the second volume.

The organizers acknowledge financial support from the Austrian Government, the Styrian Government, the City of Graz, the International Union of Pure and Applied Physics, and the International Atomic Energy Agency, and also greatly appreciate the help received from the University of Graz and the Institut für Theoretische Physik.

The editors of these proceedings express their gratitude to Miss M. Krautilik and to Mrs. E. Walter for their assistance in typing.

Finally, thanks are due to the Springer-Verlag and in particular to Dr. W. Beiglböck and to Dr. H.A. Weidenmüller for facilitating the rapid publication of these proceedings.

Graz, June 20, 1978

Harald F.K. Zingl
Michael I. Haftel
Hubert Zankel

INTERNATIONAL ADVISORY COMMITTEE

I.R.Afnan,Bedford Park,Australia
E.O. Alt, Mainz, Germany
R.D. Amado, Philadelphia, USA
G. Bencze, Budapest, Hungary
W. Breunlich, Vienna, Austria
G.E.Brown,NORDITA and Stony Brook,USA
H.E. Conzett, Berkeley, USA
J.A. Edgington, London, England
H. Fiedeldey, Pretoria, Southafrica
M. Gmitro, Rez, Czechoslowakia
W. Grüebler, Zürich, Switzerland
A. Johansson, Uppsala, Sweden
Y. Kim, West Lafayette, USA
L. Kok, Groningen, Netherlands
V.V. Komarov, Moskow, USSR
B. Kühn, Dresden, GDR
M.P. Locher, SIN, Switzerland
I. Lovas, Budapest, Hungary
P. Macq, Louvain, Belgium
J.S.C. Mc Kee, Winnipeg, Canada
D.F. Measday, Vancouver, Canada
A.N.Mitra,Delhi,India
M.J. Moravcsik, Eugene, USA
H.P. Noyes, Stanford, USA
V. Valković,Zagreb,Yugoslavia
J.M. Pniewski, Warsaw, Poland
H.G. Pugh, College Park, USA
J.R.Richardson,Los Angeles,USA
L. Rosen, LAMPF, USA
W. Sandhas, Bonn, Germany
T. Sasakawa,Sendai, Japan
E. Schmid, Tübingen, Germany
P. Signell,East Lansing, USA
A.G. Sitenko, Kiew, USSR
I. Slaus, Zagreb, Yugoslavia
R.J.Slobodrian,Quebec, Canada
W. Thirring, Vienna, Austria
J.A. Tjon, Utrecht,Netherlands
R.Vinh Mau, Paris, France
R.van Wageningen,Amsterdam, Netherlands
B.Zeitnitz, Bochum, Germany

LOCAL ORGANIZING COMMITTEE

J. Fröhlich
H. Kriesche
C. B. Lang
L. Mathelitsch
H. Mitter
F. Pauß
W. Plessas
K. Schwarz
H. Zankel
H. Zingl

TABLE OF CONTENTS

Neutron-Proton Charge Exchange Scattering between 200 and 600 MeV +)

W. Hürster, Th. Fischer, G. Hammel, K. Kern, P. Kettle, M. Kleinschmidt, L. Lehmann, E. Rössle, H. Schmitt, D.M. Sheppard ++)

Fakultät für Physik der Universität Freiburg
D - 7800 Freiburg i.Br.

The elementary process $n+p \rightarrow p+n$ is not yet well understood. No satisfying theoretical approaches could be found up to now. All existing data show a pronounced increase of the differential cross section for small proton scattering angles, but there are strong discrepancies in the relative shape of the angular distributions [1-3].

We have measured differential cross sections for the elastic charge exchange scattering between 200 and 600 MeV by detecting the scattered protons in an angular range from $0^\circ \leq \theta_{Lab} \leq 18^\circ$ with the large acceptance magnet spectrometer at the nEl-neutron beam at SIN [4]. By time-of-flight techniques a primary energy resolution of better than 1.5 percent can be reached. More than 10^7 proton events have been recorded. The absolute normalization will be obtained by comparison with known cross sections for the inelastic channel $n+p \rightarrow d+\pi^0$, which has been measured as well [5].

Fig. 1 shows differential cross sections at 360 and 540 MeV, together with a fitted curve of the form

$$d\sigma/dt = \alpha_1 \exp(\beta_1 t) + \alpha_2 \exp(\beta_2 t) \qquad (1)$$

This parametrization fits well the data in the whole energy range. In Fig. 2 the energy dependence of the parameters is plotted against the neutron momentum in the lab system, where β means the logarithmic slope at $t=0$:

$$\beta = (\alpha_1\beta_1 + \alpha_2\beta_2)/(\alpha_1+\alpha_2) \qquad (2)$$

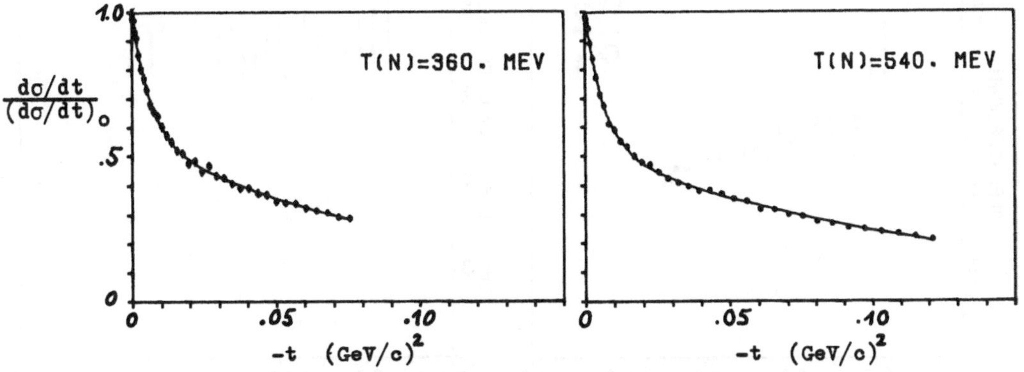

Fig. 1. Relative angular distributions at 360 MeV and 540 MeV. The curves are fits of eq. (1) to the data.

No structure can be seen in the energy dependence of β as was suggested by the Princeton-data.

+) Work supported by Bundesministerium für Forschung und Technologie
++) Present address: University of Alberta, Edmonton, Canada

[1] P. F. Shepard et al., Phys. Rev. D10 (1974), 2735
[2] G. Bizard et al., Nuclear Physics B85 (1975), 14
[3] M. L. Evans et al., Phys.Rev.Letters 36 (1976), 497; L. C. Northcliffe, private communication
[4] SIN - Jahresbericht 1977, E 25
[5] Contribution to this conference

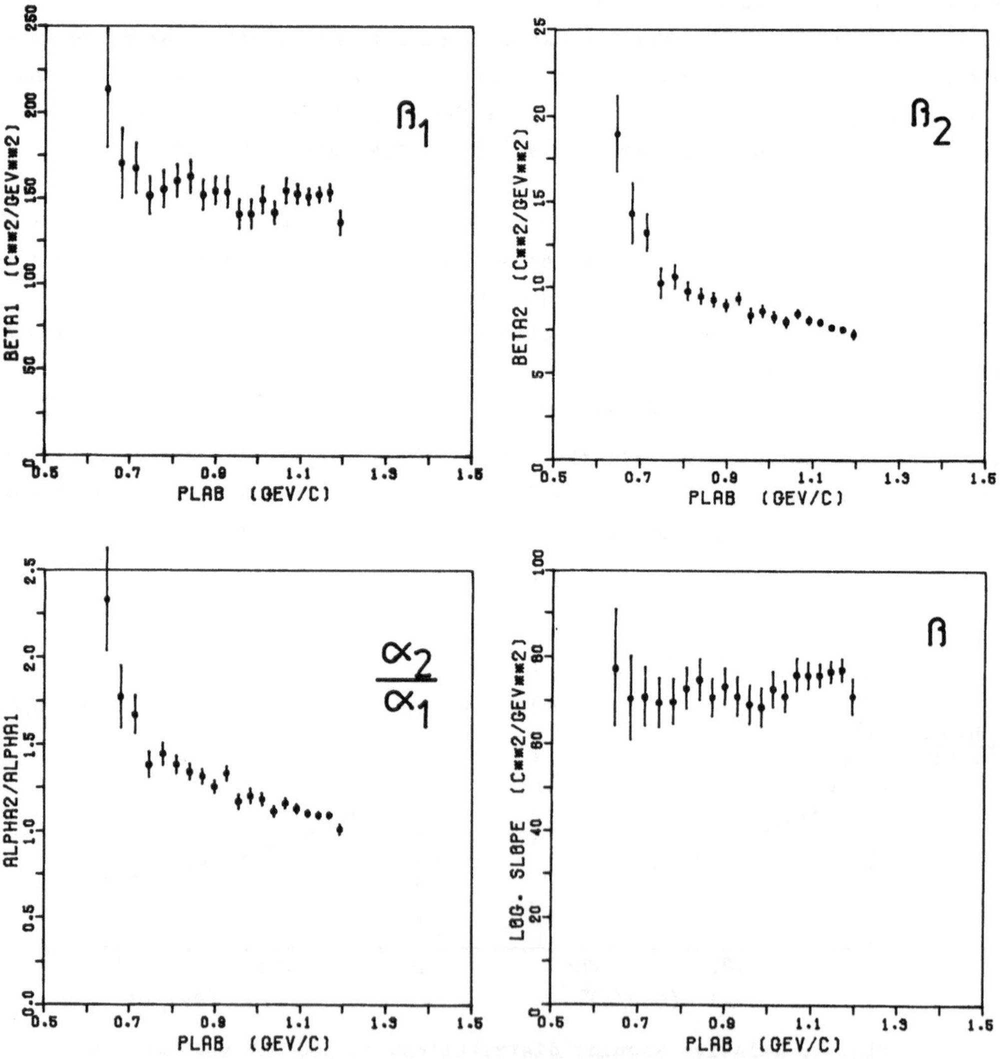

Fig.2. Parameters of eqs.(1) and (2) as functions of the incident momentum.

n-p CHARGE EXCHANGE SCATTERING FROM 150 TO 800 MeV*

B. E. Bonner and J. E. Simmons
Los Alamos Scientific Laboratory, University of California
Los Alamos, New Mexico 87545

Mahavir Jain and G. Glass
Texas A & M University, College Station, Texas 77843

C. L. Hollas, C. R. Newsom, and P. J. Riley
University of Texas, Austin, Texas 78712

The backward elastic np scattering cross section has been the subject of intense experimental and theoretical effort. It was noted[1] in the 1960's that the few extant experimental results had a similar shape, independent of incident energy, when plotted as a function of u, the square of the four-momentum transfer. This shape was fit by a sum of two exponentials: $d\sigma/du = \alpha_1 \exp(\beta_1 u) + \alpha_2 \exp(\beta_2 u)$, where the first term describes a sharp peak for small values of -u [$\lesssim 0.02(\mathrm{GeV/c})^2$] and the second a much more gradual decrease. Calculations using one pion exchange (OPE) predicted a dip instead of a peak for small values of -u. It was suggested[2] that the sharp peak could be caused by a destructive interference between the OPE amplitude and a slowly varying background term.

Two recent experiments[3,4] have provided results on the energy dependence of the np CEX cross section. Unfortunately, the data sets disagree for neutron momenta greater than about 1.2 GeV/c, but both are consistent with the broad peak in both slope and intercept reported[3] at around 900 MeV/c. The present experiment was designed to cover the region of the peak and to study in some detail the s- and u-variation of the np CEX cross section over the range $575 \leqslant P_n \leqslant 1429$ MeV/c. A description of the technique used is given in Ref. 5. The present data set consists of 29 angular distributions for neutron c.m. angles 120° to 180°. Each distribution was fit with the double exponential for all data with $|u| < 0.16\ (\mathrm{GeV/c})^2$. The quality of the fit is indicated by the overall value of chi squared ($\chi^2 = 1718$) for 1593 degrees of freedom ($\chi^2/\nu = 1.078$).

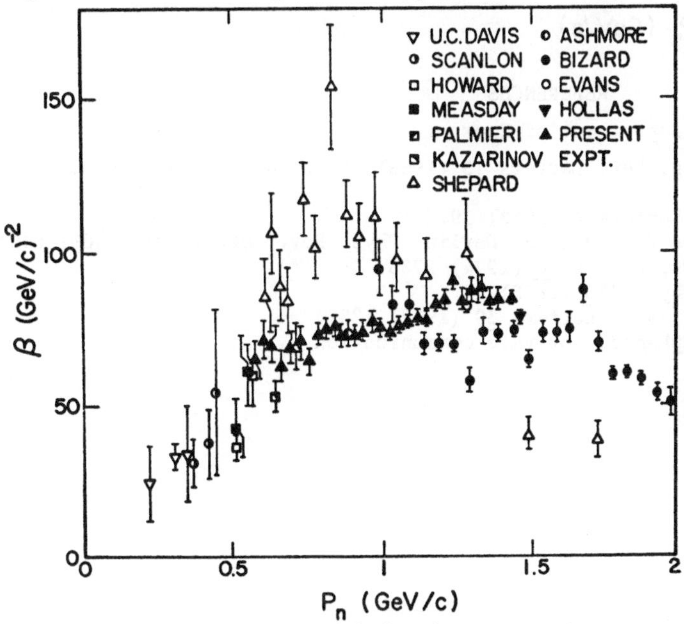

Fig. 1. Plot of the logarithmic slope vs neutron momentum for np charge exchange at intermediate energy. The slope is defined as $\beta = [d/du(\ell n\, d\sigma/du)]_{u=0}$.

Figure 1 shows the value of the logarithmic slope at u = 0 compared to previous measurements. The present results contrast sharply with the previously reported peak but agree with the older result of Ashmore at 350 MeV. In addition, we agree with the lower energy measurements. In Fig. 2 we show our results for the intercept $d\sigma/du$ (u = 0) multiplied by P_n^2. Again no evidence for the peak is seen and our data extrapolate smoothly to the lower energy results.

A recent theory due to Gibbs[6] in which both s- and p-wave pions are allowed in OPE has been applied to the present data. The remarkable agreement obtained and its possible significance will be presented.

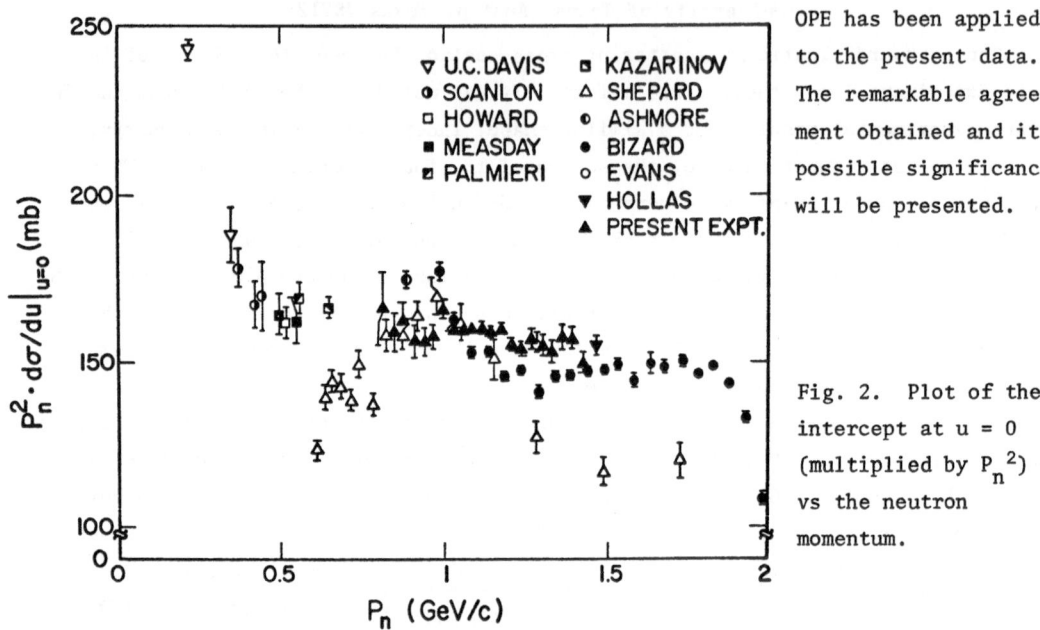

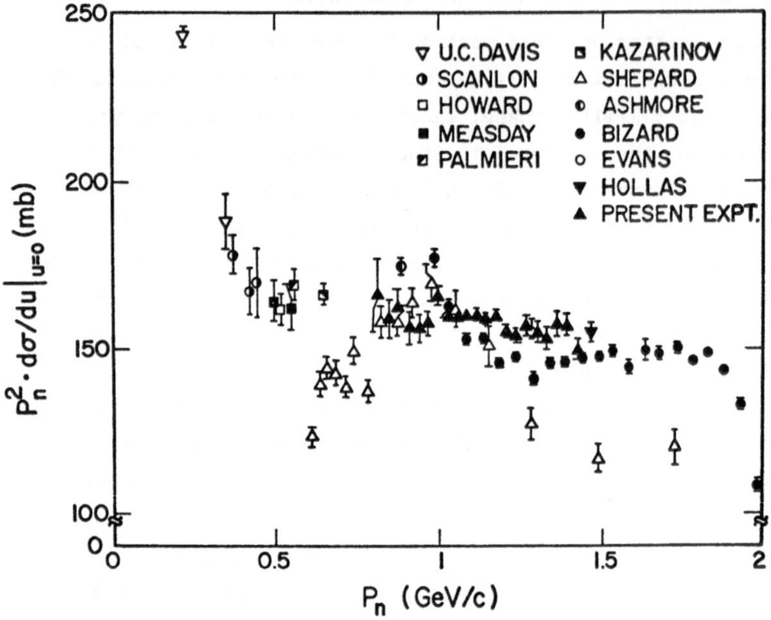

Fig. 2. Plot of the intercept at u = 0 (multiplied by P_n^2) vs the neutron momentum.

REFERENCES

*Work supported by the U. S. Department of Energy.

1. J. L. Friedes et al., Phys. Rev. Letters 15 (1965) 38; R. Wilson, Annals of Physics 32 (1965) 193.
2. R. J. N. Phillips, Phys. Letters 4 (1963) 19.
3. R. E. Mischke, P. F. Shepard, and T. J. Devlin, Phys. Rev. Letters 23 (1969) 542; P. F. Shepard et al., Phys. Rev. D 10 (1974) 2735.
4. G. Bizard et al., Nucl. Phys. B85 (1975) 14.
5. B. E. Bonner et al., Phys. Rev. Letters 39 (1977) 1253.
6. W. R. Gibbs and G. J. Stephenson, private communication.

NUCLEON-NUCLEON SCATTERING IN A NON-ABELIAN GAUGE THEORY

K. Fabricius
Department of Physics
University of Wuppertal
Wuppertal, Germany

and

J. Fleischer
Department of Physics
University of Bielefeld
Bielefeld, Germany

Introduction

Conventionally the NN-Forces are described by the exchange of π,ρ,ω and some other scalar particles [1]. Since the Born approximation does not reproduce the S-waves at all, one has to perform higher order calculations if one does not want to give up the model, but then renormalization problems arise. Models with neutral vector boson fields are renormalizable [2]. One loop calculations including π and ω have been done by Bessis et al. [3]. Renormalizable models for charged vector bosons can be constructed only in the Yang-Mills formalism [4]. Yang-Mills models differ from conventional models in their extreme complexity because a large part of the amplitudes is unphysical. Therefore in a first attempt only the π and ρ mesons have been taken into account in an earlier work of the authors [5]. Now we present here a more complete description by including in addition to π and ρ also the ω. We remark that in our model the ω mass results from spontaneous symmetry breaking as well as the ρ mass. In this point we differ from Ref. [3].

The perturbation calculations of Ref. [3] and [5] serve as a basis for Padé approximants. In Ref. [3] matrix-Padé approximants [6] were applied, whereas we confine ourselves to scalar Padé approximants because off shell calculations in Yang-Mills models are extremely laborious and because the problem of gauge invariance arises for off shell amplitudes. For scalar Padé approximants only on shell amplitudes are necessary. In our calculations we do not fix the gauge parameter. How the different unphysical contributions cancel so that a gauge invariant S-matrix results, is shown explicitely in Ref. [5]. The procedure to keep the gauge parameter free will prove to be useful for off shell matrix Padé calculations because it allows to extract the amplitude in the unitary gauge (see Ref.[7]).

The Model

As a basis of our NN calculations we take a Lagrangian which is built up by the fields of the physical particles N, π, ρ and ω. Renormalizability implies that this must be a Yang Mills model with spontaneous symmetry breaking. Therefore two Higgs multipletts χ, φ have to be introduced. The general procedure to maintain the physical symmetries in spite of spontaneous symmetry breaking is described in Ref.[8]. Here we consider only isospin invariance and nucleon number conservation. According to Ref. [8], we have to choose a $SU(2)_L \times SU(2)_G \times U(1)_L \times U(1)_G$ symmmetric Lagrangian, L and G standing for local and global, respectively. The transformation properties of the fields with respect to the four groups are given by the following table:

Field	$SU(2)_L$	$SU(2)_G$	$U(1)_L$	$U(1)_G$
Pion Φ_i i =1,2,3	1	o	o	o
Nucleon ψ	1/2	o	1	o
χ_o, χ_i i=1,2,3	1/2	1/2	o	o
φ	o	o	1	-1
φ^*	o	o	-1	1

The invariant part of the Lagrangian reads

$$\begin{aligned}\mathcal{L}_{inv} = & \frac{1}{2} D_\mu \Phi_i D^\mu \Phi_i - \frac{1}{2} m_\pi^2 \Phi_i^2 + \frac{1}{2} D_\mu \chi_i D^\mu \chi_i + \frac{1}{2} D_\mu \chi_o D^\mu \chi_o - \frac{1}{2} m_\chi^2 (\chi_o^2 + \chi_i^2) \\ & + D_\mu \varphi^* D^\mu \varphi - m_\varphi^2 \varphi^* \varphi + \frac{i}{2} \bar{\psi} \gamma_\mu D^\mu \psi - \frac{i}{2} D_\mu \bar{\psi} \gamma^\mu \psi - M \bar{\psi} \psi \\ & - \frac{1}{4} \lambda_1 (\Phi_i^2)^2 - \frac{1}{4} \lambda_2 (\chi_o^2 + \chi_i^2)^2 - \lambda_3 \Phi_i^2 (\chi_o^2 + \chi_j^2) - \frac{1}{4} \lambda_4 (\varphi^* \varphi)^2 - \lambda_5 \Phi_i^2 \varphi^* \varphi - \lambda_6 (\chi_o^2 + \chi_i^2) \varphi^* \varphi \\ & - i g_\pi \bar{\psi} \gamma_5 \tau_i \psi \Phi_i - \frac{1}{4} \rho_{\mu\nu}^i \rho_i^{\mu\nu} - \frac{1}{4} \omega_{\mu\nu} \omega^{\mu\nu},\end{aligned}$$

where the covariant derivatives D_μ and $\rho_{\mu\nu}$ and $\omega_{\mu\nu}$ are given by

$$D_\mu \Phi_i = \partial_\mu \Phi_i + g_\rho \epsilon_{ijk} \rho_{\mu j} \Phi_k \,, \quad D_\mu \psi = \partial_\mu \psi - i g_\rho \frac{\tau i}{2} \psi \rho_{\mu i} - i g_\omega \psi \omega_\mu$$

$$D_\mu \chi_i = \partial_\mu \chi_i + \frac{g_\rho}{2} \chi_o \rho_{\mu i} + \frac{g_\rho}{2} \epsilon_{ijk} \rho_{\mu j} \chi_k, \quad D_\mu \chi_o = \partial_\mu \chi_o - \frac{g_\rho}{2} \rho_{\mu i} \chi_i$$

$$D_\mu \varphi = \partial_\mu \varphi - i g_\omega \omega_\mu \varphi$$

$$\rho_{\mu\nu}^i = \partial_\mu \rho_\nu^i - \partial_\nu \rho_\mu^i + g_\rho \epsilon_{ijk} \rho_\mu^j \rho_\nu^k \,, \quad \omega_{\mu\nu} = \partial_\mu \omega_\nu - \partial_\nu \omega_\mu$$

The symmetry is broken spontaneously by putting

$$\chi_o = \hat{\chi}_o + c_1 \,, \quad \varphi_+ = \frac{1}{\sqrt{2}} (\varphi + \varphi^*) = \hat{\varphi}_+ + c_2$$

with vanishing $\langle o|\hat{\chi}_o|o\rangle$ and $\langle o|\hat{\varphi}_+|o\rangle$. The Goldstone bosons are χ_i i=1,2,3 and

$\varphi_- = -\frac{i}{\sqrt{2}}(\varphi-\varphi^*)$. By the Higgs-Kibble mechanism ρ and ω acquire their masses: $m_\rho = \frac{1}{2} g_\rho c_1$, $m_\omega = g_\omega c_2$. As $\hat{\chi}_o - \hat{\Phi}_+$ mixing occurs, we find the physical Higgs fields (σ_i, i=1,2) after diagonalisation:

$$\sigma_1 = \cos\varphi\,\hat{\chi}_o - \sin\varphi\,\hat{\Phi}_+ \text{ and } \sigma_2 = \sin\varphi\,\hat{\chi}_o + \cos\varphi\,\hat{\Phi}_+, \text{ where}$$

$$\sin 2\varphi = 2\mu^2/(m_{\sigma_2}^2 - m_{\sigma_1}^2) \text{ with } \mu^2 = 4\lambda_6 c_1 c_2 \text{ and } (m_{\sigma_1}^2 - m_{\sigma_2}^2)^2 = 4\mu^4 + (m_{\hat{\chi}_o}^2 - m_{\hat{\Phi}_+}^2)^2.$$

The complete Lagrangian now reads

$$\mathcal{L} = \mathcal{L}_{inv} + \mathcal{L}_{GB} + \mathcal{L}_{FP}$$

with the gauge breaking term ('t Hooft gauge)

$$\mathcal{L}_{GB} = -\frac{1}{2\xi_\rho}(\partial_\mu \rho_i^\mu - \xi_\rho m_\rho \chi_i)^2 - \frac{1}{2\xi_\omega}(\partial_\mu \omega^\mu + \xi_\omega m_\omega \varphi_-)^2$$

and the Faddeev-Popov part

$$\mathcal{L}_{FP} = \partial_\mu \bar{c}_i\, \partial^\mu c_i - \xi_\rho m_\rho^2\, \bar{c}_i c_i + g_\rho \epsilon_{ijk} \partial_\mu \bar{c}_i \rho_\mu^j c_k$$
$$- \frac{1}{2}\xi_\rho m_\rho g_\rho\, \bar{c}_i c_i \hat{\chi}_o - \frac{1}{2}\xi_\rho m_\rho g_\rho \epsilon_{ijk} \bar{c}_i c_j \chi_k + \partial_\mu \bar{u} \partial^\mu u + \xi_\omega m_\omega g_\omega \bar{u}\, u\, \varphi_-$$

with c_i and u the anticommuting Faddeev-Popov ghosts. ξ_ρ and ξ_ω are free gauge parameters.

So we have the physical fields $\psi, \Phi, \rho, \omega, \sigma_1, \sigma_2$ and the unphysical fields $\chi_i, \varphi_-, c_i, u$. The masses of the unphysical particles depend on the gauge parameters. In this gauge the masses of χ_i and c_i are $\sqrt{\xi_\rho} m_\rho$ and the masses of φ_- and u are $\sqrt{\xi_\omega} m_\omega$. For the vector particle propagators one obtains

$$\Delta_{F\mu\nu}^{(\varkappa)}(k) = -\frac{g_{\mu\nu}}{k^2 - m_\varkappa^2} - \frac{k_\mu k_\nu}{m_\varkappa^2}\left(\frac{1}{k^2 - m_\varkappa^2} - \frac{1}{k^2 - \xi_\varkappa m_\varkappa^2}\right), \quad \varkappa = \rho \text{ or } \omega.$$

Results

We have determined the complete one loop approximation of the tangent of the NN phase shift: $\tan\delta = K_1(g^2/4\pi) + K_2(g^2/4\pi)^2$. It is known that only matrix Padé approximants can reproduce the S-wave data in a one loop approximation [6]. Since we calculate scalar approximants: $\tan\delta = (\frac{g^2}{4\pi}) K_1^2 / (K_1 - \frac{g^2}{4\pi} K_2)$, we merely computed the triplet P-waves, which are shown in fig. 1. We point out, that in particular the ^3P_o phase shift has been considerably improved by taking into account the ω in addition to the ρ. Further improvement is not possible due to the special structure of the above [1/1] Padé approximant. The couplings of the best fit are: $(g_\rho/2)^2/4\pi = g_\omega^2/4\pi = 2.3$. The obtained phase shifts are insensitive to variations of the masses and couplings of the Higgs bosons.

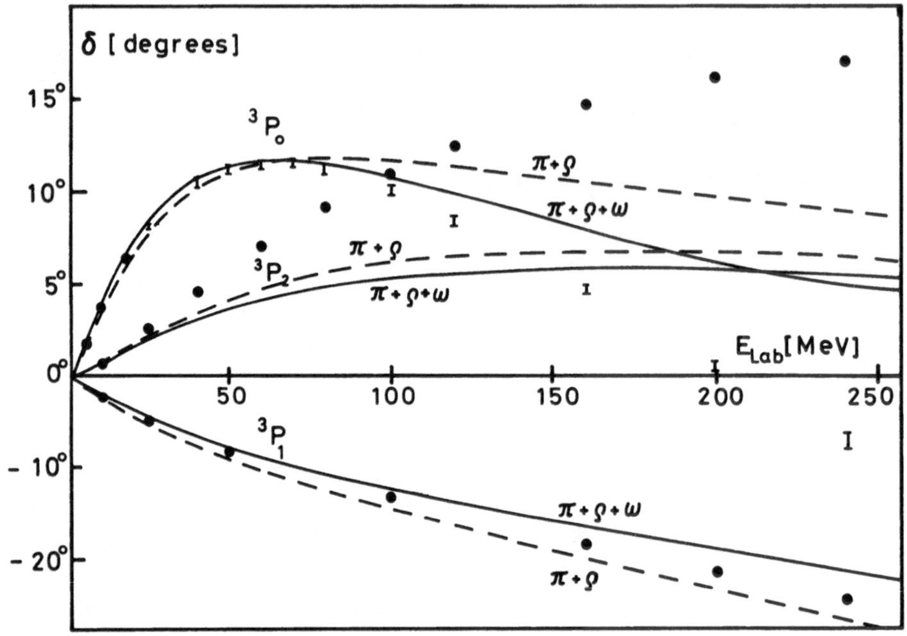

Fig. 1

Triplet P-wave phase shifts in a [1/1] scalar Padé approximation. The experimental data are taken from Ref. [9]. Solid lines: π,ρ and ω exchange, dashed lines only π and ρ. For the coupled channel wave $^3P_2$ a 2 x 2 (on shell) matrix Padé has been performed.

References

1. G. Kramer, in: Springer Tracts in Modern Physics, Vol. 55, edited by G. Höhler (Springer Verlag; Berlin Heidelberg New York, 197o)
2. K. Symanzik, DESY T 71/1 (1971)
3. D. Bessis, P. Mery and G. Turchetti, Nuovo Cimento 4oA, 198 (1977)
4. see e.g. E.S. Abers and B.W. Lee, Phys.Rep. 9,1 (1973)
5. K. Fabricius and J. Fleischer, to be published in Phys.Rev.D
6. The power of matrix Padé's for NN S-waves was demonstrated by J. Fleischer and J.A. Tjon, in: Padé and Rational Approximation,edited by E.B. Saff and R.S. Varga (Academic Press New York San Francisco London, 1977)
7. K. Fabricius and I. Schmitt, Wuppertal preprint
8. I. Bars,M.B. Halpern and M. Yoshimura, Phys. Letters 29, 969 (1972)
   B. de Wit, Nucl. Phys. B51, 237 (1973)
9. M.H. Mac Gregor, R.A. Arndt, and R.M. Wright, Phys. Rev. 182, 1714 (1969)

COUPLED CHANNEL EFFECTS ON NUCLEON-NUCLEON PHASE SHIFTS AND RESONANCES

Earle L. Lomon
Center for Theoretical Physics
Massachusetts Institute of Technology
Cambridge, MA 02139/U.S.A.

Abstract: It is shown that coupling to higher mass channels always causes an attraction which increases with energy below the inelastic threshold. Coupling of the $^3F_3$ ($^1D_2$) wave to the F(S) wave NΔ channel produces the experimental 2260 (2150) MeV resonance. In both cases the predicted lower energy phase shifts agree with recent data analyses. Other partial waves are improved by the coupling and a resonance is indicated in the ^3P_o channel.

General Effect of Coupling[1]

Elastic NN scattering proceeding through an $N^*_1N^*_2$ intermediate state (Fig. 1)

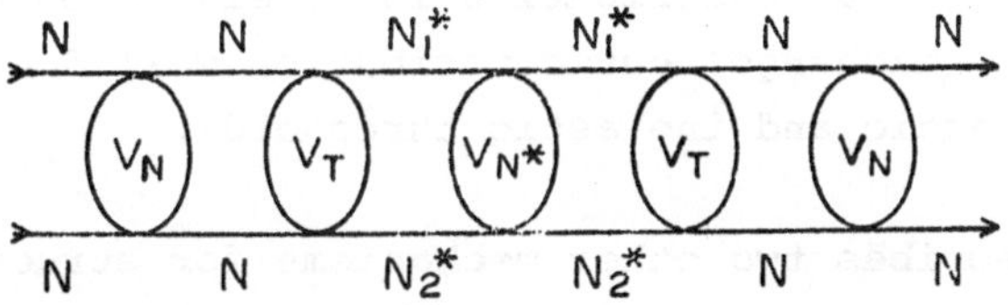

Figure 1 NN scattering through isobar intermediate states

is described by a term in the partial wave dispersion relations:

$$\Delta A_{\alpha L} = \frac{1}{\pi}\int_{s_i}^{\infty} \frac{\rho_{\alpha'L',\alpha L}(s')}{s'-s} \qquad \text{Eq.1}$$

where α represents all quantum numbers other than the orbital angular momentum L, the inelastic threshold $s_i=(M_{N^*_1} + M_{N^*_2})^2$, and

$$\rho_{\alpha'L',\alpha L}(s') = |\langle\alpha', L', s'|\alpha, L, s'\rangle|^2 \qquad \text{Eq.2}$$

is positive definite. For $s < s_i$ the denominator of the integrand is also positive definite, and the integrand is an increasing function of s for all s'. Hence the effect of the coupling, as given by the discontinuity across the elastic cut, is to add attraction in the NN channel, the amount of attraction increasing with energy up to the inelastic threshold. For normally strong transition interactions this is a large effect and may frequently result in resonances. When $s > s_i$ the denominator has contributions of both signs and the attractive effect rapidly weakens. However near inelastic threshold the angular momentum barrier keeps ρ small for

$$(s' - s_i)^{1/2} \lesssim (2L' + 1)\, M^T_{ex} \qquad \text{Eq.3}$$

where M^T_{ex} is the effective exchanged mass in the transition interaction. This extends the increasing attraction beyond s_i. Except for L'=0 the mechanism may produce partially inelastic resonances above s_i. For L'=0 the numerator varies more slowly than the denominator.

In the approximation that the integral in Eq. 1 is dominated by the peak of the integrand given by Eq. 3

$$\Delta A_{\alpha L} \sim [s_i + (2L' + 1)^2 \, (M^T_{ex})^2 - s]^{-1} \qquad \text{Eq.4}$$

so that the larger the range of V_T (i.e. the smaller M^T_{ex}) the stronger the energy dependence. The same estimate of the integral indicates that the physical cut contribution is likely to be more important than the unphysical (exchange) cut contribution well down into the region between the elastic and inelastic thresholds.

Equation 1 describes two other mechanisms for structure in the NN amplitude due to coupling to higher threshold channels. The nature of the branch point at s_i is determined by $\rho \sim (s'-s_i)^{L'}$, giving rise to a cusp in the cross section which has a discontinuous (L'+1)st derivative. The cusp is difficult to observe, and even for L'=0 requires high accuracy and resolution except in rare instances. The remaining mechanism only comes into play if V_{N*} is strongly attractive enough to bind the $N^*_1 N^*_2$ system. In the presence of coupling the bound state leaks into the NN channel causing it to resonate. This gives rise to large effects, but may be expected to be rare.

Applications to NN Partial Waves

The effect on the NN channel, as discussed above, of the coupling to isobars depends only on the range of the transition potential and on the threshold energy and orbital angular momentum of the inelastic channel. A model incorporating those features will give semi-quantitatively correct results. However the NN interaction must be realistic as a baseline against which to see the effect of turning on the interchannel coupling. For this purpose we choose the Feshbach Lomon (FL) interaction,[2] for which the medium and long range potential is constructed from one boson and two pion exchanges and which fits nucleon-

nucleon data well up to ∿300 MeV. The boundary condition core of the FL interaction can be extended into a matrix which couples channels (this is already done in FL for the tensor coupled NN states):

$$r_o \frac{dU_{\alpha L}}{dr_o} = \sum_{\alpha'L'} F(\alpha L;\ \alpha'L')\ U_{\alpha'L'}(r_o)$$

where the F's are energy independent. As $r_o = \frac{1}{2}\mu^{-1}$ the range of the transition interaction is reasonable. Transition interactions of range μ^{-1} would modify the energy dependence to a minor extent (Eq.4). Coupling to the NΔ(1232) and NN*(1470) channels are considered. The former has a lower threshold than the latter but cannot couple to all NN states. (Coupling to the ΔΔ system would produce similar effects as coupling to NN* because the value of s_i is nearly the same.)

We first look at the $^3F_3$ channel because of the rather well identified pp resonance observed in this partial wave.(3) Fig. 2 shows the effect of coupling to either NΔ or NN* with L'=3. (We use the simplified nota- $f_{\alpha L,\alpha L}=f_1, f_{\alpha'L',\alpha'L'}=f_2$ and $f_{\alpha L,\alpha'L'}=f_T$.) The choice $f_2=0$ implies a moderate, non binding attraction in the isobar channel. f_1 is adjusted for each value of f_T so that the effective core interaction at elastic threshold is the same as in the FL interaction, reproducing its fit to the low energy data. Al-

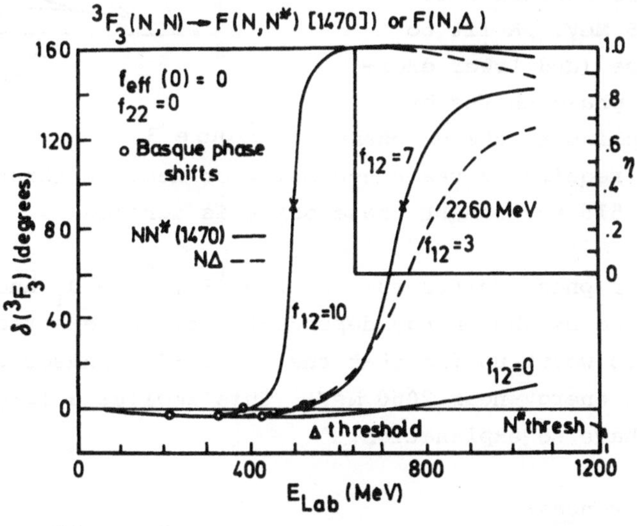

Figure 2

though the NN* coupling can produce a resonance near the experimental value of $\sqrt{s}$=2260 MeV, the width is too narrow and the resonance is completely elastic. However for NΔ coupling, when f_T is adjusted to put the resonance at 2260 MeV, Γ = 200 MeV approximately as observed. The elasticity is 80% which is larger than that observed, indicating the need for corrections due to the width of the Δ or continuum production of pions. It is interesting to note that the predicted phase shifts below the resonance follow the trend of the recent BASQUE data analysis.[4]

Another pp resonance is indicated by the data at 2150 MeV.[3] As this energy is just below the NΔ threshold it is tempting to relate it to NΔ production in an S state, which can only occur in the $^1D_2$ channel. Fig. 3 shows that when the coupling is adjusted to produce the resonance at the experimental energy, the phase shifts go through the BASQUE phase results at lower energy.

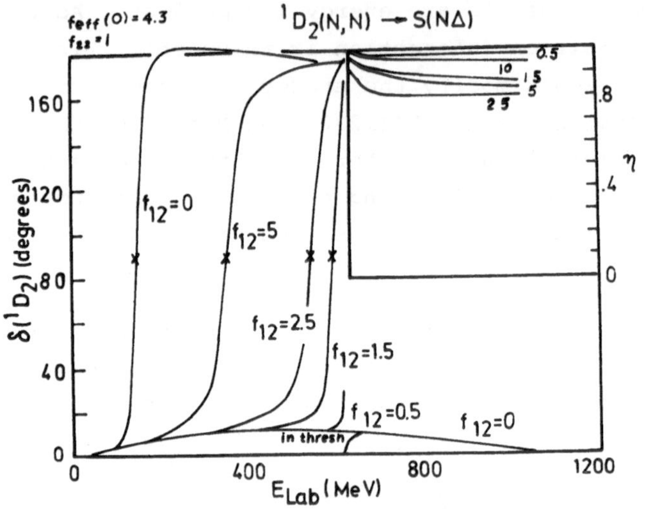

Figure 3

The BASQUE ^3P_o phase shifts indicate a possible resonance at 2115 MeV. A fit to those (and lower energy) phase shifts by coupling to the NΔ channel requires a resonance below or near NΔ threshold even if the critical 515 MeV BASQUE phase shift is ignored.

Other phase shifts such as the $^1S_0$, $^3S_1$-$^3D_1$ and $^1P_1$ are very much improved by the energy dependent attractive effect of NΔ or NN* coupling. It is worth noting that the $^1P_1$ BASQUE phases show a rapid decrease with energy near 2060 MeV. This implies a large inelasticity for which we have no explanation.

References:

(1) Earle L. Lomon, Phys. Rev. D1, 549 (1970).

(2) Earle L. Lomon and Herman Feshbach, Ann. Phys. (NY) 48, 94 (1968)

(3) H. Hidaka et al., Phys. Lett. 70B, 479 (1977); CERN Courier, Oct. 1977 and private communication re 2150 MeV resonance.

(4) D. Axen et al., Lett. Nuovo Cim. 20, 151 (1977); C. J. Oram, private communication re updated results and np analysis.

NUCLEON-NUCLEON POTENTIAL FROM REGGE-POLE THEORY IN MOMENTUM SPACE

G.J.M. Austen, T.A. Rijken, P.A. Verhoeven

Institute for Theoretical Physics, University of Nijmegen, The Netherlands

I. Introduction. Recently it has been shown that the configuration space nucleon-nucleon potential derived from Reggeon-exchange describes the low energy nucleon-nucleon data very well [1]. In this paper we give an outline of the momentum space model and report on results obtained sofar. The model is derived within the framework of S-matrix theory [2]. In the relativistic integral equations the particles in the intermediate states are on-mass-shell but off-energy-shell. These integral equations are based on: (i) the existence of a complete set of amplitudes possessing Mandelstam analyticity, (ii) the approximation of the third double spectral function ρ_{tu} by a sum of $\rho_{tu}^{(1)}$ and $\rho_{tu}^{(2)}$ which depend respectively on t and u only [2], (iii) the unitarity equation. The OBE-potentials are the leading contributions at low energy of Regge-trajectory exchanges [2]. New are the potentials due to Pomeron-, f-, and A_2-exchange. Our present treatment includes the full momentum dependence of the potentials.

II. Amplitudes and relativistic integral equations. We expand the transition matrix in terms of "derivative" amplitudes [3]:

$$\langle q_a' q_b' | M^\sigma | q_a q_b \rangle = \sum_{\alpha',\alpha} I^*_{P',\alpha'} \, I_{P,\alpha} \, M^\sigma_{\alpha',\alpha}(s,t,u)$$

where σ denotes the signature and the spinor covariants $I_{P,\alpha}$ are defined by

$$I_{P,0} = \bar{v}(q_b)\,\gamma_5\,u(q_a) \qquad\qquad I_{P,2} = i\,\bar{v}(q_b)\,\gamma_\mu\,u(q_a)\,L_P^\mu$$

$$I_{P,1} = i\,\bar{v}(q_b)\,\gamma_\mu\,u(q_a)\,P^\mu \qquad\qquad I_{P,3} = \bar{v}(q_b)\,\gamma_\mu\,u(q_a)\,T_P^\mu$$

Here $P_\mu = \frac{1}{2}(q_a-q_b)_\mu$, $W_\mu = (q_a+q_b)_\mu$, $(L_P)_\mu = i\,\varepsilon_{\mu\nu\rho\sigma} W^\nu P^\rho (\partial/\partial P_\sigma)$, $(T_P)_\mu = \varepsilon_{\mu\nu\rho\sigma} W^\nu P^\rho L_P^\sigma$, and $v = C\,\bar{u}^T$. Similar definitions hold for $I_{P',\alpha'}$, etc. corresponding to the outgoing particles. For the amplitudes $M_{\alpha',\alpha}$ we can assume Mandelstam analiticity [3]. The elastic unitarity relations read

$$M_{\alpha,\beta} - M^+_{\alpha,\beta} = -\,i\,M^+_{\alpha,\gamma}\,\rho_\gamma^2\,M_{\gamma,\beta}$$

with $\rho_0^2 = 2\,s\,n_0^2$, $\rho_1^2 = 2\,m^2(s-4m^2)\,n_1^2$, $\rho_2^2 = 2\,s^2\,n_2^2$, and $\rho_3^2 = \frac{1}{4}\,s^3(s-4m^2)n_3^2$. Here $n_0^2 = n_1^2 = 1$ and $n_2^2 = n_3^2 = L^2 = -\frac{d}{dz}(1-z^2)\frac{d}{dz}$, $z = \cos\theta$.

Because we have at low energies $\rho_{tu} \simeq \rho_{tu}^{(1)} + \rho_{tu}^{(2)}$, the $M_{\alpha,\beta}$ with the properties mentioned above can be solved with a "generalized" potential $W_{\alpha,\beta}$ (s,t,u) employing the Mandelstam iteration method. We replace that iteration scheme by the relativistic three-dimensional integral equations

$$M^\sigma_{\alpha',\alpha} = W^\sigma_{\alpha',\alpha} + \sum_{\alpha''} \int \frac{d^3q''}{\sqrt{q''^2+m^2}}\, W^\sigma_{\alpha',\alpha''}\, G_{\alpha''}\, M^\sigma_{\alpha'',\alpha}$$

where $G_{\alpha''} = \rho^2_{\alpha''}/(q''^2 - q_i^2 - i\varepsilon)$.

Partial wave integral equations are now obtained from the expansion $M_{\alpha',\alpha} = \Sigma\,(2J+1)\,M^J_{\alpha',\alpha}\,P_J(z)$ and a similar one for $W_{\alpha',\alpha}$ and the connection with the parity conserving helicity amplitudes [4] $\{f_0^J, f_1^J, f_{11}^J, f_{12}^J, f_{22}^J\}$. We find

$$f_0^J = M_{0,0}^J \quad ; \quad f_1^J = J(J+1)\, M_{2,2}^J \quad ;$$
$$f_{11}^J = M_{1,1}^J \quad ; \quad f_{12}^J = \sqrt{J(J+1)}\, M_{1,3}^J \quad ; \quad f_{22}^J = J(J+1)\, M_{3,3}^J$$

Introducing $T_{\alpha',\alpha}^J = \rho_{\alpha'}^J M_{\alpha',\alpha}^J \rho_\alpha^J$, $V_{\alpha',\alpha}^J = \rho_{\alpha'}^J W_{\alpha',\alpha}^J \rho_\alpha^J$ we obtain the partial wave equations

$$T_{\alpha',\alpha}^J(q_f,q_i) = V_{\alpha',\alpha}^J(q_f,q_i) + \int_0^\infty dq_n\, q_n^2\, V_{\alpha',\alpha}^J(q_f,q_n)\, G(q_n,\sqrt{s})\, T_{\gamma,\alpha}^J(q_n,q_i)$$

with $G(q_n,\sqrt{s}) = \left[4\, E(q_n)(q_n^2 - q_i^2 - i\varepsilon)\right]^{-1}$.

These are the integral equations for the amplitudes f_0^J, f_1^J, f_{11}^J, f_{12}^J, and f_{22}^J and we can easily transform to the LSJ-basis. From here the formalism is standard and we refer to [5] for the details and the notations we use from here.

III. OBE-potentials from Reggeon-exchange. We use the Khuri-Jones representation of the Regge Poles to define the Reggeons. The leading contributions at low energies are the crossed channel partial wave projections, with the lowest physical J [2]. The helicity partial wave potentials V_1^J, V_2^J, ... , V_6^J for pseudoscalar (π,η,η')-, scalar (ε,δ,S^*)-, and vector (ρ,ω,ϕ)-exchange can be obtained from the formulae given in [5] by substituting for $Q_J(x)$, δ_{J0}, and $\frac{1}{3}\delta_{J1}$ respectively $L_J(x)$, R_J, and S_J. The latter are

$$L_J(x) = \frac{1}{2}\int dz\, \frac{P_J(z)\, F(z)}{x-z} \;,\; R_J = \frac{1}{2}\int dz\, P_J(z)\, F(z) \;,\; S_J = \frac{1}{2}\int dz\, z\, P_J(z)\, F(z) \;.$$

The form factor $F(z) = \exp(-t/\Lambda^2)$, where $t = (\underline{q}_f-\underline{q}_i)^2 - (E_f-E_i)^2$ and Λ is a parameter. The latter is related to the trajectory parameters [2] but we treat it here as a free parameter. Furthermore $x = \left[q_f^2 + q_i^2 - (E_f-E_i)^2 + \mu^2\right]/2q_iq_f$. The appearance of $(\Delta E)^2 = (E_f-E_i)^2$ is in contrast to some models based on the Blankenbecler-Sugar equation. We need retardation in order to have Mandelstam analyticity for $M_{\alpha',\alpha}$.

The new contributions to the potentials are due to the J = 0 components of the Pomeron P, the f-, and A_2-trajectories. We find for the partial wave helicity potentials

$$V_1^J = \pi b_1\, \Delta\left[\frac{J+1}{2J+1} R_{J+1} + R_J + \frac{J}{2J+1} R_{J-1}\right]$$
$$V_2^J = \pi b_2\, \Delta\left[\frac{J+1}{2J+1} R_{J+1} - R_J + \frac{J}{2J+1} R_{J-1}\right]$$
$$V_3^J = \pi b_3\, \Delta\left[\frac{J}{2J+1} R_{J+1} - R_J + \frac{J+1}{2J+1} R_{J-1}\right]$$
$$V_4^J = \pi b_4\, \Delta\left[\frac{J}{2J+1} R_{J+1} - R_J + \frac{J+1}{2J+1} R_{J-1}\right]$$
$$V_5^J = \pi b_5 \Delta\, \frac{\sqrt{J(J+1)}}{2J+1}\left[-R_{J+1} + R_{J-1}\right]$$
$$V_6^J = \pi b_6 \Delta\, \frac{\sqrt{J(J+1)}}{2J+1}\left[R_{J+1} - R_{J-1}\right]$$

where the coefficients b_ℓ $(\ell = 1,..,6)$ are the same, except for an overall (-) sign, as those for scalar meson exchange and can be found in [5]. Furthermore $\Delta(t) = \exp(-t/4m_P^2)/m^2$ where the parameter m is in the Regge Pole model given by $m_P^2 = \frac{1}{4}\left[a + \alpha' \ln \bar{s}\right]^{-1}$ [2]. Here a comes from the exponential behavior of the Regge residue, α' is the slope of the trajectory, and $\bar{s}$ is a parameter which governs the spectral boundaries of the Khuri-Jones function. For the pomeron one finds from Regge fits to the high energy data (see [2]) $m_P \sim 250 - 300$ MeV. The results of [1] confirm this. Here we treat m_P as a free parameter.

IV. Results and discussion. We have made a fit to the nucleon-nucleon data up to 330 MeV using the χ^2 second-derivative matrices of the Livermore phase shift analysis [6]. Here we have excluded the waves: ${}^1S_0$, ${}^3S_1$, ε_1, ${}^3D_1$. (A fit including also these waves is in progress and will be published elsewhere.) In our fit we obtained sofar the satisfactory result: χ^2/data = 1.85 for 1128 data. The following 13 parameters were searched: g_π, $g_{\eta'}$, g_ρ, f_ρ, g_ω, f_ω, g_δ, g_ε, g_P, g_{A_2}, α_V^m, Λ_s, Λ_t. The last two parameters Λ_s and Λ_t denote the spin singlet and spin triplet cut off's which appear in the form factors. Furthermore we used as input: $\alpha_P = 0.361$, $\alpha_S = 1.01$, $\alpha_V^e = 1.0$, $\theta_P = -23^\circ$, $\theta_S = 37.9^\circ$, $\theta_V = 37.5^\circ$, $m_P = m_{A_2} = 307.81$ MeV (see [1] for a discussion of these parameters). The Pomeron and f,f' are treated together and denoted by P. The preliminary results for the search parameters are given in Table I. In Table II we list the corresponding phase shifts. The coupling constants are in general somewhat different than those obtained in [1]. Note however that in [1] also the S-waves are fitted.

In our work we employ the Khuri-Jones representation of the Regge Poles which, albeit convenient, is not essential in principle. Alternatively one could use Veneziano-functions. Then, for very negative intercepts ($\alpha_E \lesssim -4.0$) of the exotic trajectories, dual to the non-exotic ones, the J = 0 and J = 1 exchanges are again dominant at low energies.

Finally we mention that the recent work on bag-models [7] suggests that the quark-structure of the nucleons could play a role already at rather low energies. If this is the case a physical explanation of the Pomeron-potential could be that it is a bag-surface or -overlap effect. Then, the exchange of confined colored gluons would lead to repulsion [8]. Also, such a manifestation of the nucleons as composite particles would warrant to include somehow Regge cuts in our model, e.g. $\pi \otimes \pi$, $\pi \otimes P$, $\rho \otimes P$, etc. These cuts would in particular affect the short distance behavior of the nucleon-nucleon interaction.

Acknowledgement: We would like to thank Prof. J.J. de Swart and Dr. M.M. Nagels for discussions.

References:

1. M.M. Nagels, T.A. Rijken and J.J. de Swart, Phys.Rev. D 17, 768 (1978).
2. T.A. Rijken, Proc. of the Int.Conf. on Few Body Problems in Nuclear and Particle Physics, Quebec, 1974; T.A. Rijken, Ph.D. thesis, University of Nijmegen, 1975; T.A. Rijken and M.M. Nagels, to be published.
3. Similar amplitudes were introduced by V. DeAlfaro et al, Ann.Phys. 44, 165 (1967). See also C. Rebbi, Ann.Phys. 49, 106 (1968).
4. M.L. Goldberger et al., Phys.Rev. 120, 2250 (1960).
5. A. Gersten, P.A. Verhoeven and J.J. de Swart, Nuov.Cim. 26A, 375 (1975).
6. M.H. MacGregor, R.A. Arndt and R.M. Wright, Phys.Rev. 182, 1714 (1969).
7. See e.g. A.Th.M. Aerts, P.J.G. Mulders and J.J. de Swart, Phys.Rev. D 17, 260 (1978).
8. F.E. Low, Phys.Rev. D 12, 163 (1975).

	μ (MeV)	$g^2/4\pi$	f/g	
π	138.041	13.52		
η	548.8	4.69		
η'	957.5	7.22		
ρ	770, Γ = 146	0.47	6.67	
φ	1019.5	0.40	1.09	$\alpha_v^m = 0.32$
ω	783.9	7.74	0.33	
δ	962.0	1.28		
S*	993.0	0.89		
ε	760, Γ = 640	19.95		
P,f,f'	m_P = 307.81	7.29		
A_2	m_{A_2} = 307.81	0.45		
Λ_s	849.92			
Λ_t	953.90			

Table I: Meson-nucleon coupling constants from NN-fit. The underlined couplings are constraint via SU(3).

T_{lab} (MeV)	25	50	95	142	210	330
$^1D_2$	0.69	1.67	3.55	5.49	7.80	9.80
$^1G_4$	0.04	0.15	0.38	0.62	0.97	1.66
$^3P_0$	8.99	11.77	9.74	4.94	-2.53	-14.20
$^3P_1$	-5.16	-8.52	-12.76	-16.28	-20.59	-26.52
$^3P_2$	2.52	5.81	10.59	13.85	16.57	18.33
ε_2	-0.84	-1.79	-2.75	-3.03	-2.68	-1.17
$^3F_2$	0.11	0.34	0.74	1.01	1.06	-0.02
$^3F_3$	-0.23	-0.70	-1.49	-2.16	-2.93	-4.08
$^3F_4$	0.02	0.11	0.40	0.85	1.61	2.87
ε_4	-0.05	-0.19	-0.50	-0.81	-1.17	-1.59
$^3H_4$	0.01	0.03	0.10	0.19	0.34	0.55
$^3H_5$	-0.01	-0.08	-0.28	-0.51	-0.81	-1.23
$^1P_1$	-3.17	-10.59	-13.58	-16.44	-20.27	-25.71
$^1F_3$	-0.43	-1.14	-2.11	-2.78	-3.43	-4.37
$^1H_5$	-0.02	-0.17	-0.50	-0.83	-1.22	-1.64
$^3D_2$	3.81	9.34	17.58	22.90	26.47	26.38
$^3D_3$	0.06	0.39	1.63	3.36	5.53	7.54
ε_3	0.56	1.63	3.37	4.70	5.85	6.71
$^3G_3$	-0.05	-0.26	-0.87	-1.59	-2.59	-4.03
$^3G_4$	0.17	0.73	2.04	3.43	5.30	8.13
$^3G_5$	-0.01	-0.05	-0.16	-0.24	-0.24	-0.10
ε_5	0.04	0.20	0.67	1.18	1.86	2.80

Table II: Nucleon-bar NN phase shifts in degrees.

N-N POTENTIAL FROM REGGE-POLE THEORY

M.M.Nagels, T.A.Rijken and J.J.de Swart
Institute for Theoretical Physics, University of Nijmegen, The Netherlands

This paper contains an extension of the model of Ref. 1, consisting in the expansion of the factor M/E in the pseudoscalar meson potentials. Instead of eq. 5 of Ref. 1 we have

$$\mathcal{V}_3^{(P)} = - f_P^2 \ \Delta(\mathbb{1} - \underline{k}^2/8MM' - \underline{q}^2/2MM')/m_\pi^2 \quad .$$

In configuration space the last two terms contribute only to the nonlocal part of the potential (the local contributions cancel). In addition to eq. (31) of Ref. 1. we get

$$\tilde{V}_P(r) = \frac{f_P^2}{4\pi}\frac{m^2}{m_\pi^2}\frac{m}{4MM'}\left[\frac{1}{3}(\underline{\sigma}_1\cdot\underline{\sigma}_2)(\nabla^2\phi_C^1 + \phi_C^1\nabla^2) + (\nabla^2\phi_T^0 S_{12} + \phi_T^0 S_{12}\nabla^2)\right] \quad .$$

We describe briefly the way of solving the Schrödinger equation involving nonlocal terms of the form $-[(\nabla^2\phi(r) + \phi(r)\nabla^2)\mathbb{1} + (\nabla^2\chi(r)S_{12} + \chi(r)S_{12}\nabla^2)]/2M_{red}$ in the coupled triplet states. The radial equation

$$\{(1+2\phi)\mathbb{1} + 2\chi S_{12}\}u'' + (2\phi'\mathbb{1} + 2\chi' S_{12})u' + \left[k^2 - 2M_{red}V - \{(1+2\phi)\mathbb{1} + \chi S_{12}\}\frac{L^2}{r^2} - \frac{L^2}{r^2}\chi S_{12} + \phi''\mathbb{1} + \chi'' S_{12}\right]u = 0$$

goes under the substitution $u = A^{-\frac{1}{2}}v$, where $A \equiv \{(1+2\phi)\mathbb{1} + 2\chi S_{12}\}$, over into the radial equation for v

$$v'' + \left[k^2 - \frac{L^2}{r^2} - 2M_{red}W\right]v = 0$$

with the "potential"

$$2M_{red}W = 2M_{red}A^{-\frac{1}{2}}VA^{-\frac{1}{2}} - A^{-2}(\phi'\mathbb{1} + \chi' S_{12})^2 - (A^{-1} - \mathbb{1})k^2 + \{A^{\frac{1}{2}}[L^2, A^{-\frac{1}{2}}] + A^{-\frac{1}{2}}[L^2, A^{\frac{1}{2}}]\}/2r^2 .$$

Defining $X = (1 + 2\phi + 4\chi)^{\frac{1}{2}}$ and $Y = (1 + 2\phi - 8\chi)^{\frac{1}{2}}$ we have explicitly

$$A^{\frac{1}{2}} = \frac{1}{3}(2X + Y)\mathbb{1} + \frac{1}{6}(X - Y)S_{12} \quad \text{and} \quad A^{-\frac{1}{2}} = \{\frac{1}{3}(X + 2Y)\mathbb{1} + \frac{1}{6}(-X + Y)S_{12}\}/XY \quad .$$

Below we give the results of the present status of the fit with χ^2/data=2.17. The pion coupling, which has been fixed such as to produce the correct quadrupole moment of the deuteron, has an excellent value. Furthermore the $^3D_2$ phase shifts are already quite low at higher energies.

Results

	$g^2/4\pi$	$f^2/4\pi$	f/g		m(meV)	$g^2/4\pi$
π	(14.110)	7.805×10^{-2}		δ		1.438
η	(3.683)	2.037×10^{-2}		S^*		0.759
η'	(4.230)	2.340×10^{-2}		ε		23.597
ρ	0.800	14.973	(4.327)	P,f,f'	303.28	10.399
ϕ	0.042	0	(0)	A_2		0.260
ω	7.902	1.465	(0.431)	Λ	984.01	

Table I. Meson-nucleon coupling constants and search masses from the NN fit. Figures between parentheses give equivalent information as the neighboring columns.

	$^1S_0$	$^3S_1$	$^3P_0$	$^3P_1$	$^3P_2$	$^1P_1$
a	-7.822	5.489	-2.841	1.990	-0.294	3.023
r	2.723	1.846	2.459	-7.563	4.402	-6.895

Table II. s- and p-wave effective range parameters in units of fm.

P_d	Q	ρ(-B,-B)	N_g^2	A
5.43 %	0.2860 fm^2	1.845 fm	0.8083 fm	0.0262

Table III. Deuteron parameters.

T_{lab}(MeV)	25	50	95	142	210	330
$^1S_0$	48.98	39.23	26.35	16.28	4.93	-10.14
$^3S_1$	78.37	59.54	40.31	27.18	13.52	-3.41
ε_1	2.01	2.50	3.10	3.76	4.83	6.86
$^3P_0$	9.20	12.30	10.13	4.82	-3.67	-17.64
$^3P_1$	-5.06	-8.52	-12.99	-16.84	-21.80	-29.37
$^1P_1$	-6.17	-8.78	-11.43	-13.72	-16.87	-21.62
$^3P_2$	2.45	5.77	10.49	13.57	16.09	18.25
ε_2	-0.84	-1.80	-2.81	-3.13	-2.86	-1.56
$^3D_1$	-2.98	-6.82	-12.26	-16.22	-19.72	-21.30
$^3D_2$	4.03	9.94	19.02	25.19	29.28	29.00
$^1D_2$	0.69	1.68	3.62	5.59	8.34	11.22
$^3D_3$	0.07	0.45	1.88	3.84	6.65	10.18
ε_3	0.58	1.70	3.52	4.91	6.16	7.03
$^3F_2$	0.11	0.34	0.77	1.10	1.27	0.45
$^3F_3$	-0.23	-0.70	-1.51	-2.17	-2.89	-3.97
$^1F_3$	-0.44	-1.18	-2.18	-2.83	-3.42	-4.38
$^3F_4$	0.02	0.11	0.42	0.88	1.66	2.91
ε_4	-0.05	-0.20	-0.52	-0.83	-1.20	-1.63
$^3G_3$	-0.06	-0.28	-0.92	-1.74	-2.93	-4.61
$^3G_4$	0.18	0.76	2.13	3.60	5.67	8.41
$^1G_4$	0.04	0.15	0.39	0.63	1.00	1.71
$^3G_5$	-0.01	-0.05	-0.16	-0.25	-0.26	0.01
ε_5	0.04	0.22	0.70	1.23	1.93	2.90
$^3H_4$	0.00	0.03	0.10	0.20	0.35	0.60
$^3H_5$	-0.01	-0.08	-0.29	-0.52	-0.83	-1.23
$^1H_5$	-0.03	-0.17	-0.52	-0.86	-1.25	-1.65
$^3H_6$	0.01	0.01	0.04	0.09	0.20	0.50
ε_6	-0.00	-0.03	-0.11	-0.21	-0.36	-0.61

Table IV. Nuclear bar pp and np phase shifts in degrees.

T_{lab}(MeV)	2	4	6	8	10	12
$^1P_1$	-0.494	-1.180	-1.863	-2.502	-3.088	-3.623
^3P_C	-0.010	-0.011	0.007	0.042	0.092	0.154
^3P_T	-0.114	-0.320	-5.53	-0.793	-1.030	-1.260
$^3P_{LS}$	0.009	0.028	0.055	0.090	0.134	0.188
$^1D_2$	0.005	0.025	0.061	0.108	0.163	0.224
^3D_C	0.005	0.025	0.059	0.105	0.161	0.226
^3D_T	0.014	0.064	0.147	0.258	0.390	0.537
$^3D_{LS}$	0.002	0.007	0.015	0.024	0.035	0.047

Table V. Low energy ^1L_L, ^3L_G, ^3L_T, and $^3L_{LS}$ phase shifts in degrees for L = 1,2..

Reference: 1. M.M. Nagels, T.A. Rijken, and J.J. de Swart, Phys.Rev. D 17, 768 (1978).

ON THE POTENTIAL RECONCILIATION OF THE n-p $^3S_1-^3D_1$ and $^3D_2$ PHASE PARAMETERS AND THE QUADRUPOLE MOMENT OF THE DEUTERON

M. W. Kermode and A. McKerrell

DAMTP, University of Liverpool, PO Box 147, Liverpool L69 3BX, UK

It is well known that the Reid $^3S_1-^3D_1$ neutron-proton potential does not provide a good description of the $^3D_2$ phase shifts. In a recent paper (2), we have shown that a simple modification to the Reid potential which gives a phase equivalent nonlocal potential also describes, to a high degree of accuracy, the $^3D_2$ phase shifts from the analysis of the experimental data by Arndt <u>et al</u> (3).

However, this local plus rank one separable potential in each channel, changed the model value of the deuteron quadrupole moment Q from 0.2770 fm^2 to 0.3111 fm^2 (the experimental value is 0.2860 ± 0.0015 fm^2 (4)). We suggested that although the Reid potential gives good deuteron wave functions it gives an unsatisfactory description of the scattering phase parameters. In general, it appears that a local potential that gives a good description of, say, the Livermore energy-dependent $^3S_1-^3D_1$ phase parameters (5) produces a poor description of the quadrupole moment and vice versa. We indicated that it might be possible, but difficult, to find a simple nonlocal phenomenological potential giving a good description of all three sets of data (i.e. $^3S_1-^3D_1$, $^3D_2$ and Q).

Since we made our suggestion that it would be difficult to construct a nonlocal potential to describe only the $^3S_1-^3D_1$ phase parameters and Q, Groenenboom and Boersma (6) have provided suitable candidates. Their potential describes the Livermore $^3S_1-^3D_1$ phase parameters. Presumably, there is a similar potential that would fit those obtained by Arndt <u>et al</u>.

We take this opportunity to point out that the unusual behaviour of the bar mixing parameter obtained by Arndt <u>et al</u> is a peculiarity, not of the potentials, but of the phase parametrization. We suggest that future analyses should express the results in terms of the M-phase parameters (7) rather than the bar phase parameters. In particular, the M-mixing parameter, Δ_{12}, given by

$$\tan \Delta_{12} = \tan\bar{\varepsilon}\cos\bar{\delta}_S \cos\bar{\delta}_D - \cot\bar{\varepsilon}\sin\bar{\delta}_S \sin\bar{\delta}_D$$

has the <u>same</u> energy-dependent shape for the two different forms of $\bar{\varepsilon}$ (3,5), with that from the analysis by Arndt <u>et al</u> having a larger maximum value. The M-phase parameters are those that result from effective range theory (7) and the M-matrix elements are closely related to the asymptotic wave functions of the deuteron.

It would be of interest to know the values of the $^3D_2$ phase shifts which result from the nonlocal potential of Groenenboom and Boersma and how they compare with experiment. Although their potential does not contain local components, their results indicate that it should be possible to find a local plus separable potential that has the same properties, with the local term having the one pion tail.

When an agreed set of phase parameters is obtained from experiment, it will be worth while constructing a phenomenological nonlocal potential to describe all three sets of data. Such a potential would be of use for, say, nuclear structure calculations and also of interest for a comparison of its local and nonlocal (or momentum dependent) components with those of the potentials from more fundamental considerations; for example, the Paris potential (8) or the Nijmegen potential (9). Hopefully such a potential would be a useful substitute for the more complicated versions.

References

1) Reid R V 1968 Ann. Phys., NY 50 411-48
2) McKerrell A, Kermode M W, Mines J R and Mustafa M M 1978 J. Phys. G: Nucl. Phys. (to be published).
3) Arndt R A, Hackman R H and Roper L D 1977 Phys. Rev. C15 1002-20.
4) Reid R V and Vaida M L 1975 Phys. Rev. Lett. 34 1064.
5) MacGregor M H, Arndt R A and Wright R N 1969 Phys. Rev. 182 1714-28.
6) Groenenboom P H L and Boersma H J 1978 Phys. Lett. 74B 1-5.
7) Kermode M W 1967 Nucl. Phys. A99 605-24.
McKerrell A, Kermode M W and Mustafa M M 1977 J. Phys. G: Nucl. Phys. 3 1349-57.
8) Vinh Mau R, 'The Paris nucleon-nucleon potential' to appear in 'Mesons in nuclei'.
9) Nagels M M, Rijken T A and de Swart J J 1978 Phys. Rev. D17 768-76.

ROLE OF NON-ITERATIVE DIAGRAMS IN NN-SCATTERING

K. Bleuler, K. Holinde and R. Machleidt
Institut für Theoretische Kernphysik der Universität Bonn
Nussallee 14-16, D-5300 Bonn, W. Germany
and
M.R. Anastasio, A. Faessler and H. Müther
Institut für Kernphysik der Kernforschungsanlage Jülich
Postfach 1913, D-5170 Jülich, W. Germany

In meson theory, the intermediate-range attraction of the two-nucleon force is built up by contributions arising from 2π-exchange. In dispersion-theoretic models, these contributions are determined by an analytic continuation of empirical πN- (and $\pi\pi$-) scattering data. In OBE-models, such contributions are effectively described by the exchange of a scalar, isoscalar meson usually called σ, with a mass around 500 MeV.

Recent nuclear matter calculations suggest however that such an approach, in which one does not give the intermediate-range, attractive part of the nucleon-nucleon potential any structure, does not lead to saturation at empirical densities. Thus, in order to understand nuclear structure phenomena, it is absolutely essential to work out explicitly the contributions arising from 2π-exchange. In order to have a well-defined transition from the two-body to the many-body problem we start from a field-theoretical Hamiltonian (leaving out antinucleons) and use non-covariant perturbation theory.

A considerable part of the 2π-exchange is built up by the fourth-order iterative diagrams with $N\Delta$- and $\Delta\Delta$-intermediate states involving π-exchange, Δ being the 33-isobar with a mass of 1236 MeV. (The corresponding diagrams involving NN-intermediate states are included by iteration of OPEP in the scattering equation). In a recent paper [1], we evaluated these diagrams in momentum space without any approximation (together with those involving ρ-exchange). They replace about one third of the original σ in pure OBE-models. We should note that, due to retardation effects in the propagators, these contributions are by a factor of 3 smaller than former models using twice-iterated transition potentials with simple propagators of pion (and rho) range, i.e. $(\underline{q}'-\underline{q})^2 + m_\pi^2$ (m_ρ^2).

It has been pointed out, however, that the non-iterative diagrams, see fig. 1, involving NN-, NΔ-, and $\Delta\Delta$-intermediate states, are of equal importance [2]. Since the calculation of these contributions is quite heavy, we present here - as a first step - the results for the case with NN-intermediate states only (involving π-exchange). We use γ^5-coupling, $g_\pi^2 = 14.4$ as coupling constant, and a form factor $F_\pi = (\Lambda_\pi^2 - m_\pi^2)/(\Lambda_\pi^2 - t)$ with $t = (E'-E)^2 - (\underline{q}'-\underline{q})^2$ and $\Lambda_\pi = 2.5$ GeV.

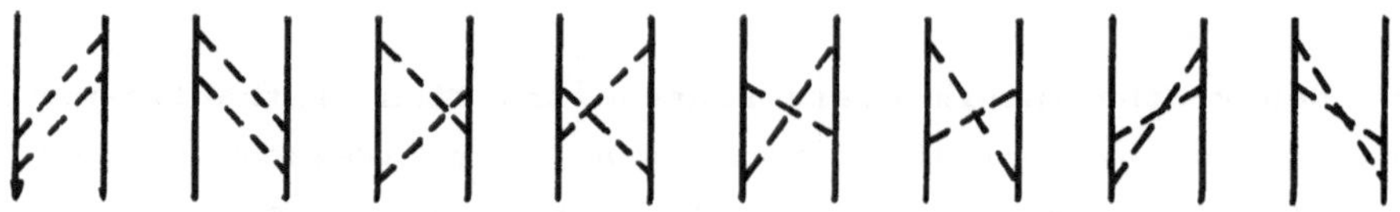

Fig. 1

In order to study the effect of the diagrams of fig. 1 on NN-scattering phase shifts, we add these contributions to an OBE-potential [3], likewise derived from non-covariant perturbation theory, and solve the corresponding scattering equation of Lippmann-Schwinger type, see ref. [3]. The results can be summarized as follows:

(i) The non-iterative box diagrams (first and second diagram of fig. 1) give a negligible contribution in all partial wave states apart from the $^3S_1$-state, in which they considerably increase the attraction.

(ii) The crossed-box diagrams (diagrams 3-8 of fig. 1) contribute in all partial waves. Their contribution is attractive or repulsive, depending on the specific partial wave. For example, they give attraction in ^1S_o, but repulsion in $^3SD_1$.

(iii) The diagrams of fig. 1 account for about 20% of the σ-contribution in pure OBE-models. Furthermore, they build up the long-range (non-resonant) part of the ρ-channel contribution to the NN-interaction.

References

1. K. Holinde, R. Machleidt, M.R. Anastasio, A. Faessler and H. Müther, Isobar Contributions to the Two-Nucleon Interaction Derived from Non-Covariant Perturbation Theory, preprint, to be published in Phys. Rev. C

2. J.W. Durso, M. Saarela, G.E. Brown and A.D. Jackson, Nucl. Phys. A278 (1977) 445

3. K. Kotthoff, K. Holinde, R. Machleidt and D. Schütte, Nucl. Phys. A242 (1975) 429

CORRELATIONS BETWEEN POTENTIALS AND OBSERVABLES IN THE NN INTERACTION*

F. Pauss and L. Mathelitsch
Institut für Theoretische Physik, Universität Graz, Austria

and

M. Lacombe, B. Loiseau, R. Vinh Mau
Division de Physique Théorique, Institut de Physique Nucléaire
91406 Orsay Cedex, France

We wish to study the effects of the components of a NN potential on the observables for energies between 20 and 350 MeV. The knowledge of these correlations might then be useful to give constraints on components of the NN force. For this purpose, we have chosen the soft core and velocity dependent potential as given in ref. [1].

We show results for pp and np obtained with: (0) the complete Paris potential - (I) the potential without the central p^2 velocity dependent term - (II) a static equivalent potential - (III) the central and spin spin potentials - (IV) the potential without $\vec{L}.\vec{S}$ and SO2 forces - (V) the potential without tensor and SO2 forces - (VI) the potential without SO2 force - (VII) (only for np) case (III) for the isospin I = 0 and the complete I = 1 potential.

We found a) for the I = 1 interaction: - the p^2 term acts strongly and is attractive at low energy and repulsive at high energy - the tensor force contribution is important at low energy and that of LS at high energy (I_o fig. 1) - the Polarization P (fig.2) (and also D, A, R, C_{NN}) are good test for the tensor, LS and to a smaller extent the SO2 potentials.

b) for the I = 0: - the p^2 term behaves like that of the I = 1 - the contribution of the tensor is attractive and very important - that of LS negligeable - that of SO2 small (I_o fig. 3) - P is mainly given by the I = 1 LS and I = 0 tensor forces (fig. 4) - C_{NN} constraints the I = 0 tensor force. It seems a reasonable fit to the data could be obtained without the SO2 component of the potential.

*Supported by the Fonds zur Förderung der wissenschaftlichen Forschung, Project 2900.

Reference

[1] M. Lacombe, B. Loiseau, J.M. Richard, R. Vinh Mau, P. Pirès and R. de Tourreil, "Parametrization of the Paris potential", Contributed paper to the 2nd International Conference on the NN interaction, June 1977, Vancouver, Canada.

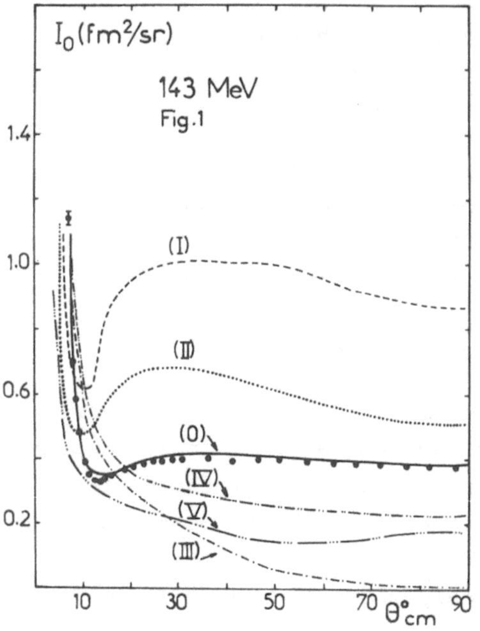

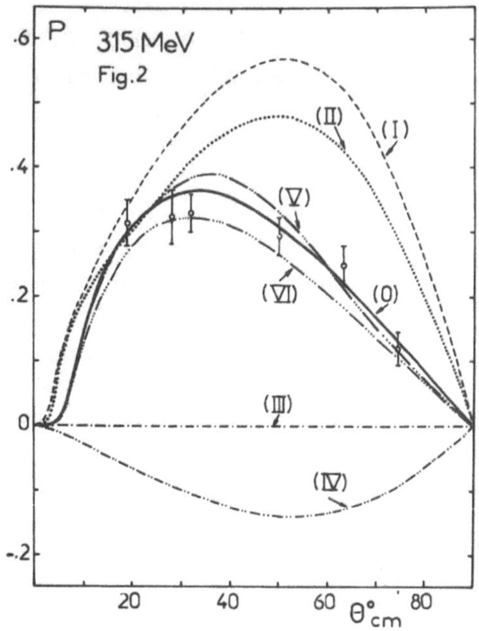

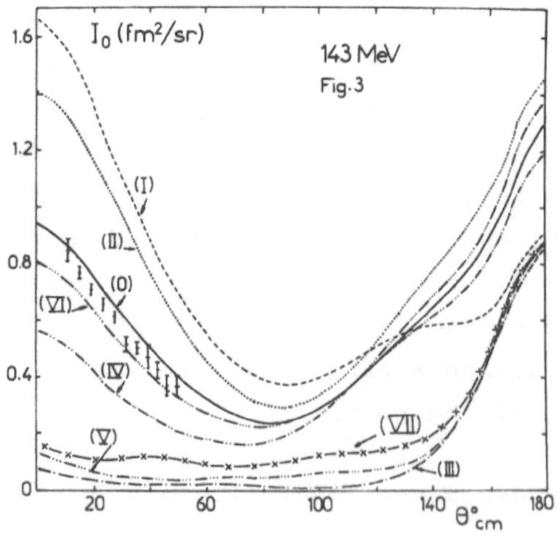

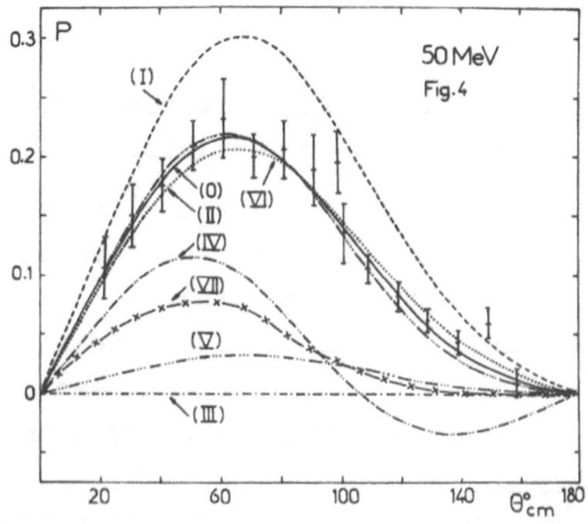

PARTICLE MIXING AND CHARGE ASYMMETRIC ΛN FORCES

Sidney A. Coon[†*] and Peter C. McNamee[††]

[†]Institut de Physique, Université de Liège, Belgium and Department of Physics, University of Arizona, Tucson, Arizona[+].

[††]Department of Physics, University of Arizona, Tucson, Arizona, and Stanford Research Institute, Menlo Park, California[+].

A nonnegligible charge asymmetry in the baryon-baryon strong interaction has been conclusively demonstrated by reliable determinations[1,2)] of the electromagnetic (EM) contribution to the binding energy difference of the mirror nuclear pair (^{3}He-^{3}H). Charge asymmetric NN potentials[3,4)] due to particle mixing ($\pi^\circ\eta$ and $\rho^\circ\omega$) in one-meson-exchange models match very well (17 keV + 56 keV) the now determined nuclear charge asymmetry (∿81 ± 29 keV) in the pair.[1)]

There is further evidence for baryon-baryon charge asymmetry in the separation energy difference for the mirror hypernuclear pair ($^4_\Lambda$He-$^4_\Lambda$H): $\Delta B_\Lambda = B_\Lambda(^4_\Lambda\text{He}) - B_\Lambda(^4_\Lambda\text{H})$ where $B_\Lambda \equiv B(^A_\Lambda Z) - B(^{A-1}Z)$. The experimental value of ΔB_Λ is large and positive (+350 ± 50 keV). The Coulomb contribution (due to ^{3}He core compression within $^4_\Lambda$He) to ΔB_Λ is small (∿10-20 keV) and negative,[2)] indicating a sizable charge asymmetry in the ΛN interaction. This paper estimates the contribution of $\pi^\circ\eta$, $\rho^\circ\omega$, $\phi\rho^\circ$, $\phi\omega$, and $\Sigma^\circ\Lambda$ mixing to ΔB_Λ in a simple 2-body "core plus lamda" model of the 4-body hypernuclei. We find that particle mixing, especially $\Sigma^\circ\Lambda$ mixing, seems to be the dominant source of charge asymmetry in the NΛ interaction.

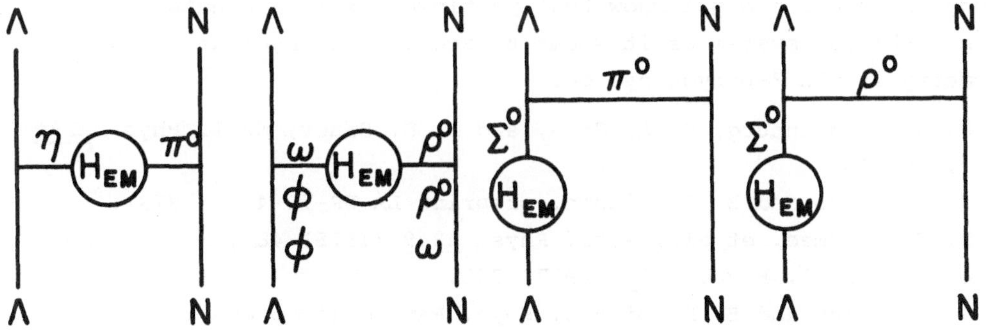

We calculate the ΛN potentials corresponding to the diagrams shown in an analogous way to the NN case.[3)] The "hadronic hamiltonian" contains an isospin-violating perturbation H_{EM} which mixes particles of different isospin;[3)] $\langle\omega|H_{EM}|\rho^\circ\rangle$ and $\langle\Lambda|H_{EM}|\Sigma^\circ\rangle$ are known from experiment. We substitute for the unknown inner region of the potentials a hard core of radius 0.4 fm.

We calculate the expectation value of these potentials with a 2-body model for the wavefunctions.[5] A crude estimate of ΔB_Λ in the excited (J=1) states can be made if we assume they have the same wavefunction as the ground states. The experimental evidence, not yet conclusive for B_Λ ($^4He^*$), is that ΔB_Λ(J=1) is -20 ± 70 keV. Contributions from each of the meson mixing diagrams are tabulated below in keV.

	J=0(ground states)	J=1(excited states)
$\Delta B^{\pi\eta}$	- 18	+ 6
$\Delta B^{\rho\omega}$	+ 22	+23
$\Delta B^{\phi\rho}$	- 1	+ 7
$\Delta B^{\phi\omega}$	+ 5	+ 9
$\Delta B_\pi^{\Sigma\Lambda}$	+ 59	-20
$\Delta B_\rho^{\Sigma\Lambda}$	+114	-62
Total (ΔB_Λ)	+181	-37

In this model ΔB_Λ ranges from 181 to 189 keV for radial compression of the core from 0% to 5%. The mass difference of charged Σ particles may also contribute to ΔB_Λ about +30 to +100 keV in a 2π exchange graph,[6] or ∿130 keV in a Λ-Σ conversion model.[7]

Adding up particle mixing and Σ mass differences we nearly match the experimental charge asymmetry. But all these estimates were made in a 2-body model of the 4-body problem. An exact 4-body model estimate of ΔB_Λ indicates that the 2-body model approach underestimates the exact model result by more than a factor of 2.[8] A more exact 4-body calculation might show that particle mixing accounts for ΔB_Λ in the 4-baryon system as it seems to account for nuclear charge asymmetry in the 3-baryon system.

1. R. A. Brandenburg, S. A. Coon, and P. U. Sauer, Nucl. Phys. A294 (1978) 305.
2. J. L. Friar and B. F. Gibson, preprint LA-UR-78-411 (1978).
3. P. C. McNamee, et al., Nucl. Phys. A249 (1975) 483, S. A. Coon, et al., Nucl. Phys. A287 (1977) 381.
4. J. L. Friar and B. F. Gibson, Phys. Rev. C (in press).
5. P. C. McNamee and R. J. Oakes, Phys. Rev. 149 (1966) 1157.
6. R. H. Dalitz and F. Von Hippel, Phys. Lett. 10 (1964) 153.
7. J. Dabrowski, Acta Physica Polonica B6 (1975) 453.
8. B. F. Gibson and D. R. Lehman, LASL preprint (1978).

*Chercheur I.I.S.N.
+Present address.

TREATMENT OF DIVERGENT EXPANSIONS IN SCATTERING THEORY

A. Gersten and S. Malin
Department of Physics,
Ben Gurion University of the Negev,
Beer Sheva, Israel

Introduction

One of the biggest obstacles in applying quantum field theory to realistic scattering problems are the divergencies of perturbation expansions for large coupling constants and the divergencies of partial wave expansions for massless particles exchanges. There exist, however methods of summation of the divergent expansions[1] which can lead to significant applications in physics[2]. In this paper we treat the problem of summing such expansions using three methods: (i) a generalization of the Padé approximation to the multivariable care. The suggested definition is unique and preserves unitarity. (ii) The summation of divergent partial waves for arbitrary spins. (iii) A successful application of a series inversion to the $^3P_1$ nucleon-nucleon phase shift up to 200 Mev.

Multivariable Padé approximants

There exists an extensive literature on the applications of Padé approximants[3]. These approximants are defined in the following way. Let f(z) have the formal expansions

$$f(z) = a_0 + a_1 z + a_2 z^2 + \dots \tag{1}$$

We define:

$$\begin{aligned} f_L(z) &= a_0 + a_1 z + \dots + a_L z^L \\ P_N(z) &= b_0 + b_1 z + \dots + b_N z^L \\ Q_M(z) &= 1 + c_1 z + \dots + c_M z^M \end{aligned} \tag{2}$$

The Padé approximant [N/M](z) is defined as

$$[N/M](z) = P_N(z)/Q_M(z) \tag{3}$$

where the polynomials $P_N(z)$ and $Q_M(z)$ are obtained from the equation

$$f_{N+M}(z)Q_M(z) - P_N(z) = 0(z^{N+M+1}) \tag{4}$$

In problems where more then one variable is treated, (like many coupling constants) there is a need to develop a generalization of the above method. Here we suggest a unique definition based on a variational principle as follows: Consider the n-variable minimum problem

$$\lim_{R\to 0} \text{Min} \frac{1}{R^N} \int_{-R}^{R} \dots \int_{-R}^{R} |f_L(z_1,\dots,z_n)Q(z_1,\dots,z_n)-P(z_1,\dots,z_n)|^2 dz_1,\dots,dz_n \tag{5}$$

and let the generalized Padé approximants be defined as $P(z_1,\dots,z_n)/Q(z_1,\dots,z_n)$, where P and Q are polynomials in $z_1,\dots,z_n$, the coefficients of which are found by solving Eq. (5). The power N is chosen to leave the functional (5) finite and different from zero. This is a generalization of the usual (one-variable) Padé approximation, where the coefficients $b_0,\dots,b_N$ and $c_1,\dots,c_n$ can be found by solving the minimum problem

$$\lim_{R\to 0} \frac{1}{(R^{N+M+1})^2} \int_{-R}^{R} |f_{N+M}(z)Q_M(z) - P_N(z)|^2 dz \tag{6}$$

In the one-variable case this variational method is equivalent solving Eq. (4). In the multivariable case the generalization of Eq. (4) runs into problems of uniqueness[4]. Eq. (5), on the other hand, defines a unique approximant, which seems to have the property that unitary amplitudes correspond to unitary approximants. Work on the details of the proof is now in progress. The solution of Eq. (5) goes via a set of linear equations for the coefficients of the polynomials P and Q. Therefore if a solution exists, it is unique. As an example let us take the two variable approximant for

$$\begin{aligned} f_{22}(z_1,z_2) &= a_0 + a_{10}z_1 + a_{01}z_2 + a_{11}z_1z_2 + a_{20}z_1^2 + a_{02}z_2^2 \\ P(z_1,z_2) &= b_0 + b_{10}z_1 + b_{01}z_2 \\ Q(z_1,z_2) &= 1 + c_{10}z_1 + c_{01}z_2 \end{aligned} \tag{7}$$

Substituting Eqs. (7) in Eq. (5) (with N=2) we obtain after lengthy calculations

$$\begin{aligned} b_0 &= a_0 \\ c_{10} &= (a_{10}a_{01}^2a_{02} - a_{10}^3a_{20} - a_{01}^3a_{11} - \tfrac{4}{5}\, a_{10}a_{01}^2a_{20})/D \\ c_{01} &= (a_{01}a_{10}^2a_{20} - a_{01}^3a_{02} - a_{10}^3a_{11} - \tfrac{4}{5}\, a_{10}^2a_{01}a_{02})/D \\ D &= a_{10}^4 + a_{01}^4 + \tfrac{4}{5}\, a_{10}^2a_{01}^2 \end{aligned}$$

Divergent Partial Wave Expansions

The scattering amplitude for particles with spin is given by the sum

$$f_{\lambda\mu}(\theta) = \sum_{j=0}^{\infty} c_j d^j_{\lambda\mu}(\theta).$$

The expansions are usually divergent if massless paricles are exchanged (e.g. in quantum electrodynamics). To sum these expansions we suggest the following generalization of Padé approximants

$$f_{\lambda\mu}(\theta) \simeq \Sigma a_J d^J_{\lambda\mu}(\theta)/\Sigma b_J P_J(\cos\theta)$$

where $P_J(\cos\theta)$ are the Legendre polynomials. Most of the details of similar approximants were worked out by Holdeman[5]. One can easily adopt the Holdeman procedure by adding the recursion relation

$$d^J_{\lambda\mu}(\theta)d^1_{00}(\theta) = \cos\theta\, d^J_{\lambda\mu}(\theta) = \Sigma < J,\lambda;1,0|\ell,\lambda >< J,\mu;1,0|\ell,\mu > d^{\ell}_{\lambda\mu}(\theta)$$

Series inversion

Series inversion can be used in some cases as a method of summing divergent expansions. We have tried this method for the nucleon-nucleon phase shifts. Denoting $\alpha = g^2/(4\pi)$, where g is the pion-nucleon coupling constant, one may have an expansion for the phase shift:

$$\delta = d_1\alpha + d_2\alpha^2 + d_3\alpha^3 + \ldots + \text{other terms.}$$

Assuming only expansion in α one can invert the series

$$\alpha = b_1\delta + b_2\delta^2 + \ldots$$

$$b_1 = 1/d_1, \quad b_2 = -d_2/d_1^3, \ldots$$

Using a recent phase shift analysis[6] and computed partial waves of one and two pion exchanges (d_1 and d_2) in the N-N interaction, we have obtained the following results for the $^3P_1$ phase shift which are depicted in fig. 1.

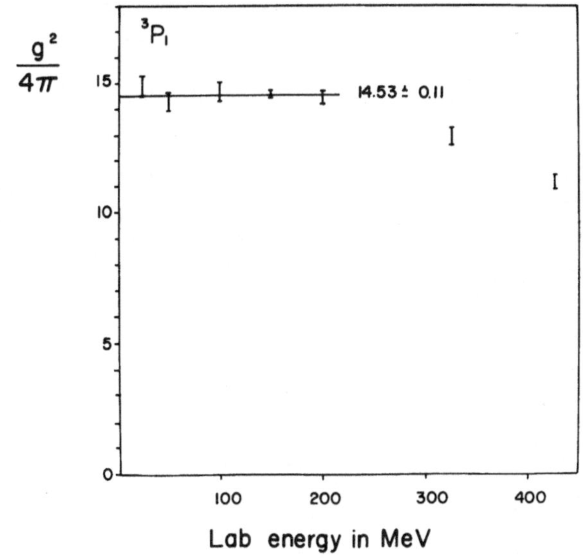

Fig 1

The pion-nucleon coupling constant for different lab energies

Up to about 200 MeV a perfact fit with a straight line $\alpha = 14.53 \pm 0.11$ is obtained. For other well determined phase shifts like $^1S_0$, $^3P_0$, $^3P_2$, $^1D_2$ we do not have such a result. An explanation to this might be the influence of the N-N core which causes a change of sign of the above phase shifts making the inversion double valued and singular.

Conclusions

The mere fact that some expansions of theoretical physics are divergent should not be discouraging since excelent methods of summing divergent expansions exist and others are being developed. These methods preserve essential physical properties of the solutions.

References

1. Hardy G.H. Divergent Series, Oxford, London (1949).
2. "The Padé approximants in theoretical physics" Ed. G. Baker and J. Gammel, Academic Press (1970).
3. Proc. of the 1970 Cargese Summer School on Padé Approximants, Ed. D. Bessis, Gordon and Breach, London (1972).
4. J.S.R. Chisholm and J. McEwan, Proc. Roy. Soc. Lond. A336, 421 (1974).
5. J.T. Holdeman, Jr., Math. Comput. 23 (1969) 275.
6. R.A. Arndt et al. Phys. Rev. C9, 555 (1974).

MEASUREMENTS ON BOUND ANTINUCLEON-NUCLEON SYSTEMS

G. Backenstoss[1], P. Blüm[3], K. Fransson[4], R.Guigas[3]
M. Izycki[1], H. Koch[3], A.Nilsson[4], P.Pavlopoulos[1],
H. Poth[3], M. Suffert[2], L. Tauscher[1] and K.Zioutas[2]
University Basel, Switzerland[1],
CERN Geneva, Switzerland[2]
Kernforschungszentrum and University Karlsruhe, Germany[3]
Research Institute for Physics, Stockholm, Sweden[4].

Recently a number of experiments became known where narrow resonances above the antinucleon-nucleon threshold and narrow bound states below threshold have been reported[1]. The great interest arisen from these experiments originates from the rather unexpectedly long lifetime of these states excluding a fast annihilation into mesons. Several theoretical attempts have been made to understand this phenomenon. One approach takes advantage of the known NN boson exchange potential[2] which is transformed by a G-parity transformation in an $N\bar{N}$ potential which thus results in a deeply binding potential. It is however by no means obvious that the states produced have the narrow widths observed[3]. A completely different approach is based upon the quark model. The narrow states are associated with baryonium[4], a diquark-antidiquark state ($qq\,\bar{q}\bar{q}$) formed by $N\bar{N}$ which is prevented from decaying into meson states ($q\bar{q}$) by a rule analogous to the Okubo-Zweig-Iizuka rule which prevents the Ø-meson to decay into pions. Regge trajectories for baryonium have been calculated[5].

The only subthreshold state so far seen is associated with the $\bar{p}n$ system as observed from $\bar{p}$ stopping in a deuterium bubble chamber as an enhancement in some pionic annihilation channels[6]. The interpretation relies however on the spectator role of the p in deuterium. A study of $p\bar{p}$ systems hence seems to be most desirable.

If antiprotons slow down in matter they are Coulomb captured and form antiprotonic atoms. The states of these atoms are determined by the pure electromagnetic interaction, except for the lowest observable states. There the influence of the strong $\bar{p}N$ interaction manifests itself in a shift ΔE and a broadening Γ of the state which is related to the complex effective scattering length A^* by

$$\Delta E - (i/2)\Gamma = - (2\pi/m)A^* \; |\psi\ (0)\ |^2 \qquad (1)$$

where m is the reduced $\bar{p}$-mass and the $\bar{p}$ wave function ψ is taken at the origine. A^* however depends on the properties of the scattering amplitude near threshold. In kaonic atoms the influence of the $Y_0^x(1405)$ resonance in the K^-p channel 25 MeV below threshold was clearly seen in heavier kaonic atoms ($_6C$,$_{15}P$..) shifting the lowest observable atomic levels as an attractive potential. A similar situation occured in heavier antiprotonic atoms[7)]. Should that be a hint for so far unknown $\bar{p}N$ bound states near threshold ? In order to avoid complications due to nuclear structure, measurements on H should be performed where formula (1) should hold for A_s the true s wave scattering length for $p\bar{p}$. Unfortunately, measurements of the 2p-1s transition in H turn out to be rather difficult. An other possibility is given by measurements on suitable isotopes where the nuclear structure effects may be compensated to a certain extent. Finally one may attempt to detect direct transitions into the strongly bound states. All three possibilities have been followed up and should shortly be discussed here.

a) Isotope effects in $\bar{p}$-Atoms.

The effective scattering length determined from a number of $\bar{p}$-atoms turned out to be $A^* = (+2.0+i.2.0)$ fm corresponding to attraction. However, the interpretation of A^* in terms of the $\bar{p}p$ scattering length is by no means straightforward. A careful measurement on the isotopes ^{16}O and ^{18}O yielded for the isotopic difference of the strong interaction shift $(\Delta E(^{18}O)-\Delta E(^{16}O))/E(^{16}O)$ $= (-1.06\pm0.65)\ 10^{-3}$ resulting in $A_{\bar{p}n} = [(-0.3\pm1.4)+i(1.0\pm1.7)]$ fm.This value is unfortunately too insensitive to determine the sign of the interaction.

b) Antiprotonic hydrogen.

The energy shift ΔE of the 1s level would directly allow to draw conclusions about the $\bar{p}p$ scattering length. Unfortunately due to the Stark effect, absorption from highly excited s states is likely and hence a suppression of the 2p-1s transition by $\sim 10^3$ must be expected. A search for very weak lines in the energy region between 5 and 15 keV is necessary where the only signature are the energy differences of the 2p-1s/3p-1s/4p-1s-transitions with unknown intensity pattern. At the sensitivity required a great number of Xray lines of target surrounding elements are present. Although statistically significant structures have been observed an unambiguous identification of the lines has not yet been established.

c) Direct Transitions.

Direct transitions between the atomic states about 10 keV below threshold to the deeply bound states may occur in the energy region of several hundred MeV by photon or pion emission. We report here on observations of high energy γ-rays associated with stopping $\bar{p}$ in H performed at the CERN-PS separated stopping $\bar{p}$ beam[8]. The main difficulty in observing γ rays of low yield is a high continuous background originating from the decay of π^0 mesons produced in the $p\bar{p}$ annihilation where 3.52 γ quanta are known to be produced. The detection was attempted with a large NaI crystal (25x30cm) surrounded with a plastic shield in order to reduce the leakage of the shower produced by the γ and hence improving on the energy resolution. A sophisticated trigger was required to select the $\bar{p}$'s stopping in a thin liquid hydrogen target. Calibration of energy and pulse shape was done with a tagged photon beam at the Bonn synchroton. Also stopping pions obtained at the CERN $\bar{p}$-beam provide by the reaction $\pi^- + p \rightarrow n + \gamma$ monoergetic γ rays of 129 MeV suitable for monitoring the apparatus. The spectra obtained show three statistically significant unknown lines betweeen 180 and 420 MeV. At 132 MeV a further line appears of comparable intensity which is fully understood as originating from stopping π^- from $p\bar{p}$ annihilation. The results are given in table I. It should be emphasized that not so much the existence of the transitions but the narrow width for which the experimental resolution is only a limit is the very new exciting and unexpected result.

Antiprotons have also been stopped in He with the intention to obtain a good spectrum for the $\bar{p}p$ annihilation background. However, also this spectrum shows some structure, possibly broader than the resolution, in the region above 350 MeV. The statistical significance must however be improved.

The present data should be considered only as a first indication of the new states. The significance of the lines and the resolution of the detection systems must be improved. A correlation of the transitions with certain annihilation channels and particularly the quantum numbers are of greatest interest.

The interpretation of the data in terms of the various theoretical ideas may lead to important results. The relation between the NN- and the $N\bar{N}$ potential could be used to derive information about the interaction at small distances where bound states are much more sensitive than scattering amplitudes. The baryonium, its relation to meson states and hence important aspects of the quark picture may be studied in some detail. Finally the knowledge of the $N\bar{N}$ interaction near threshold is of considerable interest in astrophysics and cosmology where the question of separation between matter and antimatter is discussed.

Table I

Energy (MeV)	Γ(MeV)	Conf.level (%)	Yield (10^{-3})	Mass (MeV)
132±6	< 16	0.7	5.1±2.7	$\pi^- + p \to n + \gamma$
183±7	< 19	1.0	7.2±1.7	1684
216±9	< 21	2.5	6.0±1.9	1646
420±17	< 34	1.8	8.5±2.0	1395

References

[1]) L.Montanet, $N\bar{N}$ Review, V.International Conference on Experimental Meson Spectroscopy, Boston 1977

[2]) R.A. Bryan and R.J.N. Phillips, Nucl.Phys.B5,201 (1968)

[3]) L.N. Bogdanova, O.D. Dalkarov and I.S.Shapiro,Ann.Phys.84,261(1974)
I.S. Shapiro, Phys.Rep.35,129 (1978)
C.B. Dover,Proc.IV Int.Symp.Nucl.-Antinucl.Interact.,Syracuse (1975)
J.M. Richard,M.Lacombe and R.Vinh Mau,Phys.Lett.64B,121,(1976)

[4]) G.F. Chew, Proc.III Europ.Symp.on $N\bar{N}$ Interact.,Stockholm(1976),p.515

[5]) G.C. Rossi and G. Veneziano, Nucl.Phys.B123,507 (1977)
Chan Hong Mo and H. Høgassen, Phys.Lett.72B,121 (1977)

[6]) L. Gray, P. Hagerty and T. Kalogeropoulos,Phys.Lett.26, 1491 (1971)

[7]) H. Poth,G. Backenstoss,I. Bergström,P.Blüm,J.Egger,W. Fetscher, R.Guigas,C.J.Herrlander,M.Izycki,H.Koch,A.Nilsson,P.Pavlopoulos, I.Sick,L.Tauscher, et.al. Nucl.Phys.A294,435 (1978)

[8]) P.Pavlopoulos,G. Backenstoss,P.Blüm,K.Fransson,R.Guigas,N.Hassler, M.Izycki,H.Koch,A.Nilsson,H.Poth,M.Suffert,L.Tauscher and K.Zioutas Phys.Lett.72B,415 (1978).

NONPERTURBATIVE CALCULATION OF NUCLEON-ANTINUCLEON ANNIHILATION DIAGRAMS IN A SIMPLE MODEL

M. BAWIN
Physique Nucléaire Théorique, Institut de Physique,
Université de Liège, Sart Tilman,
B-4000 Liège 1, Belgium.

As is well known, a potential for the description of the nucleon-antinucleon $(N-\overline{N})$ system can be obtained using N-N one boson exchange models and G-parity transformation. To that potential, one must then add the annihilation diagrams describing the virtual annihilation of the $N\overline{N}$ system into pions. The simplest such diagram (one pion annihilation) can be computed in a straightforward way as a perturbation. Although diagrams associated with annihilation into two (or more) pions are expected to contribute the more important part of the annihilation "potential", it is nevertheless of interest to compare a nonperturbative calculation of the one pion annihilation diagram to the perturbative one. Indeed, the former calculation implies a renormalization of the wavefunction[2], and it is of interest to know to what extent this renormalization will affect the final result.

We consider a model involving two identical neutral spinless particles ("nucleons"), which can be considered a three-dimensional analogue of the model studied by Okubo and Feldman[2] in the framework of the Bethe-Salpeter equation. Our basic equation is, in the center-of-mass (C.M.) frame :

$$(\vec{p}^2 - \vec{k}^2)\Psi(\vec{p}) = \int V(\vec{p},\vec{p}\,')\Psi(\vec{p}\,')d^3p' + \frac{\lambda}{2\pi^2}\frac{m}{2(W^2 - m_\pi^2)}\int \Psi(\vec{p}\,')d^3p' \quad . \quad (1)$$

In eq. (1), m is the "nucleon" mass, m_π is the "pion" mass, and $W = 2\sqrt{k^2+m^2}$ is the total C.M. energy of the $N-\overline{N}$ system, k being the C.M. relative momentum of one nucleon. The interaction $V(\vec{p},\vec{p}\,')$ describes the "potential" part of the $N\overline{N}$ interaction (obtained, e.g. from the N-N interaction via G-parity). The second term on the right-hand side is the one "pion" annihilation diagram, whose coupling strength is given by $\lambda \simeq 0.08$.

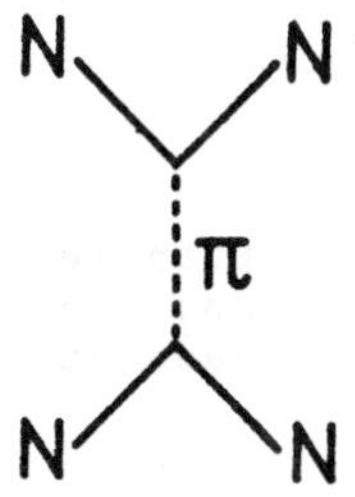

One pion annihilation diagram

The equation we use comes from a simple Klein-Gordon-like formulation of the relativistic two-body problem[3]. In order to make eq. (1) analytically soluble, we take :

$$V(\vec{p},\vec{p}\,') = \frac{\alpha}{4\pi}\,\frac{v(p)v(p')}{2m} \quad , \quad \text{with} \quad v(p) = \frac{1}{p^2 + m^2} \quad .$$

This potential yields a total energy

$$W_o = 2m\sqrt{1 - X_o^2} \quad , \quad \text{where} \quad X_o = \sqrt{\frac{\alpha\pi}{8}} - 1 \quad .$$

Solving eq. (1) and renormalizing $\Psi(\vec{p})$ in such a way that $W_R = m_\pi$ is a bound state solution, we find :

$$W_R = 2m\sqrt{1 - X_R^2} \quad , \quad \text{with} \quad X_R \simeq X_o + \frac{\lambda}{16} \quad ,$$

where we have made the approximation $W_R >> m_\pi$.

On the other hand, the perturbation result from the annihilation diagram yields :

$$W = 2m\sqrt{1 - X_o^2 - \Delta X_o^2} \quad , \quad \text{with} \quad \Delta X_o^2 = \frac{\lambda X_o(1 + X_o)\, m^2}{4\,\pi\, W_o^2} \quad .$$

For $\lambda << 1$, we then get, writing $X_R^2 = X_o^2 + \Delta X_R^2$:

$$\frac{\Delta X_R^2}{\Delta X_o^2} \simeq \frac{4\,\pi\, W_o^2}{m^2(1 + X_o)8} \quad . \qquad (2)$$

Although both ΔX_R and ΔX_o are much smaller than X_o , our simple model nevertheless suggests that, once a more fundamental description of the N-$\overline{N}$ system is arrived at, nonperturbative calculations of the one pion annihilation diagram may turn out to be important.

REFERENCES

1. I.S. Shapiro, Phys. Rep. 35C, 131 (1978).
2. S. Okubo and D. Feldman, Phys. Rev. 117, 279 (1960).
3. J.S. Kang and H.J. Schnitzer, Phys. Rev. D12, 841 (1975).

Investigation of the ΔN Interaction in the Deuteron

R. Händel, M. Dillig, M. G. Huber
Institute for Theoretical Physics, Univ. Erlangen, W-Germany

The interaction of isobars with nucleons manifests itself in nuclear reactions which involve the excitation of baryonic resonances. For $\Delta(1236)$ dominated scattering processes on the deuteron the transition amplitude reads[1)]

$$T(\vec{k},\vec{k}')=\langle\vec{k}'|H'G_{\Delta N}(\omega)H|\vec{k}\rangle \qquad (1)$$

where

$$G_{\Delta N}(\omega)=\frac{1}{\omega-H_{\Delta N}}=\sum_{\mu}\frac{|D^{*}_{\mu}\rangle\langle D^{*}_{\mu}|}{\omega-\varepsilon^{*}_{\mu}} \qquad (2)$$

The positions and strengths of the eigenmodes $|D^{*}_{\mu}\rangle$ of the ΔN system reflect the influence of the ΔN interaction in the Hamiltonian $H_{\Delta N}$. The D^{*}-spectrum has been calculated[2)] using for the ΔN interaction a one boson-exchange potential including π, σ, ρ and ω exchange (Fig. 1a). The strong spin dependence of the resulting potential is demonstrated by the splitting of the D^{*} spectrum in Fig. 1b: the two spectra correspond to two choices for the ρBB coupling constants (for further details see ref. 2).

The differential cross sections for elastic pion scattering and for coherent photopion-production have been calculated using the same parameter sets as for the spectra (A) and (B), respectively, of Fig. 1b. The results are compared with available experimental data in Fig. 2. It turns out that (i) the ΔN interaction plays an important role; (ii) the cross sections (particularly polarization data) are sensitive to the details of the interaction[2)] and (iii) the limited amount of data presently available does not allow to discriminate between the interactions (A) and (B).

The investigation of the ΔN interaction is presently only in its beginning; for a more detail understanding of the ΔN system a systematic analysis of various medium energy reactions on the deuteron (and on more complex nuclei) is required.

1) L. S. Kisslinger, W. L. Wang, Phys. Rev. Lett. 30 (1973) 1071
2) R. Händel, M. Dillig, M. G. Huber, Phys. Lett. 73 B (1978) 4
   R. Händel, Dissertation (Erlangen 1978)
3) B. Bouquet et al., Nucl. Phys. 79 B (1974) 45
   G. von Holtey et al., Zeitsch. Phys. 259 (1973) 51

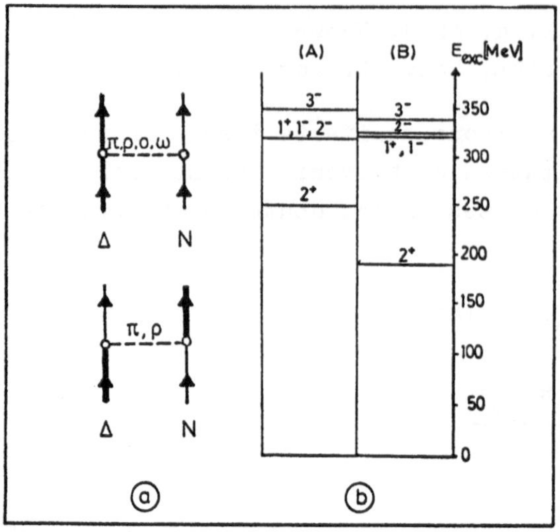

Fig. 1:

(a) Schematic representation of the direct and the exchange term in the ΔN interaction

(b) Sensitivity of the $D^*(T=1)$ spectrum on the ρBB coupling constant. Spectrum (A) refers to $f_{\rho NN}/4\pi$ = 1.89, spectrum (B) results from $f_{\rho NN}/4\pi$ = 3.32 (for details see ref. 2).

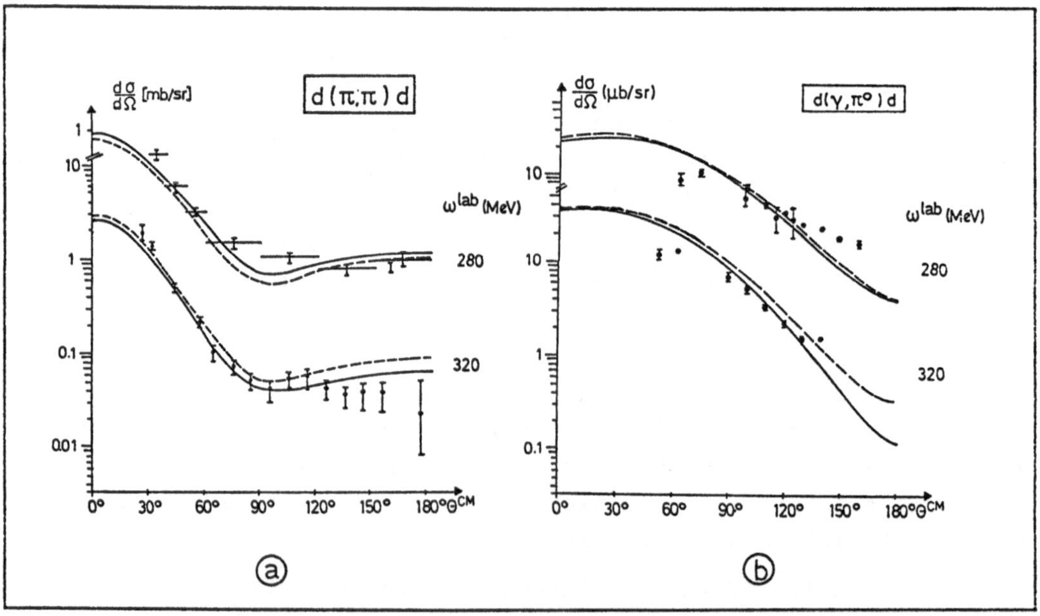

Fig. 2: Influence of the ρBB coupling constant on the differential cross section in elastic π-scattering (a) and pion-photo-production (b) on the deuteron at two different scattering energies. The full and the dashed lines correspond to the weak and the strong ρ-coupling as specified in Fig. 1b (the experimental values are from ref. 3).

Investigation of the πNN, $\pi \Delta N$ and $\pi\Delta\Delta$ Formfactor

M. Dillig
Institute for Theoretical Physics, Univ. Erlangen

M. Brack
Institute v. Laue-Langevin, Grenoble, France

The proper off shell continuation of the pion-baryon vertices requires information about their finite range structure. As the extraction of such information from experimental data is rather ambiguous and model-dependent a detailed microscopic analysis of the πNN, the $\pi N\Delta$ and the $\pi\Delta\Delta$ form factors was performed.

In a first step the contributions from first order triangle and loop diagrams were evaluated including contributions from π, σ and ρ exchange together with excitation of N, $\overline{N}$ and Δ(1236) intermediate states. For the dominating agency, the $\pi\rho$-triangle diagram the contributions were summed to all orders by solving the integral equation for $F_{\pi BB}(q)$

$$F_{\pi BB}(q^2)=1+\int G_{\pi\rho b}(\vec{q}-\vec{k})F_{\pi BB}(k^2)d\vec{k} \qquad (1)$$

the selfconsistent evaluation of the πNN, $\pi N\Delta$ and $\pi\Delta\Delta$ form factors leads to a set of coupled integral equations (Fig. 1).

The resulting q-dependence of the various form factors is demonstrated in Fig. 2; in a monopole parametrization

$$F_{\pi BB}(q^2)=\Lambda^2_{\pi BB}/(\Lambda^2_{\pi BB}+q^2) \qquad (2)$$

the cut off masses turn out to be $\Lambda_{\pi NN}\sim 7m_\pi$, $\Lambda_{\pi N\Delta}\sim 5m_\pi$ and $\Lambda_{\pi\Delta\Delta}\sim 6m_\pi$ for purely virtual pions. Increasing the pion energy leads to a decreasing cut off mass with $\Lambda_{\pi N\Delta}\sim 4m_\pi$ for $\omega_\pi\sim 2m_\pi$, accompanied by an increasing complex component (which accounts for the real π decay for energies above the π-production threshold).

Unfortunately the present findings suffer from a serious sensitivity on various model parameters (the main uncertainty is introduced by the ρBB coupling constants, which are rather poorly known). For a more detailed understanding of the finite range structure of vertices further investigations are highly requested.

1) M. Dillig, M. Brack, preprint, Erlangen 1977
(to be published in Journ. Nucl. Phys. G)

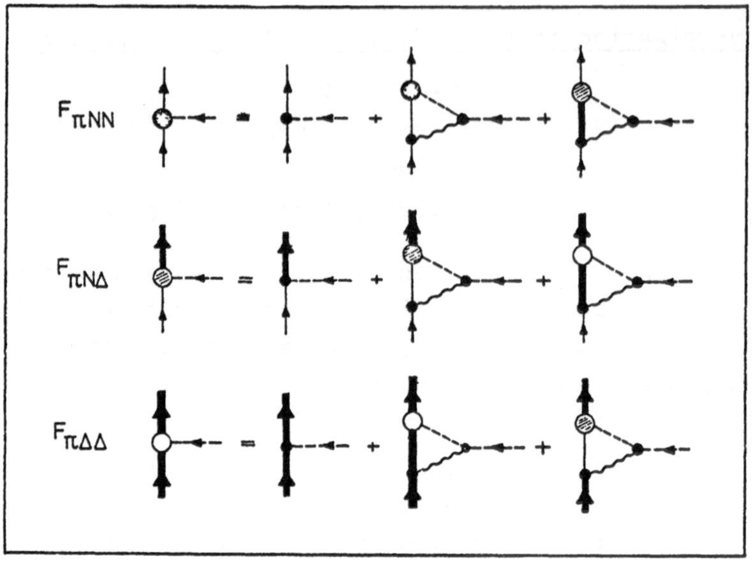

Fig. 1: Schematical representation of the set of coupled integral equations for a self consistent calculation of the πNN, $\pi N\Delta$ and $\pi\Delta\Delta$ form factors.

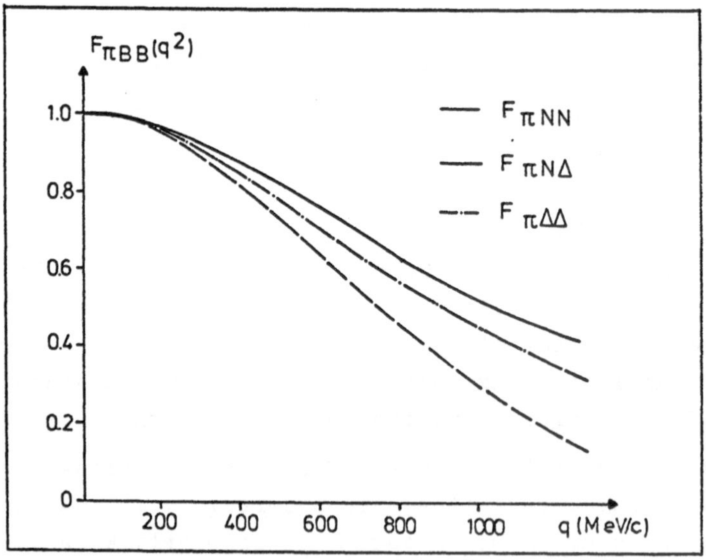

Fig. 2: Dependence of the various πBB form factors on the pion momentum q (for a pion energy $\omega = 0$).

LIGHT FRONT FIELD THEORY TREATMENT OF THE 2N PROBLEM

J.M. Namyslowski
Institute of Theoretical Physics, Warsaw University
Pl-00 681 Warsaw, ul. Hoza 69, Poland

1. Introduction

Two nucleons, moving in their CMS with a relative momentum in the range (0.1 - 0.3) GeV/c, are usually considered as being properly described by the non-relativistic dynamics, with some relativistic corrections, roughly of the order of 10% effects. The non-relativistic approximation of $(k^2 + m^2)^{1/2}$ by $m + \frac{1}{2}k^2m^{-1}$ corroborates that statement. Such reasoning is partly based on the pure kinematical argument, and partly on the phenomenological fit [1] of the 2N data by the Schrödinger equation, in which there is no attempt to explain the nature of the NN potential.

Field theory offers some insight into the structure of the 2N interaction. However, so far the field theoretic 2-body problem was either analysed using the Bethe-Salpeter equation, with its interpretational difficulties, and with the simultaneous treatment of the 2-body and only some many-body intermediate states, or, more often, using the 3-dimensional approximations of the Bethe-Salpeter equation, which are suffering from many non-unique assumptions.

The Weinberg equation [2], which is the exact formulation of the Bethe-Salpeter type equation in the light front field theory [3], considered according to the old fashioned perturbation rules [4], offers a unique opportunity to study relativistically the nature of the NN forces. We found [5] that the binding energies of a few MeV, and the phase shifts for the lab. energies (0.1 - 0.3) GeV are quite sensitive to the forces at short distances, which are directly connected to the relativistic effects. Our results differ substantially, depending on the Lippmann-Schwinger and the Weinberg treatment. In a simple spinless model calculation this difference may be as large as (60 - 90)% of the deuteron binding energy.

The large relativistic dynamical effects are due to the repulsion present in the Weinberg one-meson exchange interaction. Previously, Franz Gross [6] found the relativistic origin of a repulsion which is due to the coupling to the negative energy states. The Gross repulsion, depending on the value of the coupling constant, can be also recovered in the Weinberg scheme. However, in the Weinberg approach there is yet another repulsion, which is present for an arbitrary value of the coupling constant. Its origin lies in the non-instantaneous character of the one-particle exchange, if it is studied on the light plane.

The Weinberg repulsion we illustrate here only on the example of the spinless nucleons, exchanging a scalar meson. However, if we extend our considerations to the vector meson exchanges, analysed on the light plane, then the resulting potential is less singular at the origin than in the standard treatment.

The very large decrease of the binding energy, evaluated in the Weinberg approach for the bound states commonly considered as "non-relativistic", should have an essential impact on the extensively studied NN and quark-antiquark systems.

2. The Weinberg equation in the light front field theory

For simplicity we limit ourselves to the $\phi^2\phi_0$ field theory. In this case Weinberg [2] studied the infinite momentum limit of the old fashioned perturbation theory, and wrote a Bethe-Salpeter type equation. The essential points of this approach are: 1° the 3-dimensional integration loops, without any approximation, though the theory is fully relativistic, 2° all particles are on their mass shells, 3° the particle-antiparticle intermediate states give zero contribution, 4° there are the energy denominators, similarly as in the non-relativistic theory, and 5° the driving term in the ladder approximation is non-local and energy dependent.

The Weinberg infinite momentum rules can be also derived in the field theory quantized on the light like surfaces [3]. The light like variables are denoted as $p_\pm = p_0 \pm p_z$, and the total, and the relative 4-momentum as P, and q, respectively. From the Weinberg rules we get the following conservation laws for the 2-body t-matrix, denoted as M,

$$\langle P_+'\vec{P}_\perp' q_+'\vec{q}_\perp' | M | P_+\vec{P}_\perp = 0 q_+\vec{q}\rangle = \delta(P_+'-P_+)\,\delta^{(2)}(\vec{P}_\perp')\langle q_+'\vec{q}_\perp'|M|q_+\vec{q}_\perp\rangle . \tag{1}$$

The "minus" components of P and q are determined from the mass shell conditions. Usually, instead of q_+ one introduces the light like variable $\eta = \frac{1}{2} + q_+(P_+)^{-1}$, and then the Weinberg equation is

$$\langle \eta'\vec{q}_\perp'|M|\eta\vec{q}_\perp\rangle = \langle \eta'\vec{q}_\perp'|I^W|\eta\vec{q}_\perp\rangle -$$

$$-\frac{1}{2}(2\pi)^{-3}\int d\eta'' d^2\vec{q}_\perp'' \eta''^{-1}(1-\eta'')^{-1}\langle \eta'\vec{q}_\perp'|I^W|\eta''\vec{q}_\perp''\rangle\langle \eta''\vec{q}_\perp''|M|\eta\vec{q}_\perp\rangle (P_-''-\sqrt{s}-i\varepsilon)^{-1}s^{-1/2}, \tag{2}$$

where

$$P_-'' = (\vec{q}_\perp''^2 + m^2)s^{-1/2}\eta''^{-1}(1-\eta'')^{-1}, \quad s = P^2 = P_+ P_- .$$

3. The Plmn Vierbein, and the invariant variables

The Weinberg rules, found in the light front field theory, can be made manifestly invariant, and in this respect they resemble either the covariant Feynman rules, or the Mandelstam dispersion relations. To get the invariant Weinberg rules we have to introduce the Plmn Vierbein [4]. For any given process we have the characteristic initial total 4-momentum P, which is a time like 4-vector, and we can explicitly construct three orthonormal space like 4-vectors from the initial and final relative Wightman-Garding relative 4-momenta q and q'. We get

$$n = q(-q^2)^{-1/2}, \quad m^\mu = N\varepsilon^{\mu\nu\rho\sigma} P_\nu q_\rho q'_\sigma, \quad \ell^\mu = L\varepsilon^{\mu\nu\rho\sigma} P_\nu m_\rho n_\sigma,$$

where N and L are the appropriate normalization factors for $l^2 = m^2 = -1$. Any 4-vector can be projected on the Plmn Vierbein. In particular, we get $(P_{Plmn}) = (\sqrt{s},0,0,0)$, and $(q_{Plmn}) = (0,0,0,(\frac{1}{4}s-m^2)^{1/2})$.

The conservation laws, stated in Eq.(1), give us $P'_{lm} = P''_{lm} = 0$, $P'_p+P'_n=P''_p+P''_n=P_p=s^{1/2}$, while the mass shell conditions are equivalent to the following relations: $P'q' = P''q'' = 0$, $P'^2 = 4(-q'^2+m^2)$, $P''^2=4(-q''^2+m^2)$. Therefore, it is useful to introduce two more Vierbeins P'lmn' and P"lmn", in terms of which we have automatically $q'_{p'}=q''_{p''} = 0$, $P'^2=4(q'^2_{lmn'}+m^2)$, $P''^2=4(q''^2_{lmn''}+m^2)$. We also get $q'_m = 0$, because of our construction of m. The remaining projections of q'on the P'lmn' Vierbein can be parametrized in terms of the following invariants: $\bar{q}' \equiv (-q'^2)^{1/2}$, and

$$\cos\theta' = -q'q(q'^2q^2)^{-1/2}\, 2\left(\frac{M_o}{M'_o} + \frac{M'_o}{M_o}\right)^{-1}, \tag{3}$$

where $M_o^2 \equiv P^2$, $M'^2_o \equiv P'^2$. Thus, we get $(q'_{Plmn'}) = (0,\bar{q}'\sin\theta',0,\bar{q}'\cos\theta')$. Similarly, for the intermediate relative 4-momentum q" we get $(q''_{p''lmn''})=(0,\bar{q}''\sin\theta''\cos\phi'',\bar{q}''\sin\theta''\sin\phi'',\bar{q}''\cos\theta'')$, where $\cos\theta''$, $\cos\phi''$, and $\sin\phi''$ are shorthands for invariants of the similar type as in Eq.(3).

4. The invariant Weinberg equation, and the Weinberg one-boson exchange

The Lorentz invariants $q_p = qP(P^2)^{-1/2}$, and $q_n = qn$ can be used for defining such invariants, which have the algebraic properties of the light front variables. We denote them by the same symbols as the non-invariant light front variables, because they play the same role, however we have to remember, that they are Lorentz invariants. We get

$$q_\pm=q_p \pm q_n, \quad \eta = \frac{1}{2} + (q_p + q_n)s^{-1/2}, \quad q_\ell = q\ell, \quad q_n = qn. \tag{4}$$

The 4-dimensional scalar product q^2 can be rewritten in terms of the invariant variables as $q^2 = (q_p+q_n)(q_p-q_n)-q_\ell^2 - q_n^2$, and one recognizes the basic algebraic property of the light front variables, namely the <u>linear</u> dependence of q^2 on q_+ and q_-.

The invariant Weinberg equation has the same structure as Eq.(2). Then, passing from the variables given in Eq.(4) to the invariants $\bar{q}$, $\bar{q}'$, $\cos\theta'$, $\bar{q}''$, $\cos\theta''$, $\cos\phi''$, defined in the previous Section, we get

$$\langle\bar{q}'\cos\theta'|M|\bar{q}\rangle = \langle\bar{q}'\cos\theta'|I^W|\bar{q}\rangle -$$
$$-\tfrac{1}{4}(2\pi)^{-3}\int\bar{q}''^2 d\bar{q}'' d\cos\theta'' d\phi'' (\bar{q}''^2+m^2)^{-1/2}(\bar{q}''^2-\bar{q}^2-i\varepsilon)^{-1}\langle\bar{q}'\cos\theta'|I^W|\bar{q}''\cos\theta''\cos\phi''\rangle$$
$$\langle\bar{q}''\cos\theta''|M|\bar{q}\rangle. \tag{5}$$

The amplitude M is normalized in such a way, that the differential cross section is $\pi^{-1}\bar{q}^2 d\sigma/dt = \frac{1}{16}(2\pi)^{-2}s^{-1}|\langle\bar{q}'\cos\theta'|M|\bar{q}\rangle|^2_{\bar{q}'^2=\bar{q}^2=\frac{1}{4}s-m^2}$, where $t = -2\bar{q}^2(1-\cos\theta')$.

The Weinberg one-boson exchange is given by the ladder approximation to the irreducible kernel I^W. We take the interaction lagrangian $2mg:\phi^2\phi_o:$, to which in the non-relativistic theory, in the static limit, it corresponds in the position space the potential $V(r) = -\frac{g^2}{4\pi}\frac{e^{-\mu r}}{r}$. We denote by μ the mass of the ϕ_o field, and we get

$$\langle\bar{q}'\cos\theta'|I^W|\bar{q}''\cos\theta''\cos\phi''\rangle = -4m^2g^2\ (A+\alpha)^{-1}, \tag{6}$$

where $A \equiv \mu^2+\bar{q}'^2 + \bar{q}''^2-2\bar{q}'\bar{q}''(\cos\theta'\cos\theta''+\sin\theta'\sin\theta''\cos\phi'')$,

$$\alpha\equiv\left|\frac{\bar{q}'\cos\theta'}{\sqrt{\bar{q}'^2+m^2}} - \frac{\bar{q}''\cos\theta''}{\sqrt{\bar{q}''^2+m^2}}(\bar{q}'^2+\bar{q}''^2-2\bar{q}^2) - \frac{\bar{q}'\cos\theta'\bar{q}''\cos\theta''}{\sqrt{\bar{q}'^2+m^2}\,\sqrt{\bar{q}''^2+m^2}}(\sqrt{\bar{q}'^2+m^2}-\sqrt{\bar{q}''^2+m^2})^2.\right.$$

There should be noticed the following properties of I^W: 1° manifest invariance, 2° energy dependence through $\bar{q}^2=\frac{1}{4}s-m^2$, 3° non-locality, and 4° if $\bar{q}' = \bar{q}'' = \bar{q}$, then $I^W = -4m^2g^2A^{-1}$, as usually.

5. Relativistic repulsion, and conclusions

The Weinberg one-boson exchange contains two parts. One is the standard attractive term, and the second is a term which shows the repulsive character. We extract these parts from Eq.(6) by writing

$$(A+\alpha)^{-1} = A^{-1}\ [1-\alpha(A+\alpha)^{-1}]\ . \tag{7}$$

The repulsive term is energy dependent and nonlocal. It is the direct consequence of the old fashioned perturbation procedure, applied to the light front field theory, which leads to the P_- denominators, and to the P_+ conservation. The s-dependence of I^W is the new feature, absent in the Feynman theory, and for fixed q' and q" the fully off shell element of $I^W \to 0$, if $s \to \infty$.

The modifying factor $[1-\alpha(A+\alpha)^{-1}]$ in Eq.(7) provides a form factor to the standard expression A^{-1}. Note, that although the underlying field theory is local, the effective Weinberg interaction in Eq.(5) is nonlocal.

The physical effect of the extra term $-\alpha(A+\alpha)^{-1}$ is large, even in problems which may be classified as "non-relativistic". For example, in the spinless model of the 2N system with the exchange of the π meson, and the coupling constant adjusted in the Lippmann-Schwinger approach to give the correct deuteron binding energy, we get [5] for the difference between the non-relativistic and the Weinberg result about 60% of the deuteron binding energy. The decrease of the binding energy due to the Weinberg repulsion is present even in positronium [7], which is bound only by 7 eV. There, the effect is about 4%. In the NN system, if instead of the π meson exchange we consider an effective scalar particle exchange of the mass $\gtrsim$ 500 MeV, then we get [5] about 90% for the difference between the standard and the Weinberg result. This is expected, since for the shorter range the relativistic repulsive effect should be stronger.

In the continuum of the 2N problem the effect of the Weinberg repulsion is also large. Some crude estimates show, that for the laboratory energy in the range (0.3 - 0.6) GeV the extra term $-\alpha(A+\alpha)^{-1}$ may take over 1 in Eq.(7), changing the original attractive interaction into the repulsive one.

The large physical effects of the Weinberg repulsion in the 2N problem at low and intermediate energies, forces us to doubt in the present meson exchange analysis of the NN forces, and calls for redoing it in the Weinberg approach. This approach is as exact as the non-approximated Bethe-Salpeter equation, and yet it is free from the difficulties associated with the Bethe-Salpeter equation. The 2-body propagator is of the Schrödinger type, the formalism is manifestly invariant, and the analysis of the n-body intermediate states can be made in a systematic way. Moreover, the extension to the 3- and many-body problems is straightforward, because of the simple cluster decomposition property in the P_- variable.

Acknowledgements

The author benefited from many discussions, calculations, and criticism of Mr. P. Danielewicz.

References

[1] R.V. Reid, Ann. of Phys. 50 (1968) 411.
[2] S. Weinberg, Phys. Rev. 150 (1966) 1313.
[3] S-J. Chang, R.G. Root and T-M. Yan, Phys.Rev. D7 (1973)1133; S-J. Chang and T-M. Yan, ibid. D7 (1973) 1145; T-M. Yan, ibid. D7 (1973); T-M. Yan, ibid. D7 (1973) 1780.
[4] J.M. Namyslowski, Phys.Rev.D (1978), Warsaw Univ.preprint IFT (1977) 15.
[5] P. Danielewicz and J.M. Namyslowski, to be published.
[6] F. Gross, Phys.Rev. D10 (1974) 223.
[7] G.Feldman, T.Fulton and J.Townsend, Phys.Rev.D7 (1973) 1814.

UNITY OF RELATIVISTIC CORRECTIONS AND MESON EXCHANGE CURRENTS

Franz Gross*
Department of Physics
William and Mary
Williamsburg, VA 23185/USA

Measurements of electron-deuteron scattering at high momentum transfer [1] have stimulated considerable interest in the proper way to calculate meson exchange currents and relativistic corrections. In a recent short review given at the Zurich Conference [2], the relationship between various approaches to this problem and the interpretation of some correction terms was not clear. Today, all remaining ambiguities seem to be resolved, and in this short note I will briefly describe how exactly the same correction terms can be obtained from two very different approaches. I will limit myself to a comparison of the pion contribution to the deuteron charge form factor only.

1. The Perturbation Approach (P)

The first approach will be represented by the work of Friar [3], and Gari and Hyuga [4,5], although many others have also contributed [6]. These authors develop the subject from the standpoint of non-relativistic perturbation theory, and calculate corrections in increasing powers of (p/M), where p is some characteristic nuclear momentum and M is the nucleon mass.

In the (P) approach, the corrections can be classified as coming from:

(a) the nucleon current (Ref. [3])

(b) Lorentz contraction of the nuclear wave functions (Ref. [3])

(c) retardation, or relativistic corrections due to the combined effect of the recoil term and wave function re-orthonormalization (Ref. [4])

(d) the meson exchange "pair" term (Ref. [5])

(e) the $\rho\pi\gamma$ contribution (Ref. [5]).

The full result should be the sum of all of these corrections.

I will not discuss the $\rho\pi\gamma$ contribution, since this is essentially the same in both approaches. If we write the deuteron charge form factor as [7]

$$G_C(q^2) = G_E^S(q^2)\, D_C(q^2) + \left[2G_M^S(q^2) - G_E^S(q^2)\right] D_C^{SO}(q^2), \qquad (1)$$

where G_E^S and G_M^S are the nucleon isoscalar charge and magnetic form factors, then:

*Supported in part by the National Science Foundation.

$$D_C(q^2) = \left[1 - \frac{q^2}{8M^2} - \frac{q^4}{16M^2}\frac{d}{dq^2}\right] F_C(q^2) + X_1 + X_2$$
$$\qquad\quad (a) \qquad\quad (b) \qquad\qquad\qquad (c) \quad (d) \qquad\qquad (2)$$

$$D_C^{SO}(q^2) = \frac{q^2}{8M^2} \int_0^\infty dr[j_o(\tau) + j_2(\tau)]w^2(r) + X_2$$
$$\qquad\qquad\qquad (a) \qquad\qquad (d)$$

where the letter below each term indicates its origin, and

$$F_C(q^2) = \int_0^\infty dr\ j_o(\tau)\ [u^2(r) + w^2(r)], \tag{3}$$

and $j_\ell(\tau)$ are the spherical bessel functions, u and w the deuteron S and D state wave functions and $\tau = \frac{qr}{2}$ where q is the magnitude of the momentum transferred by the electron in the Breit frame.

The formulae given by Gari and Hyuga for the π exchange contribution to X_1 and X_2 reduce to (taking the form factor to be unity):

$$X_1 = -\frac{q}{4M} C_\pi \int_0^\infty dr\ j_1(\tau) \left\{ (1 - \frac{2}{x} - \frac{2}{x^2})e^{-x}(u^2 + w^2) + 2(1 + \frac{1}{x} + \frac{1}{x^2})e^{-x}(2\sqrt{2}uw - w^2) \right\} \tag{4}$$

$$X_2 = -\frac{q}{2M\mu^2} \frac{C_\pi}{\pi} \int_0^\infty k^2 dk \int_{-1}^{+1} dz \int_0^\infty dr \times \frac{1}{\mu^2 + k^2 + \frac{q^2}{4} + qkz}$$
$$\{(\frac{1}{2} q + kz)j_o(kr)(u^2 + w^2) - (qP_2(z) + 2kz)j_2(kr)[2\sqrt{2}uw - w^2]\} \tag{5}$$

where μ is the pion mass, $x = \mu r$, and $P_2(z)$ are Legendre polynomials, and $C_\pi = (\mu/2M)^2\ g_\pi^2/4\pi$.

To facilitate comparison, we can recast X_2 into a single integral. The method is to combine the integrals over k and z into d^3k, expand the Legendre polynomials using the addition theorem, and expand the r integration to d^3r. Then the integration over d^3k can be performed, giving derivatives of a position space potential, and the d^3r integration can subsequently be reduced to a single integral. The final result is:

$$X_2 = -\frac{q}{2M} C_\pi \int_0^\infty dr\ j_1(\tau)e^{-x}(\frac{1}{x} + \frac{1}{x^2}) \left[u^2 - w^2 + 4\sqrt{2}uw\right] \tag{6}$$

$$X_1 + X_2 = -\frac{q}{4M} C_\pi \int_0^\infty dr\ j_1(\tau) \left\{ e^{-x}(u^2 + w^2) - 2e^{-x}(1 + \frac{3}{x} + \frac{3}{x^2}) \times (w^2 + 2\sqrt{2}uw) \right\} \tag{7}$$

2. The Non-Perturbative Approach (NP)

This approach is based on the evaluation of the relativistic Feynman diagram for the impuse approximation, in which the spectator nucleon is restricted to its mass shell [8,9]. The deuteron structure is described by a relativistic wave function with 4 components. The usual S and D state wave functions describe the structure of the deuteron when the virtual nucleon is in a positive energy state, but when it is in a negative energy state two P state wave functions, denoted v_t and v_s, are needed [10]. To compare with the (P) approach, the exact results can be expanded in powers of p/M, and the leading order terms compared. In this case, the corrections come from only 2 sources:

(a) terms proportional to u^2, uw, or w^2 (Ref. [8])

(b) terms proportional to uv_t. wv_t, wv_s or uv_s (Ref. [9]-reported in Ref. [7])

One obtains Eq. (2) with X_1 replaced by X_1' and X_2 by X_2' where

$$X_2' = +\frac{q}{2M} \int_0^\infty dr\; j_1(\tau)\left\{\frac{v_s}{\sqrt{3}}(u - \sqrt{2}w) - \sqrt{\tfrac{2}{3}}\, v_t\left(u + \frac{w}{\sqrt{2}}\right)\right\} \tag{8}$$

$$X_1' + X_2' = -\frac{q}{4M^2} \int_0^\infty dr\; r\; j_1(\tau)[u\hat{u} + w\hat{w}] \tag{9}$$

$$\hat{u} = \left(-\frac{d^2}{dr^2} + M\varepsilon\right)u\,, \qquad \hat{w} = \left(-\frac{d^2}{dr^2} + \frac{6}{r^2} + M\varepsilon\right)w$$

and ε is the deuteron binding energy.

To compare Eq. (9) with (P) it is sufficient to use the Schrödinger equation in the OPE approximation [11]:

$$\hat{u} = -(U_c u + U_t\, 2\sqrt{2}\, w); \qquad \hat{w} = -(U_c w - 2U_t w + U_t\, 2\sqrt{2} u) \tag{10}$$

$$U_c = -M\mu C_\pi \frac{e^{-x}}{x}\;; \qquad U_t = -M\mu C_\pi \frac{e^{-x}}{x}\left(1 + \frac{3}{x} + \frac{3}{x^2}\right) \tag{11}$$

Substituting (10) and (11) into (9) gives the remarkable result that $X_1 + X_2 = X_1' + X_2'$.

To compare equation (8) with (6), use the relativistic wave equation in the OPE approximation (see the first of Ref. [10]) to relate v_t and v_s to u and w:

$$v_t \cong \sqrt{3}\, C_\pi \frac{e^{-x}}{x}\left(1 + \frac{1}{x}\right)(\sqrt{2}u + 2)$$

$$v_s \cong \sqrt{3}\, C_\pi \frac{e^{-x}}{x}\left(1 + \frac{1}{x}\right)(u - \sqrt{2}w) \tag{12}$$

(The parameter λ has been set equal to unity to correspond to the γ^5 coupling used by Gari and Hyuga.) Substituting (12) into (8) gives $X_2 = X_2'$.

3. Discussion

When all of the corrections are included from either approach, the result to order M^{-3} is identical. One can therefore have confidence in the results of either method, and is justified in regarding the relativistic corrections and meson exchange currents as "known". The "potential" correction, Eq. (9), which I obtained some years ago (and which has been the subject of some uncertainty) is now interpreted as a combination of corrections due to pair currents and retardation. Comparison of the two approaches shows that one cannot make an unambiguous distinction between pair currents and relativistic effects.

It is a pleasure to acknowledge helpful conversation with Ray Arnold and Ben Chertok, who drew my attention to Ref. [4].

REFERENCES

1. R. G. Arnold et al., Phys. Rev. Lett. 35, 776 (1975).

2. F. Gross, Proc. 7th Int'l. Conference on High Energy Physics and Nuclear Structure, Zurich, 1978 (ed. by M. Locher-Birkhäuser Verlag Basel), p. 329.

3. J. L. Friar, Ann. Phys. (N.Y.) 81, 332 (1973); Phys. Rev. C12, 695 (1975).

4. M. Gari and H. Hyuga, Zeitschrift fur Physik A277, 291 (1976).
H. Hyuga and M. Gari, Nuclear Physics A274, 333 (1976).
M. Gari and H. Hyuga, Nuclear Physics A278, 372 (1977).

5. M. Gari and H. Hyuga, Nuclear Physics A264, 409 (1976).

6. L. L. Foldy, Phys. Rev. 122, 275 (1961); D15, 3044 (1977).
H. Osborn, Phys. Rev. 176, 1514 (1968); 176, 1523 (1968).
F. E. Close and H. Osborn, Phys. Rev. D2, 2127 (1970).
M. Chemtob and M. Rho, Nuclear Physics A163, 1 (1971).
D. O. Riska and G. E. Brown, Phys. Letters 38B, 193 (1972).
R. A. Krajcik and L. L. Foldy, Phys. Rev. D10, 1777 (1974).
W. Kloet and J. Tjon, Phys. Letters 49B, 419 (1974).
A. D. Jackson, A. Lande, and D. O. Riska, Phys. Letters 55B, 23 (1975).
F. Coester and A. Ostabee, Phys. Rev. C11, 1836 (1975).

7. F. Gross, Proc. 7th Int'l. Conference on Few Body Problems, Delhi, 1976 (ed. by Mitra et al., - North Holland Publishing Co.), p. 523.

8. F. Gross, Phys. Rev. 142, 1025 (1966); 152, 1517E (1966).
B. M. Casper and F. Gross, Phys. Rev. 155, 1607 (1967).

9. R. G. Arnold, C. E. Carlson and F. Gross, Phys. Rev. Lett. 26, 1516 (1977); and another in preparation.

10. J. Hornstein and F. Gross, Phys. Lett. 47B, 205 (1973).
W. W. Buck and F. Gross, Phys. Lett. 63B, 286 (1976). The $\lambda = 0$ solution contains a numerical error and should not be used.

11. R. J. Reid, Ann. Phys. (N.Y.) 50, 411 (1968).

NUCLEON - NUCLEON PHASE SHIFTS CALCULATED USING THE BLANKENBECLER - SUGAR EQUATION

K. Schwarz, H. F. K. Zingl, L. Mathelitsch
Institut für Theoretische Physik der Universität Graz
Universitätsplatz 5, A-8010 Graz

The inclusion of relativistic kinematics yields effects in phase shifts which corresponds to a reduction of the attraction in the interaction. This was shown for local potentials and S-waves by several authors [1-4].

We investigated relativistic effects on phase shifts δ^L, up to L = 2, with the aid of the Lippmann-Schwinger equation (LS) and the Blankenbecler-Sugar equation (BBS) using separable interaction models. For S-waves we chose two models: model 1 is of the Yamagucchi type [5] and model 2 consists of a sum of Yamagucchi formfactors as given by Alt [6]. For higher partial waves we used the potential of Doleschall [7].

To show the relativistic influence at low energies we calculated also the scattering length for the S-waves. The results are given in the following tables where δ^L_{RE} means the difference $\delta^L_{LS}-\delta^L_{BBS}$. To see the relative effect we give these values in percentage to δ^L_{LS} also.

We can see that the addition of relativistic kinematics using separable potentials leads to similar results as from local ones. The effect is not only seen in S-waves but it is evident in the same order also in higher partial waves. For low energies (see table 3) our results are similar to [2] . It is interesting to note that δ^L_{RE} strongly depends on the interaction model (e.g. for Reid SC $\delta^1_{RE} \sim 10^o$ at 10 MeV [3]).

Therefore we conclude : if one uses relativistic equations with a given interaction model one has to readjust the potential parameters in order to obtain the observed phase shifts.

Table 1: δ_{RE} for S-waves in degrees (in percents)

Labor-energy (MEV)	$^1S_0$		$^3S_1$	
	model 1	model 2	model 1	model 2
10	1.9 (3.2)	3.3 (5.0)	4.0 (3.8)	3.3 (3.2)
50	1.1 (3.0)	2.0 (4.0)	2.7 (4.1)	2.0 (3.6)
100	1.0 (3.8)	1.5 (4.9)	2.3 (4.9)	1.5 (4.4)
200	0.9 (5.7)	0.8 (7.3)	2.0 (6.6)	0.8 (6.5)

Table 2: δ_{RE} for higher partial waves in degrees (in percents)

Labor-energy (MEV)	$^1P_1$	$^3D_2$	$^3P_1$	$^3P_2$
10	0.2 (8.7)	0.1 (4.8)	0.3 (15.0)	0.1 (4.3)
50	o.6 (7.8)	0.6 (5.5)	1.2 (14.0)	0.4 (5.9)
100	0.9 (6.9)	1.4 (7.9)	1.7 (13.1)	0.8 (7.4)
200	1.3 (5.7)	2.3 (9.3)	2.6 (12.0)	1.5 (9.3)

Table 3: Scattering lengths

a (fm)	$^1S_0$		$^3S_1$	
	model 1	model 2	model 1	model 2
a_{LS}	-23.71	-23.59	5.41	5.48
a_{BBS}	-19.03	-16.08	6.09	6.04

References :

[1] G.E. Brown, A.D. Jackson and T.T.S. Kuo; Nucl.Phys. A 133 (1969) 481

[2] A.D. Jackson and J.A. Tjon; Phys.Lett. Vol. 32B (1970) 9

[3] M. Fortes and A.D. Jackson; Nucl.Phys. A 175 (1971) 449

[4] M. Bawin and J.P. Lavine; Nucl. Phys. B 49 (1972) 610

[5] H. Zankel and H.F.K. Zingl; Acta Phys.Austr. 44 (1976) 245

[6] E.O. Alt and P.L.G. Bakker; Z. Physik A273 (1975) 37

[7] P. Doleschall; Nucl.Phys. A 220 (1974) 491.

QUASI-ELASTIC ELECTRON SCATTERING AND THE MOMENTUM DISTRIBUTION OF THE DEUTERON

D. Royer, M. Bernheim, A. Bussière, J. Mougey, D. Tarnowski, S. Turck
DPh-N/HE, CEN Saclay, BP 2, 91190 Gif-sur-Yvette, France

G.P. Capitani, E. de Sanctis
INFN, Frascati, Italy

and

S. Frullani
INFN, Istituto di Sanità, Roma, Italy.

The coincidence cross section for the d(e,e'p)n reaction in the quasifree kinematics has been measured at 500 MeV incident energy with the high duty cycle electron linac at Saclay.

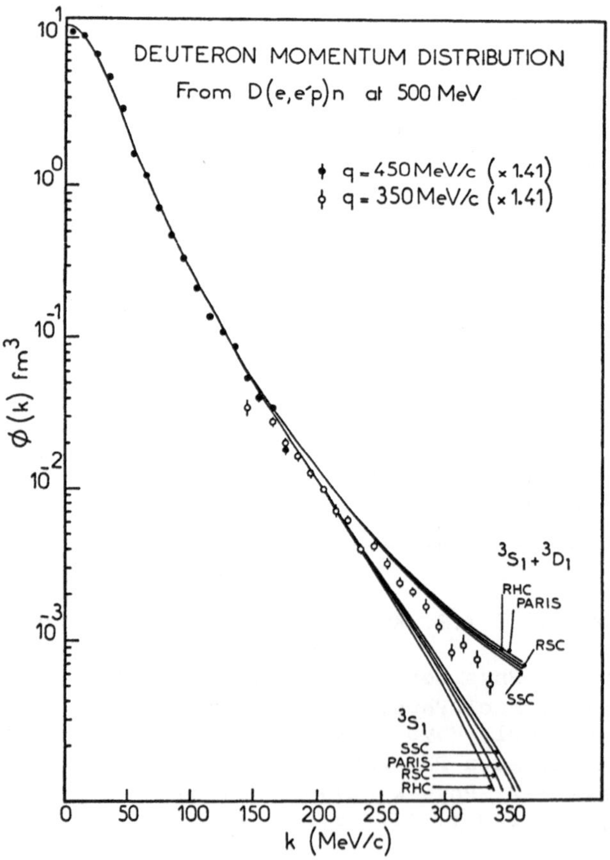

Fig. 1 Momentum distribution of the deuteron from the present data as compared with S and S + D theoretical momentum distributions (see text).

In the plane wave impulse approximation, the cross section factorizes into the off shell elastic electron-proton cross section and the deuteron momentum distribution $\phi(k)$ [1]. In this framework, we determined $\phi(k)$ up to 340 MeV/c, in a region where the amplitude of the process is mainly due to the D state admixture (Fig. 1). Two kinematics with momentum transfers of 450 and 350 MeV/c were used to cover respectively the low k and high k parts of the distribution. The corresponding center of mass energies of the final n-p system were kept constant at 51 and 115 MeV. The cross section in the missing energy spectrum is shown in Fig. 2 after removal of the accidentals, for the highest k of this experiment.

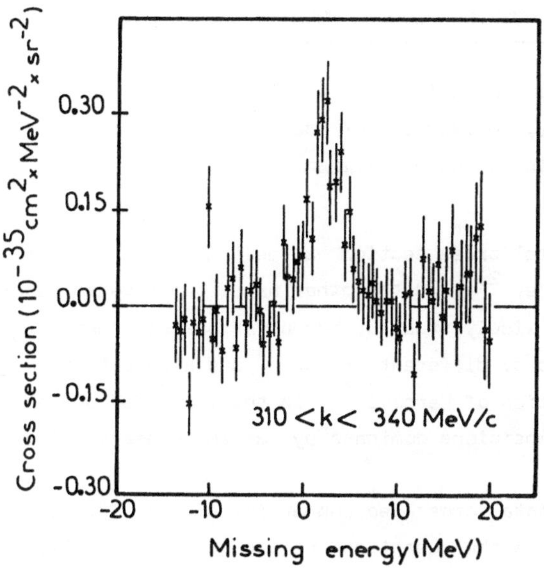

Fig. 2 Missing energy spectrum for the reaction d(e,e'p)n at k ≤ 340 MeV/c.

The improvement in the determination of $\phi(k)$ over the previous measurements consists in the absence of higher order effects which dominates the cross section in (p,2p) for k > 200 MeV/c [2,3], and in the greatly enlarged k range compared to the existing (e,e'p) [4].

First evaluations lead to small final state interaction effects [5] then allowing us to compare directly the PWIA results with the theoretical momentum distributions. On Fig. 1 are plotted the S and S + D distributions as obtained from four nucleon-nucleon potentials : Reid hard core (RHC) Reid soft core (RSC), Paris and De Tourreil super soft core (SSC) [6-8]. They have D state probabilities ranging from 5.5 to 7 % and are all consistently above the data.

Before drawing definitive conclusions from these data, the overall normalization and liquid deuterium target thickness have to be checked.

References

[1] J. Mougey et al., Nucl. Phys. A262, 461 (1976).
[2] T.R. Whitten et al., Nucl. Phys. A254, 269 (1975).
[3] R.D. Felder et al., Nucl. Phys. A264, 397 (1976).
[4] P. Bounin, Ann. de Physique 10, 475 (1965).
[5] G. Kingma and A.E.L. Dieperink, preprint and private communication.
[6] R.V. Reid , Ann. Phys. 50, 411 (1968)
[7] M. Lacombe et al., Phys. Rev. D12, 1495 (1975).
[8] R. de Tourreil and D.W.L. Sprung, Nucl. Phys. A201, 193 (1973).

NEUTRON-PROTON CAPTURE TOTAL CROSS SECTION BETWEEN 38 AND 73 MEV

M. Bosman, A. Bol, J.F. Gilot, P. Leleux, P. Lipnik, P. Macq
Cyclotron laboratory, University of Louvain
B - 1348 Louvain-la-Neuve, Belgium

Below 100 MeV, the n-p capture total cross section had been measured only at thermal neutron energy [1], and at 14.4 MeV [2]. On the other hand, the inverse reaction (deuteron photodisintegration) was widely studied, but unfortunately there exist discrepancies between measurements by different groups [3] and also between measurements and the exhaustive calculation of Partovi [4], in the 15-25 MeV photon energy region, where the "simple" E1 transitions dominate by far the cross section.

We have measured the n-p capture total cross section at 38, 43, 48, 53, 58, 63 and 73 MeV ; this energy range overlaps the questioned region of the photodisintegration. Neutrons are produced by the ^{7}Li(p,n) reaction at 0°. The collimated neutron beam has been described elsewhere [5]. A liquid hydrogen target is situated at 3 m from the production target. Deuterons from capture processes are emitted in a narrow forward cone (less than ± 7°). The time of flight between two NE102 scintillators (base length : 80 cm), as well as ΔE and E signals from those scintillators are used to select deuterons. High-energy "monokinetic" incident neutrons are also selected by time-of-flight between the production target and the first thin ΔE scintillator. The hydrogen target is sandwiched between two gas proportional counters : the first one, in anticoïncidence, supresses charged particles contaminating the neutron beam, while the second one, in coïncidence, triggers on charged particles from the hydrogen target. Protons from n-p elastic scattering at 3° lab are also detected. The capture total cross section is thus normalized to the well-known n-p differential cross section at backward c.m. angles [6].

Until now, only the high energy data have been analysed and they fit Partovi's calculations ; further results will be presented at the conference.

1) A.E. Cox et al., Nucl. Phys. 74 (1965) 497
2) J. Tudoric-Ghemo, Nucl. Phys. A92 (1967) 233
3) J. Ahrens et al., Phys. Lett. 52B (1974) 49 ; J.E.E. Baglin et al., Nucl. Phys. A201 (1973) 593 ; B. Weissmann and H.L. Schultz, Nucl. Phys. A174 (1971) 129
4) F. Partovi, Ann. Phys. (N.Y.) 27 (1964) 79
5) M. Bosman et al., Nucl. Instr. Meth. 148 (1978) 363
6) R.A. Arndt et al., Phys. Rev. D8 (1973) 1397 ; R.A. Arndt et al., Phys. Rev. C15 (1977) 1002

THE NEUTRON ELECTRIC FORM FACTOR AND QUASI-ELASTIC ELECTRON-DEUTERON SCATTERING

H. Arenhövel and W. Fabian
Institut für Kernphysik
Universität Mainz
D-6500 Mainz

B.A. Craver and Y.E. Kim
Department of Physics
Purdue University
W. Lafayette, Indiana

We consider electron-neutron coincidence measurements in quasi-elastic electron-deuteron scattering as an alternative to most previous experiments for a more accurate determination of the neutron electric form factor $G_{En}(q_\mu^2)$. Aside from certain kinematic factors the differential cross section may be expressed in terms of the functions[1] $\rho_{\mu\mu'}$, $f_{\mu\mu'}$, $\mu = 0, 1$ and $\mu' = 0, \pm 1$. Here $\rho_{\mu\mu'}$ is the virtual photon density matrix and the $f_{\mu\mu'}$ are essentially products of matrix elements of the nuclear current. To indicate the sensitivity to G_{En} we evaluate the relative difference of the calculated cross section with respect to the prediction for $G_{En} = 0$. For $G_{En} \neq 0$ we use a representative fit to the existing data, here taken to be $G_{En} = -\mu_n\tau(1 + 4\tau)^{-1} G_{Ep}$, where $\tau = q_\mu^2/4M^2$, G_{Ep} is the proton electric form factor and M the nucleon mass[2]. Our main results are summarized in table 1. The magnitude of the cross sections is of the order of $(5.0\text{-}0.1)10^{-32}\text{cm}^2/\text{MeV-sr}^2$.

Table 1: Per cent difference in cross section relative to $G_{En} = 0$. All cases at the quasi-elastic peak and for an electron laboratory scattering angle of 28°. The nuclear potential is that of Hamada-Johnston.

$q_\mu^2(\text{fm}^{-2})$	0.48	0.965	2	6	10	12	30
% difference	-4.1	-5.4	-2.2	2.3	3.1	3.2	2.6

For scattering at the quasi-elastic peak the amplitudes f_{01} and f_{1-1} vanish, and the cross section depends only upon f_{00} and f_{11}. One may then determine f_{00}, which is very sensitive to G_{En} and independent of the neutron magnetic form factor, by the usual Rosenbluth plot.

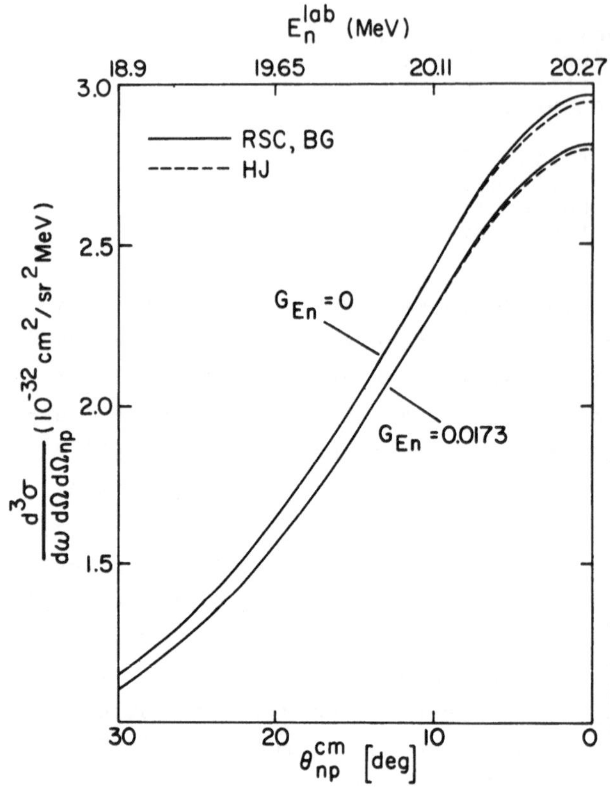

Fig. 1: The e-n coincidence angular distribution near the quasi-elastic peak for $q_\mu^2 = 0.965\ fm^{-2}$. The outgoing electron solid angle is $d\Omega$, the solid angle of the n-p relative momentum as seen in their CMS is $d\Omega_{np}$ and the energy range of the outgoing electron is $d\omega$.

In fig. 1 we show the differential cross section near the quasi-elastic peak and for $q_\mu^2 = 0.965\ fm^{-2}$, as calculated with three different nucleon-nucleon potentials. There is a slight model dependence only at the quasi-elastic peak ($\theta_{np}^{cm} \approx 0^o$), and even there it is no more than 1%. This is also the case at $q_\mu^2 \approx 10\ fm^{-2}$. In addition, contributions from pionic and pair excitation currents and from isobar configurations[3], calculated with the Hamada-Johnston potential at $q_\mu^2 \approx 1$ and $10\ fm^{-2}$, give a net contribution of less than 0.5% to the calculated cross section.

In view of these results it seems hopeful that if one can measure the coincidence cross sections to within 1-2% one might be able to extract more accurate measurements of G_{En}. This would be especially so for $q_\mu^2 < 5\ fm^{-2}$. Here many possibly complicating factors which we have considered are not very important; this also appears to be the case for relativistic corrections[4].

References:

1. H. Arenhövel, Y.E. Kim, B.A. Craver, W. Fabian and D.P. Saylor, "Quasi-Elastic Electron-Deuteron Scattering and the Electric Form Factor of the Neutron", preprint KPH 9/18;
   H. Arenhövel and W. Fabian, "Electro- and Photodisintegration of Deuterium", in the Proceedings of Workshop on Few Body Systems and Electromagnetic Interactions, Frascati (1978), Italy
2. S. Galster et al., Nucl. Phys. B32 (1971) 221
3. W. Fabian and H. Arenhövel, Nucl. Phys. A258 (1976) 461
4. J.L. Friar, Nucl. Phys. A264 (1976) 455

THE ELECTRON-DEUTERON TENSOR POLARIZATION AND THE SHORT RANGE BEHAVIOUR OF THE DEUTERON WAVE FUNCTION

L.J. Allen and H. Fiedeldey

Department of Physics, University of South Africa,
P O Box 392, Pretoria, Republic of South Africa

Hockert and Jackson [1] suggested that a single measurement of the tensor polarization of recoil deuterons in electron-deuteron scattering (P_e) at a momentum transfer of $q^2 = 19.52\ fm^{-2}$ ($q \approx 4.5\ fm^{-1}$) could, in the absence of meson exchange currents, distinguish between the "hardest" and "softest" models of the nucleon-nucleon interaction. Similarly Moravcsik and Ghosh [2] suggested that measurement of P_e in the region $q = 6\text{-}10\ fm^{-1}$ is supersensitive to the deuteron wave function within 0.5 fm. We have shown by means of examples that if the suggested measurements are made to within an accuracy of 10% the ambiguity in the wave function in the core region would remain much larger than expected.

We take as reference interaction the SSC potential of De Tourreil and Sprung [3] and take as pseudodata for P_e that calculated non-relativistically from this interaction and assigned an error of 10%. From the SSC potential we generate phase equivalent interactions by the method of finite range unitary transformations [4]. The following form factor for the unitary transformation has been employed

$$g(r) = \begin{cases} C(R-r)^{\alpha}(1-\beta r), & r \leqslant R \\ 0, & r > R \end{cases} \tag{1}$$

The constant C is determined by the unitarity of the transformation.

Table 1. Interactions varying in the core region obtained using eq. (1) with $R = 0.7$ fm, $\alpha = 2.1$ and for various β. Interactions I1 to I4 are transformed only in the $^3D_1$ partial wave, I5 to I8 only in the $^3S_1$ partial wave and I9 to I12 have the same transformation applied in both partial waves. For each interaction P_e at $q = 4.5\ fm^{-1}$ is given. All the interactions have the same percentage D state (5.45%) and quadrupole moment (.279 fm^2) as that of the SSC potential.

Int.	β	P_e	Int.	β	P_e	Int.	β	P_e
I1	-5.0	.627	I5	-5.0	.659	I9	-5.0	.660
I2	2.2	.626	I6	2.2	.625	I10	2.2	.625
I3	3.1	.627	I7	3.1	.637	I11	3.1	.638
I4	3.6	.627	I8	3.6	.667	I12	3.6	.668
Pseudodatum		.626± 10%						

In table 1 we list 12 interactions obtained using eq. (1). We take as pseudodata for the deuteron electric form factor A(q) that obtained non-relativistically from the SSC deuteron wave function and assigned an error of 10%. (In calculating A(q) we have used the nucleon electric form factors of Janssens et al. [5] throughout.) All the interactions of table 1 fit the pseudodata for A(q) to at least $q \gtrsim 6.5\ fm^{-1}$. Furthermore it is clear that the value of P_e at $q = 4.5\ fm^{-1}$ is always within the assumed error on the pseudodatum. The deuteron partial wave components which suitably combined give the deuteron wave functions of table 1 are plotted in fig. 1 and show a wide variety of behaviour inside 0.7 fm.

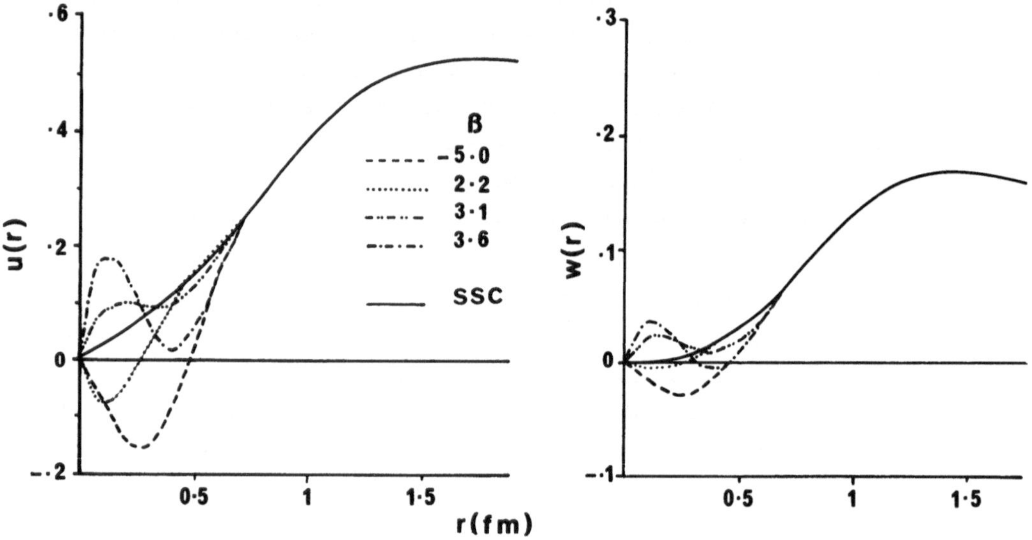

Fig. 1. The $^3S_1$ and $^3D_1$ deuteron partial wave components which suitably combined give the deuteron wave functions of table 1.

The interactions I2, I6 and I10 fit A(q) to within 4% and P_e to within 1% out to $q = 10\ fm^{-1}$ while I3 likewise fits A(q) and P_e out to $q = 8.5\ fm^{-1}$. These wave functions are rather different from the SSC wave function inside 0.7 fm.

We conclude that a rather complete measurement of P_e out to $q = 10\ fm^{-1}$ to within 10% error would not allow us to distinguish between "hard" and "soft" interactions. Furthermore meson exchange currents could be expected to complicate the analysis considerably in the region $q = 6\text{–}10\ fm^{-1}$ as well as (to a lesser degree) relativistic effects.

References

1. J. Hockert and A.D. Jackson, Phys. Lett. 58B (1975) 387
2. M.J. Moravcsik and P. Ghosh, Phys. Rev. Lett. 32 (1974) 321
3. R. de Tourreil and D.W.L. Sprung, Nucl. Phys. A201 (1973) 193
4. M.W. Kermode, J.R. Mines and M.M. Mustafa, J. Phys. G: Nucl. Phys. 2 (1976) L113
L.J. Allen, H. Fiedeldey and N.J. McGurk, J. Phys. G: Nucl. Phys. 4 (1978) 353
5. T. Janssens, R.H. Hofstadter, E.B. Hughes and M.R. Yearian, Phys. Rev. 142 (1966) 922

DEUTERON FORMFACTOR CALCULATIONS WITH THE BETHE-SALPETER EQUATION

J.A. Tjon and M.J. Zuilhof
Institute for Theoretical Physics, University of Utrecht, The Netherlands

The elastic em formfactors (FF) for the deuteron are calculated in an one boson exchange model using the Bethe-Salpeter (BS) equation. Relativistic and pair current contributions to the FF are investigated in the impulse approximation represented by the diagrams in fig. 1, where the four-point function satisfies the homogeneous BS equation

fig. 1

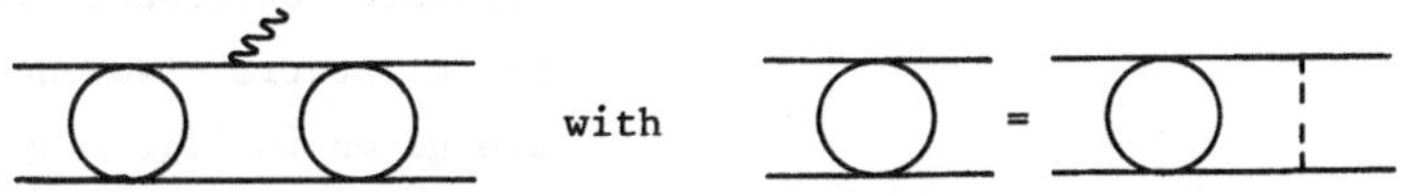

In calculating the FF for non-zero momentum transfer q the BS wavefunction is needed in Lorentz-frames with total momentum $\vec{P} = \pm \vec{q}/2$. These wavefunctions can in principle be obtained from the cm BS states by applying the boost operators Λ for spin $\frac{1}{2}$ particles

$$|P,p\rangle_D = \Lambda^{(1)}(L)\ \Lambda^{(2)}(L)\ |\ L^{-1}p\rangle_{cm} \text{ with } L^{-1}P = (M_D,\vec{o})$$

where p is the relative momentum between the two nucleons. The cm states have been constructed from the homogeneous solutions of the Wick rotated BS equations. Due to the presence of the negative energy states the solution contains 8 components. For the driving force the exchange of π, η, ε, δ, ρ and ω mesons were used, where the coupling constants are such that the experimental NN phase shifts are reproduced [1]. Pseudovector coupling has thereby been used for the pion-nucleon interaction.

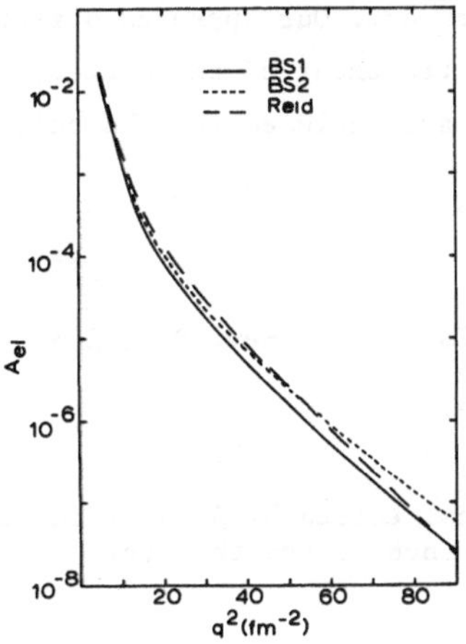

fig. 2

The basic approximation which can be made in calculating the FF is to neglect the effects due to the cm motion of the deuteron i.e. $\Lambda = 1$, $L = 1$ and in addition to restrict the wavefunction to the positive energy state components ${}^3S_1^+$, ${}^3D_1^+$. The result for the electric FF is shown in fig. 2 (BS1). For comparison the FF for the Reid potential is also shown in the figure. In all the calculations dipole nucleon FF were used. Since the arguments of the BS wavefunction become complex in the complete calculation due to L, the latter dependence was

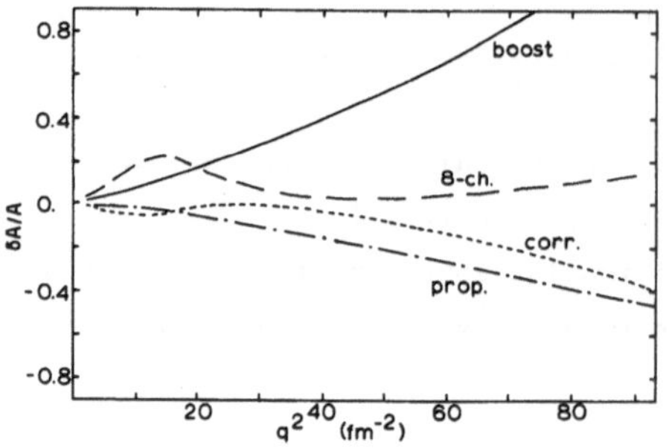

fig. 3

accounted for in an approximate way by expansion to lowest order. The corrections due to the cm motion of the deuteron can be distinguished as coming from the inclusion of the additional 6 channels, the lowest order correction in the arguments of the wave function and the boost transformation Λ acting on the em nucleon vertex operator and the propagator. The separate contributions, relative to the basic "static" two-channel approximation are given in fig. 3 by the curves labeled 8-CH, CORR, BOOST and PROP respectively. From this figure we see that the inclusion of the negative energy states is non-negligible even at low momentum transfer. However, if the static two-channel approximation is corrected for the boost transformation on the em nucleon vertex, the corresponding FF being shown in fig. 2 (BS2), one finds that the effects of the negative energy states becomes negligible. Up to momentum transfer of $q^2 = 100\ \mathrm{fm}^{-2}$ the correction amounts to less than 5%. From this it follows that these two corrections are not additive, but that there exists a strong interference effect. In this light the conventional approach of estimating exchange current effects [2-4] should be considered critically. Adding all the additional contributions together we find, due to cancellations of the various corrections, that the static two-channel approximation is not significantly changed up to a momentum transfer of $50\ \mathrm{fm}^{-2}$. Relativistic corrections to the deuteron formfactor have been considered by various groups [5], starting from a non-relativistic potential theory. Our approach differs basically from this because of the genuine relativistic treatment of the deuteron in its cm system. Moreover, some off mass shell effects are included in this calculation.

References

1. J. Fleischer and J.A. Tjon, Nucl. Phys. B84, 375 (1975); Phys. Rev. D15, 2537 (1977); and to be published.
2. W.M. Kloet and J.A. Tjon, Phys. Lett. 49B, 419 (1974).
3. A.D. Jackson et al, Phys. Lett. 55B, 23 (1975).
4. M. Gari and H. Hyuga, Nucl. Phys. A264, 409 (1976).
5. See for a review F. Gross in Few Body Dynamics, edited by A.N. Mitra et al (North-Holland, Amsterdam, 1976), p. 523 and references cited therein.

RELATIVISTIC CALCULATION OF THE PHOTON CIRCULAR POLARIZATION IN $np\to\gamma d$ PROCESS*

Shin-ichi MORIOKA and Tamotsu UEDA[†]

School of Physical Sciences, The Flinders University of South Australia, Bedford Park, S.A. 5042, AUSTRALIA.

[†]Faculty of Engineering Science, Osaka University, Toyonaka, Osaka, JAPAN.

We propose that the present discrepancy between the experimental value[1] [$(-1.30 \pm 0.45) \times 10^{-6}$], and the theoretical result[2] (10^{-7} to 10^{-9}) for the circular polarization of γ-rays in $np\to\gamma d$ may be removed by employing a relativistic deuteron wave function.

The covariant amplitude for $np\to\gamma d$ is given by (see Fig. 1)

$$M^{\lambda} = \bar{u}^{c}(p)\varepsilon\Gamma \frac{\gamma k + M}{k^2 - M^2} e\gamma A^{\lambda}(\ell)u(n), \qquad (1)$$

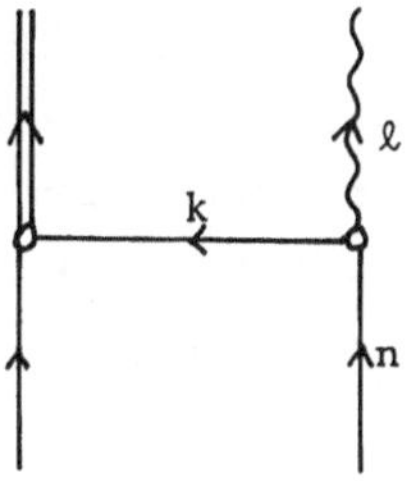

Fig. 1

where u^c, ε, and A^{λ} are the charge-conjugated spinor, the deuteron polarization vector, and the photon polarization vector with helicity λ respectively. Here p, n, k, and ℓ are the four momenta as defined in Fig. 1, while M is the nucleon mass, and Γ is a form factor given by

$$\Gamma_{\mu} = S(q^2)\gamma_{\mu} + W(q^2)\gamma_5\gamma_{\mu}, \qquad (2)$$

with $q = (p - k)/2$, and $S(q^2)$ and $W(q^2)$ are the Lorentz-invariant dnp vertex function corresponding respectively to the parity conserving and parity-violating interaction in the deuteron. We may decompose the intermediate nucleon propagator into its positive and negative energy component[3] as

$$\gamma k + M = \frac{2M}{1+\delta}\Big(\sum_{\zeta} u^{\zeta}(\bar{k})\bar{u}^{\zeta}(\bar{k}) - \delta\sum_{\xi} v^{\xi}(p)\bar{v}^{\xi}(p)\Big), \qquad (3)$$

where $\bar{k} = k + \delta(p + k)$, $\delta = (M^2 - k^2)/M_D^2$ with M_D the deuteron mass. Here u^{ζ} and v^{ξ} represent the Dirac Spinor for the positive and negative energy states respectively. Using Eq. (3) we can write the amplitude as

$$M^{\lambda} \propto \psi_{+}\bar{u}(\bar{k})\gamma A^{\lambda}(\ell)u(n) + \psi_{-}\bar{v}(p)\gamma A^{\lambda}(\ell)u(n). \qquad (4)$$

Here ψ_+ and ψ_- are defined in the deuteron rest frame as

$$\psi_{+} = \phi_0^S(q)P_0 + \phi_2^S(q)P_2 + \phi_{11}^W(q)P_{11}, \qquad (5)$$

$$\psi_{-} = \phi_{10}^S(q)P_{10} + \phi_0^W(q)P_0 + \phi_2^W(q)P_2, \qquad (6)$$

where $q = |\vec{q}|$, and ϕ_0, ϕ_2, ϕ_{10}, and ϕ_{11} denote the $^3S_1$, $^3D_1$, $^1P_1$, and $^3P_1$ components of the deuteron wave function, with the P's the corresponding projection operators. In Eqs. (5) and (6) the superscript S and W refer to the vector and axial vector component of Γ_{μ} in Eq. (2). The first term in Eq. (4) corresponds to the positive energy part of the nucleon propagator and reduces to the usual non-relativistic amplitude. The second term on the r.h.s. of Eq. (4) corresponds to the negative energy part of the intermediate nucleon propagator and is purely a relativistic contribution.

In the ordinary non-relativistic formulation of the deuteron, the main contribution to P_γ arises from the interference term $a_+\phi_0^S\,\phi_{11}^W$. However, in the present formulation there is another contribution to P_γ from the second term on the r.h.s. of Eq. (4) and of the form $b_-\phi_0^S\phi_0^W$. Here a_+ and b_- express the γNN vertex's contribution to the positive and negative energy part respectively (see Eq. (7) below). The ratio of these two contributions to P_γ i.e. $P_\gamma(b_-\phi_0^S\phi_0^W)/P_\gamma(a_+\phi_0^S\phi_{11}^W) \simeq 10^3$, is quite large.

This large contribution to P_γ from the negative energy part is due to two factors. <u>(i) Wave function</u>: Since ϕ_0^W is the relativistic correction to the usual component ϕ_{11}^W, it is expected in general to be smaller. However, at low energies the wave function with angular momentum L goes as q^L, and thus making the S wave ϕ_0^W and P-wave ϕ_{11}^W of the same order of magnitude. <u>(ii) γNN vertex</u>: Since the photon's circular polarization is defined[4] as $\vec{A}^{\pm} = \mp\frac{1}{\sqrt{2}}(\vec{A}^1 \pm \vec{A}^2)$ the contributions from the γNN vertex in Eq. (4) are

$$a_+ = \bar{u}_{\nu_1}(k)\vec{\gamma}\cdot\vec{A}^\lambda u_{\nu_2}(n) = \chi^\dagger_{\nu_1}[\vec{A}^\lambda\cdot(\tilde{p} - \tilde{n}) + i\vec{\sigma}\cdot(\tilde{p} + \tilde{n}) \times \vec{A}^\lambda]\chi_{\nu_2} \tag{7}$$

$$b_- = \bar{v}_{\nu_1}(p)\vec{\gamma}\cdot\vec{A}^\lambda u_{\nu_2}(n) = \chi^\dagger_{\nu_1}[-\vec{\sigma}\cdot\vec{A}^\lambda - (\vec{A}^\lambda\cdot\tilde{n})(\tilde{\sigma}\cdot\tilde{p}) - (\vec{A}^\lambda\cdot\tilde{p})(\tilde{\sigma}\cdot\tilde{n}) + (\tilde{n}\cdot\tilde{p})(\vec{\sigma}\cdot\vec{A}\,) - i\tilde{p}\cdot(\vec{A}^\lambda \times \tilde{n})]\chi_{\nu_2}$$

where $\tilde{p} = \vec{p}/(E_p + M)$ and so on. In the above we have dropped the normalization factors. At very low energies one has $b_-/a_+ \simeq 10^3$. This ratio shows the dominance of the negative energy part. We note that the above values of a_+ and b_- are independent of the dnp vertex which can have additional terms to those in Eq. (2) such as $T(q^2)q_\mu$, $v(q^2)\gamma^5 q_\mu$.

Although we recognize the possibility of other contributions to the amplitude other than those discussed in this note such as the weak interaction existing before photon emission, we believe it worthwhile to emphasize the importance of treating the deuteron relativistic.

One of us (S.M.) wishes to thank I.R. Afnan and B.H.J. McKellar for their useful comments on this work.

* A part of this research supported by ARGC and The Flinders University Research Budget.

<u>REFERENCES</u>:

1. V.M. Lobashov et al., Nucl. Phys. A197, 241 (1972).
2. R.J. Blin-Stoyle and H. Feshbach, Nucl. Phys. 27, 395 (1961).
   K. Ohya et al., Prog. Theor. Phys. 56, 875 (1976).
   B.H.J. McKellar, Nucl. Phys. A254, 349 (1975).
3. E.A. Remler, Nucl. Phys. B42, 56 (1972).
4. See, for example, H. Muirhead, "The Physics of Elementary Particles", (Pergamon Press, 1968).

THERMAL n-p RADIATIVE CAPTURE

M. W. Kermode and A. McKerrell
DAMTP, University of Liverpool, PO Box 147, Liverpool L69 3BX, UK

In his invited talk at the Laval conference, Sprung (1) briefly mentioned the agreement between the theoretical and experimental values for the capture cross section of thermal neutrons (v = 2200 m/s) by protons. In the ensuing discussion, Noyes (2) retracted his value of 302.5 mb (3) for the magnetic dipole cross section in the single particle approximation, on which the agreement was based, in favour of the value of 321 mb calculated by Bosco et al (4). Noyes used the Bethe-Longmire shape-independent formula to calculate the matrix element whereas Bosco et al used dispersion theory.

We have calculated the cross sections, together with meson exchange and nucleon isobar effects using a neutron-proton $^1S_0$ wave function from a local potential which describes to a high degree of accuracy the phase shifts of (1) MacGregor et al (5) (labelled M in the table) or (2) Arndt et al (6) (labelled A). It has been the lack of knowledge of this wave function that has limited the accuracy of the calculation of the matrix elements. Both the $^1S_0$ potentials (A,M) give (by construction) a scattering length of -23.72 fm. These potentials do not have a hard core.

For the one pion potential, we used the same form as Reid (7); for consistency since we used the Reid deuteron wave functions (hard core (H) and soft core (S)). Then the pion mass $0.7\,\hbar c$ = 138.13 MeV and the coupling constant $f^2/4\pi = 14.94714/\hbar c$ = 0.07575 are determined and these values were inserted in the formulae of Riska and Brown (8) (see also, Colocci et al (9)).

Our results for the various combinations of potentials are given in the table. In this table, σ_1 is the cross section including meson exchange effects and σ_2 is the cross section including meson exchange and nucleon isobar (Δ) effects. The δ's are percentage increases with respect to the magnetic dipole matrix element (8).

$\psi(^1S_0)$	M	M	A	A
$\psi(^3S_1-^3D_1)$	H	S	H	S
σ_{dip} (mb)	305.6	305.2	304.2	304.2
δ^{SS} (%)	1.71	1.70	1.70	1.69
δ^{SD} (%)	1.30	1.30	1.26	1.28
σ_1 (mb)	324.3	323.7	322.5	322.6
δ_Δ^{SD} (%)	1.84	1.94	1.74	1.99
σ_2 (mb)	335.9	336.1	333.5	335.2
Experimental σ (mb)		334.2 ± 0.5		

A small change in the potential to give a singlet scattering length of -23.74 fm (-23.70 fm) increases (decreases) the cross section by 0.4 mb.

The Reid hard core potential gives a deuteron quadrupole moment, Q, of 0.2770 fm^2. A nonlocal potential that is phase equivalent to this Reid potential but giving Q = 0.2883 fm^2 (10), which is closer to the experimental value of 0.2860 ± 0.0015 fm^2 (11), reduces the value of σ_2 by 4.3 mb (for M).

We see that the values for σ_2 are very close to the experimental value, without the Stranahan reduction suggested by Riska and Brown. Clearly this point requires further investigation. Also, the value of σ_1 is very close to Noyes' value of 302.5 mb.

References

1) Sprung D W L 1974 Proc. Int. Conf. on few body problems in nuclear and particle physics, Quebec 482.
2) Noyes H P ibid. 494.
3) Noyes H P 1965 Nucl. Phys. 74 508-32.
4) Bosco B, Ciocchetti G and Molinari A 1963 Il. Nuo. Cim. 28 1427-36.
5) MacGregor M H, Arndt R A and Wright R N 1969 Phys. Rev. 182 1714-28.
6) Arndt R A, Hackman R H and Roper L D 1977 Phys. Rev. C15 1002-20.
7) Reid R V 1968 Ann. Phys., NY 50 411-48.
8) Riska D O and Brown G E 1972 Phys. Lett. 38B 193-5.
9) Colocci M, Mosconi B and Ricci P 1973 Phys. Lett. 45B 224-6.
10) McKerrell A, Kermode M W and Mustafa M M 1977 J. Phys. G: Nucl. Phys. 3 1349-57.
11) Reid R V and Vaida M L 1975 Phys. Rev. Lett. 34 1064.

SPECIAL COHERENT STATES FOR π-N SCATTERING*

M. Bolsterli
T Division, Los Alamos Scientific Laboratory
Los Alamos, NM 87545/USA

The static model for pion-nucleon interaction provides a basis for understanding p-wave pion-nucleon scattering phase shifts. That reasonable phase shifts can be computed by using the dispersion method of Chew and Low[1] is well known; the direct method used by Friedman, Lee, and Christian[2] (FLC) to obtain equally interesting phase shifts is less well known. Recent work[3,4] has clarified the connection between static and nonstatic models for meson-nucleon interaction: for weak coupling the static model is unrelated to a corresponding nonstatic model, but for strong coupling the static model solutions are a useful first step toward constructing solutions of the corresponding nonstatic model.

The direct method of FLC is based on the idea of Tomonaga[5] that a useful approximate state vector is one in which the mesons are in a single mode described by the function $b(k)$; the only creation operator needed to construct such a single-mode state of the mesons is

$$A^\dagger = \int b(k) a^\dagger(k)\, dk \quad ,$$

where $A^\dagger$ and $a^\dagger(k)$ may have spin and isospin indices. The Tomonaga-FLC choice for the single-mode state $|b\rangle_T$ is the one that minimizes the expectation of the Hamiltonian. Then states $a^\dagger(k)|b\rangle_T$ are used to compute scattering systems. However, the states $a(k)|b\rangle_T$ have been neglected by FLC, although they are clearly as important as the states $a^\dagger(k)|b\rangle_T$.

An alternative choice[4] of the basic single-mode state $|b\rangle_S$ that avoids the difficulty with the states $a(k)|b\rangle_T$ is to choose $|b\rangle_S$ to be an eigenvector of the annihilation operators $a(k)$, that is, $a(k)|b\rangle_S = b(k)|b\rangle_S$. If the meson field has internal degrees of freedom, then $a(k)$ must be replaced by a suitable scalar annihilation operator; for the pion case, where a is a vector in spin and isospin, the single-mode state $|b\rangle_S$ is chosen to satisfy

$$\sum_{i\lambda=1}^{3} \sigma_i \tau_\lambda a_{i\lambda}(k)|b\rangle_S = b(k)|b\rangle_S .$$

As in the case of isovector mesons, it turns out that the special coherent state $|b\rangle_S$ for the pion case gives energy expectation that is the same as that given by $|b\rangle_T$ in both the strong and weak coupling limits. Moreover, the one-meson subspace contains only the states

$$\sum_{i\lambda=1}^{3} \sigma_i \tau_\lambda a^\dagger_{i\lambda}(k)|b\rangle_S$$

and the states neglected by FLC are now included.

So far, calculations have been carried out only in the P_{11} states. An interesting general result has been obtained for the P_{11} phase shift: in the static model the P_{11} phase shift is zero in an energy range Δ_0 above the scattering threshold. In the one-meson approximation this is obtained by solving the scattering problem, which is like that coming from a separable two-body interaction. When more mesons are added, higher thresholds appear, but not lower ones, so the one-meson result holds in all approximations. Again, this result is like that obtained previously for isovector mesons; FLC noted that this phase shift was very nearly zero in their approximation. Experimentally, the P_{11} phase shift is nearly zero until the kinetic energy reaches about $2m_\pi$, at which point it resembles a normal p-wave phase shift. It is interesting that the static model gives a simple explanation of this striking effect, as well as accounting for the P_{33} resonance.

A second general result comes from the fact that there is a countable infinity of special coherent states $|b\rangle_S$ for each value of spin and isospin, rather than the single such state for each isospin that arises in the case of isovector field. In the P_{11} partial wave, one of these states is the ground state, and the others, as well as the bare source state, lie in the scattering continuum. Each of these discrete states in the continuum couples to the ground state, and therefore each can lead to a resonance in the scattering in the P_{11} partial wave. The behavior in other partial waves is expected to allow a similar description. Thus, the special coherent states $|b\rangle_S$ provide a natural mechanism for resonances in the one-Fermion sector in a simple meson field theory.

---

*Work performed under the auspices of the United States Department of Energy.

[1] G. F. Chew and F. E. Low, Phys. Rev. 101, 1570 (1956).

[2] M. H. Friedman, T. D. Lee, and R. Christian, Phys. Rev. 100, 1494 (1955).

[3] M. Bolsterli, Phys. Rev. D 13, 1727 (1976).

[4] M. Bolsterli, Phys. Rev. D 16, 1749 (1977); Phys. Rev., to be published.

[5] S. Tomonaga, Progr. Theoret. Phys. (Japan) 2, 6 (1947).

ON THE FIELD-THEORETIC APPROACH TO LOW-ENERGY π-N INTERACTION

R. F. Alvarez-Estrada
Departmento di Fisica Teorica
Universidad Complutense, Madrid-3, Spain

Let the low-energy π-N interaction be described approximately by the following hamiltonian, which generalizes the static models of Chew-Low (P-wave) [1] and Drell-Friedman-Zachariasen (S-wave) [2] by including nucleon recoil:

$$H = H_o + H_I \ , \qquad H_o = \frac{\bar{p}^2}{2M_o} + \sum_{j=1}^{3} \int d^3\bar{k}\,\omega(|\bar{k}|)\,a^+(\bar{k}j)\,a(\bar{k}j) \tag{1}$$

$$H_I = \sum_{i=1}^{3} H_{I,i}, \quad H_{I,1} = \frac{f_o}{\mu} \sum_{j=1}^{3} \int d^3\bar{k}\, \frac{v(|\bar{k}|)}{[(2\pi)^3 2\omega(|\bar{k}|)]^{1/2}} \{ i[\bar{k}(1+\lambda_1)+\lambda_2\bar{p}]\bar{\sigma}\tau_j a(\bar{k}j) \cdot$$

$$\cdot \exp i\bar{k}\bar{x} + \text{h.c.}\} \tag{2}$$

$$H_{I,2} = \frac{g_1}{\mu} \sum_{j=1}^{3} \int d^3\bar{k}_1 d^3\bar{k}_2 \frac{v(|\bar{k}_1|)v(|\bar{k}_2|)}{(2\pi)^3[4\omega(|\bar{k}_1|)\omega(|\bar{k}_2|)]^{1/2}} \{a(\bar{k}_1 j)a(\bar{k}_2 j)\exp i\bar{x}(\bar{k}_1+\bar{k}_2)+$$

$$+ a^+(\bar{k}_1 j)a(\bar{k}_2 j)\exp i\bar{x}(\bar{k}_2-\bar{k}_1) + \text{h.c.}\} \tag{3}$$

$$H_{I,3} = \frac{g_2}{\mu^2} \sum_{\alpha,\beta,\gamma=1}^{3} \varepsilon_{\alpha\beta\gamma}\, \tau_\alpha \int d^3\bar{k}_1 d^3\bar{k}_2 \frac{i\omega(|\bar{k}_2|)v(|\bar{k}_1|)v(|\bar{k}_2|)}{(2\pi)^3[4\omega(|\bar{k}_1|)\omega(|\bar{k}_2|)]^{1/2}}$$

$$\{a^+(\bar{k}_1\beta)a^+(\bar{k}_2\gamma)\exp[-i\bar{x}(\bar{k}_1+\bar{k}_2)] + a^+(\bar{k}_2\gamma)a(\bar{k}_1\beta)\exp i\bar{x}(\bar{k}_1-\bar{k}_2) + \text{h.c.}\} \tag{4}$$

Here, M_o, $\bar{x}$ and $\bar{p}$ are the nucleon bare mass, position and threemomentum operators, respectively, while $\varepsilon_{\alpha\beta\gamma}$ is the fully antisymmetric tensor. Other notations and dependences are standard [3,4], and f_o, g_1, g_2, λ_1, λ_2 are dimensionless coupling constants (λ_1,λ_2 either vanish or are of order μ/M_o). Generalizing the bounds presented in [5], one can show that for any ket ψ:

$$\| H_I\psi \| \leq e_1 \| H_o\psi \| + e_2 \| \psi \| \tag{5}$$

$$e_1 = \frac{f_o}{\mu}[1.10|1+\lambda_1|x_4 + 1.91\left(\frac{M_o}{\mu}\right)^{1/2}|\lambda_2|x_1^{1/2} + 0.55\frac{\lambda_2}{\mu}x_2^{1/2}] + 0.35\frac{g_1}{\mu^2}x_1 + 0.49\frac{g_2}{\mu^3}(x_1x_3)^{1/2}$$

$$e_2 = [0.55|1+\lambda_1|x_2^{1/2} + 0.96(M_o\mu)^{1/2}|\lambda_2|x_1^{1/2}]\frac{f_o}{\mu} + 0.087\frac{g_1}{\mu}x_1 + 0.12\frac{g_2}{\mu^2}(x_1x_3)^{1/2}$$

$$x_i = \int_0^{+\infty} dk \frac{k^{2i}[V(k)]^2}{\omega(k)}, \quad i = 1,2$$

$$x_3 = \int_0^{+\infty} dk.k^2\omega(k)[V(k)]^2 \qquad x_4 = [\int_0^{+\infty} dk \frac{k^4[V(k)]^2}{\omega(k)^3}]^{1/2}$$

We shall concentrate on the dressed nucleon state $\psi_+(\bar{\pi}\sigma\tau)$ with three-momentum $\bar{\pi}(\bar{\pi}^2/2M_o < \mu)$, third components of spin σ and isospin τ and physical energy $E((H-E)\psi_+ = 0)$.

Let $\psi(\bar{\pi}\sigma\tau)$ be the bare nucleon state and $Q_{\bar{\pi}}$ be the projector on the fourdimensional subspace spanned by all $\psi(\bar{\pi}\sigma'\tau')$ as $\sigma'\tau'$ vary, for fixed $\bar{\pi}$. The Brillouin-Wigner equation for $\psi_+(\bar{\pi}\sigma\tau)$, which generalizes the one for static models [3,5] reads:

$$\psi_+(\bar{\pi}\sigma\tau) = \psi(\bar{\pi}\sigma\tau) + (1 - Q_{\bar{\pi}})(E - H_o)^{-1} H_I\psi_+(\bar{\pi}\sigma\tau) \qquad (6)$$

Generalizing the treatment in [5] for static models, one shows that the Neumann series for $\psi_+(\bar{\pi}\sigma\tau)$, formed by all successive iterations of Eq.(6) converges in norm when $e_1[1+|E|/\rho(E)]+e_2/\rho(E) < 1$. Here, $\rho(E)$ is the smallest distance from E to the set of all eigenvalues of H_o corresponding to eigenvectors with threemomentum $\bar{\pi}$, the smallest eigenvalue $\bar{\pi}^2/2M_o$ being excluded from that set. The physical energy can be rigourosly obtained through the contraction mapping principle for suitably small f_o,g_1,g_2 as in [5]. For the usual V(k) [3], the above convergence condition holds for values of f_o,g_1,g_2 which are one or two orders of magnitude smaller than the actual physical values [3,4].

Furthermore, for the case $g_1=g_2=\lambda_1=\lambda_2=0$ and any $f_o \neq 0$, we have been able to establish the general validity of the Tamm-Dankoff approach (see [4] for an introduction to it). In particular, we have been able to show that, for $\bar{\pi} = 0$, the contribution of n pions to $\psi_+(\bar{\pi}\sigma\tau)$ can be safely neglected provided that $0.076\ (f_o/\mu)^2.x_4^2$ be small compared to n. For the physical case [3,4] $(f_o^2/4\pi = 0.22)$ we conclude that only a few pions contribute to $\psi_+(\bar{\pi}\sigma\tau)$. This purely theoretical result is essentially consistent with other estimates based upon sum rules and low-energy experimental data [4].

References

[1] G.F. Chew and F.E. Low, Phys.Rev. 101 (1956) 1571.
[2] S.D. Drell, M.Friedman and F.Zachariasen, Phys.Rev.104 (1956) 236.
[3] S.S. Schweber "An Introduction to Relativistic Quantum Field Theory", Sect. 12d,Evanston and Row, Illinois,1961.
[4] E.M. Henley and W. Thirring "Elementary Quantum Field Theory" Part III, McGraw-Hill, New York, 1962.
[5] R.F. Alvarez-Estrada, Il Nuovo Cimento 27A, (1975) 37.

S-WAVE PION-NUCLEON PHASE SHIFTS IN PADÉ APPROXIMATION

P. Achuthan, T. Chandramohan and K. Venkatesan
Department of Mathematics, I.I.T., Madras - 600 036

Abstract

The two S-wave pion nucleon phase shifts $\delta^{(1)}$ $(I = 1/2)$ and $\delta^{(3)}$ $(I=3/2)$ have been calculated in the Padé approximation using $\varepsilon(700)$, $\rho(770)$, $f(1260)$, $\Delta(1236)$ and $N(938)$ for the energy range W = 1085 MeV - 1820 MeV in the centre of mass. Contributions from suitable resonance combinations which agree nearest with the $\delta^{(3)}$ experimental values are given.

Though the two S-wave πN phase shifts $\delta^{(1)}$ and $\delta^{(3)}$ in the isospin states 1/2 and 3/2 respectively, belong to the set of small phase shifts in the low energy pion-nucleon scattering, this area of hadron physics has always been problematic. $\delta^{(3)}$ is repulsive whereas $\delta^{(1)}$ is attractive [1]. None of the earlier theories adequately explains this feature [2]. Sakurai's theory of strong interactions [3] qualitatively explains the opposite signs of $\delta^{(1)}$ and $\delta^{(3)}$ by appealing to the rule that like charges repel and unlike charges attract. In recent years good fits to one ($\delta^{(3)}$) or both the phase shifts have been claimed [4-8]. The results of [5] have been questioned [9]. The fits of [6] and [7] are obtained at the cost of unacceptable coupling constants, like $g^2_{\pi NN}/4\pi \approx 7$, $g^2_{\pi N\Delta}/4\pi \approx 0.48$ and those of [4] and [9] with unusual scalar meson mass.

In the present paper we have recalculated the S-wave phase shifts with ε, ρ, f exchanges in the t-channel and N and Δ exchanges in the u-channel. The partial wave projection of the contributions from these particles together with the contribution of the s-channel nucleon is the first approximation $(f^{(2I)}_{1\pm})_1 = R^{(2I)}_1$, real due to our zero width approximation. The second approximation is computed by initially finding the imaginary part using unitarity condition

$$T^+ T = \frac{i}{2} (T^+ - T) \tag{1}$$

in the one meson approximation and the real part by using fixed-t dispersion relations. No subtraction has been performed in any of the amplitudes following arguments of [10-13]. The [1,1] Padé approximant for the partial wave amplitude is

$$(f^{(2I)}_{1\pm})_{[1,1]} = (f^{(2I)}_{1\pm})^2_1/[(f^{(2I)}_{1\pm})_1 - (f^{(2I)}_{1\pm})_2] \tag{2}$$

where $(f_{1\pm}^{(2I)})_2 = R_2^{(2I)} + i\,D_2^{(2I)}$. The expression for the phase shift is given by

$$\delta_{1\pm}^{(2I)} = 1/2 \text{ arc tan } \frac{2q\, {R_1^{(2I)}}^2 (R_1^{(2I)} - R_2^{(2I)})}{(R_1^{(2I)} - R_2^{(2I)})^2 + D_2^{2(2I)} - 2q\, R_1^{(2I)}\, D_2^{(2I)}} . \qquad (3)$$

The results of our numerical calculation done on IBM 370/155 Computer, are displayed in fig. 1. As seen from the figure ε by itself seems to be the best candidate for explaining the $\delta^{(3)}$. Though the signs turn out to be incorrect for $\delta^{(1)}$ we have displayed the corresponding contributions for completeness. In conclusion we emphasize that the problem of S-wave πN phase shifts, in respect of sign and magnitude, is still an open problem and would need better methods for its solution.

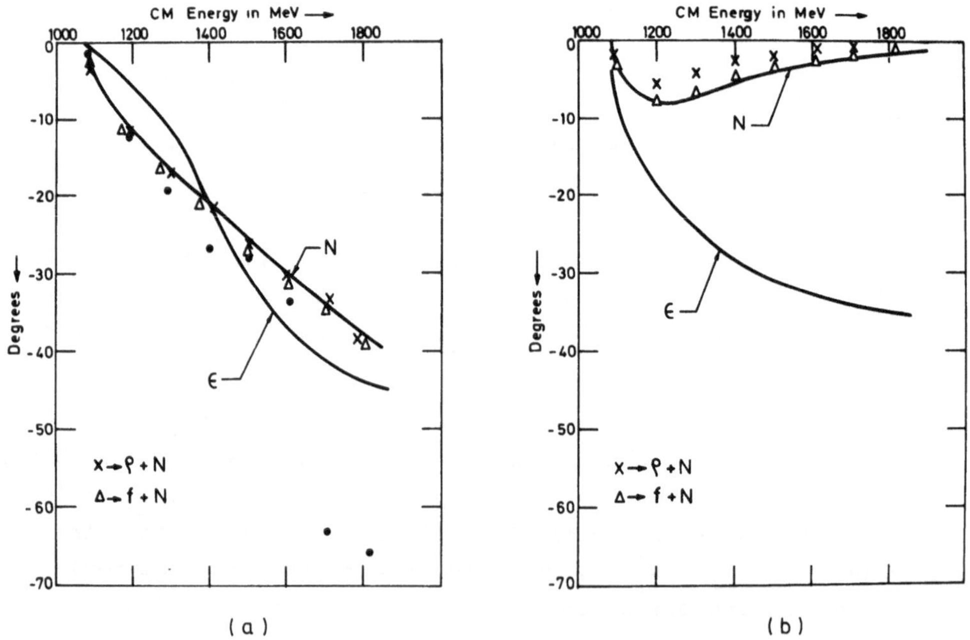

FIG - 1

Fig. 1(a) $\delta^{(3)}$ phase shift; (b) $\delta^{(1)}$ phase shift. The points indicate the experimental data of CERN [14].

References

[1] H.A. Bethe and F.D. Hoffmann, 'Mesons and Fields', Vol.I, p.70.
[2] A good account of early theories is given in ref. 1.
[3] J.J. Sakurai, Ann. Phys. 11 (1960) 1.
[4] J.A. Mignaco and E. Remiddi, Nuovo Cimento, A1 (1971) 395.
[5] L.V. Filkov and B.B. Palyushev, Nucl. Phys. B42 (1972) 541.
[6] T. Yanagida and M. Yonezawa, Prog. Theor. Phys. 54 (1976) 184.
[7] N. Hiroshige and T. Tsugimure, Prog. Theor. Phys. 57 (1977) 901.
[8] T. Yenagida and M. Yonezawa, Prog. Theor. Phys. 56 (1976) 1199.
[9] M.C. Bergere and J.M. Drouffe, Nuovo Cimento A11 (1972) 121; Nucl. Phys. B53 (1973) 191.
[10] P. Achuthan, H. Schlaile and F. Steiner, Nucl. Phys. B24 (1970) 398.
[11] G. Höhler and R. Straus, Z.Physik, 232 (1970) 205.
[12] V.Barger and R.J.N. Phillips, Phys.Rev., 187 (1969) 2210.
[13] H.Nielsen, Nucl. Phys. B33 (1971) 152.
[14] D.J. Herndon, A.Barbaro-Galtieri and A.H. Rosenfeld, πN Partial-Wave Amplitude - A Compilation, UCRL 20030 (1970).

SOME CONSIDERATIONS OF THE NUCLEON AS A BOUND TRIQUARK

E.P. Harper+
Dept. of Physics
The George Washington University
Washington, D.C. 20052, USA

I. Introduction

In this paper we show how it is possible, on the basis of a non-relativistic triquark model of the nucleon, to simultaneously reproduce the nucleon magnetic moment ratio and the weak interaction coupling constant ratio. The usual SU(6) model gives for the magnetic moment ratio, -3/2, and for the weak coupling constant ratio, 5/3, which are to be compared with the experimental values -1.46 and 1.26 respectively. It occurred to a number of authors [1] that a possible source of the discrepancies between theory and experiment (small in the case of the magnetic moment ratio) might be the failure to take into account possible non-central forces in the qq interaction giving rise to L = 2 states of the triquark. The bound triquark is in many ways similar to the bound trinucleon but there are some differences such as the apparent symmetry of the wave function under space-spin-isospin exchange and the fact that all constituents are charged. These differences are explicitly taken into account in our calculation.

In addition to the ratios just discussed we also calculate a matrix element for the photo-transition $\gamma + N \rightarrow \Delta(1236)$ [2] and compare the value with the experimental finding. This latter calculation should be viewed with some circumspection since it is no means clear that the D-state probabilities for the nucleon and $\Delta(1236)$ are related in any simple way.

II. The Nucleon Wave Function, Magnetic Moments and Weak Coupling Constants

We can represent the wave function of the bound triquark as follows [3]

$$|\Psi_N\rangle = |W^S_{\frac{1}{2}\frac{1}{2}}\ \Psi_S^{1/2}(^2S_{1/2})\rangle + |W^+_{\frac{1}{2}\frac{1}{2}}\Psi_+^{1/2}(^2S_{1/2}) + W^-_{\frac{1}{2}\frac{1}{2}}\Psi_-^{1/2}(^2S_{1/2})\rangle$$

$$+ \sum_{m_s} \langle 2\ 1/2{-}m_s \tfrac{3}{2} m_s | \tfrac{1}{2}\tfrac{1}{2}\rangle |W^+_{\frac{3}{2}m_s}\ \Psi_+^{m_s}(^4D_{1/2}) + W^-_{\frac{3}{2}m_s}\ \Psi_-^{m_s}(^4D_{1/2})\rangle \qquad (1)$$

where each term corresponds to the spatially symmetric, mixed symmetry (L = 0) and mixed symmetry (L = 2) states familiar to trinucleon theorists. The spin-isospin functions have the following permutation properties

$P(i)W^S = +W^S$

$P(1)W^+ = W^+$ $\quad$ $P(2)W^+ = -\frac{1}{2}W^+ - \frac{\sqrt{3}}{2}W^-$ $\quad$ $P(3)W^+ = -\frac{1}{2}W^+ + \frac{\sqrt{3}}{2}W^-$

$P(1)W^- = W^-$ $\quad$ $P(2)W^- = -\frac{\sqrt{3}}{2}W^+ + \frac{1}{2}W^-$ $\quad$ $P(3)W^- = \frac{\sqrt{3}}{2}W^+ + \frac{1}{2}W^-$

(2)

where P(i) is the operator which exchanges the spin-isospin coordinates of particles k, j ≠ i. The spatial wave functions behave similarly under space exchange.

The spin-isospin functions are constructed thus,

$$W^A_{\frac{1}{2}m_s} = \frac{1}{\sqrt{2}}[\chi_2^{m_s}\eta_1 - \chi_1^{m_s}\eta_2]$$

$$W^+_{\frac{1}{2}m_s} = \frac{1}{\sqrt{2}}[\chi_2^{m_s}\eta_2 - \chi_i^{m_s}\eta_1]$$

$$W_{\frac{1}{2}m_s} = \frac{1}{\sqrt{2}}[\chi_1^{m_s}\eta_2 + \chi_2^{m_s}\eta_1]$$

$$W^-_{\frac{3}{2}m_s} = \chi_3^{m_s}\eta_1 \qquad W^+_{\frac{3}{2}m_s} = -\chi_3^{m_s}\eta_2 \tag{3}$$

where

$$\chi_1^{m_s} = |(0\ 1/2)1/2\ m_s>_1$$

$$\chi_2^{m_s} = |(1\ 1/2)1/2\ m_s>_1$$

$$\chi_3^{m_s} = |(1\ 1/2)3/2\ m_s>_1 \tag{4}$$

in the usual notation [3]. A similar representation holds for the isospin functions η.

In terms of the wave function given by equation (1) the nucleon magnetic moment is

$$\mu_N = <\Psi_N|\vec{J}\cdot\vec{M}_N|\Psi_N> \quad \frac{J_z}{J(J+1)} \tag{5}$$

with the magnetic moment operator given by

$$\vec{M}_N = \sum_{i=1}^{3} 1/2(1 + \tau_z(i))[\mu_u\vec{\sigma}(i) + \tfrac{2}{3}\vec{\ell}(i)] + \tfrac{1}{2}(1-\tau_z(i))$$
$$[\mu_d\,\vec{\sigma}(i) - \tfrac{1}{3}\vec{\ell}(i)] \tag{6}$$

in units of quark magnetons $\frac{e\hbar}{2m_q c}$ where m_q is the quark mass. $\vec{\ell}(i)$ is the orbital angular momentum of quark i in the nucleon center of mass.

μ_u and μ_d are the magnetic moments of (in quark magnetons) of the u and d quarks. J is the total angular momentum of the triquark system.

Splitting equation (6) into isoscalar and isovector parts we obtain the following, using equations 1, 2 and 3

$$\mu_N(\text{isoscalar}) = \frac{\mu_+}{2}[P_S + P_{S'} - P_D] + P_D/6$$

$$\mu_N(\text{isovector}) = \pm\frac{1}{6}[\mu_-(SP_S + P_{S'} - 8P_{SS'} - P_D) + P_D] \qquad (7)$$

where $\mu^+ = \mu_u + \mu_d$ and $\mu^- = \mu_u - \mu_d$ and the probabilities P_S, $P_{S'}$, $P_{SS'}$ and P_D are, respectively

$$P_S = \langle\Psi_S^{1/2}(^2S_{1/2})|\Psi_S^{1/2}(^2S_{1/2})\rangle$$

$$P_{S'} = \langle\Psi_+^{1/2}(^2S_{1/2})|\Psi_+^{1/2}(^2S_{1/2})\rangle + \langle\Psi_-^{1/2}(^2S_{1/2})|\Psi_-^{1/2}(^2S_{1/2})\rangle$$

$$P_{SS'} = \langle\Psi_S^{1/2}(^2S_{1/2})|\Psi_+^{1/2}(^2S_{1/2})\rangle$$

$$P_D = \langle\Psi_+^{m_s}(^4D_{1/2})|\Psi_+^{m_s}(^4D_{1/2})\rangle + \langle\Psi_-^{m_s}(^4D_{1/2})|\Psi_-^{m_s}(^4D_{1/2})\rangle \quad . \qquad (8)$$

The +, - signs in equation (7) refer to $T_z = \pm 1/2$, respectively i.e. to the proton or neutron.

We now make the simple assumption that the quark magnetic moment has no anomalous part so that

$$\mu_u = 2\mu_d = 2/3. \qquad (9)$$

In this model the anomalous magnetic moments of the nucleon are due to its being a composite of constituents which are themselves pure Dirac particles as is the electron.

Adding the isoscalar and isovector parts we get for the ratio of the proton to neutron magnetic moments

$$-\frac{\mu_p}{\mu_n} = \frac{3-2(P_{S'} + 2P_{SS'}) - 3P_D}{2-2\ (P_{S'} + 2P_{SS'}) - 2P_D} \qquad (10)$$

Notice that if $P_{S'} = 0$, this implying $P_{SS'} = 0$, that we get precisely the same prediction for the magnetic moment ratio as the obtained from the model neglecting states other than symmetric S-states. That is, inclusion of the D-state alone gives no improvement.

Turning to the weak interactions of the triquark we get

$$G_V = g_V$$

$$g_A = g_A \ \langle\Psi_p|\sum_{i=1}^{3}\vec{\sigma}(i)\tau_+(i)|\Psi_n \qquad (11)$$

where the first equation is a consequence of CVC. $G_{V,A}$ are the weak coupling constants for the nucleon while $g_{V,A}$ are the corresponding equations for the quark. The assumption corresponding to no anomalous quark magnetic moments is in the weak interaction case [4]

$$g_A = g_V \tag{12}$$

Evaluating the second of equations (11) in a manner similar to the evaluation of the isovector magnetic moment we arrive at

$$\frac{G_A}{G_V} = \frac{1}{3}\,[5 - 4(P_{S'} + 2P_{SS'}) - 6\,P_D] \tag{13}$$

Calling the negative of the experimental magnetic moment ratio R and the experimental weak coupling constant ratio S we have the following equations to solve for P_D and $P_X = P_{S'} + 2P_{SS'}$

$$2R = \frac{3 - 2P_X - 3P_D}{1 - P_X - P_D}$$

$$3S = 5 - 4P_X - 6\,P_D \tag{14}$$

The solution is $P_D = 0.247$ and $P_X = -0.066$. It is easily seen that the omission of either the D-state or the S'-state leads to no possibility of obtaining the experimental results. The small negative value of P_X means that the probability $P_{S'}$ is small but non-zero if we assume that the symmetric S-state is dominant.

Using these probabilities we can get the quark mass from the proton magnetic moment. We get $m_q = 267.4$ MeV.

III. The Photo-transition $\gamma + N \rightarrow \Delta(1236)$

In addition to the ratios considered in the previous section another experimentally measurable parameter of the non-strange triquark system is the matrix element $\langle \Psi_N^{1/2} | \sum_i m_i^z | \Psi_\Delta^{1/2} \rangle$ whose empirical value can be extracted from measurements of the process [2] $\gamma + P \rightarrow \Delta^+ \rightarrow P + \pi^0$. Experimentally its value is found to be $(1.25 \pm 0.02)\,\frac{2\sqrt{2}}{3}\,\mu_p$ where μ_p is the proton magnetic moment.

In our model we write the $\Delta(1236)$ wave function as follows

$$|\Psi_\Delta^{1/2}\rangle = |\chi^S_{\frac{3}{2}\frac{1}{2}}\,\Phi_S^{1/2}(^4S_{3/2})\rangle + \sum_{m_s} \langle 2\; 1/2-m_s\; 3/2\; m_s | 3/2\; 1/2\rangle$$

$$|\chi^S_{\frac{3}{2}\frac{1}{2}}\,\Phi_S^{1/2}(^4D_{3/2})\rangle + \langle 2\; \tfrac{1}{2}-m_s\; 1/2\; m_s | 3/2\; \tfrac{1}{2}\rangle |\chi^+_{1/2\,m_s}\Phi_+^{m_s}(^2D_{3/2})$$

$$+ \chi^-_{\frac{1}{2}m_s}\,\Phi_-^{m_s}(^2D_{3/2})\rangle \tag{15}$$

where

$$\chi^{S}_{\frac{3}{2}\frac{1}{2}} = \chi_3^{1/2}\,\eta_3$$

$$\chi^{+}_{\frac{1}{2}m_s} = -\,\chi_2^{m_s}\,\eta_3$$

$$\chi^{-}_{\frac{1}{2}m_s} = \chi_1^{m_s}\,\eta_3 \qquad (16)$$

Using this wave function for $\Delta(1236)$ and equation (1) for the proton we get for $\langle\Psi_N^{1/2}|\sum_i M_i^z|\Psi_\Delta^{1/2}\rangle$ the expression

$$\langle\Psi_N^{1/2}|\sum_{i=1}^{3} M_i^z|\Psi_\mu^{1/2}\rangle = \frac{2\sqrt{2}}{3}\,\mu_q\,[\langle\Psi_S^{1/2}(^2S_{1/2})|\Phi_S^{1/2}(^4S_{1/2})\rangle + \langle\Psi_+^{1/2}(^2S_{1/2})|\Phi_S^{1/2}(^4S_{1/2})\rangle + \langle\Psi_+^{1/2}(^4D_{1/2})|\Phi_+^{1/2}(^2D_{3/2})\rangle + \frac{2}{5}\langle\Psi_+^{1/2}(^4D_{1/2})|\Phi_S^{1/2}(^4D_{3/2})\rangle] \qquad (17)$$

where μ_q is the quark magneton, $\frac{e\hbar}{2m_qc}$.

As it stands equation (17) introduces four new quantities. If we make the assumption that the states of a given total orbital angular momentum and spatial symmetry type are the same in both the nucleon and $\Delta(1236)$ we can then make the approximation

$$\langle\Psi_N^{1/2}|\sum_{i=1}^{3} M_i^z|\Psi_\Delta^{1/2}\rangle \sim \frac{2\sqrt{2}}{3}\,\mu_q\,[1 - \frac{1}{2}(P_D - P_X)] \qquad (18)$$

where P_D and P_X are given by the solution of equations (14). Using these solutions and the mass previously obtained for the quark we find

$$\langle\Psi_N^{1/2}|\sum_{i=1}^{3} M_i^z|\Psi_\Delta^{1/2}\rangle = \frac{2\sqrt{2}}{3}\,\mu_p\,(1.06) \qquad (19)$$

The number in parentheses is to be compared with the empirical value 1.25 ± 0.02 and the SU(6) value of unity.

IV. Discussion

In the foregoing we have calculated the ratio of the nucleon magnetic moments and weak coupling constants and found that the experimental values could be reproduced provided the nucleon has a large D-state component and a small mixed symmetry S-state. These results were obtained on the basis of the assumption that the electromagnetic and weak charges of the constituent quarks were unrenormalized by their coupling to the "gluon" field [4]. In this model the anomalous magnetic moments and axial vector coupling constant of the nucleon are due not to its coupling with the meson field mediating nuclear interactions but to the fact that it is a composite system of three quarks. The results obtained for the nucleon configuration probabilities are then used to obtain an estimate for the matrix element describing a $\gamma + N \rightarrow \Delta(1236)$ phototransition. The result of this calculation gives a prediction which is marginally better than that given by SU(6) model [5].

References

[1] J. Franklin, Phys. Rev. 172, 1807 (1968).
Y. Gell and D. Lichtenberg, Nuovo Dimento 1A, 29 (1969).

[2] See, for example, J.J.Kokkedee "The Quark Model" W.A. Benjamin (1969).

[3] E. Harper, Y.E. Kim and A. Tubis, Phys. Rev. C6, 126 (1972).

[4] N.N. Bogoliubov, V.A. Matvejev and A.N. Tavkhelidze, Nuovo Cim.48A, 132 (1967).

[5] R.H. Dalitz and D.G. Sutherland, Phys.Rev. 146, 1180 (1966).

+Work supported by the George Washington University Committee on Research.

MULTIQUARK DIBARYON RESONANCES

A.Th.M. Aerts, P.J.G. Mulders and J.J. de Swart
Institute for Theoretical Physics, University of Nijmegen, The Netherlands

Recent calculations of six quark states in the bag model show many dibaryon resonances [1]. These are positive parity resonances, when all the quarks are in 1s states of the (spherical) bag. The splitting of states is caused by the color magnetic interaction. About the widths of the predicted states not much is known. Calculations by DeTar [2] show that the $^3S_1$ and $^1S_0$ resonances in the NN channel, predicted at 2.16 and 2.24 GeV., are probably very unstable, because they can spontaneously decay in two nucleons in relative S-waves. An interpretation as soft core in the NN interaction might be possible. This could be the more fundamental explanation of the soft core due to Pomeron exchange as obtained and successfully used in the NN potential by the Nijmegen group [3]. The other q^6 resonances predicted in the NN channel, a $^1D_2$ and a $^3D_3$ state both at 2.36 GeV., probably are less unstable due to the angular momentum barrier and could have an acceptable width. The pp resonance at 2.39 GeV. [4] and the peak in deuteron photodesintegration around 2.38 GeV. [5] could be due to these resonances.

The lowest positive parity resonances in the YN channels are a 1^+ resonance at 2.17 GeV. and a 2^+ resonance at 2.24 GeV., both with $I = \frac{1}{2}$. The 2^+ resonance decays in a D-wave and therefore might be less unstable. A candidate is the resonance at 2.26 GeV., found by Shahbazian [6].

Negative parity resonances must come from L excitations of the bag. The bag now becomes a stringlike object, with at each end a certain number of quarks coupled to conjugate color irreps [7]. The mass of the states is given by the simple mass formula

$$M = A_L + B_1 + B_2$$

where A_L is the mass of an L-excited bag neglecting the color magnetic interaction. These masses A_L are supposed to lie on linear trajectories, i.e.

$$A_L^2 = A_0^2 + (1/\alpha')L$$

where A_0 is the mass of a spherical six quark bag neglecting color magnetic interactions, L is the angular momentum and $(1/\alpha')$ is the slope which is proportional to the square root of the quadratic Casimir of the color irreps, to which the quarks at the ends couple. For color $\underline{3}$-$\underline{3}^*$ configurations $(1/\alpha')$ equals the universal slope $(1/\alpha') = 1.1\ \text{GeV}^2$. B_1 and B_2 are the color magnetic energy contributions of the ends of the bag. SU(3) breaking comes in from the dependence of A and B on the number of strange quarks. In Tables I and II the masses of six quark states for S = 0, I = 0,1 and S = -1, I = 1/2, 3/2 have been given. The first column gives the L = 0 six quark states (positive parity), the other columns the L = 1 excited states (negative parity). We distinguish four types of resonances: q^5-q ($\underline{3}^*$-$\underline{3}$), q^4-q^2 ($\underline{3}$-$\underline{3}^*$), q^4-q^2 ($\underline{6}^*$-$\underline{6}$) and q^3-q^3 ($\underline{8}$-$\underline{8}$) resonances, indicating the number of quarks at the ends and the color configuration. n and s indicate nonstrange respectively strange quarks.

In the decays of these resonances the four simplest mechanisms we expect to be:
1. Spontaneous rearrangements of quarks at one end (q^4 or q^5) such that a color singlet is formed. This is only possible for $\underline{3}$-$\underline{3}^*$ configurations above the BB^* threshold, B^* indicating the lowest L = 1 excited baryon resonances. For S = 0 the important threshold is $M(NN^*) \simeq 2.46$ GeV. If no angular momentum barrier is present (i.e. for 0^-, 1^- and 2^- states) these resonances will be very unstable.
2. Tunneling followed by spontaneous decay. In this way q^4-q^2 ($\underline{3}$-$\underline{3}^*$) states can couple to BB P-waves. For S = 0 this mechanism is possible for I = 1, $J^P = 0^-, 1^-, 2^-$ and I = 0, $J^P = 1^-$ (P-waves), it is impossible for F-waves, e.g. 3^- and 4^- states.
3. Pair creation followed by spontaneous decay to BBM channels. This is possible for all types, if after pair creation the ends are in relative S-waves (also D-waves are possible). Except for the S = 0, I = 0, $J^P = 1^-$ state, spontaneous decay of the 0^- and 1^- states in BBM channels, where all particles are in relative S-waves, is possible. The thresholds are for S = 0: $M(NN\pi) = 2.02$ GeV. and for S = -1: $M(N\Lambda\pi) = 2.20$ GeV., $M(N\Sigma\pi) = 2.28$ GeV., $M(NN\bar{K}) = 2.37$ GeV.
4. For q^3-q^3 states decay is possible due to the color mixing between the q^3-q^3 ($\underline{8}$-$\underline{8}$) configuration and the q^3-q^3 ($\underline{1}$-$\underline{1}$) configuration, which also occur. This mixing is also present between q^4-q^2 ($\underline{6}^*$-$\underline{6}$) and q^4-q^2 ($\underline{3}$-$\underline{3}^*$) configurations. We do not expect that this mixing has considerable influence on the masses.

These decay possibilities considerably reduce the number of dibaryon resonances which we expect to be prominent. The mechanisms 1 and 3 are largely inelastic. For the elastic BB reactions only the higher spin states (3^- and 4^-) of the q^4-q^2 ($\underline{3}$-$\underline{3}^*$) type might be most prominent, although even they couple only weakly. For S = 0, I = 1 two 3^- resonances lie around 2.34 GeV., which can explain the structure found by Kroll and Grein in one of the uncoupled $^3P_1$ or $^3F_3$ waves [4]. Also noteworthy is the existence of some states with quantum numbers forbidden in the NN system (extraneous states). We mention two low-lying extraneous states in the I = 0 channel with $J^P = 0^-$ and 2^- at $M \simeq 2.11$ GeV. The 2^- state will be the most prominent because the 0^- state easily decays via mechanism 3.

In the S = -1 channels we mention states with I = 1/2 at 2.11 GeV (1^-) and at 2.15 GeV (0^-, 1^- and 2^-). Except the beforementioned 2^+ resonance, also $J^P = 0^-, 1^-$ and 2^- (I = 1/2) resonances occur in the region around 2.26 GeV. [6].

References:

1. R.L. Jaffe, Phys.Rev.Lett. 38, 195 (1977) and Errata 38, 617 (1977).
A.Th.M. Aerts, P.J.G. Mulders and J.J. de Swart, Phys.Rev. D 17, 260 (1978).
P.J.G. Mulders, A.Th.M. Aerts and J.J. de Swart, Nijmegen report no. THEF-NYM-78.1.
V. Matveev and P. Sorba, Lett.Nuov.Cim. 20, 435 (1977), Fermilab PUB 77/56.
2. C. DeTar, Phys.Rev. D 17, 323 (1978)
3. M.M. Nagels, T.A. Rijken and J.J. de Swart, Phys.Rev. D 17, 768 (1978).
4. W. Grein and P. Kroll, University of Wuppertal preprint WUB 77-6 and refs. therein.
5. T. Kamae et al., Phys.Rev.Lett. 38, 468 (1977).

6. B.A. Shahbazian and A.A. Timonina, Nucl.Phys. B 53, 19 (1973);
B.A. Shahbazian et al., in Proceedings of the XVIII International Conference on High Energy Physics, Tblisi, 1976, edited by N.N. Bogolubov et al., (JINR, Dubna, 1977), Vol. I, p. C 35.

7. K. Johnson and C.B. Thorn, Phys.Rev. D 13, 1934 (1976);
R.L. Jaffe, MIT Report no. CTP 657.

	n^6		n^5-n $(\underline{3}^*-\underline{3})$		n^4-n^2 $(\underline{3}-\underline{3}^*)$		n^4-n^2 $(\underline{6}^*-\underline{6})$		n^3-n^3 $(\underline{8}-\underline{8})$	
S=0,I=0	1	2.16	0,1	2.37	1	2.11	1	2.39	0,1	2.43
	3	2.36	1,2	2.43	1	2.43	1,2,3	2.52	1,2	2.50
			2,3	2.54	0,1,2	2.47	1	2.68	0,1,2,3	2.58
					1,2,3	2.56			1,2	2.87
S=0,I=1	0	2.24	0,1	2.37	0	2.20	0	2.50	0,1	2.43
	2	2.36	1,2	2.43	1	2.25	0,1,2	2.57	1,2	2.50
			2,3	2.54	2	2.34	2	2.64	0,1,2,3	2.58
			0,1	2.56	0,1,2	2.34	1	2.68	0,1	2.65
			1,2	2.62	1	2.43	0	2.91	1,2	2.72
					0,1,2	2.47			1,2	2.87
					1,2,3	2.56				
					0,1,2	2.74				

TABLE I: The S = 0 q^6 resonances for I = 0 and I = 1. The total spin has been given. Except for the first column (L = 0), the spin S has to be combined with L = 1 to find J. The second part of every column gives the mass in GeV.

	n^5s		n^5-s	$(\underline{3}^*-\underline{3})$	n^4s-n	$(\underline{3}^*-\underline{3})$	n^3s-n^2	$(\underline{3}-\underline{3}^*)$	n^4-ns	$(\underline{3}-\underline{3}^*)$	n^3s-n^2	$(\underline{6}^*-\underline{6})$	n^4-ns	$(\underline{6}^*-\underline{6})$	n^2s-n^3	$(\underline{8}-\underline{8})$
S=-1,I=1/2	1	2.17	0,1	2.52	0,1	2.30	0	2.11	1	2.29	0,1,2	2.48	1	2.56	0,1	2.37
	2	2.24	1,2	2.59	1,2	2.35	1	2.15	0	2.38	1	2.56	0	2.65	1,2	2.44
	1	2.33	2,3	2.69	0,1	2.41	1	2.27	1	2.43	1	2.59	1,2,3	2.69	0,0,1,1	2.56
	0	2.40			1,2	2.47	1	2.34	0,1,2	2.48	0	2.67	0,1,2	2.74	1,1,2,2	2.63
	2	2.51			0,0,1,1	2.52	0	2.35	2	2.52	1,2,3	2.68	2	2.79	1,1,2,2	2.64
	3	2.51			1,1,2,2	2.58	0,1,2	2.38	1	2.57	0,1,2	2.72	1	2.83	0,0,1,1,2,2,3,3	2.70
					2,2,3,3	2.67	1	2.39	0,1,2	2.62	2	2.79			0,1	2.75
					0,1	2.69	2	2.47	1,2,3	2.71	1,1	2.83			1,2	2.78
					1,2	2.75	0,1,2	2.50			0	3.03			1,2	2.83
							1,1	2.58							0,1,2,3	2.85
							0,0,1,1,2,2	2.62							1,2	2.98
							1,1,2,2,3,3	2.70								
							0,1,2	2.86								
S=-1,I=3/2	1	2.33	0,1	2.71	0,1	2.41	1	2.34	0	2.38	1	2.59	0,1,2	2.74	0,1	2.56
	0	2.40	1,2	2.77	1,2	2.47	0	2.35	1	2.43	0	2.67	1	2.83	1,2	2.59
	2	2.51			0,1	2.52	0,1,2	2.38	2	2.52	0,1,2	2.72	1	2.96	1,2	2.63
	1	2.66			1,2	2.58	1	2.39	1	2.57	2	2.79	0	3.06	1,2	2.64
					2,3	2.67	2	2.47	0,1,2	2.62	1,1	2.83			0,1,2,3	2.70
					0,0,1,1	2.69	0,1,2	2.50	1	2.70	1	2.92			0,1	2.76
					1,1,2,2	2.75	1,1	2.58	1,2,3	2.71	0	3.03			1,1,2,2	2.78
					0,1	2.98	0,0,1,1,2,2	2.62	0,1,2	2.89					1,2	2.83
							1	2.63							0,0,1,1,2,2,3,3	2.85
							1,1,2,2,3,3	2.70							1,2	2.98
							1,2,3	2.86								

TABLE II: S=-1 q^6 resonances for I = 1/2 and I = 3/2. The total spin has been given. Except for the first column (L = 0), the spin S has to be combined with L = 1 to find J. The second part of every column gives the mass in GeV.

HIGH ENERGY p-p ELASTIC SCATTERING AND CONSTITUENT MULTIPLE SCATTERING MODEL

S. Wakaizumi
Department of Physics
Hiroshima University
Hiroshima 730/JAPAN

Recent measurements of the differential cross section for p-p elastic scattering have exhibited the following features in the range of momentum-transfer squared $1 \leq -t \leq 10$ GeV^2 at center-of-mass energy $\sqrt{s} = 53$ GeV[1]; (i) it has a pronounced sharp dip at $t = -1.34$ GeV^2, (ii) the dip is followed by a second peak of structureless single exponential with a very small slope $b_2 = (1.81 \pm 0.02)$ GeV^{-2} for $2 \leq -t \leq 6$ GeV^2 and (iii) another change of slope occurs near $t = -6.5$ GeV^2. The small value of slope b_2 in (ii) has not been able to be reproduced by the eikonal models such as Chou-Yang model, which have long been considered to describe well the high energy scattering[2], because the convolution integral involved in those models yields the slope of half the first one for the second peak, i.e. $b_2 = b_1/2$. Slope of the first peak, b_1, is measured to be 10.3 GeV^{-2} for $0.2 \leq -t \leq 0.6$ GeV^2[3], so those models give 5 GeV^{-2} to the second slope, which is too large as compared with the experimental value.

Here, we attempt to explain the above-mentioned structure stated in (i)∿(iii) by the multiple scattering model of hypothetical "constituents" originated by Takada and by Harington and Pagnamenta. The proton is assumed to be composed of N constituents, which could be identified with quarks, partons or something like that. So, the proton-proton elastic scattering, regarded as a collision of two composite particles, can be described by a successive series of single, double, triple, $\cdots\cdots$ scatterings of constituents as follows,

$$T_{pp}(s,t)/R_0^2 = \binom{N}{1}^2 t_{qq}(t^0)[f_1(t^0)]^2 + \binom{N}{2}^2 2! \frac{1}{2\pi}\int dq_1^0 dq_2^0 t_{qq}(q_1^0) t_{qq}(q_2^0) \delta(q_1^0 + q_2^0 - q^0)$$
$$\times [f_2(q_1^0, q_2^0)]^2 - \binom{N}{3}^2 3! \frac{1}{(2\pi)^2}\int dq_1^0 dq_2^0 dq_3^0 t_{qq}(q_1^0) t_{qq}(q_2^0) t_{qq}(q_3^0) \delta(q_1^0 + q_2^0 + q_3^0 - q^0)$$
$$\times [f_3(q_1^0, q_2^0, q_3^0)]^2 + \cdots\cdots, \qquad (1)$$

where $q_1^0, q_2^0, \cdots$ are non-dimensional momenta transfered between the two colliding constituents, $t^0 = R_0^2 t \equiv -q^{0^2}$ is a scaling variable of the same quantity squared between the two protons, t_{qq} is the constituent scattering amplitude. In our model is incorporated the important property of hadron reactions at high energy, the "geometrical scaling". The first term in eq.(1) represents the single scattering, the second term the double one and so on. The functions $f_1(t^0)$, $f_2(q_1^0, q_2^0)$, $f_3(q_1^0, q_2^0, q_3^0)$, etc. are the hadronic form factors related to each multiple scattering, and are calculated with

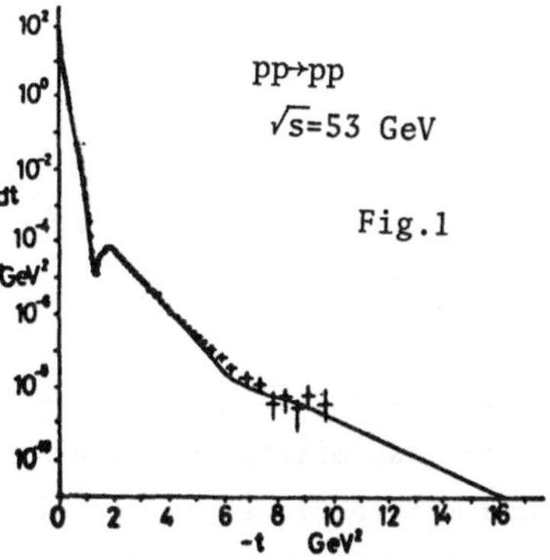

Fig.1

the wavefunction of proton constructed from N constituents. If a Gaussian form is taken for t_{qq} and the wavefunction for simplicity, then is obtained the slope of each multiple scattering peak which leads to the forward, second and thid peaks in dσ/dt as follows; $b_1=a+\frac{N-1}{N}<r^2>$, $b_2=\frac{1}{2}\times(a+\frac{N-2}{N}<r^2>)$ and $b_3=\frac{1}{3}(a+\frac{N-3}{N}<r^2>)$, where a is constituent scattering slope and $\sqrt{<r^2>}$ is the root mean square distance between a constituent and the center-of-mass of the proton. The experimental ratio of the forward peak slope to the second peak one, $R\equiv b_2/b_1=1.8/10.3\simeq 1/5$, is reproduced only when $<r^2>\gg a$ and N=3 [4] , which leads to $R\simeq 1/4$. Consequently, we can say that the proton is an object with an extended distribution of very small <u>three</u> constituents(quarks). To be more interesting, the slope of constituent scattering directly appears as a slope of third peak in dσ/dt, i.e. $b_3=a/3$ for N=3. In order to obtain an overall fit to dσ/dt for $0\leq|t|\leq 10$ GeV^2, a more realistic wavefunction is adopted which leads to $f_1=1/(1-t/\mu_1{}^2)(1-t/\mu_2{}^2)$,known to reproduce well the forward peak. Moreover, it is necessary to introduce, besides the multiple scattering amplitude, an additional term which represents the very forward steep peak before the break at $t\simeq -0.2$ GeV^2. The origin of the steep peak is not known, so it is phenomenologically given as $T_0/R_0{}^2=iA_0e^{b_0t/2}$. The calculated result is given at $\sqrt{s}=53$ GeV in Fig.1. Values of $A_0=2.05$ and $A_1=1.81$(A_1:the magnitude of t_{qq} at t=0) are taken to fit $\sigma_T=42.5$mb and the height of the second peak. The break is eventually derived well at t=-0.22 GeV^2. And, the first dip is obtained at t=-1.37 GeV^2 in agreement with the observed value $t_{min}=-(1.34\pm 0.02)$ GeV^2 and the slope b_2 is 2.13 GeV^{-2}, a little larger than $b_2{}^{exp}=(1.81\pm 0.02)$ but not inconsistent. Another change of slope is predicted to occur near t=-7 GeV^2. To test the composite model of hadrons, $\pi^-p\to\pi^-p$ is also calculated with the same parameters of t_{qq} as in pp → pp. For comparison is used the relation $P^{\pi}_{LAB}/P^{p}_{LAB}=m_\pi/m_p$ [5] and π^-p elastic scattering at 200 GeV/c is calculated with p-p parameters of $\sqrt{s}=53$ GeV and is compared with the data on dσ/dt obtained at FNAL[6]. The result is shown in Fig.2. The pion constituent number N'=2 is more favored than N'=3.

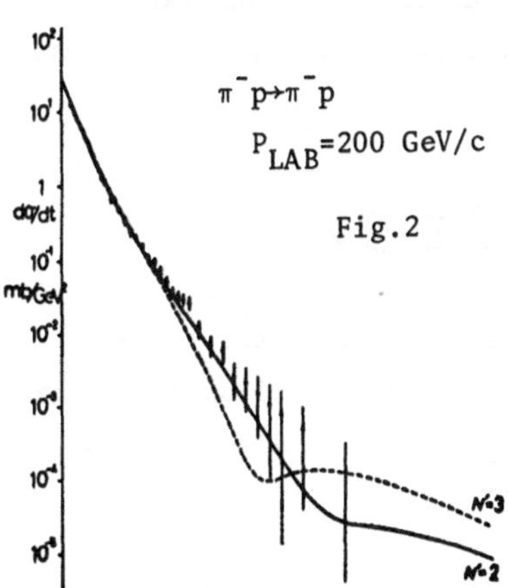

Fig.2

1. De Kerret,H.,et al.:Phys. Letters62B, 363(1976). —De Kerret,H.,et al.:Phys. Letters 68B, 374(1977).
2. Chou,T.T.,Yang,C.N.:Phys. Rev. 170, 1591(1968).
3. Kwak,N.,et al.:Phys. Letters 58B, 233(1975).
4. Wakaizumi,S.,et al:Phys. Letters 70B, 55(1977).
5. James,P.B.,et al.:Phys. Rev. Letters18,179(1967).
6. Akerlof,C.W.,et al.:Phys. ReV. 14, 2864(1976).

A NEW STABLE DIBARYON

H. Høgaasen and P. Sorba
CERN -- Geneva

Today, experiments give strong evidence for hadron constituents (quarks) and for at least five different flavours of quarks. Moreover, the possibility that quarks carry a degree of freedom called colour is taken more and more seriously; especially in the framework of quantum chromodynamics (QCD), where colour is the source of the gluon field that propagates the strong interactions. Each quark q can exist in three colour states, the quarks (antiquarks) transform under the $3(\bar{3})$ fundamental representation of the supposedly exactly conserved SU_3^{colour} group, whereas the gluons transform under the eight-dimensional adjoint representation.

The masses of the s wave baryons (qqq states) and mesons $(q\bar{q})$ are very well explained by the one-gluon exchange interaction, which is analogous to the hyperfine splitting in electrodynamics due to photon exchange.

The Hamiltonian for the colourmagnetic interaction between two quarks in a relative s wave state is given[1)] by

$$H' = -\frac{8\pi}{3}\alpha_s \sum_{\alpha=1}^{8} \lambda_1^\alpha \underline{\sigma_1} \cdot \lambda_2^\alpha \underline{\sigma_2}\, \delta^3(\underline{r}_1 - \underline{r}_2) M_{12}$$

where α_s is the running strong interaction "constant" for the coupling of gluons to quarks, λ_1^α and λ_2^α are the usual Gell-Mann matrices and $\underline{\sigma}_1$ and $\underline{\sigma}_2$ the Pauli spin matrices for quarks 1 and 2. The constant M_{12} is dependent on the masses of the two quarks. The mean value of this two-particle Hamiltonian in a physical state is

$$\langle H' \rangle = -C_{12} \left\langle \sum_{\alpha=1}^{8} \lambda_1^\alpha \lambda_2^\alpha \underline{\sigma_1} \cdot \underline{\sigma_2} \right\rangle \tag{1}$$

where C_{12} is proportional to the product of M_{12} and the square of the space wave function in the point $\underline{r}_1 = \underline{r}_2$. Symbolically we shall, in the following, write

$$H' = -C_{12} \lambda_1^\alpha \lambda_2^\alpha \underline{\sigma_1} \cdot \underline{\sigma_2} .$$

Then, for a system of n quarks

$$H' = -\sum_{i>j}^{n} C_{ij} \lambda_i^\alpha \lambda_j^\alpha \underline{\sigma_i} \cdot \underline{\sigma_j} . \tag{2}$$

The mass spectrum of all s wave (non-charmed) baryon states is correctly reproduced[2] if we write the mass as

$$M = \sum_i \bar{m}_{q_i} + \langle H' \rangle \qquad (3)$$

The effective mass of the non-strange quarks n and d is: $\bar{m}_n = \bar{m}_d = 360$ MeV $\equiv \bar{m}_n$, while for the strange quark: $\bar{m}_s = \bar{m}_n + \Delta m$ with $\Delta m = 175$ MeV; the "constants" are given by $C_{uu} = C_{dd} = C_{nd} \equiv C_{nn} = 20$ MeV, $C_{ns} = 12.5$ and $C_{ss} = 9.5$ MeV. Here, C_{ns} denotes the effective interaction between non-strange quarks and stranges ones, and so on. The mass formula is of comparable quality to the one deduced from the MIT bag model when non-charmed quarks are excluded and probably even better if the hadron contains heavy quarks like charmed, top or bottom quarks. As hadronic states are colour singlets, one can ask whether the colour degree of freedom of quarks will have any impact on the description of nuclear matter. Apart from our belief that nuclear forces will, one day, be deduced from the interquark forces, there are arguments showing that even a simple nucleus as the deuteron[3] has a 5% admixture of a six-quark dibaryonic component with non-trivial hidden colour.

Moreover, Jaffe[4] and later Aerts, Mülders and de Swart[5] have shown that the colourmagnetic force (1) leads to a bound $\Lambda\Lambda$ state with a binding energy of around 30 MeV and many dibaryon states that could be resonant.

In the flavour-symmetric limit of H' when all C_{ij} are equal $C_{ij} \equiv C$, the calculation of the binding energy due to the one-gluon exchange for n quarks in a relative s state is very simple[6], namely

$$\langle H' \rangle = C\left(8n - \frac{1}{2} C_6 + \frac{4}{3} S(S+1)\right) \qquad (4)$$

Here C_6 is the quadratic Casimir operator for the colour-spin group $SU_6^{CS} \supset SU_3^{colour} \times SU_2^{spin}$ and S is the total spin of the system. Clearly, the greater C_6, the stronger the binding; the more symmetry we have in colour-spin, the stronger the attraction. Now as the Pauli principle demands antisymmetric states in the product of flavour × colour × spin, a given colour spin state uniquely fixes the flavour multiplet to which the n quark state can belong. By assumption, only colour singlets can exist in nature, for a six-quark state this means that the Young tableau of the SU_3^{colour} representation must be 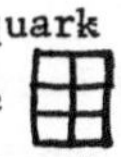 . The Pauli principle then gives[3)-5)] the Young tableau 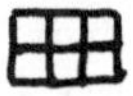 for the usual flavour spin group $SU_{2F}^{FS} \supset SU_2^{spin} \times SU_F^{flavour}$, where F flavours are assumed. The dimension of the flavour spin representation 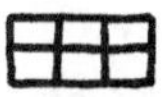 is 50 with two flavours 490 with three, 2590 with four, 26026 with six, and so on.

With two flavours where a deuteron-like state can be realized, the smallest repulsion between the six quarks is in the colour-spin representation of dimension 175, Young tableau 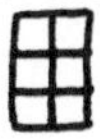 and value of the Casimir operator $C_6[175] = 96$. Increasing the number of flavours to three gives the 490 dimensional representation 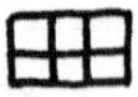 with $C_6[490] = 144$ as the state with the biggest attraction. For the number of flavours equal to or greater than four[7], the 1134 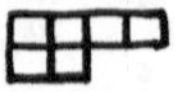 dimensional representation of colour-spin with the Casimir operator $C_6[1134]$ can always be realized. The flavour multiplicity of this state is six when we have four flavours, its $SU_4^{flavour}$ Young tableau is  .

Among these six flavour states (with spin 1) the lightest will be the one consisting of ududsc quarks. In the flavour-symmetric limit, the minimal value of H^0 given by (4) is 8/3C when we have two flavours, -24C with three, and -29,33C with four or more flavours. The last situation, therefore, makes it plausible that the 1134 colour-spin representation leads to more strongly bound dibaryons than the 490 dimensional, where Jaffe found his $\Lambda\Lambda$ bound state. We should note that the six-quark attraction is very big, as a comparison the nucleon binding is -6C. However, flavour-symmetric breaking is important[2),4),5),8)] in the interaction Hamiltonian: in the 1134 representation we find for the ududcs (I = 0, S = 1) state

$$\langle H' \rangle = -4\, C_{nn} - 12\, C_{ns} - 12\, C_{nc} - \frac{4}{3}\, C_{sc} \qquad (5)$$

The lightest state this could decay into is ΛC^0 (by the charm and strangeness conserving strong interactions) where Λ = uds with a hyperfine binding of $\langle\Lambda|H^0|\Lambda\rangle =$ $= -8/3\ (C_{ns} + 2C_{nn})$ and C^0 is the $I = 0,\ S = 1/2^+$ udc state with binding $-8/3\ (C_{nc} + 2C_{nn}) = \langle C^0|H'|C^0\rangle$. Since $C_{ns} \simeq 5/8\ C_{nn}$ as mentioned above, we find for the difference of binding between the six-quark system ududcs and the sum of the binding of Λ and C^0 which form the lightest possible strong interaction decay product

$$\langle 6q|H'|6q\rangle - \langle\Lambda|H'|\Lambda\rangle - \langle C^0|H'|C^0\rangle = \frac{2.5}{3} C_{nn} - \frac{28}{3} C_{ns} - \frac{4}{3} C_{sc} \qquad (6)$$

Unfortunately, we have no well-established charmed baryon state to determine C_{ns} and C_{sc} from; however, a fairly good guess of C_{ns} can be obtained from charmed mesons where the D^*D and F^*F mass differences give $C_{nc} \simeq 6$ MeV, $C_{sc} \simeq 5$ MeV. With $C_{nn} = 20$ MeV this gives a net binding energy of the $C^0\Lambda$ system in (6) of 46 MeV. It can, therefore, only decay weakly. In conclusion, we have proved the stability of at least one of the six states corresponding to the Young tableau 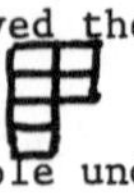 of $SU_4^{flavour}$ and established the existence of a new dibaryon state that is stable under strong interaction and is a ΛC^0 bound state.

ACKNOWLEDGEMENTS

We would like to thank Chan Hong-Mo for asking the questions that led us into this investigation and G. Girardi for reading the manuscript.

REFERENCES

1) A. De Rujula, H. Georgi and S.L. Glashow, Phys. Rev. D12 (1975) 147.

2) Tsou Shen Tsun, Oxford Mathematical Institute preprint (1978).

3) V.A. Matveev and P. Sorba, Lettere al Nuovo Cimento 20 (1977) 145; and Fermilab Pub. 77/56, to appear in Nuovo Cimento A.

4) R.L. Jaffe, Phys. Rev. Letters 38 (1977) 195.

5) A.Th.M. Aerts, P.J.G. Mülders and J.J. de Swart, Phys. Rev. D17 (1978) 260 and THEF-NYM 78-1 preprint.

6) R.L. Jaffe, Phys. Rev. D15 (1977) 281.

7) H. Høgaasen and P. Sorba, forthcoming CERN preprint (1978).

8) Chan Hong-Mo et al., Rutherford preprint RL-78-027 (to be published in Phys. Letters B, 1978).

PROTON-PROTON BREMSSTRAHLUNG*

Adam Szyjewicz and A.N. Kamal
Theoretical Physics Institute and Department of Physics
University of Alberta
Edmonton, Alberta, Canada T6G 2J1

1. INTRODUCTION

A year ago we presented the results of a field theory calculation for pp bremsstrahlung at the Second International Conference on Nucleon-Nucleon Interaction held in Vancouver[1]. The purpose of this report is to present the current status of our calculation. We undertook a field theory calculation for its inherent advantages, namely, (i) it is fully relativistic, (ii) gauge invariance is imposed from first principles and (iii) particle excitations (like Δ excitation) can be handled relatively easily. The disadvantage is that taking strong interactions into account to all orders is a task which can only be achieved in an approximate way. For example, use of strong form factors is commonly made in regions of high momentum transfers. This introduces higher order strong interaction corrections to the vertices. Such corrections demand introduction of new gauge terms i.e. the photon can be emitted from one of the internal lines in the vertex modification. A complete calculation taking all such effects into account would be forbiddingly difficult. Our calculation takes strong interactions into account to second order only. Form factors which are higher order strong interaction corrections are not used. The potential model calculations, on the other hand, have the distinct advantage that they take strong interactions into account to all orders but they suffer from the fact that they are (i) non-relativistic and cannot be extended to high incident energies, (ii) gauge invariance, which can be imposed for the long range part of the interaction, cannot be imposed so unambiguously for the shorter range part of the potential which is purely phenomenological and (iii) absorption effects like Δ excitation cannot be taken into account in any simple way short of doing a coupled channel calculation.

2. MODEL

All calculations reported here were done in the laboratory system. Chronologically, in our program, the calculations were done in the following order,

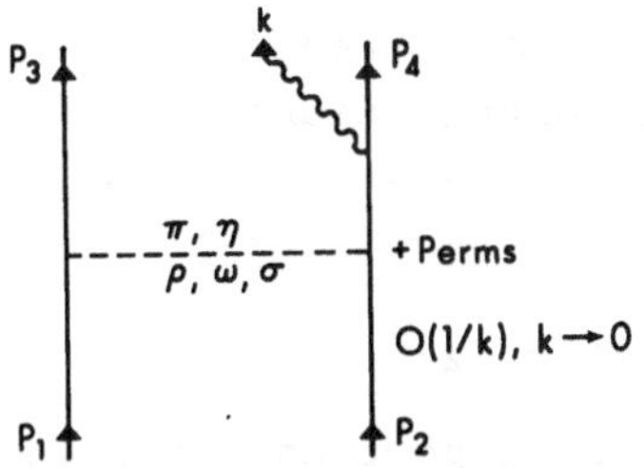

Fig. 1. OBE external emission graph.

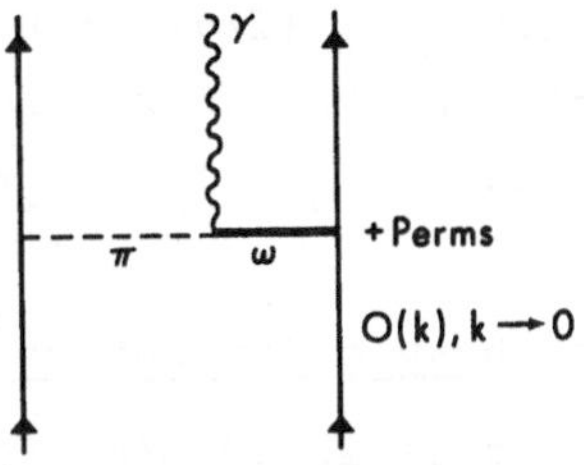

Fig. 2. ω radiative decay graph.

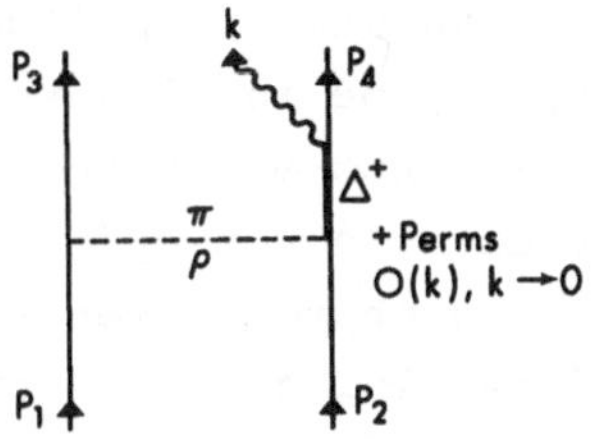

Fig. 3. Δ excitation graph.

(i) OBE external emission[2] (Fig. 1)

(ii) ω-radiative-decay contribution[3] (Fig. 2)

(iii) Δ excitation[1] (Fig. 3)

The results reported at the Second International Conference on Nucleon-Nucleon Interaction were for symmetric and coplanar geometry only. Since then we have extended our calculations to asymmetric and non-coplanar geometries and have worked out both the photon angle and the photon energy-spectra.

<u>The OBE external emission[1,2]</u> (Fig. 1)

The amplitude from this process is of O(1/k) for soft photons. The Lagrangian density used was[1,2],

$$\mathcal{L}_{int} = \sqrt{4\pi}\ \{g_\pi \bar\psi\gamma_5 \vec\tau\cdot\vec\phi_\pi\psi + g_\sigma\bar\psi\psi\phi_\sigma + g_\omega\bar\psi\gamma^\mu\psi\omega_\mu + \bar\psi[g_\rho\gamma^\mu\vec\tau\cdot\vec\rho_\mu + \frac{f_\rho}{4M}\sigma^{\mu\nu}\vec\tau\cdot\vec\rho_{\mu\nu}]\psi\} \quad (1)$$

where $\vec\rho_{\mu\nu} = \partial_\mu\vec\rho_\nu - \partial_\nu\vec\rho_\mu$.

The strong coupling parameters chosen were those given by Arndt, Bryan and MacGregor[4] which were chosen to fit the peripheral ($L \geq 1$) partial waves in the pp-elastic scattering. They were

$$\begin{array}{llll} g_\pi^2 = 12.74 & & m_\pi = 136.5\text{ MeV} & \\ g_\sigma^2 = 2.33 & & m_\sigma = 421\text{ MeV} & \\ g_\rho^2 = 0.92\,, & (f/g)_\rho = 4 & m_\rho = 763\text{ MeV} & \\ g_\omega^2 = 3.06\,, & (f/g)_\omega = 0 & m_\omega = 783\text{ MeV}\;. & \end{array} \qquad (2)$$

We have neglected the η exchange contribution[1]. The electromagnetic vertex used was $e(\gamma_\mu - i\sigma_{\mu\nu}k^\nu \frac{\kappa}{2M})$ with $\kappa \simeq 1.79$. No form factors were used.

ω-radiative-decay contribution[3] (Fig. 2)

This process is of O(k) for soft photons. For ideally mixed ω and ϕ SU(3) suggests that $g^2_{\omega pp} = 9g^2_{\rho pp}$. Furthermore SU(3) suggests that $\Gamma(\omega\to\pi\gamma) = 9\Gamma(\rho\to\pi\gamma)$. Experimentally[5] $\Gamma(\omega\to\pi\gamma) \simeq (25\pm7)\Gamma(\rho\to\pi\gamma)$. On both counts one anticipates that the ω radiative decay amplitude will dominate over the ρ radiative decay amplitude. This argument, based on pure vector couplings, may be invalidated due to the tensor coupling of the ρ-meson. We have not evaluated the ρ radiative decay contribution. The $\omega\to\pi\gamma$ vertex is given by[3] $g_{\omega\pi\gamma}\varepsilon_{\mu\nu\rho\sigma}\varepsilon^\mu_{(\omega)}\varepsilon^\nu_{(\gamma)}p^\rho k^\sigma$ where p and k are the pion and the photon momenta respectively. $g^2_{\omega\pi\gamma}/4\pi = 0.864\times 10^{-3}\ m_\pi^{-2}$ was used resulting in $\Gamma(\omega\to\pi\gamma) = 880$ KeV.

Δ excitation[1] (Fig. 3)

This process results in an amplitude of O(k) for soft photons. At larger incident energies we expect this process to become important simply because of the enhancement coming from the Δ propagator[1]. The interaction lagrangian used was[6],

$$\mathcal{L}_{\Delta N\pi} + \mathcal{L}_{\Delta N\gamma} = G^* \sum_i \bar\psi^i_\mu \psi \partial^\mu \phi^i_\pi - ec\bar\psi^3_\mu \gamma_\nu \gamma_5 \psi F^{\mu\nu} \qquad (3)$$

with $G^{*2}/4\pi = 18.4\times 10^{-6}\ \text{MeV}^{-2}$ and $cm_\pi = 0.315$.
The width of Δ was taken into account by using $s - m_\Delta^2 + im_\Delta\Gamma f(s)$ in the denominator of the Δ propagator where f(s) is a threshold factor = $(s-\Lambda/m_\Delta^2-\Lambda)^{\frac{1}{2}}$ with $\Lambda = (m_p + m_\pi)^2$. A full width of 115 MeV and $m_\Delta = 1232$ MeV were used. We have also done a preliminary calculation where Δ excitation by ρ-exchange is included (see Fig. 3). The $\rho N\Delta$ coupling can be obtained from vector meson dominance or from the quark model[7]. The results are not reproduced here. The effect of the ρ-exchange is to suppress the contribution of the π-exchange process. We return to this point briefly in the next section.

3. RESULTS

200 MeV data

Fig. 4 shows the plot of $d\sigma/d\Omega_3 d\Omega_4 d\theta_\gamma$ in microbarns/(sr^2-rad) vs θ_γ at 200 MeV in symmetric ($\theta_3 = \theta_4 = 16.3°$) and coplanar geometry. The data are that of Beveridge et al[8] obtained at Triumf. The data are preliminary and some points have since moved as a result of reanalysis[9]. We have shown our OBE external emission prediction. If the ω-radiative-decay amplitude and the Δ excitation processes are taken into account the forward ($\theta_\gamma < 30°$) and the backward ($\theta_\gamma > 150°$) cross sections change by ±5% depending on the relative signs of the amplitudes. We have also shown the soft photon calculation by Fearing[10,11] and potential model calculations by Bohannon[9,11]. Given the accuracy of the experiments we note that all calculations fare well in representing the data.

Fig. 5 shows $d\sigma/d\Omega_3 d\Omega_4 d\theta_\gamma$ in microbarns/(sr^2 - rad) vs θ_γ at 200 MeV in symmetric ($\theta_3 = \theta_4 = 12°$) and coplanar geometry. The Triumf data

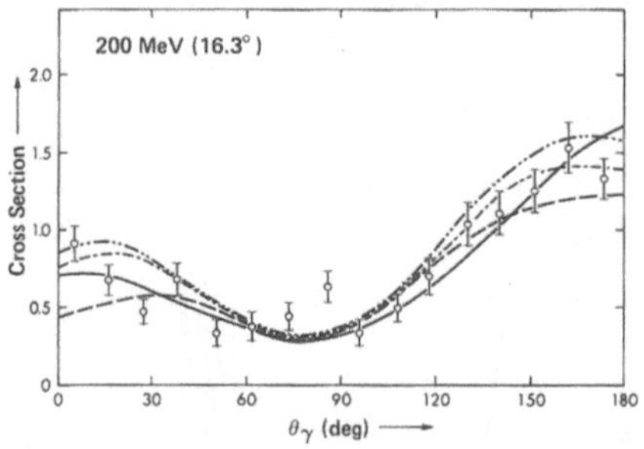

Fig. 4. $d\sigma/d\Omega_3 d\Omega_4 d\theta_\gamma$ in μb/(sr)^2-rad vs θ_γ for symmetric ($\theta_3 = \theta_4 = 16.3°$) and coplanar geometry.
——— : OBE external emission model.
- - - : Soft photon approximation[10].
-·-·- : Hamada-Johnston potential model[11].
-··-··- : Reid potential model[11].

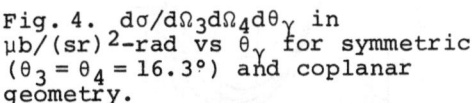

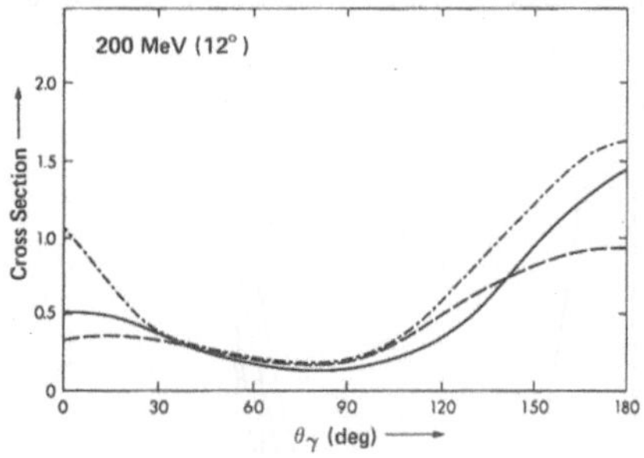

Fig. 5. $d\sigma/d\Omega_3 d\Omega_4 d\theta_\gamma$ in μb/(sr)^2-rad vs θ_γ for symmetric ($\theta_3 = \theta_4 = 12°$) and coplanar geometry.
——— : OBE external emission model.
- - - : Soft photon approximation[10].
-·-·- : Hamada-Johnston potential model[11].

at $\theta_3 = \theta_4 = 13°$ are unfortunately not yet available and are not likely to have the accuracy of the $\theta_3 = \theta_4 = 16.3$ data to discriminate between different calculations[9]. The results of soft photon approximation calculation of Fearing[10,11] and potential model calculation of Bohannon[9,11] are also shown for comparison.

730 MeV data

The OBE external emission model which uses parameters determined from elastic data below 350 MeV lab. kinetic energy is not expected to work at 730 MeV as unitarity and inelasticity are expected to play an important role. Nevertheless, we repeated our calculation for the UCLA G7 and G8 geometries[12] with the parameters of eq. (2). In order to normalize the data[12] at smaller values of k we found that the OBE amplitude had to be suppressed by a factor of 4. The same suppression of the OBE contribution can be secured by use of form factors[13]. In Figs. 6 and 7 we have shown the UCLA data for G7 and G8 geometries. The curve marked (d) in these figures represents the result of using the OBE amplitude suppressed by a factor of 4. It follows the data quite well up to $k \simeq 60$ MeV. Curve marked (e) is the contribution of Δ excitation by π exchange alone. Curves marked (a) and (c) are results of coherent addition of OBE external emission (suppressed by a factor of 4) and the Δ excitation process with either relative sign. Curve (b) is the result of incoherent addition of the two cross sections. Given the accuracy of the data the incoherent addition of the

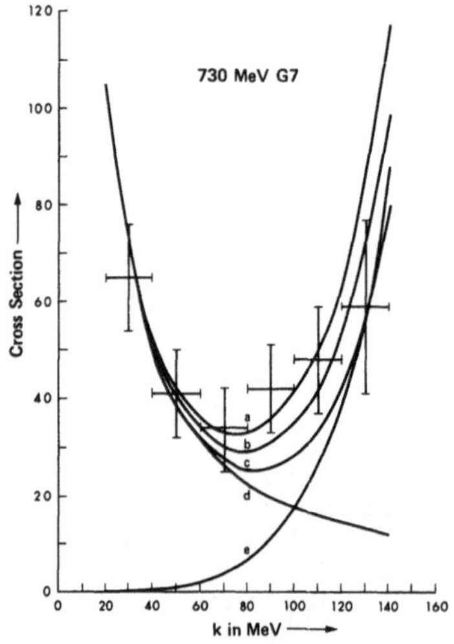

Fig. 6. $d\sigma/d\Omega_3 d\Omega_\gamma dk$ in nb/MeV-sr^2 vs k for UCLA G7 geometry[12].
Curve d : OBE amplitude $\div 4$.
Curve e : Δ excitation by π exchange.
Curve a and c : Coherent addition of d and e with either relative sign.
Curve b : Incoherent addition of d and e.

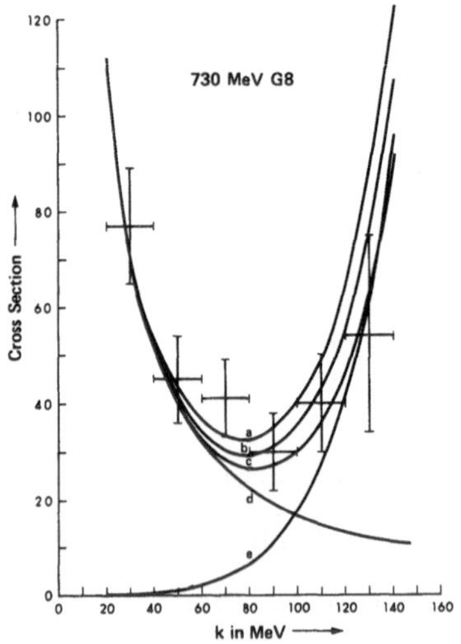

Fig. 7. $d\sigma/d\Omega_3 d\Omega_\gamma dk$ in nb/MeV-sr^2 vs k for UCLA G8 geometry[12].
Legend : Same as that for Fig. 6.

cross section fares as well as the coherent addition. What is relevant is that the interference effects are as big as ~20% at $k \gtrsim 100$ Mev.

We did a preliminary calculation of including the Δ excitation by ρ exchange. At 400 MeV, symmetric ($\theta_3 = \theta_4 = 20°$) and coplanar geometry the effect of including ρ exchange in Fig. 3 was to suppress the π exchange contribution (Fig. 3) by almost a factor of 2 in the θ_γ-spectrum for forward and backward photon angles. The suppression which was in effect for all photon angles was not so large for intermediate photon angles.

ACKNOWLEDGMENTS

We thank Harold Fearing for letting us use some of his computer programs for kinematics and several discussions. Thanks are also due to Joel Rogers for discussions and providing us with the results of George Bohannon's potential model calculations.

*Research partly supported by the National Research Council of Canada.

REFERENCES

1. Adam Szyjewicz and A.N. Kamal, Nucleon-Nucleon Interactions - 1977, H. Fearing, D. Measday and A. Strathdee eds., A.I.P. Conference Proceedings No. 41, 502 (1978).
2. R. Baier, H. Kühnelt and P. Urban, Nucl. Phys. B11, 675 (1969). It is fitting that this work was done at Graz.
3. A.N. Kamal and Adam Szyjewicz, Nucl. Phys. A285, 397 (1977).
4. R.A. Arndt, R.A. Bryan and M.H. MacGregor, Phys. Letters 21, 314 (1966).
5. Particle data group, Rev. Mod. Phys. 48, S1 (1976).
6. M.G. Olsson and E.T. Osypowski, Nucl. Phys. B87, 399 (1975).
7. P. Haapakoski, Phys. Letters 48B, 307 (1974). A.M. Green and P. Haapakoski, Nucl. Phys. A221, 429 (1974).
8. J.L. Beveridge et al, Nucleon-Nucleon Interactions - 1977, H. Fearing, D. Measday and A. Strathdee eds., A.I.P. Conference Proceedings No. 41, 446 (1978).
9. J.G. Rogers, Private communication (1978).
10. H.W. Fearing, Private communication and Nucleon-Nucleon Interactions - 1977, H. Fearing, D. Measday and A. Strathdee eds., A.I.P. Conference Proceedings No. 41, 506 (1978).
11. J.G. Rogers, Invited talk at the A.P.S. meeting, Washington, April 24-27, 1978. Bull. Am. Phys. Soc. 23, 534 (1978).
12. B.M.K. Nefkens, O.R. Sander and D.I. Sober, Phys. Rev. Lett. 38, 876 (1977).
13. L. Tiator, H.J. Weber and D. Drechsel, Mainz Preprint (1978).

PROTON-PROTON BREMSSTRAHLUNG WITH POLARIZED PROTONS

Harold W. Fearing
TRIUMF, Vancouver, B.C., Canada V6T 1W5

Abstract. The asymmetry in proton-proton bremsstrahlung is given for a variety of kinematic conditions in the soft photon approximation. It is argued that the asymmetry may be more sensitive to the unknown higher-order terms than the cross section and is shown that several models giving similar cross sections give quite different asymmetries, which could thus be used to distinguish among models.

Proton-proton bremsstrahlung (ppγ) provides in principle a straightforward way of investigating off-shell aspects of the nucleon-nucleon force. In practice however, significant results have been slow in coming since most data are reproduced at least qualitatively by purely on-shell soft photon calculations.[1] Furthermore model calculations have proved difficult and only fairly recently has the importance of some corrections, e.g. relativistic corrections,[2] been appreciated.

Renewed interest in ppγ has been generated recently by several new experiments, including a 200 MeV equal-angle experiment at TRIUMF,[3] a 730 MeV experiment at UCLA,[4] and the somewhat older, but high statistics, 42 MeV experiment at Manitoba.[5] The UCLA experiment seems to show deviations from the on-shell soft photon approximation (SPA) calculations at the highest photon energies, but because of the high incident energy it is difficult to make contact with the standard potential models available at lower energies. The TRIUMF experiment also shows some deviations from SPA at $\theta_\gamma=0°$ and 180° but the available model calculations, i.e. a relativistic one-boson-exchange model[6] (OBE) and a relativistically corrected potential model,[7] do not differ drastically from SPA, and the net result is that all three calculations differ from each other and the data by amounts of the order of the error bars on the data. The particular relativistic corrections put into the potential model are crucial, however, particularly near $\theta_\gamma=0°$, 180°, and bring the potential model calculations much closer to SPA. One thus asks whether other ppγ experiments might be more sensitive to off-shell effects or more readily distinguish among models. One possibility is a measurement of the ppγ asymmetry using polarized protons. Such experiments are particularly suited to the new meson facilities where relatively high intensity polarized beams are available. In this note we examine the asymmetry as calculated in SPA and compare the results with the few available model calculations. We hope to understand the contributions to the asymmetry of the various terms of the SPA as was done for the cross section[1] and to see how such measurements may distinguish among models which give roughly the same cross section.

First define the asymmetry as $A_i = (d\sigma_+ - d\sigma_-)/(d\sigma_+ + d\sigma_-)$ where $d\sigma_\pm$ are the ppγ cross sections with incoming proton spin parallel or anti-parallel to $\hat{i}$. For elastic scattering and also for coplanar ppγ only A_y, the component perpendicular to the scattering plane, is non-zero by parity arguments. In the non-coplanar case, however, the extra vector available allows A_x and A_z to be non-zero. We will, however, defer discussion of these 'forbidden' asymmetries and of the interesting suggestion of Moravcsik[8] regarding them and restrict our attention here to the coplanar case.

To estimate the asymmetry we use the SPA. Such an approach is very simple, making it possible to look at a larger variety of kinematic situations than in more complicated models. It also presumedly gives something like the correct answer since at least for the cross section the SPA results are in rough agreement with the data. One must interpret SPA results with the usual caution, however. Since it is purely an on-shell calculation it gives no direct information about the off-shell aspects of the interaction. One gets information from SPA only as it differs from data or model calculations, and even then such differences are not purely due to off-shell effects.[1] An understanding of the relative importance of the various terms in SPA can be very useful though for the clues it provides to the kinematic regions where more detailed model calculations may be most fruitful.

In the usual SPA the ppγ amplitude is expanded in powers of the photon momentum k as $M_{pp\gamma} = A/k + B + Ck$ where A and B are determined by gauge invariance in terms of elastic information. C contains the unknown off-shell information, but also higher-order on-shell contributions, ambiguities arising from the freedom in the choice of energy and angle at which to evaluate the elastic information, etc. The cross section is then given by $d\sigma \sim k|M_{pp\gamma}|^2 \sim A^2/k + 2AB + (B^2+2AC)k + \ldots$ where again the A^2, AB, and B^2 terms are known. The Kroll-Burnett theorem[9] and its generalization to the polarized case[10] tell us that for the first two terms only the square of the elastic amplitude, i.e. the cross section, is required. For the B^2 term, however, a knowledge of the elastic amplitude itself is necessary. Here, the amplitudes A and B were constructed using phase shifts[11] taken at an average energy and momentum transfer [Low prescription] and the result squared. Thus the A^2, AB, and B^2 terms are included. The AC term is formally of the same order in k as the B^2 terms, but as shown before,[1] at least for the equal-angle geometry at 200 MeV and above, is expected to be small since A is small. Hence keeping the B^2 term makes sense.

Some results are shown in Figs. 1-5 for a variety of kinematic situations. One observes immediately that A_y can be 20-30% and that there is a fair amount of interesting structure in these curves. Much of this structure comes from interference effects. This is particularly true at 42 MeV where, for the cross section, the A^2+2AB and B^2 contributions are both significant, with B^2 dominating at θ_γ=0°, 180° and ~80° and the other terms dominating in between. At higher energies, where B^2 dominates for all θ_γ, the effects of the first two terms show up as ripples on the main B^2 contribution. In any case since A_y involves interferences it may be particularly sensitive to changes in the amplitude produced by the unknown C term.

More important is the drastic difference, even as to sign, in A_y calculated with and without the B^2 term. This indicates that the different terms in the soft photon expansion have different intrinsic asymmetries. Once a region is found where higher-order or off-shell terms are significant, these terms may also lead to different

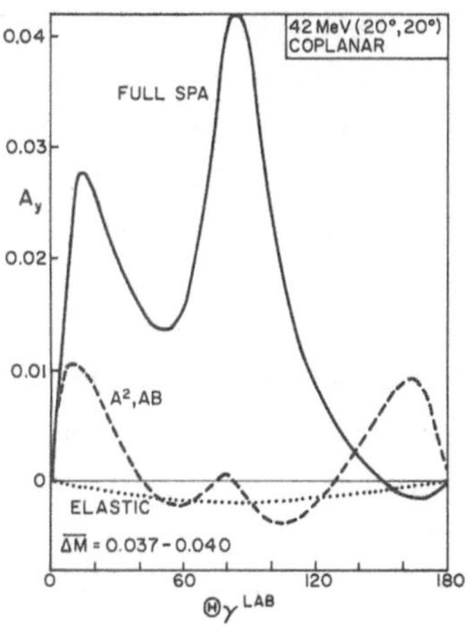

Fig. 1

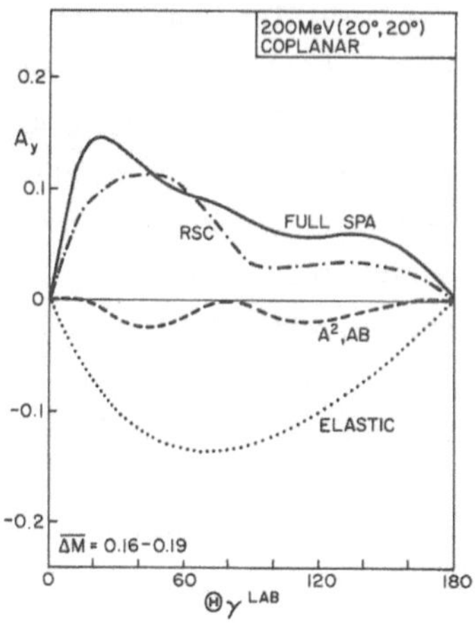

Fig. 2

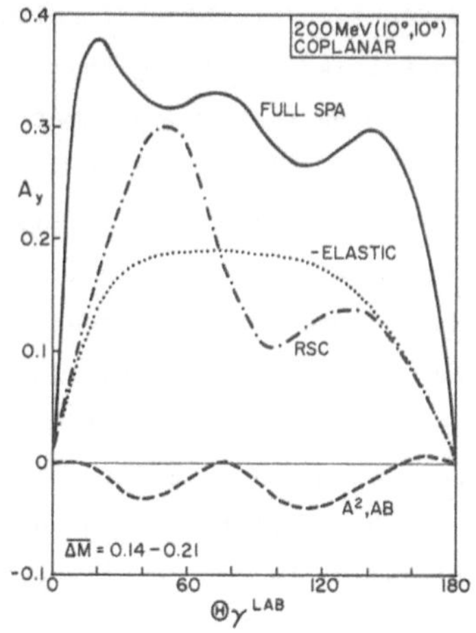

Fig. 3

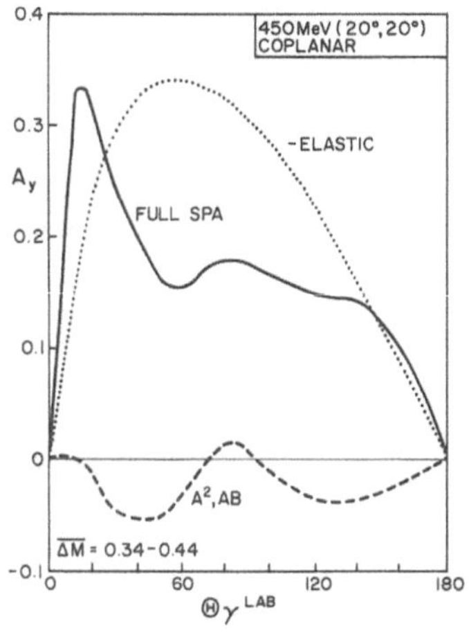

Fig. 4

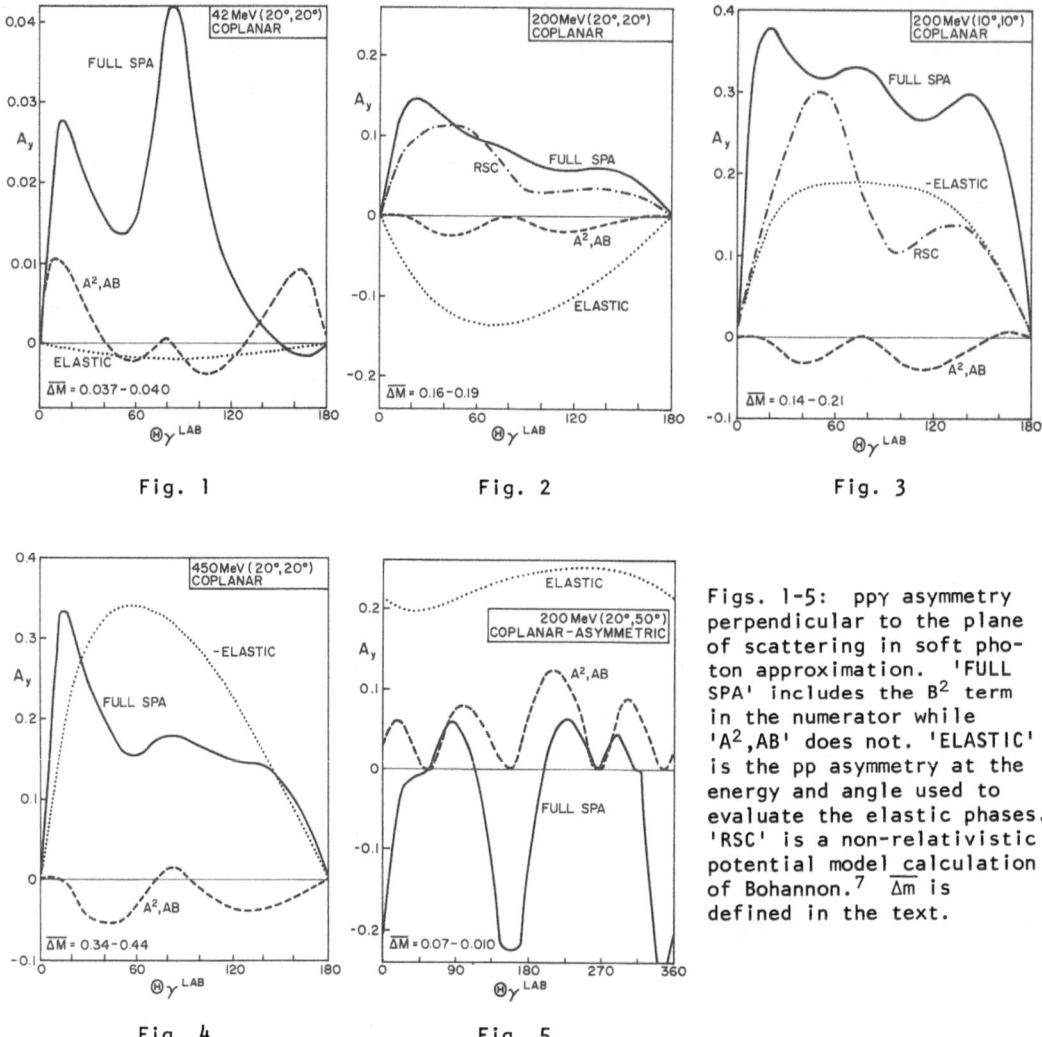

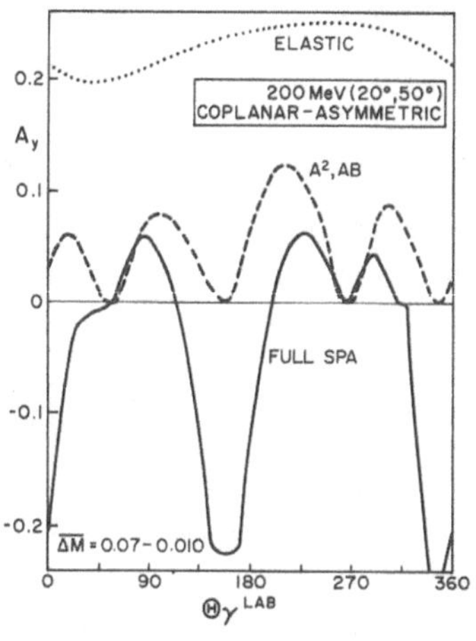

Fig. 5

Figs. 1-5: ppγ asymmetry perpendicular to the plane of scattering in soft photon approximation. 'FULL SPA' includes the B^2 term in the numerator while 'A^2,AB' does not. 'ELASTIC' is the pp asymmetry at the energy and angle used to evaluate the elastic phases. 'RSC' is a non-relativistic potential model calculation of Bohannon.[7] $\overline{\Delta m}$ is defined in the text.

intrinsic asymmetries. Thus this also suggests that the asymmetry may be a more sensitive test than the cross section of the character of the higher-order terms.

Shown also is the elastic asymmetry, i.e. A_y for elastic scattering, evaluated at the energy and angle used for the elastic information for the SPA. In the k=0 limit only A^2 contributes and the pp and ppγ polarized cross sections are proportional, the proportionality factor being a kinematic one which vanishes at θ_γ=0°, 180° and ∿80°. This factor times the smooth elastic asymmetry produces the characteristic two-humped character of the 'A^2,AB' curve in this geometry which can be seen particularly clearly in Figs. 2-4.

Of special interest are Figs. 2 and 3 which show both SPA results and results calculated in a standard non-relativistic potential model using a Reid soft core potential.[7,12] The two calculations give quite different predictions for A_y both in magnitude and qualitative shape, particularly in the (10°,10°) case, whereas the predictions for the cross section are quite similar over most of the angular range. Thus the asymmetry seems to be distinguishing between two models which give similar cross sections. One reservation should be noted, however. For the cross section, relativistic spin corrections[2,7] which are not included in the potential model A_y results, were important near θ_γ=0° and 180° and served to bring the non-relativistic potential model in much closer agreement with the SPA than before. Since these corrections depend on spin, the asymmetry might be affected by such corrections. Preliminary indications are, however, that they are not large enough to remove the qualitative difference between the two results.[12] The OBE model[6] also gives cross sections at 200 MeV similar to those of SPA. In its present form,[6] however, it is a Born approximation calculation and so gives $A_y \equiv 0$. Thus until some form of unitarity corrections are included, it really cannot be used to predict the asymmetry.

Finally, to give some idea of the off-shell region being probed we have included in each figure the value of an off-shell parameter $\overline{\Delta M}$ defined as $\overline{\Delta M}=\sum_i |P_i^2-m^2|/4m^2$ where P_i^2 are the squares of the four-momenta of the off-shell legs corresponding to the four possible external radiation diagrams. Thus $\overline{\Delta M}$=0 is the on-shell point and, for orientation, $\overline{\Delta M}$=0.74 corresponds to an intermediate Δ(1236) in each leg. The maximum possible $\overline{\Delta M}$ increases with incident energy but the actual value achieved depends on the geometry.

In summary, it appears that the asymmetry in ppγ measured with polarized protons may be sensitive to higher-order terms in the soft photon expansion since individual contributions of the terms which can be calculated are drastically different and since the asymmetry shows structure depending on interferences among terms. Furthermore models giving similar cross sections show quite different asymmetries. Thus the asymmetry may distinguish models in situations where the cross sections cannot. It should thus be very profitable to study such asymmetries more carefully, both theoretically in models more sophisticated than the SPA and experimentally.

References

1. H.W. Fearing, Nucleon-Nucleon Interactions-1977, edited by D.F. Measday, H.W. Fearing, and A. Strathdee, AIP Conf. Proc. No. 41 (AIP, N.Y., 1978) p. 506.
2. L.S. Celenza, M.K. Liou, M.I. Sobel and B.F. Gibson, Phys. Rev. C8, 838 (1973); M.K. Liou and M.I. Sobel, Ann. Phys. (N.Y.) 72, 323 (1972).
3. J.L. Beveridge *et al.*, N-N Interactions-1977 *op. cit.* p. 446.
4. B.M.K. Nefkens, O.R. Sander, and D.I. Sober, Phys. Rev. Lett. 38, 876 (1977); B.M.K. Nefkens, O.R. Sander, D.I. Sober, and H.W. Fearing, to be published.
5. J.V. Jovanovich, N-N Interactions-1977 *op. cit.* p. 451, and references therein.
6. A. Szyjewicz and A.N. Kamal, N-N Interactions-1977 *op. cit.* p. 502.
7. G.E. Bohannon, N-N Interactions-1977 *op. cit.* p. 482 and private communication.
8. M. Moravcsik, N-N Interactions-1977 *op. cit.* p. 515 and Phys. Lett. 65B, 409 (1976)
9. T.H. Burnett and N.M. Kroll, Phys. Rev. Lett. 20, 86 (1968).
10. H.W. Fearing, Phys. Rev. D7, 243 (1973).
11. R.A. Arndt, R.H. Hackman and L.D. Roper, Phys. Rev. C15, 1002 (1977).
12. Other potentials give similar results. G.E. Bohannon, private communication.

ISOBAR EXCITATION IN PROTON-PROTON BREMSSTRAHLUNG

L. Tiator and D. Drechsel
Institut für Kernphysik
Universität Mainz

D-6500 Mainz

Recently, Nefkens et al.[1] reported on experiments on proton-proton bremsstrahlung at an incident proton laboratory energy of 730 MeV and different photon angles. Up to photon energies of $\omega \approx 80$ MeV the experimental results are in good agreement with the predictions of external-emission dominance (EED), which are based on the lowest term in a soft photon expansion. Above 80 MeV the photon spectra rise markedly, a feature which cannot be explained in a soft photon theory. Since it does not seem likely that this off-shell effect can be described in a potential model with external and rescattering diagrams, we have studied the contribution of isobar ($\Delta_{3,3}$) excitation and subsequent radiative decay. We have shown that this gives rise to a resonant term, which dominates in the energy region of $\omega \gtrsim 120$ MeV[2]. Our calculations have been performed in a relativistic one-boson-exchange model including π- and ρ-mesons. Since our present results do not yet contain a dynamical description of the nucleon-nucleon force, which is difficult to obtain at the proton energies of the experiments, we have added the contributions of the external and resonant diagrams incoherently.

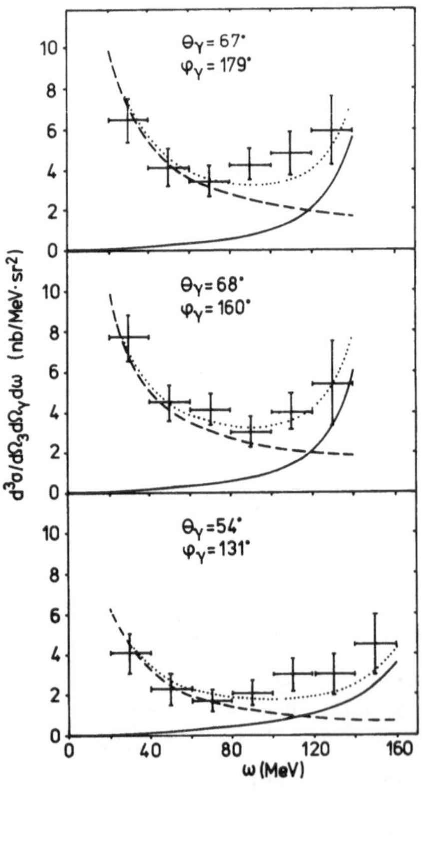

Fig.1: pp → ppγ laboratory differential cross-section at 730 MeV, proton angles $\theta_p = 50.5°$ and $\phi_p = 0°$. The experimental data of Nefkens et al. are compared to their EED-calculations (dashed line), the Δ-contributions (full line) and the addition of both contributions (dotted line).

Fig. 1 shows the experimental results, the EED predictions of Nefkens et al., the isobar contribution and the incoherent superposition of these amplitudes in different kinematical situations. For our Δ-contributions we have taken the ρNN-coupling constants of Höhler and Pietarinen[3], which correspond to a markedly stronger tensor coupling than in our previous calculation. This reduces our earlier results slightly at the lower energies and by up to 20% at the highest photon energies. In view of the crude approximations in our treatment of the external scattering terms as well as the neglect of off-shell effects and interference terms between external and isobar contributions, this reduction is not significant. There is, however, a certain arbitrariness of the cut-off masses of the meson propagators. Using π-exchange only, we have taken the usual long range cut-off Λ_π = 650 MeV, whereas in calculations with π- and ρ-exchange much of the uncertainties cancel and the results become less sensitive to the choice of the cut-off. Following ref. 4 we choose $\Lambda_\pi = \Lambda_\rho$ = 1.5 GeV. Particularly in the transition region between external and resonance dominance ($\omega \approx$ 80-120 MeV) a proper coherent superposition of the various amplitudes would be necessary. At this point we are presently developing a more realistic description of the external diagrams using the partial wave decomposition of the five independent helicity amplitudes.

References:

1. B.M.K. Nefkens, O.R. Sander and D.I. Sober, Phys. Rev. Lett. 38 (1977) 876
2. L. Tiator, D. Drechsel and H.J. Weber, to be published in Nucl. Phys.
3. G. Höhler and E. Pietarinen, Nucl. Phys. B95 (1975) 210
4. A.M. Green and J.A. Niskanen, Nucl. Phys. A249 (1975) 493

CRITERIA FOR THE CHOICE OF p-p BREMSSTRAHLUNG EXPERIMENTS

J.V. Jovanovich
Cyclotron Laboratory, Department of Physics,
University of Manitoba, Winnipeg, Manitoba, Canada R3T 2N2

It has been demonstrated[1,2] that accurate p-p bremsstrahlung (ppB) experiments can be performed using conventional wire chamber technology. In addition, it has been pointed out[3] that it is feasible to perform very precise experiments, with either polarized or unpolarized beams, in which several hundred data points can be measured with, say, 2-3% accuracy. Since these experiments are rather expensive and elaborate, one would like to choose them so as to maximize the amount of useful information obtainable on off-energy-shell effects in the N-N interaction, as well as to choose the most favourable experimental conditions. It is usually considered[4] that bremsstrahlung experiments will be most informative in the 200-300 MeV range and at small ($\lesssim 10^{\circ}$) scattering angles, where they are quite difficult and that the experiments can be of little use if they are performed at lower energies ($\lesssim$ 100 MeV) and at larger angles (20°-40°), where they are much easier to do. It is the purpose of this note to present evidence which raises doubts about the validity of this point of view.

Comparing The Model Dependent With The Soft Photon Calculations

If one were able to compute quantitatively the contributions from the off-energy-shell (OFES) and the on-energy-shell (ONES) parts of the N-N interaction separately, then the experimentalist would have a firm guide as to how to choose his experimental conditions. There exists no known way to compute separately the OFES contribution but, in the soft photon approximation (SPA), the major contribution due to the ONES effects can be computed. Where the difference between the SPA and the model dependent calculation is appreciable, one can then reasonably hope to be able to observe the OFES effects.

This kind of comparison was first summarized by Signell[5]. The numbers presented[5] in Table IX show that at 30°-30° there exists a fairly constant 20% to 25% difference between the exact Hamada-Johnston (HJ) cross sections as calculated by Brown[6] and the soft photon calculations performed by Nyman[7] in the energy range from 46 MeV to 204 MeV. This result is not very comprehensive and the difference between the two calculations may not be entirely due to the OFES effects, but the result is consistent with the notion that the contribution due to OFES effects does not change greatly in this energy range.

In a recent paper[8] Fearing has presented the results of a soft photon calculation and has compared them at 42 MeV and 200 MeV with model dependent calculations performed using the Hamada-Johnston potential. It is most interesting to note that around 17°-

17° proton angles there exists a much larger difference between the HJ and SPA calculations at 42 MeV than at 200 MeV. Furthermore, according to Fearing, at $\Theta_\gamma \approx 80^{o}$ the first term of the Low expansion goes to zero. At this point the model dependent calculation is about 50% larger than the SPA prediction at 42 MeV indicating that the contribution of the OFES-dependent Ck term is not negligible in comparison to the B term. This kind of difference is not observed at 200 MeV, thus one is led to an apparently paradoxial conclusion that in that particular region of phase space it would be more profitable to study OFES effects at 42 MeV than at 200 MeV! It is also of interest to note that a significant discrepancy between the experimental data and the HJ predictions was observed at 42 MeV (see Fig. 2 in ref. 2 and Fig. 9 in ref. 3) at ψ_γ (that is Θ_γ) angles around 80°. This is exaclty where, the first term of Low's expansion goes to zero[8)]. At present, it is not clear whether this result is just coincidental, or whether the ppB cross section is quite sensitive to OFES effects at that point, thus indicating that OFES effects may not be properly described by the Hamada-Johnston potential.

Comparing Phase Equivalent Potentials

There exists an infinite number of phase equivalent potentials which, in general, have different OFES, but identical ONES behaviour. Heller[9)] studied the effects of one particular transformation and he found that the $d^2\sigma/d\Omega_1 d\Omega_2$ cross sections changed by relatively easily observable amounts only at high proton energy (158 MeV) and at small scattering angles ($\Theta_1 = \Theta_2 = 10^{o}$), but not otherwise. This particular quantitative prediction was the main motivation for doing the 200 MeV experiment at TRIUMF[10)]. Bohannon[11)] studied the consequences of another transformation at 42 MeV and 200 MeV and found effects of various magnitudes, the effects at 200 MeV being again larger.

It is interesting to reflect on the general usefulness of these calculations and the incentive they give to the experimentalists. Heller's predictions were quanitattive and the results were welcomed by experimentalists. On the other hand, both Heller's and Bohannon's results, although quantitative in terms of the cross sections computed, were conceptually only qualitative. Each of them has taken only one out of infinitely many transformations as an illustration of what can happen when these transformations are applied. It is questionable if these calculations should be used as a firm guide in choosing the ppB experiments.

A Comment On The Range Of Validity Of Low's Theorem

It has been pointed out[12)] that one can introduce a dimensionless OFES parameter $\varepsilon_I = k/k_I$, where k is the emitted photon energy and k_I is a scale factor which depends on some very general assumptions about the nature of the nuclear force. If one assumes that the effective nuclear force is sufficiently singular so that the characteristic length for the nuclear interaction producing bremsstrahlung is equal to the wave length of the incident particle, then $k_I = k_L = vp$ (v and p are the velocity and momentum of the incident particle; units $\hbar = c = 1$ are used). On the other hand, if one assumes that the range of the nuclear force is equal to $1/\mu_B$ where μ_B is the mass of the exchanged boson, then $k_I = k_B = \mu_B v$. In terms of this parameter, the condition for the validity of the SPA simply becomes $\varepsilon_I \ll 1$. From the kinematics of the bremsstrahlung process, it can be readily seen that ε_I has an upper bound $\varepsilon_{I,max} = k_{max}/k_I$, where k_{max} is the maximum photon energy in the lab system. In Figure 1, the upper bounds for ε_L, ε_π and ε_ω are shown (ε_π and ε_ω are obtained by setting μ_B equal to the pion and the ω_o mass respectively) together with the ranges of the ε_I's covered in some recent experiments[1,2,11,13,14,15]. From this figure and the definitions of ε_I, it can be seen that ε_L and ε_B have qualitatively different behaviour as a function of the incident kinetic energy, T. As T increases from zero to infinity, $\varepsilon_{L,max}$ changes slowly from 0.25 to 1, while $\varepsilon_{B,max}$ increases from 0 to ∞.

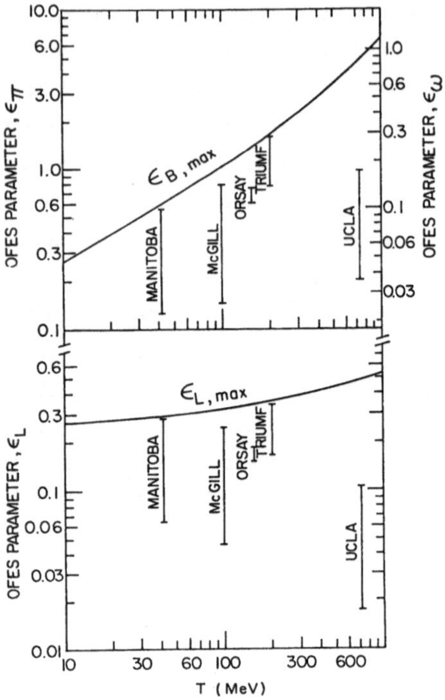

Figure 1

The full physical significance of these parameters is not quite clear. One can argue[12] that while the overall range of the nuclear force is determined by the pion mass, the force becomes stronger and has a shorter range as the incoming particle penetrates deeper into the nucleon. Thus, the characteristic range, which is really the important parameter in the case of bremsstrahlung, decreases with increasing energy (or decreasing wave length) of the incoming particle. In this case, one would expect Low's theorem to be valid for $\varepsilon_L \ll 1$, rather than for $\varepsilon_B \ll 1$. This picture of the (effective) nucleon-nucleon force being sufficiently singular seems to be consistent with the emerging picture that nucleons are composed of point-like quarks.

Calculational results described in the papers by Signell[5] and Fearing[8] are consistent with the interpretation that the relevant dimensionless parameter in Low's expansion is ε_L, rather than ε_B, although they, obviously, cannot be said to prove it. However, there exists experimental evidence which is also consistent with this point of view. This evidence is (a) the qualitative agreement between the Manitoba[1] and Orsay[14] results in their common disagreement with the HJ calculations, and (b) the excellent agreement of the UCLA results[15] with the SPA calculations[8].

Conclusion

The "common wisdom" that bremsstrahlung experiments are interesting only at higher energies and smaller angles seems to be inconsistent with the results reviewed in this paper. Lower energy experiments performed at medium scattering angles, experiments which are much easier to do, might contain just as much OFES information. It would be very interesting and useful to perform more theoretical/numerical calculations (considering polarization effects as well) that would give experimentalists a more detailed guide in choosing bremsstrahlung experiments. Finaly, an experiment of the UCLA type performed at a few GeV might be able to settle conclusively whether the expansion parameter should be ε_L or ε_B, since, at that energy, ε_π and ε_L differ by nearly two orders of magnitude.

The author is thankful to L. Heller, J.S.C. McKee, T.A. Osborn, C.A. Smith, and J.P. Svenne for reading the manuscript of this paper and making some useful comments.

References

1. L.G. Greeniaus et al., Phys. Rev. Lett. 35, 696 (1975).
2. J.V. Jovanovich et al., Phys. Rev. Lett. 37, 631 (1976).
3. J.V. Jovanovich, Proceedings of the Second International Conference on the Nucleon-Nucleon Interaction, Vancouver, Canada, June 27-30, 1977, edited by H.W. Fearing (American Institute of Physics, 1978) (in print).
4. M.K. Srivastava and Donald W.L. Sprung, Adv. Nucl. Phys. 8, 121 (1975).
5. P. Signell, Advances in Nuclear Physics, Vol. 2, 223 (1969). Edited by M. Baranger and E. Vogt (Plenem Press).
6. V. Brown, Phys. Letters 25B, 506 (1967).
7. E.M. Nyman, Phys. Rev. 170, 1628 (1968).
8. H.W. Fearing, ppB Workshop, Nucleon-Nucleon Interaction Conference, Vancouver, June 27th to July 1st, 1977 (in print).
9. G.E. Bohannon, Proceedings of the Second International Conference on the Nucleon-Nucleon Interaction, Vancouver, Canada, June 27-30, 1977, edited by H.W. Fearing (American Institute of Physics, 1978) (in pirnt).
10. L. Heller, Bull. Am. Phys. Soc. 17, 480 (1972). See also P. Signell, in Proceedings of the International Conference on Few Particle Problems and the Nuclear Interaction, Los Angeles, California, 1972, edited by I. Slaus et al. (North-Holland, Amsterdam, 1973), pp. 1-25.
11. J.L. Beveridge et al. Proceedings of the Second International Conference on the Nucleon-Nucleon Interaction, Vancouver, Canada, June 27-30, 1977, edited by H.W. Fearing (American Institute of Physics, 1978) (in print).
12. J.V. Jovanovich, preprint, see also Manitoba Cyclotron Annual Report 1975/76, p. 77.
13. F. Sannés et al. Nucl. Phys. A164, 438 (1970).
14. A. Willis et al., Phys. Rev. Lett. 28, 1063 (1972).
15. B.M.K. Nefkens et al., Phys. Rev. Lett. 38, 876 (1977).

THE INTERFERENCE OF INITIAL AND FINAL STATE AMPLITUDES IN p-d BREMSSTRAHLUNG NEAR THE BREAK-UP THRESHOLD

R.J. Slobodrian
Laboratoire de Physique Nucléaire
Département de Physique, Université Laval
Québec, G1K 7P4, Canada

The subject of the emission of low energy quanta in nuclear reactions was started a good number of years ago by Eisberg et al.[1] and Feshbach and Yennie [2]. The nucleon-deuteron bremsstrahlung is of particular interest to the three body problem [3], and the region near the break-up threshold may show effects due to the quasi two-body nature of the three body channel, and to a possible rapid variation of both on-shell and off shell amplitudes [3]. Most bremsstrahlung measurements with particles of comparable masses are carried out by the detection of the two final state particles, which is a kinematically complete experimental condition. In the approximation of emitted photons with negligible momentum with respects to the particle momenta, the cross-section for the emission of dipole radiation is given by

$$\frac{d^2\sigma}{d\Omega_d d\Omega_p} = \frac{1}{(2j_d+1)(2j_p+1)} A \frac{p_d}{m_d} \frac{e^2}{\pi} \left(Z_d - \frac{m_d}{m_p} Z_p\right)^2 \mathrm{Tr}\, \{B|t_i|^2 - |t_f - Ct_i|^2\} \qquad (1)$$

p_d is the deuteron momentum, m_d its mass, Z are charges, m_p is the proton mass, $\hbar = c = 1$, t_i, t_f are elastic amplitudes, A, B and C are kinematical coefficients involving laboratory angles of the final state particles. Low energy experimental data from the p-d system have become available above threshold [4] and around threshold [5]. Both experiments were performed using a deuteron beam on a hydrogen target. The experiments of Hall et al [4] produced data at 7, 8 and 9 MeV in the laboratory system, and the geometry was chosen such as to make small the interference term of expression (1), $C(t_i t_f^\dagger + t_i^\dagger t_f)$ with $\langle\Theta_{cm}\rangle = 88^0$ permitting the substitution

$$\frac{1}{(2j_d+1)^{\frac{1}{2}}(2j_p+1)^{\frac{1}{2}}} \mathrm{Tr}|t_{i,f}| = \frac{d\sigma}{d\Omega_{if}} \qquad (2)$$

in expression (1) without appreciable error. The results of Hall et al. [4] were largely independent of energy. The experiments of ref. 5 were carried out in a Harvard geometry and scanned the region around threshold at five energies between 6.3 and 7.1 MeV. An interesting cross section anomaly was observed in proceeding through the deuteron break-up threshold. However, the theoretical computation of expression (1) is here complicated considerably, as the average C.M. scattering angle is 48^0 and hence the interference term is not small. The complication arises from the complexity of spin space and the large number of phase shifts and mixing parameters involved.

The importance of treating the general case resides precisely in the possibility of testing the conjecture of Feshbach and Yennie [2] about the determination of a relative phase of the scattering amplitudes and fluctuations of the cross section related to it. The trace of the interference term is given by

$$\mathrm{Tr}[t_i t_f^{\dagger} + t_i^{\dagger} t_f] = 2R[A_i^* A_f + C_i^* C_f] + \frac{4}{3} R[D_i^* D_f + B_i^* B_f + E_i^* E_f + F_i^* F_f] + \frac{4}{9} R\,[G_i^* G_f +$$

$$H_i^* H_f + I^* I_f + J_i^* J_f] + \frac{2}{6} R[K_i^* K_f + L_i^* L_f] - \frac{2}{9} R[G_i I_f^* + G_f I_i^* + H_i J_f^* + H_f J_i^*] \qquad (3)$$

where A, B, C.... are the t-matrix coefficients of the twelve independent operators in 6x6 matrix form [6]. The computation has been carried out using the phase shifts and mixing parameters of Schmelzbach et al.[7] with smooth interpolation as a function of energy whenever necessary. Figure 1 shows the results and indicates a weak energy dependence. Hence the experimental anomaly reported in ref. 5 cannot be accounted for by expression 1, and in particular not by the interference term mentioned above, on the basis of presently existing elastic scattering data and phase shifts. However, it should be noted that the latter have been obtained in steps of about 100 KeV in the C.M. system, whereas the bremsstrahlung data have been measured at steps of 30 and 50 KeV. Hence, a reinvestigation of elastic scattering in small energy steps around the break-up threshold seems to be indicated. It is of course possible to represent the bremsstrahlung data around threshold by a rapidly varying phase and a term of the form $2.\cos\phi.|t_i|.|t_f|$.

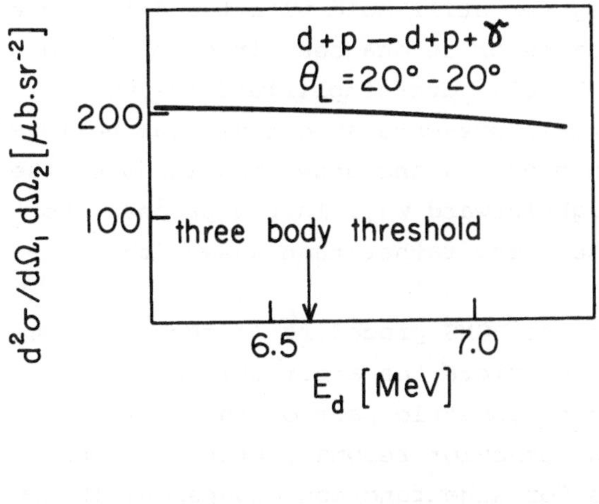

FIGURE 1 - Cross section as a function of energy from the interference term $\mathrm{CTr}|t_i t_f^{\dagger} + t_i^{\dagger} t_f)$

1. R.M. Eisberg, D.R. Yennie and D.H. Wilkinson, Nucl. Phys. 18, 338 (1960).
2. H. Feshbach and D.R. Yennie, Nucl. Phys. 37, 150 (1962).
3. A.M. Green, Few Body Problems in Nuclear and Particle Physics, Eds. R.J. Slobodrian, B. Cujec, K. Ramavataram, Les Presses de l'Univ. Laval, Québec (1975),refs.therein.
4. J. Hall, W. Wölfli and R. Müller, Phys. Lett. 37B, 53 (1971).
5. R.J. Slobodrian, R. Roy, C. Rioux and B. Frois, to be published.
6. R.G. Seyler, Nucl. Phys. A124, 253 (1969).
7. P.A. Schmelzbach et al. Nucl. Phys. A197, 273 (1972).

OFF-ENERGY-SHELL CONTINUATION OF THE TWO-BODY T-MATRIX IN THE PRESENCE OF RESONANCES

A. V. Lagu
Physikalisches Institut der Universität, Bonn, West Germany

and

O. Zohni
Physics Department and Cyclotron Institute
Texas A&M University, College Station, Texas 77843, U.S.A.

Some time back Baranger et al.[1] (BGMS) had demonstrated how the two nucleon, non-relativistic T matrix may be continued off-the-energy shell without explicitly introducing a potential when the uncoupled partial wave eigenchannel has only scattering states. Haftel[2] had subsequently extended the method when a bound state was also present. More recently Sauer and Sevgen[3] (SS) have tackled the same problem in the presence of inelasticities. In their introduction, SS have discussed the relevance of inelasticities to pion-nucleon system. Our aim is slightly different in the present work. We wish here to discuss again a situation like pion-nucleon interaction but in the spirit of BGMS-Haftel and hence assuming again that all the conditions relevant to the nucleon-nucleon system still hold with the difference that the eigenchannel has a resonance or resonances (in case of π-N, this is expected). We have, undoubtedly, oversimplifield the whole problem but having a resonance itself presents some novel (we believe) points which need to be discussed properly and once this is done, the rest follows in a generally straightforward way. In the process, however, we have posed additional questions rather than answer the existing ones.

As is well known the crux of the BGMS procedure is the existence of a completeness relation which restricts the arbitrariness in the full off-shell T matrix to only the symmetric part of the half-shell function. Because of their violent behavior resonant states are not usually used, however, as a basis for eigenfunction expansion. The main problem is therefore the construction of appropriate completeness relations which bring out explicitly the presence of resonant states. A number of studies have been made to construct appropriate completeness relations involving resonant states. Various restrictive conditions, however, are usually associated with such completeness relations.[4] This leads therefore to nonuniqueness in defining completeness relations involving resonant states. As pointed out by Zohni,[6] a generalized completeness relation for resonant states can be defined by following the work of Meetz[5] and using a symmetric kernel K related to the two-body Green function $G_0\ (E+i\varepsilon) = (E-H_0+i\varepsilon)^{-1}$, H_0 being the kinetic energy operator, by:

$$K^{\pm} = |V|^{1/2}\, G_0^{\pm}\, |V|^{1/2} \quad , \tag{1}$$

such that a biorthonormal system $(\xi, \tilde{\xi})$ can be constructed in L^2 with:

$$<\xi|\tilde{\xi}> = <\tilde{\xi}|\xi> = 1 \tag{2}$$

It follows that the biorthonormal system $\{\xi, \tilde{\xi}\}$ is complete in the sense of Hilbert space and any element f of L^2 can be expanded in terms of these states.[5] This completeness is further verified[6] by the existence of a generalized Parseval's identity which applies to any biorthonormal system:

$$\|f\|^2 = \sum_n < f|\xi_n> <\tilde{\xi}_n| f > \tag{3}$$

The fact that the biorthonormal system $\{\xi, \tilde{\xi}\}$ spans the space L^2 removes the restriction associated, e.g., with the regularization and contour deformation methods.[7,8] This allows to construct a generalized completeness relation involving those resonant states (assuming for simplicity the absence of any bound states and other defects[3]):

$$\sum_c \int dE |\psi_c^0(E)> <\psi_c^0(E)| + \sum_n \int dE_r |\tilde{\xi}_n(E_r)> <\xi_n(E_r)| = 1 \tag{4}$$

where $\psi_c^0(E)$'s are real stationary scattering states for channel c and $\{\tilde{\xi}_n, \xi_n\}$ satisfy:

$$\begin{aligned} \xi_n &= \lambda_n PK\xi_n \\ \tilde{\xi}_n &= \tilde{\lambda}_n P\tilde{K}\tilde{\xi}_n \end{aligned} \tag{5}$$

with $\tilde{\xi} = \xi^*$, $\tilde{\lambda}_n^* = \lambda_n$, P denotes the principal value,[9] and K as defined in (1) with $H = H_0 + V$.

One might now follow Haftel's procedure, defining a model Hamiltonian $H_r = K + V_r$ with V_r giving the same resonance states as the original underlying potential V, then applying the completeness of the eigenfunctions of H and H_r to construct appropriate half-shell functions (since the eigenfunctions of H and H_r are complete, it should presumably be possible to proceed parallel to Haftel's paper with details, however, depending on the biorthonormal properties characteristic of the resonant wave functions as will be given elsewhere).

However, unlike Haftel, we face here a problem, viz. whereas the boundstate wave function was assumed attainable from experiment and also because of the knowledge of binding energy, a model potential V_m could be constructed which could lead to χ_m's and δ_m's. In the present case of resonance, such possibility, to our knowledge, does not exist. Thus we have no option but to treat it as an additional degree of freedom and test the sensitivity of various nuclear processes to off-shell properties of T and the resonance wave function. In the present instance, therefore, we feel that it would be closer in spirit of BGMS to follow Amado's[10] prescription because of the availability of experimental information about resonance energies and widths which give, in turn, the phase shifts through well known relations and then the phase shifts δ and δ_r corresponding to V and V_r, respectively, become identical which is requisite for the applicability of Amado's method.

AVL whishes to express his gratitude to Prof.Dr.W.Sandhas for his hospitality, to A.V.Humboldt Foundation for a fellowship and to Banaras Hindu University, Varanasi, India for leave of absence.

References

1. M.Baranger et al., Nucl.Phys. A138, 1 (1969).
2. M. I. Haftel, Phys. Rev. Lett. 25, 120 (1970).
3. P. U. Sauer and A. Sevgen, Phys. Rev. C13, 720 (1976).
4. See, e.g. G. Garcia-Calderón, Nucl. Phys. A265, 443 (1976).
5. K. Meetz, J. Math. Phys. 3, 690 (1962).
6. O. Zohni, Bull. Am. Phys. Soc. 22, 1030 (1977).
7. R. M. More and E. Gerjuoy, Phys. Rev. A7, 1288 (1973).
8. T. Berggren, Nucl. Phys. A109, 265 (1968).
9. T. Sasakawa, Nucl. Phys. A160, 321 (1971).
10. R. D. Amado, Phys. Rev. C2, 2439 (1970).

THE COULOMB CONTRIBUTION TO PROTON-PROTON SCATTERING PHASE SHIFTS

L. Streit
Fakultät für Physik, Universität Bielefeld
D-48 Bielefeld, Germany
J. Fröhlich, H. Zankel, H. Zingl
Institut für Theoretische Physik, Universität Graz
A-8010 Graz, Austria

The usual way to compare potential model calculations with proton-proton experiments is to replace the strong scattering amplitude by the Coulomb distorted one or the plane waves by Coulomb wave functions, as an improvement over the inclusion of some arbitrary set of residual phases.

We use the Lippmann-Schwinger formalism to calculate appropriate residual phases and are able to do so in a model-independent way with the help of the pure nuclear half-shell T matrices. It is well-known that these transition matrices can be obtained directly from experiments, such as bremsstrahlung measurements. However a consistent on-shell approximation suffices in most cases.

Splitting the scattering matrix $S_\ell(p)$ as well as the scattering amplitude $T_\ell(p)$ into strong, Coulomb, and mixed parts

$$S_\ell(p) = S_{S,\ell}(p) S_{C,\ell}(p) S_{R,\ell}(p) \qquad T_\ell(p) = T_{S,\ell}(p) + T_{C,\ell}(p) + T_{R,\ell}(p)$$

we obtain the following expression for the "mixed" S-matrix $S_{R,\ell}(p)$ resp. the residual phase shift $\delta_{R,\ell}(p)$:

$$S_{R,\ell}(p) - 1 = e^{2i\delta_{R,\ell}(p)} - 1$$

$$= -\lim_{\varepsilon\to 0} \frac{2i\pi\mu p e^2}{S_{S,\ell}(p)} \Big\{ \frac{1}{2}(S_{S,\ell}(p)+1) \frac{4\mu}{\pi p} \, \mathrm{P}\!\int_0^\infty dp' Q_\ell\Big(\frac{p^2+p'^2+\varepsilon^2}{2pp'}\Big) \frac{p'}{p^2-p'^2} \times$$

$$T_{S,\ell}(p',p;E) - \pi\mu^2 T^2_{S,\ell}(p) Q_\ell\Big(\frac{2p^2+\varepsilon^2}{2p^2}\Big) + \frac{4\mu^2}{\pi} \, \mathrm{P}\!\int_0^\infty dp' \, \mathrm{P}\!\int_0^\infty dp'' T_{S,\ell}(p,p';E) \times$$

$$\frac{p'}{p^2-p'^2} Q_\ell\Big(\frac{p'^2+p''^2+\varepsilon^2}{2p'p''}\Big) \frac{p''}{p^2-p''^2} T_{S,\ell}(p'',p;E) \Big\} + O(e^4) \, . \quad (1)$$

For the derivation (of an approximate version) see ref.[1], also ref.[2]. For a given strongly interacting system this reduces the calculation of Coulomb corrections in the "lowest order regime"

$$p_{CM} \gg \mu e^2 \qquad (= 3{,}4\ \frac{\text{MeV}}{c} \text{ for } \mu = \frac{1}{2} m_p)$$

to the computation of the above integrals. A class of examples for which the calculation is particularly simple is furnished by separable models with a rank 1 strong interaction potential. Details of these calculations will be published elsewhere [2]. The main contributions are those of the external Coulomb interactions

for which our formula simplifies to

$$S_{R,\ell}(p)-1 \approx \frac{8i\mu^2 e^2}{S_{S,\ell}(p)} \, ⨍_0^\infty dp' \, Q_\ell\left(\frac{p^2+p'^2+\varepsilon^2}{2pp'}\right) \frac{p'}{p^2-p'^2} \, T_{S,\ell}(p',p;E) . \tag{2}$$

Note that we are free to consider the unscreened Coulomb potential ($\varepsilon = 0$) since the energy shell singularity is integrable (see also [3]). A consistent on-shell approximation of this integral as in ref.[1] leads to the model independent expression

$$\delta_{R,\ell}(p) = \frac{1}{2} \operatorname{arctg} \frac{\alpha_\ell}{1+\alpha_\ell} \frac{\operatorname{Im}(z)}{\operatorname{Re}(z)}$$

with $\quad z = 2i\delta'_{S,\ell}(p) + p^{-1}\left(1-e^{-2i\delta_{S,\ell}(p)}\right)$

and

$$\alpha_\ell = \frac{2\mu e^2}{\pi} ⨍_0^\infty dx \, \frac{x}{1-x^2} \, Q_\ell\left(\frac{1+x^2}{2x}\right) = \begin{cases} -\frac{\pi}{2}\,\mu e^2 & \ell = 0 \\ -\frac{2}{\pi}\,\mu e^2 & \ell = 1 \\ \frac{\mu e^2}{2\ell+1} & \ell \geq 2 \end{cases} . \tag{3}$$

The large ℓ behaviour is in keeping with the WKB predictions of [4].

To display the accuracy of eq.(3) we invoke one of the most realistic nucleon-nucleon interactions presently available - the meson theoretical potential of the Paris group [5]. In figures 1-5 we present

- the residual phases calculated explicitly by solving the dynamical problem both with and without the Coulomb force [6],
- the approximate residual phases which one obtains immediately from an application of our eq.(3),
- and for completeness the strong phases.

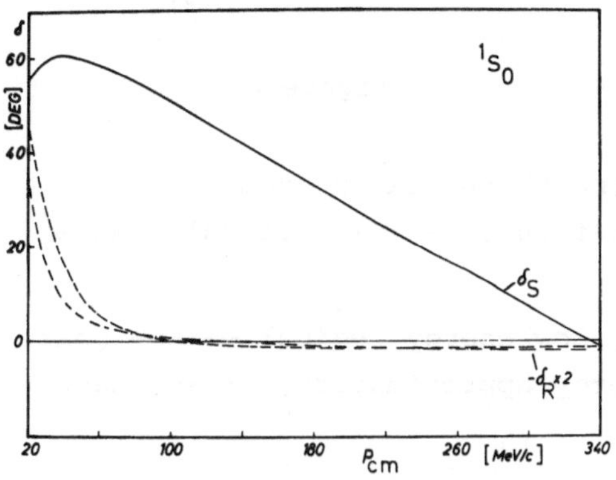

figure 1

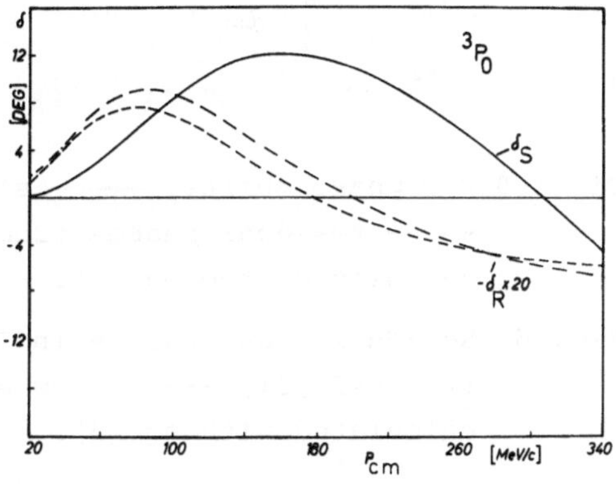

figure 2

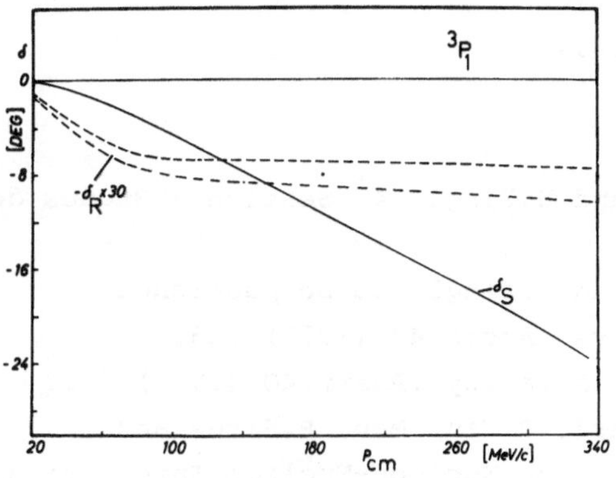

figure 3

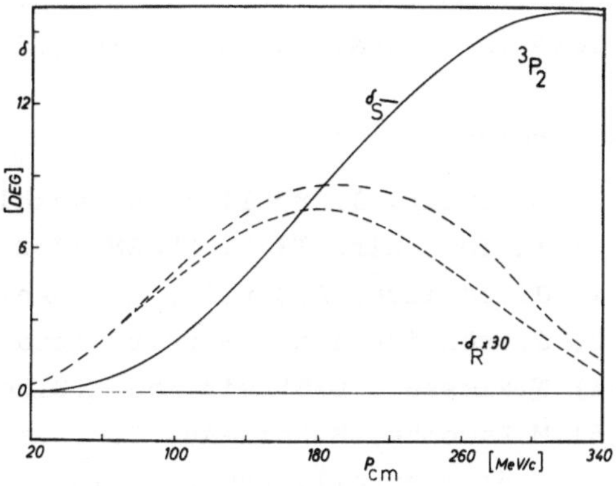

figure 4

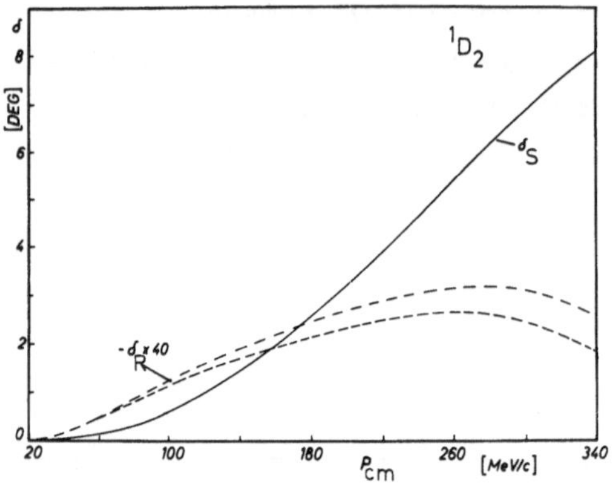

figure 5

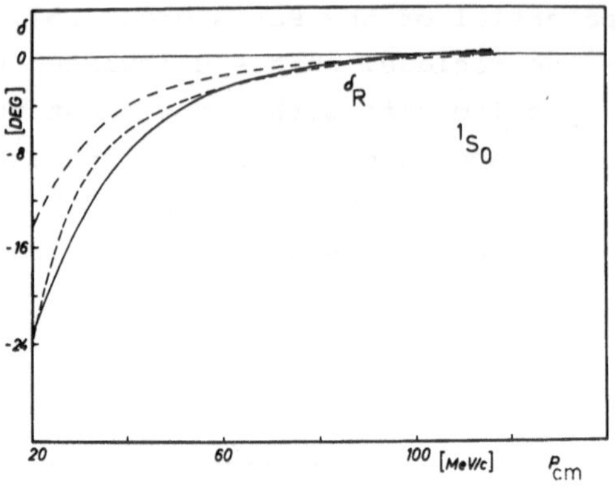

figure 6

Fig.1-5 P-P phase shifts: ——— strong phases taken from ref.[5], ----- residual phases from ref.[6], -·-·-· residual phases calculated from eq.(3).

Fig. 6 Residual phase shifts in ^1S_o low energy region: ——— phase from ref.[6], ------ low energy approximation, -·-·- phase calculated with eq.(3).

The agreement of the residual phases is quite satisfactory, with the exception of the low energy region of the S-wave where off-shell effects must be taken into account. As a consequence we use the simple model independent low energy approximation for the half off-shell extension of the T-matrix ("Noyes function") given by Fuda [7] to restore the overall accuracy (Fig.6) [2].

References

[1] L.Streit, J.Fröhlich, H.Zankel and H.Zingl, 4° Session d'Etudes de la Toussuire 1977, LYCEN 77o2.

[2] J.Fröhlich, L.Streit, H.Zankel and H.Zingl, to be published.

[3] J.Fröhlich and L.Streit, Acta Phys. Austr.47 (1977) 125.

[4] W.Plessas, L.Streit and H.Zingl, Acta Phys.Austr.40 (1974) 272.

[5] M.Lacombe, B.Loiseau, J.M. Richard, R.Vinh Mau, P.Pires and R.de Tourreil, Proc.II. Int. Conf. on Nucleon-Nucleon Interaction, Vancouver 1977 (to be published).

[6] B.Loiseau, privat communication.

[7] M.G. Fuda, Phys. Rev. C1 (1970) 1910.

THE COULOMB CONTRIBUTION TO π^+-p PHASE SHIFTS

J. Fröhlich and H. Zankel
Institut für Theoretische Physik, Universität Graz
A-8010 Graz, Austria

We apply the Blankenbecler-Sugar formalism to derive Coulomb corrections for π^+-p elastic scattering phase shifts up to $T_\pi^{Lab} \approx 250$ MeV. Starting with the splitting of the S-matrix and the T-matrix as given in ref.[1]:

$$S_\ell(p) = S_{S,\ell}(p)\, S_{C,\ell}(p)\, S_{R,\ell}(p) \qquad T_\ell(p) = T_{S,\ell}(p) + T_{C,\ell}(p) + T_{R,\ell}(p)$$

we obtain for the residual phase $\delta_{R,\ell}(p)$, which originates from the interference between the strong and Coulomb force:

$$\begin{aligned} S_{R,\ell}(p) - 1 &= e^{2i\delta_{R,\ell}(p)} - 1 \\ &= -(\lim_{\varepsilon\to 0}) \frac{2i\pi\mu p e^2}{S_{S,\ell}(p)} \{ \tfrac{1}{2}(S_{S,\ell}(p)+1) \frac{2}{\pi p} P\!\!\int_0^\infty dp' Q_\ell\left(\frac{p^2+p'^2+\varepsilon^2}{2pp'}\right) \frac{p'(E_1'+E_2')}{2E_1'E_2'[s-(E_1'+E_2')^2]} \times \\ &\times T_{S,\ell}(p,p';s) - \frac{\pi}{16s} Q_\ell\left(\frac{2p^2+\varepsilon^2}{2p^2}\right) T^2_{S,\ell}(p) + \frac{1}{\pi} P\!\!\int_0^\infty dp' P\!\!\int_0^\infty dp'' T_{S,\ell}(p,p';s) \times \\ &\times \frac{p'(E_1'+E_2')}{2E_1'E_2'[s-(E_1'+E_2')^2]} Q_\ell\left(\frac{p'^2+p''^2+\varepsilon^2}{2p'p''}\right) \frac{p''(E_1''+E_2'')}{2E_1''E_2''[s-(E_1''+E_2'')^2]} T_{S,\ell}(p'',p;s) \} \\ &+ O(e^4) \qquad \text{with } s=(E_1+E_2)^2 \text{ and } E_1=\sqrt{p^2+m_\pi^2},\ E_2=\sqrt{p^2+m_p^2}\,. \end{aligned} \tag{1}$$

Whenever the hadronic interaction is weak we neglect the term of equ.(1) which is of second order in the strong T-matrix:

$$S_{R,\ell}(p)-1=-\frac{4i\mu e^2}{S_{S,\ell}(p)} P\!\!\int_0^\infty dp'\, Q_\ell\left(\frac{p^2+p'^2}{2pp'}\right) \frac{p'(E_1'+E_2')}{2E_1'E_2'[s-(E_1'+E_2')^2]} T_{S,\ell}(p',p;s)\,. \tag{2}$$

Note that we are free to take the limit $\varepsilon \to 0$ because the on-shell singularity is integrable (see also ref.[2]). Using an on-shell approximation [3] for the remaining part of equation (1) we finally obtain:

$$\delta_{R,\ell}(z) = \frac{1}{2} \operatorname{arctg} \frac{\operatorname{Im}(z)}{1+\operatorname{Re}(z)}$$

with

$$z = \alpha_\ell \quad [2i\delta'_{S,\ell}(p) + p^{-1}(1-e^{-2i\delta_{S,\ell}(p)})(1 + \frac{p^2}{E_1E_2})] \tag{3}$$

and

$$\alpha_\ell = \frac{4\mu\sqrt{s}e^2}{\pi} \int_0^\infty \!\!\!\!\!\!\!\!P\ \ dp'\, Q_\ell\left(\frac{p^2+p'^2}{2pp'}\right) \frac{p'(E_1'+E_2')}{2E_1'E_2'[s-(E_1'+E_2')^2]} \quad ,$$

which is model independent in the sense that $\delta_{R,\ell}(p)$ is roughly a function of the hadronic phase shift only. To test the quality of equation (3) we have calculated the residual phases of the S_{31} and P_{31} wave, which are displayed in Fig. 1 and Fig. 2. For comparison the electromagnetic corrections obtained with a relativistic formalism [4] and the Coulomb corrections calculated with the nonrelativistic formula (3) of ref.[1] are shown.

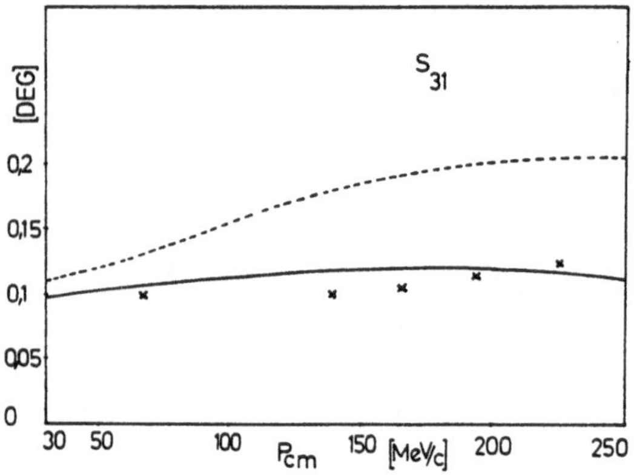

figure 1

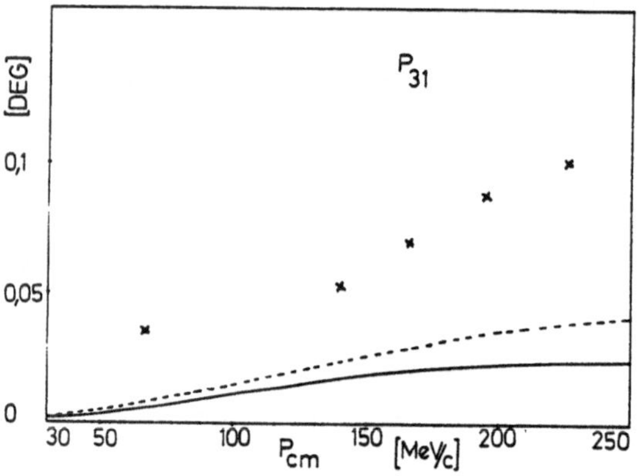

figure 2

Fig.1-2: Residual phase shifts:
----- nonrelativistic calculation, ——— calculation using equation (3), x x x x relativistic calculation from ref.[4].

The residual phases resulting from an application of the Blankenbecler-Sugar technique differ significantly from those one gets from the non-relativistic formalism. Furthermore one can see that the Coulomb contribution accounts for the main part of the electromagnetic corrections in the S_{31}, but it is less important in the P_{31}. This is in fair agreement, at least qualitatively, with the results derived by Tromberg et al.[4] in a full relativistic framework.

[1] L.Streit, J.Fröhlich, H.Zankel and H.Zingl, Inv.Contr.to this Conf..

[2] J.Fröhlich and L.Streit, Acta Phys.Austr. 47 (1977) 125.

[3] L.Streit, J.Fröhlich, H.Zankel and H.Zingl, 4° Session d'Etudes de la Toussuire 1977, LYCEN 7702.

[4] B.Tromberg, S.Waldenstrøm and I.Øverbø, Ann. of Phys.100 (1976) 1.

THE ONE PHOTON EXCHANGE POTENTIAL IN CONFIGURATION SPACE

G.J.M. Austen, J.J. de Swart

Institute for Theoretical Physics, University of Nijmegen, The Netherlands

Recently we derived the potential corresponding with relativistic one photon exchange, starting from the Bethe-Salpeter equation in a way simular to the one described elsewhere [1] for the nuclear potential. Despite the fact that there are some slight ambiguities the central part of the one photon exchange potential becomes:

$$V_C(r) = \frac{\alpha}{r}\left(1 + \frac{1}{4}\frac{k_s^2}{M^2}\right) - \frac{5}{8M^2}\left(\vec{\nabla}^2 \frac{\alpha}{r} + \frac{\alpha}{r}\vec{\nabla}^2\right) + \delta \text{ function} \qquad (1)$$

where $\frac{1}{2}\sqrt{s} = \sqrt{M^2 + k_s^2}$, M nucleon mass. Potential (1) holds for NN scattering. The usual combinations of the spin dependent potentials are also found in this way (see e.g. [2]). These potentials are to be used in a Schrödinger equation with the relativistic value of the momentum (compare [1], page 224). Furthermore it contains as new elements: a. an "energy dependent coupling": $\alpha\ (1 + \frac{1}{4} k_s^2/M^2)$ and b. nonlocal terms: $(\vec{\nabla}^2 \frac{\alpha}{r} + \frac{\alpha}{r}\vec{\nabla}^2)$.

For $r \to \infty$ the $(-\vec{\nabla}^2)$ operator in connection with the Schrödinger equation can be replaced by k_s^2 leading to the potential described by Naisse [3] for small momentum k_s:

$$V_C(r) = \frac{\alpha\ (1 + \frac{3}{2} k_s^2/M^2)}{r} \qquad (2)$$

The substitution $\alpha \to \alpha\ (1 + \frac{3}{2} k_s^2/M^2)$ and $k^2 \to k_s^2$ corresponds with the usual procedure to simulate relativistic corrections in the Coulomb amplitude: $\eta \to \eta_{rel}$, $k \to k_s$ (see [2]). However, if the total potential energy becomes comparable with the energy this is not a good approximation and one should use (1). Techniques to handle such potentials are well known [4].

The most important consequences are:

- The Coulombamplitude corresponding with (1) becomes:

$$f_C(\theta) = -\frac{\eta}{2k_s \sin^2 \frac{1}{2}\theta}\, e^{-2i\eta \ln \sin \frac{\theta}{2}} + \frac{3}{16}\,\frac{\alpha^2 \pi}{k_s \sin \frac{1}{2}\theta} \qquad (3)$$

$$\eta = \frac{M\,\alpha(1 + \frac{3}{2} k_s^2/M^2)}{k_s}$$

- The Coulombphases modify according to:

$$\sigma_\ell \to \sigma_\ell + \frac{3}{8}\,\frac{\pi\alpha^2}{(2\ell+1)} + \ldots \qquad (4)$$

  The α^2 terms in (3) and (4) are in the order of 5 - 10% times the corresponding vacuumpolarization quantities.

- As far as the use of (1) in the Schrödinger equation is concerned, applying Greens trick [4] leads to "effective" local potentials:

$$V'_c = \frac{\alpha\,(1 + \frac{3}{2}k_s^2/M^2)}{r + d} \quad ; \quad d = \frac{3}{4}\frac{\alpha}{M} \tag{5a}$$

$$V'_{nuclear} = \frac{V_{nuclear}}{1 + \frac{d}{r}} \tag{5b}$$

(only if local nuclear potentials are used).

The asymptotic potential is now (5a) instead of $\frac{\alpha}{r}$: therefore one must take as s-wave asymptotic Coulombfunctions:

$$\text{regular: } \mathcal{F}_o = \{-\sin\alpha\, G_o(\eta,k_s(r+d)) + \cos\alpha\, F_o\,(\eta,k_s(r+d))\}$$

$$\text{irregular: } \mathcal{G}_o = \{\cos\alpha\, G_o(\eta,k_s(r+d)) + \sin\alpha\, F_o\,(\eta,k_s(r+d))\} \tag{6}$$

$$\text{tg}\,\alpha = \frac{F_o(\eta,kd)}{G_o(\eta,kd)}$$

- Due to the nonlocal terms in (1) the pp-effective range produced by the Nijmegen OBE model [5] raised by 0.02 fm. Furthermore there is a charge symmetry breaking between the "nuclear" scattering lengths: $a_{pp} - a_{nn} = + 0.3$ fm.

References:

1. J.J. de Swart, M.M. Nagels, Fortschritte der Physik 28, 215 (1978)
2. M.S. Sher, P. Signell, Annals of Physics 58, 1 (1970)
3. J.P. Naisse, Bull.Cl.Sc.Ac.R.Belg. 61, 589 (1975)
4. A.M. Green, Nucl.Phys. 33, 218 (1963)
5. M.M. Nagels, T.A. Rijken, J.J. de Swart, Phys.Rev. D 17, 768 (1978)

THE COULOMB INTERFERENCE TERM IN POLARIZATION ANALYSIS AND THE DETERMINATION OF THE TRANSITION MATRIX

Albrecht Lindner
1.Institut für Experimentalphysik der Universität Hamburg

If elastic scattering experiments with charged particles are analyzed with a partial wave decomposition, it is important to separate the rather slow convergent effect of the Coulomb field. Due to Gell-Man & Goldberger we have the separation

$$\langle f|T|i\rangle = \langle f|T_C|i\rangle + {}^{(-)}\langle f_C|\tilde{T}|i_C\rangle^{(+)} \tag{1}$$

with the well-known result for Coulomb scattering

$$\langle E\Omega_f s_f m_f|T_C|E\Omega_i s_i m_i\rangle = \frac{\eta}{2}\left[\frac{\exp i(\sigma_o - \eta \ln \sin\frac{\theta}{2})}{2\pi\ \sin\frac{\theta}{2}}\right]^2 \langle s_f m_f|s_i m_i\rangle\ . \tag{2}$$

(The interaction of electric charges with magnetic moments is included in $\tilde{T}$.) Thus in

$$\frac{d\sigma}{d\Omega_f} = \frac{(2\pi)^4}{k^2}\left(|T_C|^2 + 2\,\mathrm{Re}\,T_C^*\tilde{T} + |\tilde{T}|^2\right) \tag{3}$$

the first summand is the Rutherford cross section. This is also true for polarized particles; polarization affects only the remaining terms. They have been given by Heiß[1]) using the conventional expansion in terms of associated Legendre-polynomials. However, this expansion is not rotational invariant and introduces redundant parameters[2]). This can be avoided by using correlation functions as has been shown[3,4]) for the last term of eq.(3) and will now be given for the interference term. Moreover, we find the expansion coefficients of the spin-correlation experiment $\vec{a}+\vec{b}\rightarrow a+b$ to determine all other coefficients.

In the case $\vec{a}+\vec{b}\rightarrow a+b$ we find

$$\frac{d\sigma^I}{d\Omega_f} = 2\,\mathrm{Re}\left(\langle E\Omega_f|T_C|E\Omega_i\rangle^* \sum_{l_i l_f n n_a n_b} A^I_{l_i l_f(n)n_a n_b}\, t^{(n_a)} t^{(n_b)} P_{l_i l_f(n)n_a n_b}(\Omega_i\Omega_f\Omega_a\Omega_b)\right) \tag{4}$$

where the sum resembles the last term of eq.(3)- cf. ref.4):

$$\frac{d\sigma^N}{d\Omega_f} = \sum_{n_i n_f n n_a n_b} A_{n_i n_f(n)n_a n_b}\, t^{(n_a)} t^{(n_b)} P_{n_i n_f(n)n_a n_b}(\Omega_i\Omega_f\Omega_a\Omega_b)\ . \tag{5}$$

The equation for the (complex) coefficients

$$A^I_{l_i l_f(n)n_a n_b} = i^{l_i+l_f-n_a-n_b}\,\frac{4\pi^3\hat{n}}{k^2\hat{s}_a\hat{s}_b}\sum_{s_i J s_f}(-)^{J+s_f}\,\hat{s}_i\hat{s}_f\hat{J}^2\begin{Bmatrix} s_a & s_b & s_f\\ s_a & s_b & s_i\\ n_a & n_b & n\end{Bmatrix} \tag{6}$$

$$\times\begin{Bmatrix} s_i & s_f & n\\ l_f & l_i & J\end{Bmatrix}{}^{(-)}\langle E(l_f(s_a s_b)s_f)J|\tilde{T}|E(l_i(s_a s_b)s_i)J\rangle^{(+)}$$

can be inverted yielding the transition matrix $\tilde{T}$:

$${}^{(-)}\langle E(l_f(s_a s_b)s_f)J|\tilde{T}|E(l_i(s_a s_b)s_i)J\rangle^{(+)} = (-)^{J+s_f}\,\frac{k^2}{4\pi^3}\,\hat{s}_a\hat{s}_b\hat{s}_i\hat{s}_f \tag{7}$$

$$\times\sum_{n n_a n_b} i^{n_a+n_b-l_i-l_f}\,\hat{n}\hat{n}_a^2\,\hat{n}_b^2\begin{Bmatrix} s_a & s_b & s_f\\ s_a & s_b & s_i\\ n_a & n_b & n\end{Bmatrix}\begin{Bmatrix} s_i & s_f & n\\ l_f & l_i & J\end{Bmatrix} A^I_{l_i l_f(n)n_a n_b}.$$

Thus the (real) coefficients in eq.(5) can be calculated from those in eq.(4):

$$
\begin{aligned}
A_{n_i'' n_f'' (n'') n_a'' n_b''} &= (-)^{2s_a+2s_b}\, i^{n_i''+n_f''-n_a''-n_b''} \frac{k^2}{16\pi^4} \hat{s}_a \hat{s}_b \hat{n}_i'' \hat{n}_f'' \hat{n}'' \\
&\times \sum_{\substack{l_i l_f n\, n_a n_b \\ l_i' l_f' n' n_a' n_b'}} i^{l_i+l_f-n_a-n_b+l_i'+l_f'-n_a'-n_b'}\, \hat{l}_i \hat{l}_f \hat{n} \hat{n}_a^2 \hat{n}_b^2 \hat{l}_i' \hat{l}_f' \hat{n}' \hat{n}_a'^2 \hat{n}_b'^2 \\
&\times \begin{pmatrix} l_i & l_i' & n_i'' \\ 0 & 0 & 0 \end{pmatrix} \begin{pmatrix} l_f & l_f' & n_f'' \\ 0 & 0 & 0 \end{pmatrix} \begin{Bmatrix} n_a & n_a' & n_a'' \\ s_a & s_a & s_a \end{Bmatrix} \begin{Bmatrix} n_b & n_b' & n_b'' \\ s_b & s_b & s_b \end{Bmatrix} \\
&\times \begin{Bmatrix} l_i & l_i' & n_i'' \\ l_f & l_f' & n_f'' \\ n & n' & n'' \end{Bmatrix} \begin{Bmatrix} n_a & n_a' & n_a'' \\ n_b & n_b' & n_b'' \\ n & n' & n'' \end{Bmatrix} A^I_{l_i l_f (n) n_a n_b} A^{I*}_{l_i' l_f' (n') n_a' n_b'}
\end{aligned} \tag{8}
$$

This should be accounted for when fitting experiments. Parity conservation demands l_i+l_f to be even and time reversal invariance

$$
A^I_{l_i l_f (n) n_a n_b} = (-)^{l_i+l_f+n+n_a+n_b} A^I_{l_f l_i (n) n_a n_b} \, . \tag{9}
$$

Restricting eqs.(4)-(6) to $n_b=0$ covers the analyzation experiment $\vec{a}+b \to a+b$ which would not be sufficient to determine the transition matrix (7) and the analogue of eq.(8) if $s_b \neq 0$. The same is true for $a+b \to \vec{a}+b$ and even for the polarization transfer $\vec{a}+b \to \vec{a}+b$, because we get for the product of cross section and component of the polarization tensor along Ω_c

$$
\sigma^{(n_c)}(\Omega_c)^I = 2\,\mathrm{Re}\left(\langle E\Omega_f | T_c | E\Omega_i \rangle^* \sum_{l_f l_i n_a n} B^I_{l_i n_a (n) l_f n_c}\, t^{(n_a)} P_{l_i n_a (n) l_f n_c}(\Omega_i \Omega_a \Omega_f \Omega_c) \right) \tag{10}
$$

with

$$
B^I_{l_i n_a (n) l_f n_c} = (-)^{l_f+n+2s_a} \hat{n} \hat{s}_a \sum_{n'} i^{n_c-n_a-n'} \hat{n}'^2 \begin{Bmatrix} n_a & n_c & n' \\ l_f & l_i & n \end{Bmatrix} \begin{Bmatrix} n_a & n_c & n' \\ s_a & s_a & s_a \end{Bmatrix} A^I_{l_i l_f (n') n' 0} . \tag{11}
$$

For $a+b \to a+\vec{b}$ however, we have the coefficients

$$
B^I_{l_i n_a (n) l_f n_c} = \sum_{n'} (-)^{n_a+l_f+n+n'} \hat{n} \hat{n}' \begin{Bmatrix} l_i & n_a & n \\ n_c & l_f & n' \end{Bmatrix} A^I_{l_i l_f (n') n_a n_c} \, . \tag{12}
$$

This expression would simplify to one term when coupling in another order- cf. eq.(8) of ref.4). Thus the experiment $\vec{a}+b \to a+\vec{b}$ is equivalent to $\vec{a}+\vec{b} \to a+b$ and determines the transition matrix as well.

1) P. Heiß: Z.Physik 251(1972)159
2) A. Lindner: Nucl.Phys. A261(1976)253; M. Dohrendorf and A. Lindner, in Jahresbericht 1976/77, 1.Inst.f.Exp.Phys.Univ.Hamburg
3) A. Lindner: Nucl.Phys. A230(1974)477
4) A. Lindner: Nucl.Phys. A268(1976)369

POLARIZED PROTON-PROTON SCATTERING AT INTERMEDIATE ENERGIES

W. M. Kloet* and R. R. Silbar**

*Rutgers University, New Brunswick, New Jersey 08903/USA

**Theoretical Division, Los Alamos Scientific Laboratory
University of California, Los Alamos, New Mexico 87545/USA

The study of nucleon-nucleon scattering in the region below pion production has been quite extensive, both experimentally and theoretically. The interest for the region above pion-production threshold has been relatively dormant until the recent discovery of dramatic structure for specific spin-dependent cross sections as a function of energy.[1] As a result of these experiments even speculations about di-proton resonances have been put forward.[2]

We have studied some aspects of the proton-proton spin-dependent cross sections in a model which is capable of describing elastic nucleon-nucleon scattering and pion production in a unitary way. The model is a three-body model for a system of two nucleons and a pion where the interaction occurs via the formation of quasiparticles. We allow for pion-nucleon interaction in the P_{11} and P_{33} states because these are the most important pion-nucleon interactions. In the present version we do not include a quasiparticle interaction for the two nucleons. This means that from the two-body point of view the model includes only a one-pion exchange force between two nucleons. For a diagrammatic representation of the coupled equations see Ref. 4. The general ingredients of the model are described by Aaron, Amado, and Young[3] and Ref. 4. It suffices here to say that the analytic structure of the driving terms and the quasiparticle propagators is determined by the requirement of two- and three-body unitarity. In addition, the Born term in the elastic nucleon-nucleon scattering is constructed in such a way that in the static approximation its strength is equal to the one-pion exchange force.

The P_{33} pion-nucleon interaction is fitted to the experimental phase shifts. The P_{11} interaction contains as its main feature the nucleon pole. The phase shift starts out negative, but does not turn positive at higher energy as experimentally required. To remedy this we could include an additional quasiparticle interaction but have not done so. We do not expect to describe the details of elastic scattering nor pion production but the model serves as an instructive exercise and several dynamical features can be studied. The main ingredients of the model are that it describes elastic scattering and pion production in a unitary way; it takes account of all spin complications, and it is relativistic. The model uses as an input the pion-nucleon interaction, which can be related to experimental data, and therefore does not contain any free parameters.

Applying the model for proton-proton scattering in the region for laboratory energies of 400-1500 MeV, we are able to address the following questions. Is there a strong energy dependence of the polarized total cross sections? Is the energy dependent behavior due to elastic or inelastic scattering? Which partial waves are responsible for the strong energy dependence?

Experimentally the most prominent energy dependence has been observed in the total cross sections, where the spins of both beam and target were oriented along the beam direction. Therefore we have used the unitary model described above to calculate these specific cross sections. In doing so we look at each partial wave and distinguish its elastic and inelastic contribution.

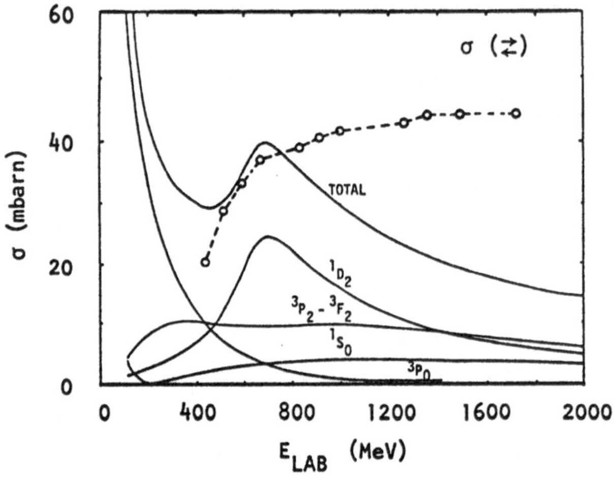

Fig. 1. Total p-p cross section for antiparallel longitudinal spins. Important partial waves are given separately. The line through the data points (Ref. 1) serves to guide the eye.

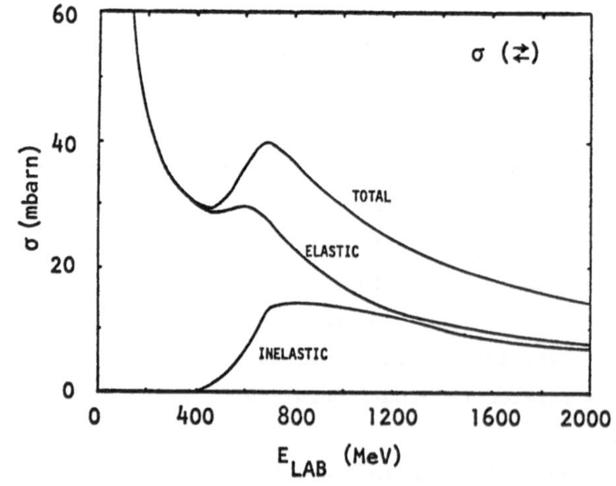

Fig. 2. Elastic and inelastic total p-p cross sections for antiparallel spins.

In figure 1 the total cross section is shown for the case where beam and target are polarized longitudinally and opposite to each other. In the model the main structure is due to the $^1D_2$ partial wave. This wave couples to an S-wave state in the nucleon-delta system. Both the elastic and inelastic parts of this partial wave show a strong energy dependence. It is interesting that the peak in the $^1D_2$ elastic scattering occurs at a lower energy (about 620 MeV) than the peak in the inelastic scattering (about 760 MeV). None of the other partial waves show any significant structure. In the $^3P_2 - ^3F_2$ wave the inelastic contribution is relatively small.

In the experimental data there is no definite indication for a similar overall structure. At best one can assume that the rapid rise between 450 and 650 MeV is due to a rapid increase of the $^1D_2$ amplitude. Above 650 MeV, however, the experimental cross section does not fall off as in the model but even shows a slight increase over a wide energy range. We conclude therefore that either the $^1D_2$ amplitude does not decrease as fast as in the model or that the other partial waves become much larger above 700 MeV. Figure 2 shows the total elastic and inelastic cross sections for this spin combination. The rapid rise between 450 and 650 MeV is due to a shoulder in the elastic cross section and a steep rise of the inelastic cross section.

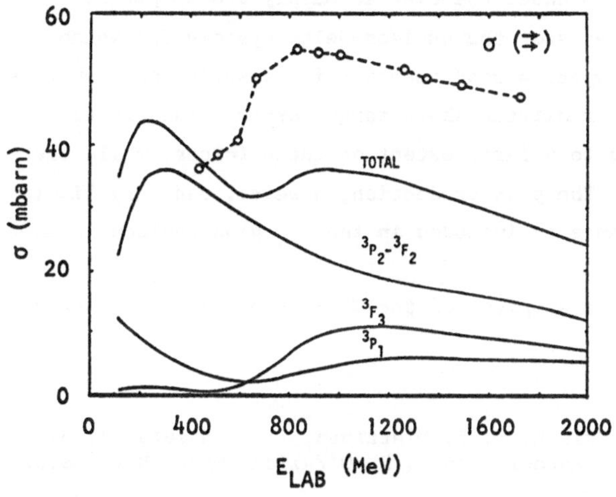

Fig. 3. Total p-p cross section for parallel longitudinal spins. Important partial waves are given separately. The line through the data points (Ref. 1) serves to guide the eye.

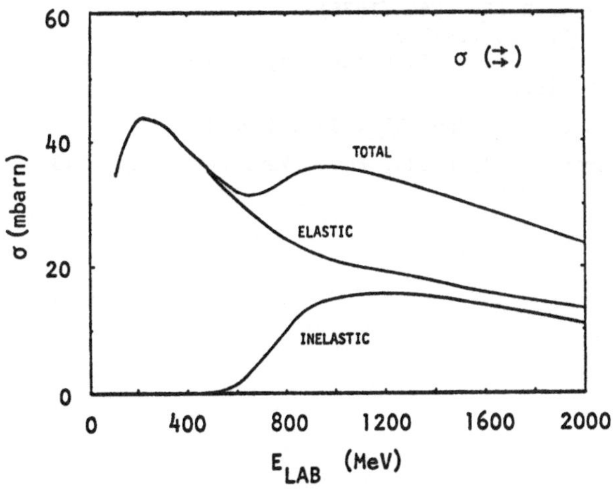

Fig. 4. Elastic and inelastic total p-p cross sections for parallel spins.

In figure 3 the total cross section is shown for the case where beam and target are polarized longitudinally and parallel. In the model the structure is due to a rise in the $^3F_3$ partial wave and to a lesser extent of the $^3P_1$. Both $^3F_3$ and $^3P_1$ partial waves couple to P-states in the nucleon-delta system. The $^3P_2$ - $^3F_2$ contribution shows a decrease over the region of 400-1500 MeV. As a result the rise in the total cross section between 650 and 950 MeV is rather modest (about 4 mbarn) as compared to the dramatic increase in the experimental cross section of at least 14 mbarn from 650 to 900 MeV. We conclude that in comparison to this model the $^3F_3$ and $^3P_1$ amplitudes should rise much faster and starting at a lower energy, or that the $^3P_2$ - $^3F_2$ should not decrease as fast. We note that the rise of the $^3F_3$ and $^3P_1$ contributions between 650 and 950 MeV accounts for about 9 mbarn which is still too small compared to the experimental rise. In this model the $^3F_3$ and $^3P_1$ waves are mainly inelastic above 650 MeV while the $^3P_2$ - $^3F_2$ amplitude is mainly elastic. In figure 4 we show the elastic and inelastic cross sections for the parallel spins. In this case the elastic contribution does not have a shoulder and the rise is only due to the inelastic cross section.

The unitary model used here gives qualitatively the rise in total cross section for both parallel and anti-parallel longitudinal spin orientation, which is also observed experimentally. In both cases the rise is mainly due to partial waves that couple to S- or P-wave states of the nucleon-delta systems and which are strongly inelastic. We do not expect a good quantitative description from this model because of the obvious lack of realistic short range forces. The elastic scattering cross section will depend to a large extent on these forces, while the pion production probably will not.[4] The pion production, however, can very likely also be modified if the Roper resonance is included in the P_{11} pion-nucleon interaction.

This work was performed under the auspices of the U. S. Department of Energy and the National Science Foundation.

---

1. W. de Boer, R. C. Fernow, A. D. Krisch, H. E. Miettinen, T. A. Mulera, J. B. Roberts, K. M. Terwilliger, L. G. Ratner, and J. R. O'Fallon, Phys. Rev. Lett. 34, 558 (1975);
A. Yokosawa, preprint ANL-HEP-PR-CP-47, Argonne (1977);
I. P. Auer, E. Colton, D. Hill, K. Nield, B. Sandler, H. Spinka, Y. Watanabe, A. Yokosawa, and A. Beretvas, Phys. Lett. 67B, 113 (1977);
I. P. Auer, A. Beretvas, E. Colton, D. Hill, K. Nield, H. Spinka, D. Underwood, Y. Watanabe, and A. Yokosawa, Phys. Lett. 70B, 475 (1977);
I. P. Auer, talk at Nucleon-Nucleon Symposium, Argonne, March 1978.

2. G. L. Kane, G. H. Thomas, Phys. Rev. D 13, 2944 (1976).
N. Hoshizaki, Prog. Theor. Phys. 58, 716 (1977);
H. Hidaka, A. Beretvas, K. Nield, H. Spinka, D. Underwood, Y. Watanabe, and A. Yokosawa, Phys. Lett. 70B, 479 (1977).

3. R. Aaron, R. D. Amado, J. E. Young, Phys. Rev. 174, 2022 (1968).

4. W. M. Kloet, R. R. Silbar, R. Aaron, R. D. Amado, Phys. Rev. Lett. 39, 1643 (1977).

POLARIZATION MEASUREMENT IN PROTON-PROTON SCATTERING AT 6.14 MeV*

G. Bittner, W. Kretschmer, H. Löh, W. Schuster, W. Stach
Tandemlabor der Universität Erlangen-Nürnberg, Erlangen, W. Germany

For a model independent determination of low energy proton-proton p-wave splitting which is caused by tensor and spin-orbit interactions, the analyzing power in p-p scattering was measured at 6.14 MeV. The very precise cross section measurements at[1] or near[2,3] this energy, which are partly contradictory, allow in principle only a determination of the $^1S_0$ phase shift and of the central p-wave combination Δ_C. The analyzing power in low energy p-p scattering depends primarily on the tensor p-wave combination Δ_T and the spin orbit p-wave combination Δ_{LS},[4] whereas the cross section is nearly insensitive to these magnitudes.[5]

At this low energy the nuclear scattering is mostly due to s-wave scattering and therefore the analyzing power is expected to be only a few tenth of a percent and peaked at extreme forward angles $\theta_{lab} \approx 10°$.[1] The measurement was performed by bombarding the supersonic H_2-gas jet of a windowless gas target[6] with the polarized proton beam of the Erlangen Lambshift source and observing the left-right asymmetry of the scattered protons in symmetrically arranged detectors. By the use of the windowless gas target the scattering center is very well defined and a scattering on entrance and exit foils of a closed gas cell is avoided. The H_2-gas jet enters the vacuum through a laval type nozzle (smallest diameter 0.6 mm) and after a region of free expansion (3 mm) it is caught by an appropriately formed tube, pumped away and going into a recycling system. The target thickness obtained for an entrance pressure of 20 bar was 10 $\mu g/cm^2$. The analyzing power was measured between $\theta_{lab} = 5°$ and 22.5° in steps of 2.5° with 8 detectors simultaneously. To avoid false asymmetries due to variations in position and direction of the incident beam, the polarization was switched on and off with a frequency of 100 Hz. This was achieved by the use of a weak magnetic field between the Sona coils of the Lambshift source, which has no effect on the beam position. The beam polarization was continuously monitored with a ^{4}He-polarimeter behind the Faraday cup.

The preliminary results are shown in Figure 1 together with theoretical curves. The curves are calculated with three different sets of $^1S_0$ and $^3P_{0,1,2}$ phase shifts, which describe the cross section measurement of Hegland et al.[2] at 5.96 MeV equally good with the same χ^2. The physical interpretation of these different p-wave sets is very different, which easily can be seen by a comparison of sign and relative magnitude of the corresponding phase shift combinations Δ_T and Δ_{LS}. The dashed curve ($\Delta_T = 0.27°$, $\Delta_{LS} = -1.11°$) corresponds to a repulsive, the dashed-dotted curve ($\Delta_T = 0.53°$, $\Delta_{LS} = 1.31°$) to an attractive dominant spin orbit interaction, the tensor interaction in both cases being attractive. The solid line, describing as well the cross section as our analyzing power data, corresponds to a dominating repulsive tensor interaction ($\Delta_T = -0.41°$) and a small attractive spin orbit interaction ($\Delta_{LS} = 0.14°$) which is consistent with the one pion exchange (OPE) model ($\Delta_T = -0.68°$, $\Delta_{LS} = 0.14°$). From the best fit (solid line) we also obtained the central p-wave parameter $\Delta_C = -0.02°$, which was corrected due to the effect of vacuum polarization according to Sher et al.[5] For all calculations the D-phase shift, obtained by the OPE-model, was used. A comparison of our results with multi-energy analyses supports the 0-50 MeV analysis of Sher et al.[5] which exclude all Berkeley and Wisconsin data, and also supports the 1-27.6 MeV analysis of Arndt et al.[7] where the normalization of the cross section data below 10 MeV was allowed to be floated. It should be mentioned that this experiment is still in progress and that further measurements are planned to increase the statistical accuracy of the data.

The authors wish to thank Prof. W. Watari (Kyoto University) for furnishing the computer code for the phase shift analysis.

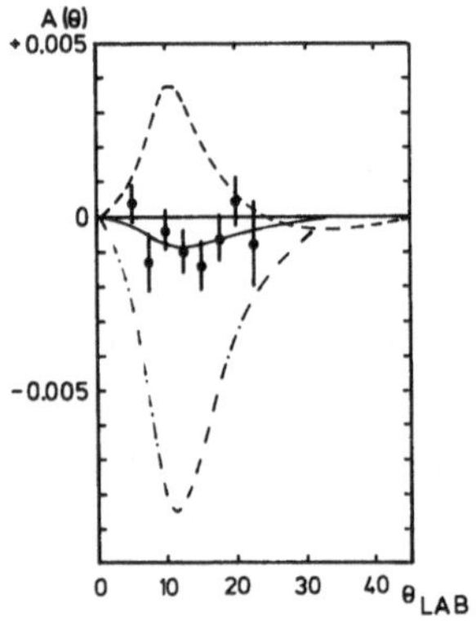

Figure 1. Angular distribution of p-p analyzing power at 6.14 MeV with theoretical curves (see text).

References

*Work supported by the Deutsche Forschungsgemeinschaft.

[1] R. J. Slobodrian, H. E. Conzett, E. Schield, and W. F. Tivol, Phys. Rev. 174 (1968) 1122

[2] P. M. Hegland, R. E. Brown, J.S. Lilley, and J. A. Koepke, Phys. Rev. Lett. 39 (1977) 9, and P. M. Hegland, Ph.D. thesis 1976.

[3] K. Imai, K. Nishimura, N. Tamura, and H. Sato, Nucl. Phys. A246 (1975) 76.

[4] J. D. Hutton, W. Haeberli, L. D. Knutson, and P. Signell, Phys. Rev. Lett. 35 (1975) 429.

[5] M. S. Sher, P. Signell, and L. Heller, Ann. Phys. 58 (1970) 1.

[6] G. Bittner, Ph.D. thesis, Erlangen 1978, and W. Schuster, Zulassungsarbeit, Erlangen 1977.

[7] R. A. Arndt, R. H. Hackmann, and L. D. Roper, Phys. Rev. C9 (1974) 555.

MEASUREMENT OF SPIN INTERACTIONS IN HIGH P_t PROTON-PROTON ELASTIC SCATTERING*

L. G. Ratner and P. F. Schultz
Argonne National Laboratory
Argonne, IL 60439

and

J. R. O'Fallon
Argonne Universities Association
Argonne, IL 60439

and

K. Abe, D. G. Crabb, R. C. Fernow, P. H. Hansen, A. D. Krisch, T. A. Mulera, A. J. Salthouse, B. Sandler, and K. M. Terwilliger
Randall Laboratory of Physics, The University of Michigan
Ann Arbor, MI 48109

and

A. Lin
Abadan Institute of Technology
Iran

I. Introduction

Since the first operation of the polarized proton beam at the Argonne National Laboratory Zero Gradient Synchrotron (ZGS), it has become increasingly clear that there is a strong spin dependence in high energy strong interactions. During the last year, we have been measuring the spin correlation parameter A_{nn} and the analyzing power at high transverse momentum and we find that in particular A_{nn} exhibits strikingly large effects. The high intensity polarized beam and a polarized target provide a very powerful tool for precise investigation of both the spin-orbit forces parameterized by A and the spin-spin forces parameterized by A_{nn}. At the normal beam momentum of 11.75 GeV/c, the extracted beam of 2 x 10^{10} protons with 70% polarization allows us to measure the pure spin state proton-proton interaction down to a level of $\lesssim$ 0.1 Fermi. This indeed probes very deeply into the inner structure of the nucleon.

In the next section we will give a short description of the Argonne Polarized Beam Facility and the experiments that are currently taking data. In Section III we will describe our current experimental set-up and in Section IV we will review our results to date.

II. Argonne Polarized Beam Facility

The polarized proton beam is produced in an atomic beam type source which is commercially available from ANAC in New Zealand. Operating in a pulsed mode, it produces 80 μA of 20 keV protons with a polarization of 75% ± 5%. This beam is then accelerated to 750 keV in a Cockcroft-Walton and injected into the linac which

* Work supported by the U.S. Department of Energy.

accelerates it to 50 MeV. A 500 μsec pulse length is then injected into the ZGS for further acceleration. To date beam has been extracted from the ZGS at energies ranging from about 400 MeV to about 11 GeV or 11.75 GeV/c.

Until its closing in October 1979, the ZGS will be used only for the acceleration of polarized protons. The current experiments, which are now running simultaneously in 6 different beam lines, cover the following topics: A_{nn} at large P^2, P↑P↑ Inclusive, Parity Violation in P- nucleus scattering, $\Delta\sigma_L$, A_{LL}, A_{LS}, A_{SS}, $\Delta\sigma_T$, A_{nn}, D_{nn}, K_{nn}, P↑P→ PππP. In August the ZGS will accelerate polarized deuterons at low energy and in late September or October it will accelerate deuterons to high energy.

III. High P_t Experimental Set-up

Figure 1 shows the layout of our experiment. The polarized beam from the ZGS first passes through a liquid hydrogen target and the left-right asymmetry from the elastic scattering is measured by the four arm spectrometer. Left scattering for the forward particle in L1,2,3 and the left scattering recoil particle in L4,5,6. Right scattering is similarly measured in R1,2,3 and R4,5,6. The beam polarization is typically 70%. The beam then continues downstream to the polarized proton target (PPT). This is a close copy of a CERN target. The PPT is maintained at 0.5°K in a magnetic field of 25 kG. $C_2H_6O_2$ beads doped with $K_2Cr_2O_7$ are contained in a flask 4.13 cm long by 2.9 gm in diameter. The target polarization, P_T, has been as high as 85%, but the high polarized beam intensity causes radiation damage to the ethylene glycol beads which reduces the average P_T. The target is annealed about every 12 hours to reduce some of the radiation damage, but the beads usually have to be changed every two days.

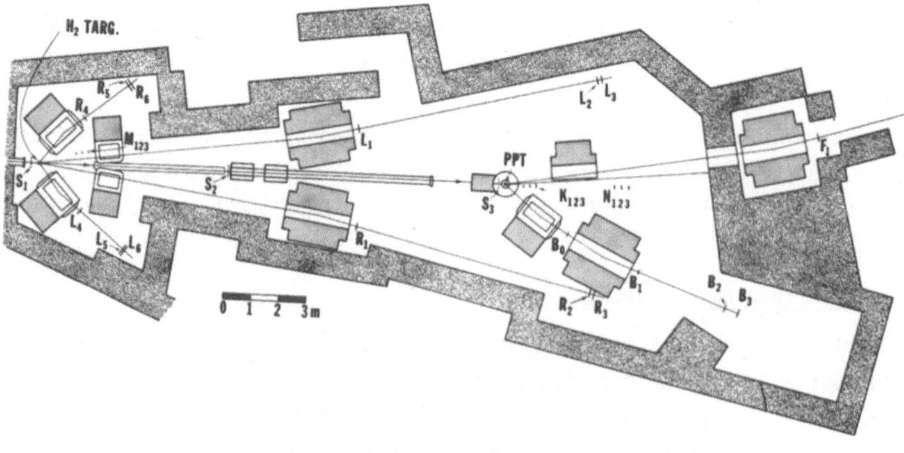

Fig. 1. Layout of High $P_\perp^2$ Experiment

The differential cross section for the elastic scattering of the polarized proton beam on the polarized target is then measured by the double arm spectrometer FB. The angle and momentum of both the forward scattered particle (F) and the backward recoil (B) is measured by the four magnets and 6 scintillation counters F123 and B123. By varying the currents in the four magnets and by occasionally reversing the PPT magnet, we can cover a large range of $P_{\perp}^2$ without moving the counters. The momentum bite defined by the F3 counter, is typically $\Delta P/_P = \pm 7\%$ and the defining solid angle $\Delta\Omega_{cm} \approx 10^{-3}$ sr. Other counters are overmatched to allow for various experimental difficulties such as: beam size, beam divergence, magnet variations, and multiple coulomb scattering. Inelastic and non-hydrogen event rates in the PPT were measured by replacing the $C_2 H_6 O_2$ beads by Teflon beads which contain no hydrogen. This correction was typically 10%. Accidentals were continuously monitored and found to be always less than 1%. The relative beam intensity I_o was monitored by the scintillation telescopes M,N, and K which were calibrated by aluminum foil irradiations. The size, position, and angle of the beam are continuously monitored by segmented wire ion chambers. The beam size at the PPT was about 20 mm FWHM and the beam movement less than 1 mm. To reduce systematic biases the beam polarization is reversed every pulse and the target polarization about every 8 hours.

We obtained the four normalized elastic event rates

$$N_{ij} = E(ij)/I_o(ij) \tag{1}$$

by simultaneously measuring the number of elastic events in the FB spectrometer, E(ij), and the number of incident protons $I_o(ij)$ in each of the four initial spin states (ij = beam, target = ↑↑, ↑↓, ↓↑ and ↓↓). The spin-spin correlation parameter, A_{nn}, was obtained from

$$A_{nn} = \frac{N_{\uparrow\uparrow} - N_{\uparrow\downarrow} - N_{\downarrow\uparrow} + N_{\downarrow\downarrow}}{P_B P_T \Sigma N_{ij}} \tag{2}$$

The analyzing power, A, was obtained by averaging over either the target or beam polarization

$$A_B = \frac{N_{\uparrow\uparrow} + N_{\uparrow\downarrow} - N_{\downarrow\uparrow} - N_{\downarrow\downarrow}}{P_B \Sigma N_{ij}}$$

$$A_T = \frac{N_{\uparrow\uparrow} - N_{\uparrow\downarrow} + N_{\downarrow\uparrow} - N_{\downarrow\downarrow}}{P_T \Sigma N_{ij}} \tag{3}$$

The equality of A_B and A_T, required by rotational invariance, gave a consistency check which held within errors, and we averaged A_B and A_T to obtain A. We obtained the four pure two-spin cross sections, $d\sigma/dt(ij)$, from the equations

$$d\sigma/dt(\uparrow\uparrow) = \langle d\sigma/dt\rangle \left[1 + 2A + A_{nn}\right]$$

$$d\sigma/dt(\downarrow\downarrow) = \langle d\sigma/dt\rangle \left[1 - 2A + A_{nn}\right] \tag{4}$$

$$d\sigma/dt(\uparrow\downarrow) = d\sigma/dt(\downarrow\uparrow) = \langle d\sigma/dt\rangle \left[1 - A_{nn}\right]$$

where $\langle d\sigma/dt\rangle$ is the measured spin average cross section.

IV. Results

Figure 2 shows the measurements of A and A_{nn} at 11.75 GeV/c as a function of P^2. We notice an increase in A_{nn} as we cross the region of the second break in the differential cross-section $P_\perp^2 = 3.5\ (GeV/c)^2$. We also see that A becomes relatively small at high $P_\perp^2$ after reaching a maximum at the first break in the differential cross-section at $P_\perp^2 = 1.5\ (GeV/c)^2$. In Figure 3 we show some preliminary results in our present run at the ZGS. Here we have plotted the ratio of the parallel to the anti-parallel cross-section and we have gone to $P_\perp^2 = 4.8\ (GeV/c)^2$. In this region A is essentially zero, but A_{nn} increases dramatically so that the parallel cross-section is almost six times larger than the anti-parallel. It appears that the spin-spin interaction A_{nn} at high $P_\perp^2$ and thus small impact parameter becomes very large while the spin-orbit parameter A is essentially zero. Large angle scattering appears to predominately occur in the triplet state. These measurements will be extended to $P_\perp^2 = 5.1$ at 90° c.m. shortly and we hope to be able to operate at still higher values at a momentum of 13 GeV/c before the ZGS is turned off.

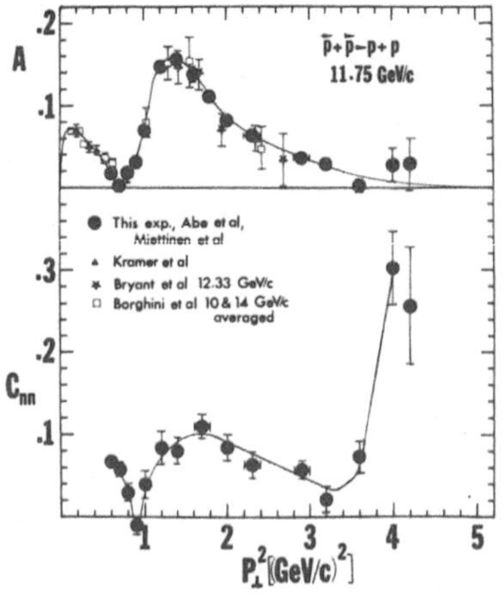

Fig.2. Analyzing power A and spin-spin correlation parameter C_{NN} (notation was changed to A_{NN} at Ann Arbor Workshop on Higher Energy Polariced Beams[5]) plotted against $P_\perp^2$ for p-p elastic scattering at 11.75 GeV/c.

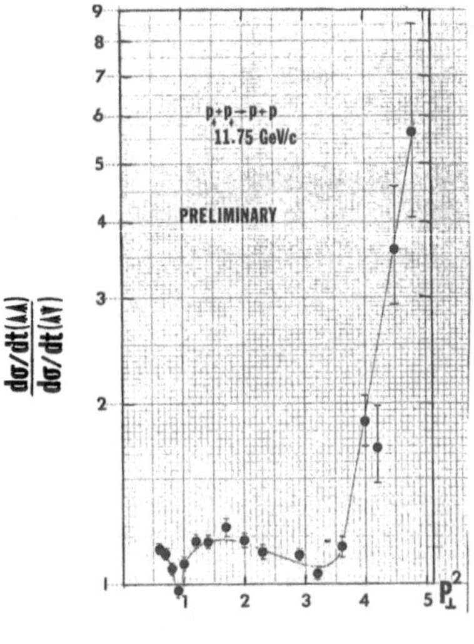

Fig.3. Ratio of parallel to antiparallel cross-sections as a function of $P_\perp^2$.

D,R,A and P FOR SMALL ANGLE p-p ELASTIC SCATTERING AT 312,392 AND 493 MeV

D. Besset, Q-H. Do, B. Favier, L.G. Greeniaus*, R. Hess, C. Lechanoine, D. Rapin, D.W. Werren**

DPNC, University of Geneva, Geneva, Switzerland

Ch. Weddigen

Kernforschungszentrum und Universität Karlsruhe, Institut für Experimentelle Kernphysik, Karlsruhe, Federal Republic of Germany.

Abstract

Results are given for the Wolfenstein parameters D,R,A and the polarization parameter P for elastic p-p scattering in the Coulomb-nuclear interference region between 3°cm and 33° cm. The experiment was performed at SIN using a 41.7% polarized beam.

1. Introduction

Measurements of complex polarization parameters in p-p scattering, unlike differential cross-sections which are sensitive only to the sum of the modulus squared of all the scattering amplitudes,provide information about the relative magnitudes and phases of individual amplitudes[1]. Between 300 and 600 MeV, only a few complex polarization measurements have been performed (see ref 2. for existing results). In particular for the D,R and A parameters, only 90 data points are available representing 8% of the overall existing p-p data. Almost none of these points however are in the Coulomb-nuclear interference region. Predictions at these small angles rely essentially on extrapolations made by means of phase shifts and indicate a strong energy dependence. In this report, 117 new data points are given for the D,R and A parameters and 39 for P. The statistical errors are typically $\simeq \pm$ 0.08 on D,R and A and $\lesssim \pm$ 0.02 on P. Data for D,R,A and P at 575 MeV, as well as for the parameter A_{oonn}[3] in the angular region $30^{o} \lesssim \theta_{CM} \lesssim 90^{o}$ at eight energies, will soon be available.

II. Experimental techniques

A detailed description of the apparatus and beam transport is given in ref 4. The 595 MeV proton beam is polarized through a scattering of 8° on a thin Be target ($|P_B|$ = 0.417 ± 0.004). Its energy can be lowered to 300 MeV by means of a Cu degrader without

* Present address : University of Alberta, Edmonton, Canada
** Present address : Landis & Gyr Ag, Zug, Switzerland

MEASUREMENT OF THE COMPONENT A_{yy} OF THE SPIN-CORRELATION TENSOR FOR p-p ELASTIC SCATTERING AT 10 MEV

K. Frank, H. Kuiper, H. Obermeyer and B. Seidler
Physikalisches Institut der Universität
D 8520 Erlangen, W. Germany

A 10 MeV polarized proton beam from the Erlangen tandem accelerator has been scattered by a polarized proton target especially designed for low energy experiments [1]. Scattered and recoil protons from pp elastic scattering events have been detected in coincidence for parallel as well as for antiparallel orientation of beam polarization and target polarization vectors. While the beam polarization was measured using a N_2-cooled ^4He polarimeter, the target polarization was measured by PIF [2], a method utilizing the internal magnetic field produced by the polarized protons in the target. In this way it has been possible to measure A_{yy}, one of the non-vanishing components of the spin-correlation tensor, at 10 MeV and at a center-of-mass angle $\theta_{cm} = 90°$. It is for the first time that A_{yy} for pp elastic scattering at low energy could be measured directly. Previous experiments performed by other authors [3] at higher energies (11.4 - 26 MeV) provided only for a direct measurement of the ratio A_{yy}/A_{xx}. In order to draw conclusions, e.g. about the relative strengths of the spin-orbit and tensor parts of the nucleon-nucleon force, it is necessary, in fact, to determine A_{yy} and A_{xx} separately. Since an experimental value of the above mentioned ratio is known, it is now sufficient to measure A_{yy} alone.

The Erlangen polarized target consists of a 0.1 mm thick LMN single crystal (doped with 1 % Nd) the protons of the water of hydration being dynamically polarized by the solid effect. Microwaves of a frequency of 70 GHz are used and the necessary magnetic field of about 1.8 T is produced by a superconducting magnet. The extremely good time stability of such a magnet system is a necessary requirement for the application of the PIF method for target polarization measurement

Despite its sensitivity against radiation damage LMN has been chosen as a polarized target material. Frozen butanol with its much more favourable radiation damage behaviour has never as yet been used for a polarized target for low energy experiments, obviously because of difficulties in preparation. We have studied the effect of radiation damage in LMN on the target polarization and its influence on the target polarization measurement by PIF. It turned out that it is possible to appropriately take these effects into account.

Our determination of A_{yy} includes the elimination of a number of systematic errors (e.g. time fluctuations of beam current and beam polarization), which are to be discussed in detail. Experimental data and numerical results of our measurements, which are still in progress, will be presented.

1) M. Eisend et al., Nucl. Instr. Meth. 140, (1977), 227
2) K. Kutschera et al., Nucl. Phys. A183, (1972), 593 and 601
3) P. Catillon et al., Nucl. Phys. B2, (1967), 93

any decrease in polarization. The transverse component of the beam polarization can be rotated using a solenoid while a deflecting magnet is used to rotate horizontal polarization components into the beam direction (see fig 1.).

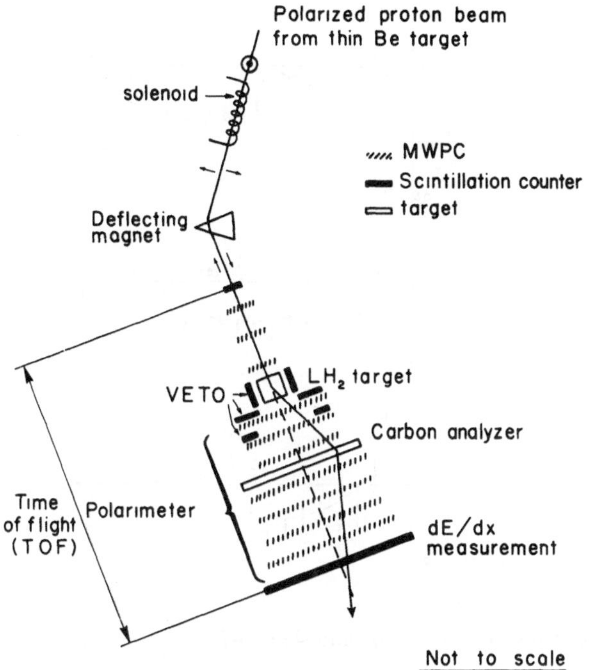

Figure 1 Experimental layout.

The p-p scattering occurs on a 14 cm long liquid hydrogen target, and the outgoing polarization components are analyzed by a second scattering on a 5 cm thick carbon analyzer (4 cm at 312 MeV). All particle trajectories are observed in multiwire proportional chambers (MWPC). A fast decision logic attached to the MWPC's selects events with scattering angles $\theta_H \gtrsim 1.5^o$ and $\theta_C \gtrsim 5^o$. Information on TOF and dE/dX, as well as VETO counters placed around the LH_2 target, allow a large fraction of the inelastic events ($pp \to \pi^+d$, $pp \to \pi^+np$, $pp \to \pi^opp$) to be tagged for off-line rejection. At all three energies, data were taken with a longitudinally and with a transversally polarized beam. Data using an empty target were also recorded in order to subtract the background due to target vessel. In order to calibrate the Carbon analyzing power and the beam polarization, data with single scattering on Carbon were recorded with the transversally polarized beam.

III. Data analysis and results

A detailed description of the data analysis is given in ref 5. The asymmetries of scattering from the Carbon analyzer are described by functions which depend on the parameters D,R,A and P, and on the azimuthal angle in the p-p scattering Φ_H. For one single energy and for each scattering angle θ_H, these functions are fitted by a leastsquare fit method to the measured asymmetries, thus providing the values of D,R, A and P. The beam polarization $|P_B|$ and the Carbon analyzing power P_C[6] are obtained after combining data with single scattering on Carbon with the asymmetry observed after the LH_2 scattering.

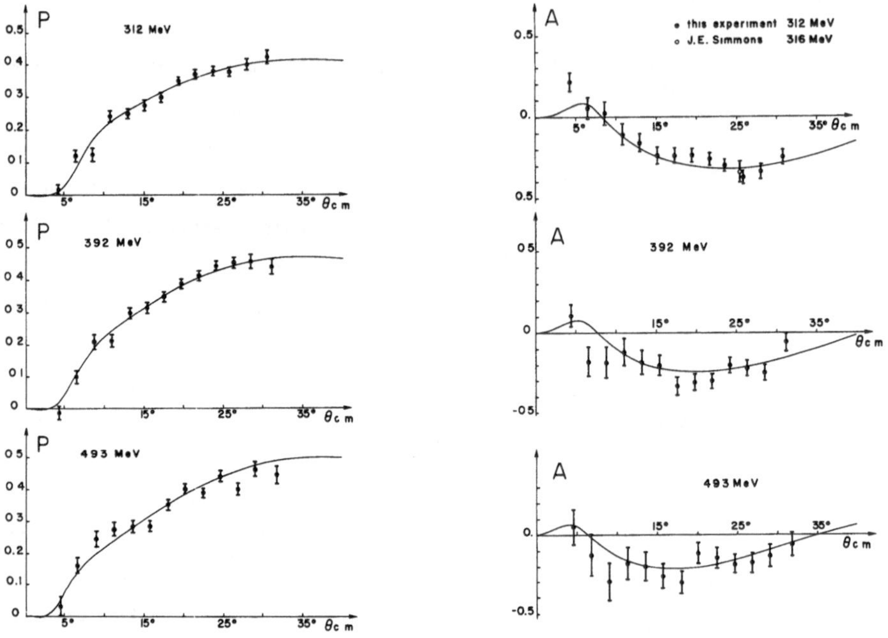

Figure 2 P and A parameters at 312, 392 and 493 MeV.

The results for P and A are shown in fig 2., and in fig 3. for D and R as a function of the c.m. scattering angle. The error bars represent statistical fluctuations only. As expected the values for the parameters D and R increase significantly with energy. For the D,R and A parameters our data are also compared with all other available data[2]. For the parameter P a good agreement with our previous CERN results[7] is observed. The curves shown correspond to predictions of the Saclay phase shift analysis[8] which include these present data.
The data have been corrected for binning effects, angular resolution and geometrical misalignments of the system. These corrections are smaller than the statistical error bars, typically < 3% on D and R and < 0.5% on P and A.
The effects of systematic errors are as follows. The normalization error due to the Carbon calibration is less than 1%. Contamination due to the three body inelastic reactions is small: results obtained with different cuts on the inelastic events give identical results. Most of the systematic errors due to asymmetries in the detectors are eliminated by rotating the spin. Moreover a good consistency has been observed

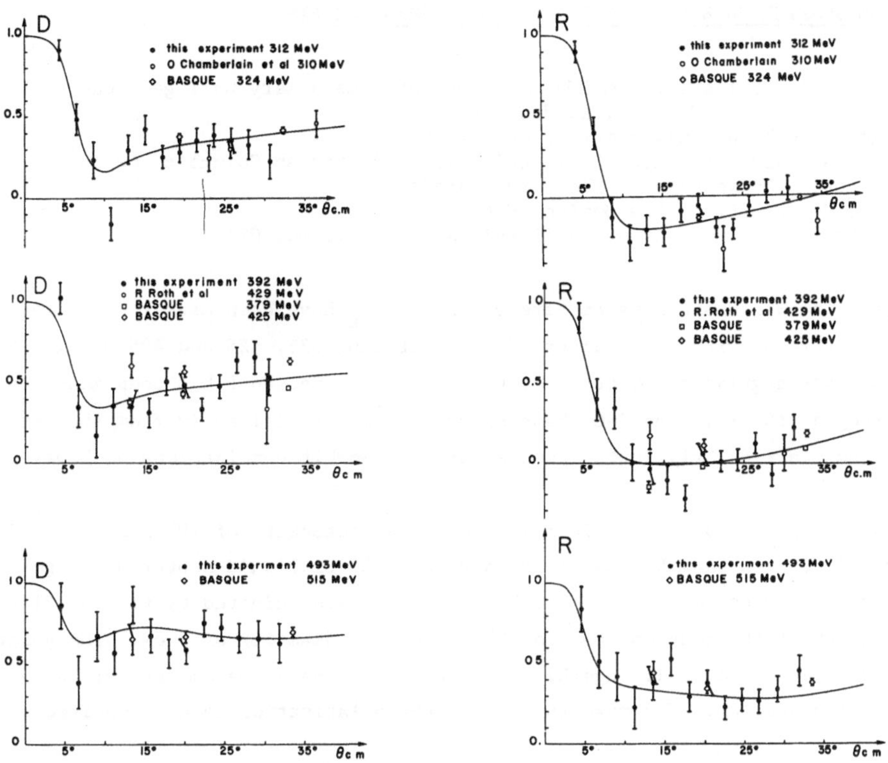

Figure 3 D and R parameters at 312 , 392 and 493 MeV.

between data taken with different beam polarization directions, indicating that systematic errors are less than the statistical ones.

We would like to thank the following organizations for having helped us at different stages of the experiment: Schweizerisches Institut für Nuklearforschung (SIN), the Swiss National Science Foundation, CERN, the University of Neuchâtel, the computer centers of Geneva Hospital and of the Federal Institute of technology of Lausanne (EPFL).

1. D. Besset et al. Nucl. Instr. & Meth. 148, 1978, 129.
2. J. Bystricky et al. ZAED Physiksdaten-physicsdata (1978).
3. D. Besset University of Geneva thesis (1978)(to be published)
4. D. Besset et al. Nucleon-Nucleon conference, Vancouver, June 1977, to appear in the American Institute of Physics, Conference Proceeding series.
   D. Besset et al Journées d'Etudes de Saturne II, Aussois, October 1977 (to be published).
5. B. Favier University of Geneva thesis 1978 (to be published).
   D. Rapin University of Geneva thesis 1978 (to be published).
6. D. Besset et al (to be published)
7. D. Aebischer et al. Nucl. Phys. A 276, 1977, 445.
8. J. Bystricky et al. Nucl. Phys. B 285, 1977, 469.

NP TRIPLE SCATTERING EXPERIMENTS AND I=0 NN PHASE SHIFTS

D.V. Bugg, J.A. Edgington, W.R. Gibson, N. Wright, Queen Mary College, London, UK
N.M. Stewart, Bedford College, London, UK
A. Clough, D. Gibson, University of Surrey, UK
D.A. Axen, G.A. Ludgate, C.J. Oram, University of British Columbia, Canada
L.P. Robertson, University of Victoria, Canada
C. Amsler, University of New Mexico, USA
J.R. Richardson, University of California at Los Angeles, USA

Abstract: The Wolfenstein parameters D_t, R_t and A_t have been measured in free np elastic scattering with an accuracy of ±0.05 at 220, 325, 425 and 495 MeV in the centre of mass angular range 60° to 160°. The polarisation P has been measured with an accuracy of ±0.015. The I=0 phase shifts differ significantly from theoretical predictions, particularly in the central and spin-orbit combinations of D waves.

The polarised proton beam at TRIUMF is used at an intensity of 100 namp to produce a polarised neutron beam by charge exchange on a 20 cm liquid deuterium target. Monoenergetic neutrons emerging at a lab angle of 9° are selected by time of flight. The intensity of the neutron beam is 10^6/sec over a diameter of 9 cm. The polarisation is given in Table 1, together with values of the R_t parameter for charge exchange from deuterium. Two magnets precess the polarisation to any required orientation.

Proton Energy (MeV)	Neutron Energy (MeV)	Neutron Polarisation (%)	R_t (9° Lab)
237	220	64	-0.81 ± 0.04
343	325	56	-0.75 ± 0.04
443	425	56	-0.78 ± 0.04
578	495	49	-0.69 ± 0.06

Table 1: Neutron polarisation as a function of energy and corresponding values of the R_t parameter for pd → n at 9° lab.

The neutron beam is scattered from a liquid hydrogen target 55 cm long. Scattered neutrons are detected in an array of scintillators 100 x 100 x 30 cm^3 with a positional accuracy of ±3.5 cm horizontally x ± 7.5 cm vertically, and a threshold of 15 MeV. Protons are detected and scattered in a polarimeter consisting of a 53 x 53 x 6 cm^3 block of carbon and an array of 12 multi-wire proportional chambers. This polarimeter has a useful energy range of 110 to 500 MeV and an average analysing power of about 35%.

The relative polarisation P(θ) as a function of angle is determined from the left-right asymmetry using the polarised neutron beam. The absolute normalisation is obtained by comparison with the data of Cheng et al.[1]. The polarisation is also determined absolutely with lower statistical accuracy from the polarisation of the recoil proton using unpolarised incident beam. The two normalisations agree.

The Wolfenstein parameters D_t, R_t and A_t are measured with a precision of ±0.05 at 10° centre of mass steps within the range determined by the kinematic constraints on the neutron counter and polarimeter.

Preliminary values of the I=0 phase shifts are shown in Table 2. This analysis uses our results for P and D_t together with earlier P and $d\sigma/d\Omega$ data. The I=1 phase shifts are taken from our previous analysis[2] of pp data, and g^2 is fixed at 14.25. At 425 and 495 MeV, the χ^2 contribution from earlier data is high, probably indicating systematic errors. A more complete analysis including our R_t and A_t data will be presented at the conference.

Bryan[3] has emphasized that I=0 D waves probe the medium-range forces where one boson exchange (other than π) should dominate. A comparison is made in Fig. 1 of the central, spin-orbit and tensor combinations of D waves with theoretical prediction of Vinh Mau et al.[4]. There is a conspicuous discrepancy in the central and spin-orbit combinations, and a smaller one in the tensor combination.

Lab Energy (MeV)	210	325	425	495
3S1	17.0 ± 1.2	- 2.4 ± 1.1	- 4.3 ± 1.6	(-12.0)
$\bar{\epsilon}1$	5.3 ± 0.5	8.4 ± 0.6	8.0 ± 0.7	11.8 ± 1.7
3D1	-18.6 ± 1.1	-27.4 ± 0.4	-26.0 ± 0.5	-27.8 ± 3.3
1P1	-24.5 ± 2.2	-27.4 ± 0.8	-38.1 ± 2.3	-43.6 ± 1.8
3D2	26.9 ± 2.0	23.9 ± 0.8	24.8 ± 1.1	23.3 ± 1.5
3D3	4.0 ± 0.6	1.9 ± 0.5	5.7 ± 0.7	4.0 ± 1.1
$\bar{\epsilon}3$	6.0 ± 0.4	7.2 ± 0.4	7.3 ± 0.5	8.5 ± 0.9
3G3	- 2.6 ± 0.4	(-4.74)	- 5.1 ± 0.6	(-6.00)
1F3	- 2.9 ± 0.9	- 7.0 ± 0.4	- 4.9 ± 0.4	-10.0 ± 1.3
3G4	5.3 ± 0.9	(-7.90)	8.2 ± 1.0	(11.05)
3G5	0.6 ± 0.4	(-0.45)	- 0.5 ± 0.5	(0.10)
χ^2	66.8	255.2	198.3	282.4
Degrees of Freedom	74	258	106	190

Table 2: I=0 phase shifts from an analysis including our P and D_t data. Values of G waves in parentheses are taken from OPE plus heavy boson exchange contributions calculated by Vinh Mau et al. At 495 MeV, 3S1 is fixed to obtain a stable solution.

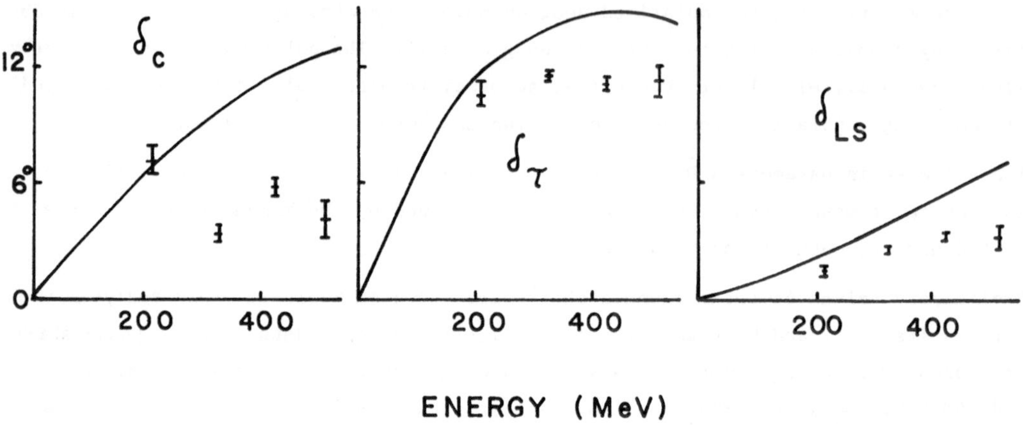

Figure 1: Central, Tensor and spin-orbit combinations of I=0 D Wave phase shifts, compared with the theoretical predictions of Vinh Mau et al.

REFERENCES

1. D. Cheng et al., Phys. Rev. 163 (1963) 1470.
2. D.V. Bugg et al., Journal of Physics G, to be published (1978).
3. R. Bryan, preprint ORO-5223-10.
4. R. Vinh Mau et al., Phys. Lett. 44B (1973) 1.

STATUS OF n-p MEASUREMENTS AND PHASE SHIFT ANALYSES NEAR 50 MeV

F. Paul Brady*

Crocker Nuclear Laboratory and Department of Physics

University of California Davis, CA 95616

ABSTRACT

Recent measurements of the n-p observables near 50 MeV and their impact on phase shift parameters, $\bar{\delta}(^1P_1)$ and $\bar{\varepsilon}_1$, are discussed.

INTRODUCTION

Several years ago, phase shift analyses[1,2] of n-p scattering data near 50 MeV revealed that ambiguous and/or anomalous values were obtained for two important phase parameters. The $^3S_1$ - $^3D_1$ mixing parameter $\bar{\varepsilon}_1$ was undetermined in the range -10° to 3°, and the $^1P_1$ phase shift disagreed both with values expected from models and with any smooth interpolation of $\bar{\delta}(^1P_1)$ values from adjacent energies.[2]

In order to help clarify the situation, we measured the n-p differential cross section at 50.0 MeV.[3] About the same time 60.9 MeV differential cross section data[4] from ORNL became available. Subsequently a phase shift analysis[5] showed that the $^1P_1$ phase depended strongly on which set of differential cross section data were used. Including only the Harwell data[6] resulted in $\bar{\delta}(^1P_1) = 0.3 \pm 1.3^\circ$ while the ORNL data alone gave a value of $-15.7 \pm 3.6^\circ$. The Davis data[3] alone produced $-7.0 \pm 1.8^\circ$, and including all data gave $-4.1 \pm 1.0^\circ$. In these analyses $\bar{\varepsilon}_1$ was fixed at 2.78° because there were insufficient n-p data to provide a fix on $\bar{\varepsilon}_1$.

Analysis[2] showed that $\bar{\varepsilon}_1$ was sensitive to the spin correlation parameters. Thus in order to help fix $\bar{\varepsilon}_1$ we made measurements of $A_{yy}(\theta)$ at 50 MeV and a phase shift analysis was carried out including these measurements.[7] This phase shift fit to p-p and n-p data near 50 MeV led to a value of $\delta(^1P_1)$ near -7° and produced a well-defined minimum in χ^2 vs $\bar{\varepsilon}_1$ and the value, $\bar{\varepsilon}_1 = 0.3 \pm 1.7^\circ$. This value for $\bar{\varepsilon}_1$ is well below that predicted by various models (2°-3°). In addition, the best-fit value for $\bar{\varepsilon}_1$ was found to depend rather strongly on the 60.9 MeV data.[8] Removal of these data from the n-p data set caused $\bar{\varepsilon}_1$ to increase by about 1.5°.

Another analysis,[9] which used different higher phase parameter restraints, obtained $\bar{\varepsilon}_1 = -0.92^\circ$ and $\bar{\delta}(^1P_1) = -4.2^\circ$ as "most reasonable" solutions. However the authors (ref. 9) noted that in their analysis a rather complex χ^2 hypersurface existed at 50 MeV which invalidated conventional error analysis.

The unsatisfactory situation noted with regard to $\bar{\varepsilon}_1$ and $\bar{\delta}(^1P_1)$ indicated the need for additional measurements of n-p observables near 50 MeV. Thus recently the n-p differential cross section at 63.1 MeV was measured[10] and additional measurements of A_{yy} at 50 MeV were carried out[11] in order to provide improved data to which both $\bar{\varepsilon}_1$ and $\bar{\delta}(^1P_1)$ are sensitive. These recent measurements,[10,11] and their impact on $\bar{\delta}(^1P_1)$ and $\bar{\varepsilon}_1$ are discussed here. ($A_y(\theta)$ measurements[12] have also been made but they have little effect on these parameters.)

EXPERIMENTAL RESULTS AND PHASE SHIFT ANALYSES

The recent 63.1 MeV cross section measurements[10] were carried out in the unpolarized neutron beam line at Crocker Nuclear Laboratory at U.C. Davis. The experimental facility and beam production techniques were similar to those which have been described earlier.[3,13,14] A neutron beam was obtained from the ^{7}Li (p,n)^{7}Be reaction and collimated at 0° to 24 mm high by 12 mm wide with an intensity of $\simeq 10^6$ n/sec in the full energy peak at 63.1 MeV. The neutron beam was incident on a CH_2 or H_2 gas target in a scattering chamber $\simeq$ 3.5 m from the ^{7}Li target. Recoil protons were detected in ΔE-E telescopes using silicon ΔE and NaI E dectectors. Time-of-flight from the ^{7}Li target was used to select the neutron beam peak of $\simeq$ 1.9 MeV FWHM. A number of corrections were applied to the data, the largest being the effect of nuclear interactions[15] in the NaI.

The differential cross sections, obtained by normalizing to a fit to total cross sections based on earlier U.C. Davis data,[16] are plotted as the solid circles in Fig. 1. The crosses and open squares are the Harwell data[6] at a mean energy of 62.5 MeV, and the open triangles are data from ORNL.[4] When allowance is made for the energy dependence, these new Davis cross sections agree well with the earlier measurements[3] at 50.0 MeV which span $20 \leq \theta_{cm} \leq 173.^o$ The ORNL data rise more steeply from 90° on backwards than do the present data, while the forward angle Harwell data (as near 50 MeV) appear to be too large.

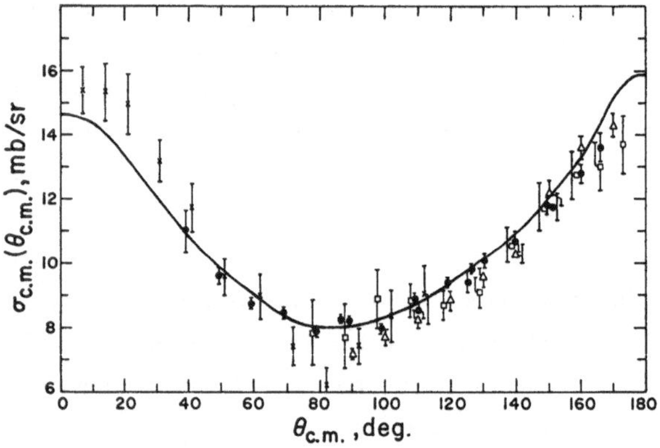

Fig. 1. Differential cross sections near 60 MeV.

The $A_{yy}(\theta)$ measurements were carried out in the polarized neutron line[17] at CNL where a 50.0 MeV beam of 0.476 ± .017 polarization[18] is produced by the reaction $d+T \rightarrow \vec{n}+{}^4He$ at 29.6^o lab. The polarized proton target consisted of a Nd-doped $La_2Mg_3(NO_3)_{12} \cdot 24H_2O$ crystal 2 mm thick and 25 mm diameter whose hydrogen is polarized dynamically. Recoil protons were detected in four ΔE-E NE102 plastic scintillator telescopes. To minimize the effect of systematic asymmetries, data were taken for all four possible combinations of neutron and proton spins. (See ref. 11 for details.)

The resulting values of $A_{yy} = A_{yy}^{(II)}$ are plotted <u>vs</u> θ_n(cm) in Fig. 2. The agreement with the earlier measurements,[7] $A_{yy}^{(I)}$, is good except that the new values at largest θ_n(cm) are consistently lower. It should be noted that the absolute normalization uncertainty, due to target and beam polarization uncertainty, is 7.8% for the more recent $A_{yy}^{(II)}$ results, while for $A_{yy}^{(I)}$ it is 26%. $A_y(\theta)$ values obtained from $A_{yy}^{(II)}$ data agreed well, within statistics, with earlier $A_y(\theta)$ measurements.[12]

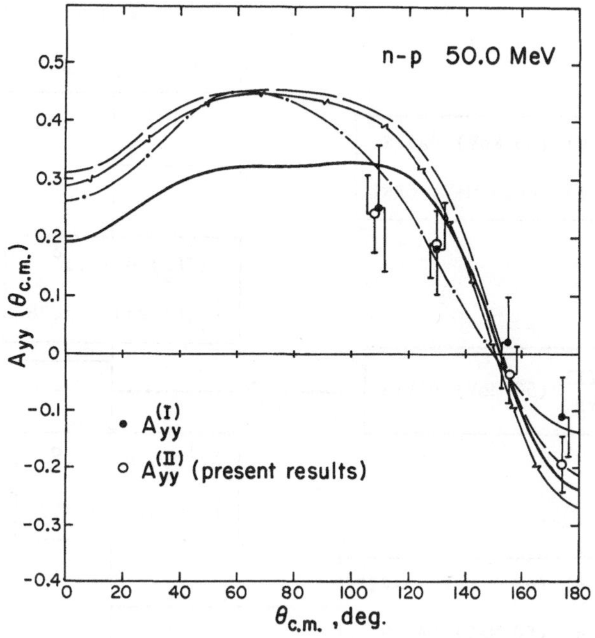

Fig. 2. A_{yy} measurements and predictions.

	Ref.	$\bar{\delta}(^1P_1)$	$\bar{\varepsilon}_1$
— — —	19	-8.8^o	2.8^o
—•—	20,21	-9.5^o	1.8^o
———	8	-6.5^o	2.9^o
—∧—	22	-10.2^o	2.4^o

The solid curve in Fig. 2 is the A_{yy} prediction of a phase shift analysis[8] incorporating essentially all new n-p data near 50 MeV and yielding $\bar{\varepsilon}_1 = 2.9^o$ and $\bar{\delta}(^1P_1) = -6.5^o$. The predictions of several model-dependent calculations are also shown. These are: (a) a one-boson-exchange-potential (OBEP) model of Bryan and Gersten[19] ("Model C": $\bar{\varepsilon}_1 = 2.78^o$, $\bar{\delta}(^1P_1) = -8.76^o$), shown as a dashed curve; (b) a separable potential model of Pauss and Zingl[20], using the potential parameters of Doleschall[21] ($\bar{\varepsilon}_1 = 1.8^o$, $\delta(^1P_1) = -9.5^o$), shown as a dot-dashed curve; and (c) a OBEP model, with asymptotic power law energy dependence of Ueda and Green[22] ("Model IV", $\bar{\varepsilon}_1 = 2.40^o$, $\bar{\delta}(^1P_1) = -10.24^o$), shown as a broken curve.

Figure 3 gives a rough idea of how the phase parameters $\bar{\delta}(^1P_1)$ and $\bar{\varepsilon}_1$ change as data from various measurements are added or removed from the n-p data set. As shown, the Harwell data were removed from the analyses[8] early. However, it turns out that when all recent n-p data are included the impact of the back angle data is small.

Fig. 3 shows that the addition of the Davis 63.1 MeV differential cross section data reduces the uncertainty in $\bar{\varepsilon}_1$ and $\delta(^1P_1)$ and also makes the value of $\bar{\delta}(^1P_1)$ less dependent on whether the ORNL data are included or not. The solid curve in Fig. 1 is a phase shift prediction based on essentially all available n-p data and having $\bar{\delta}(^1P_1) = -6.5^o$ and $\bar{\varepsilon}_1 = 2.9^o$. The prediction at back angles decreases a little ($\tilde{-}1\%$) if phase parameters $\bar{\delta}(^1P_1) = -6.4$ and $\bar{\varepsilon}_1 = 3.6^o$, obtained from a fit excluding the ORNL data, are used.

The inclusion of $A_{yy}^{(II)}$ data also has a stabilizing effect on $\bar{\varepsilon}_1$ and $\delta(^1P_1)$ (bottom of Fig. 3) with the net result that the dependence of both $\bar{\delta}(^1P_1)$ and $\bar{\varepsilon}_1$ on the ORNL data is reduced.

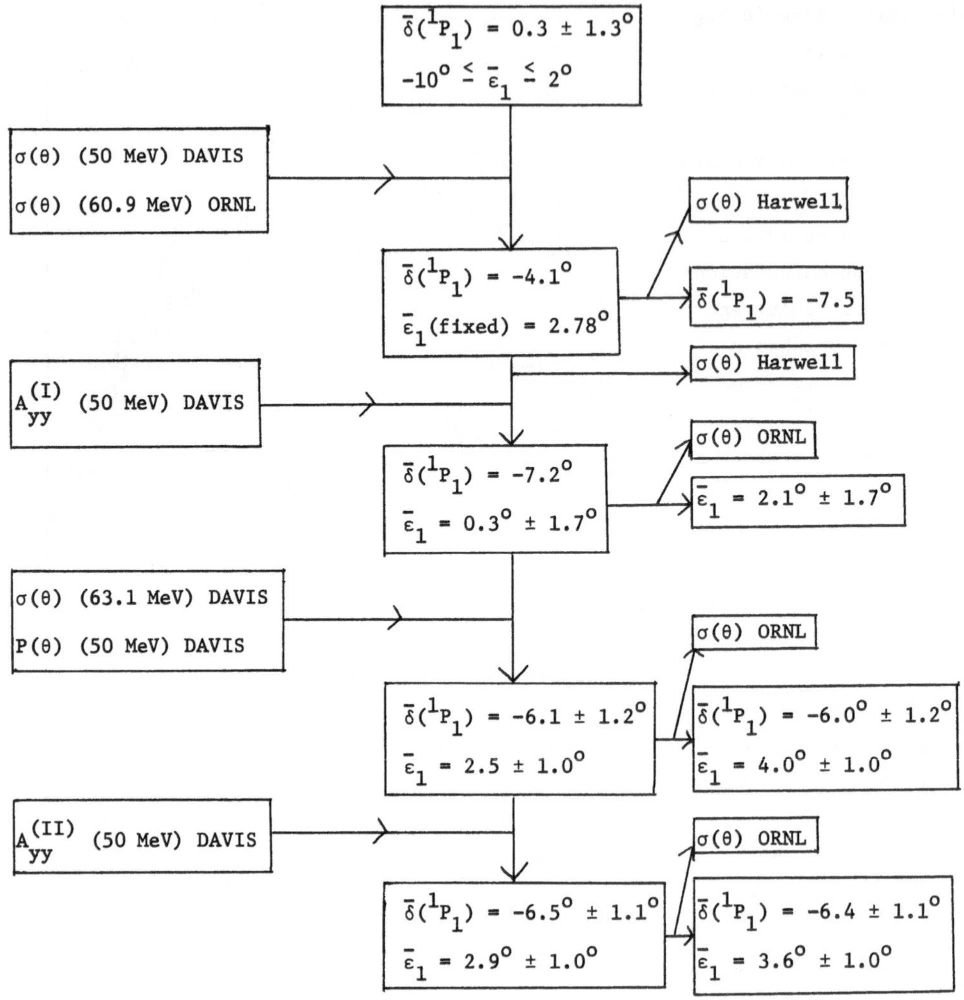

Fig. 3. Effect of measurements on $\bar{\delta}(^1P_1)$ and $\bar{\varepsilon}_1$.

One concludes that the ambiguities and uncertainties in $\delta(^1P_1)$ and $\bar{\varepsilon}_1$ which have persisted for some time[23] are considerably reduced. An independent measurement of good accuracy ($\sim$2%) of the differential cross section near 60 MeV could further reduce some of the remaining uncertainties.

We are grateful for the support of the National Science Foundation grants PHY 71-03499 and PHY 77-05301.

FOOTNOTES AND REFERENCES

* The experimental work described herein was carried out in collaboration with D.H. Fitzgerald, R. Garrett, S.W. Johnsen, N.S.P. King, M.W. McNaughton, J.D. Reber, J.L. Romero, T.S. Subramanian, J.L. Ullmann, and J.W. Watson. Tom Burt and Peter Signell carried out the phase-shift analyses.

1. R.A. Arndt, J. Binstock and R. Bryan, Phys. Rev. D8, 1397 (1973).
2. J. Binstock and R. Bryan, Phys. Rev. D9, 2528 (1974).
3. T.C. Montgomery, F.P. Brady, B.E. Bonner, W.B. Broste, and M.W. McNaughton, Phys. Rev. Letters 31, 640 (1973) and T.C. Montgomery, B.E. Bonner, F.P. Brady, W.B. Broste and M.W. McNaughton, Phys. Rev. C, (1977).
4. M.J. Saltmarsh, C.R. Bingham, M.L. Halbert, C.A. Lindemann, and A. van der Woude, Oak Ridge National Laboratory (unpublished). See Ref. 1 for table of this data.
5. R.A. Bryan and J. Binstock, Phys. Rev. D10, 72 (1974).
6. J.P. Scanlon, G.H. Stafford, J.J. Thresher, P.H. Bowen and A. Langsford, Nucl. Phys. 41, 401 (1963).
7. S.W. Johnsen, F.P. Brady, N.S.P. King, M.W. McNaughton, and Peter Signell, Phys. Rev. Lett. 38, 1123 (1977) and references mentioned therein.
8. P. Signell (details to be published).
9. R.A. Arndt, R.H. Hackman and L.D. Roper, Phys. Rev. C15, 1021 (1977), and earlier communications.
10. N.S.P. King, J.D. Reber, J.L. Romero, T.S. Subramanian, J.L. Ullmann, D.H. Fitzgerald and F.P. Brady, to be published.
11. D.H. Fitzgerald, S.W. Johnsen, F.P. Brady, R. Garrett, J.L. Romero, T.S. Subramanian, J.L. Ullmann, and J.W. Watson, to be published.
12. J.L. Romero, M.W. McNaughton, F.P. Brady, N.S.P. King, T.S. Subramanian and J.L. Ullmann, Phys. Rev. C17, 469 (1978).
13. J.A. Jungerman and F.P. Brady, Nucl. Instrum. and Meth. 89, 167 (1970).
14. F.P. Brady, N.S.P. King, M.W. McNaughton, J.F. Harrison and B.E. Bonner, NBS SP425 1, 103 (1975) Ed. R.A. Schrack and C.D. Bowman.
15. D.F. Measday and C. Richard-Serre, Nucl. Instrum. Meth. 76, 45 (1969). C.A. Goulding and J.G. Rogers, TRIUMPF rep TRL-753, 1975
16. F.P. Brady, W.J. Knox, J.A. Jungerman, M.R. McGie, and R.L. Walraven, Phys. Rev. Lett. 25, 1682 (1970).
17. A.L. Sagle, M.W. McNaughton, N.S.P. King, F.P. Brady and B.E. Bonner, Nucl. Instrum. and Methods 129, 345 (1974).
18. A.L. Sagle et al (to be published).
19. R.A. Bryan and A. Gersten, Phys. Rev. D6, 341 (1972).
20. F. Pauss and H.F.K. Zingl, Phys. Rev. C15, 1231 (1977).
21. Pauss and Zingl (Ref. 23) use the Doleschall potential parameters from Nucl. Phys. A220, 491 (1974), for the $^1P_1$ partial wave and the Doleschall "3T4" potential parameters (to be published, and U. of Graz report, 1976) for ε_1.
22. T. Ueda and A.E.S. Green, to be published.
23. See M. H. McGregor, R.A. Arndt, and R.M. Wright, Phys. Rev. 173, 1272 (1968).

THE PRODUCTION OF PIONS BY POLARIZED PROTONS INCIDENT ON HYDROGEN AND DEUTERIUM

G. Jones
Dept. of Physics, University of British Columbia
Vancouver, B.C., V6T 1W5, Canada

INTRODUCTION

The $p+p \rightleftarrows d+\pi^+$ reaction has long been regarded as the basic testing ground for exploring our understanding of the production and absorption of pions in nucleon systems. In 1955, Gell-Mann and Watson published their now-famous phenomenological analysis[1] of this reaction in the near-threshold region where the number of partial waves contributing to the reaction is small in number. Most of their considerations were confined to the energy region where only two partial waves (s- and p-wave pions) play a significant role. If the differential cross-section for the production of pions by a beam of polarized protons incident on hydrogen is written in the form:

$$\frac{d\sigma}{d\Omega} = \gamma_0 + \gamma_2\cos\theta + \gamma_4\cos^4\theta + \ldots + \vec{p}\cdot\vec{n}\,\sin\theta\,(\lambda_0 + \lambda_1\cos\theta + \lambda_2\cos^2\theta + \lambda_3\cos^3\theta + \ldots)$$

where $\vec{p}$ is the polarization of the beam of protons, and $\vec{n}$ is a unit vector in the direction of $\vec{k}_p \times \vec{k}_\pi$ then the restriction to only s- and p-wave pions results in only three non-zero coefficients, γ_0, γ_2 and λ_0. Conversely, the presence of a non-zero value for any of the other coefficients is evidence for the contribution of higher partial waves. In the years since 1955, many experimental measurements of the reaction have been performed, but very few of these at energies below 425 MeV have had enough precision to answer such questions. In fact, the extent of the "near threshold" region described above where only s- and p-wave pions contribute is still unresolved, although many authors have regarded the existing limited experimental evidence as indicating that most of the energy region below 425 MeV is consistent with such a "near threshold" interpretation. (For a proton energy of 425 MeV, the value of η in the expression for the cms pion momentum ($\eta m_\pi c$) is close to unity.) The measurements reported in this paper provide additional knowledge of the asymmetry in pion production using a polarized proton beam. Such measurements show a clear contribution from d-wave pi^ s for proton energies well below 425 MeV, even at 350 MeV. The deuteron is the simplest nuclear target that can be employed for studying the production of pions from nuclei. Differential cross-sections at a number of incident energies above 340 MeV have been reported[2,3]. Recent theoretical estimates based on impulse approximation[4] reproduces the data quite well in the forward hemisphere. Our measurements provide additional differential cross-section information as well as the asymmetries in pion production resulting from use of polarized beams of protons at 305 and 330 MeV.

EXPERIMENT

The experimental arrangement consisted of a 50 cm Browne-Buechner magnetic spectrograph mounted on beam-line 1 of the TRIUMF cyclotron. The spectrograph was instrumented with a 24 element scintillation counter hodoscope along the focal plane for determining pion momentum. For the measurements of pion production from hydrogen, both polyethylene (CH_2) and liquid hydrogen targets were used. The hydrogen gas system was simply replaced by deuterium for the measurements of pion production from deuterium. The angular range covered by this spectrograph was 35° to 145° (or alternatively the kinematic limit imposed by the pion detection threshold of 12 MeV). Both the polarization and the intensity of the proton beam were monitored continuously during the experiment by simultaneously measuring the p-p elastic scattering (at 26°) to the left and to the right from a 5 mg/cm^2 polyethylene target mounted a short distance downstream of the primary pion production target (typically, 150 mg/cm^2 of CH_2). The proton beam current was typically 1-2 nA with a polarization greater than 60%. The group responsible for these measurements consisted of: E. Auld, R. Feenstra, A. Haynes, R. Johnson, G. Jones, T. Masterson, E. Mathie, D. Ottewell, M. Sivertz, P. Walden, C. Winter and B. Tatischeff (Orsay).

RESULTS AND DISCUSSION: $p+p \rightarrow {}^2H+\pi^+$

A sample of the data obtained at a bombarding energy of 350 MeV is shown in Fig. 1. On the abscissae are shown points 2 MeV below the peaks in order to provide an indication of the resolution of the system. This figure also displays the analysing power for the pion production reaction as a function of pion energy. The analysing power, A_π is related to the asymmetry ε by: $A_\pi(\theta) = \frac{1}{P}\,\varepsilon(\theta)$ where P is the magnitude of the polarization of the proton beam, and the signs are consistent with those of the Madison convention[5].

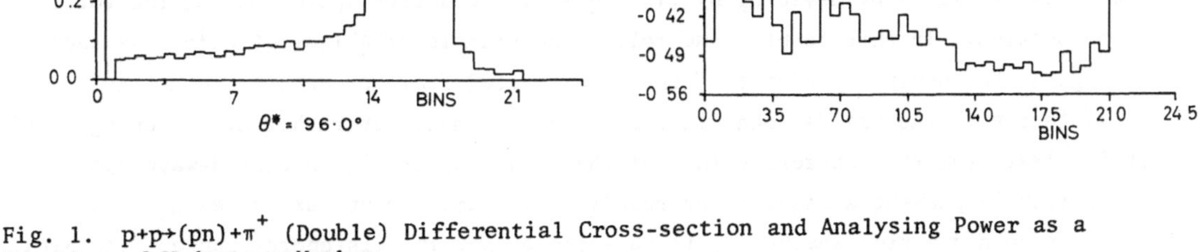

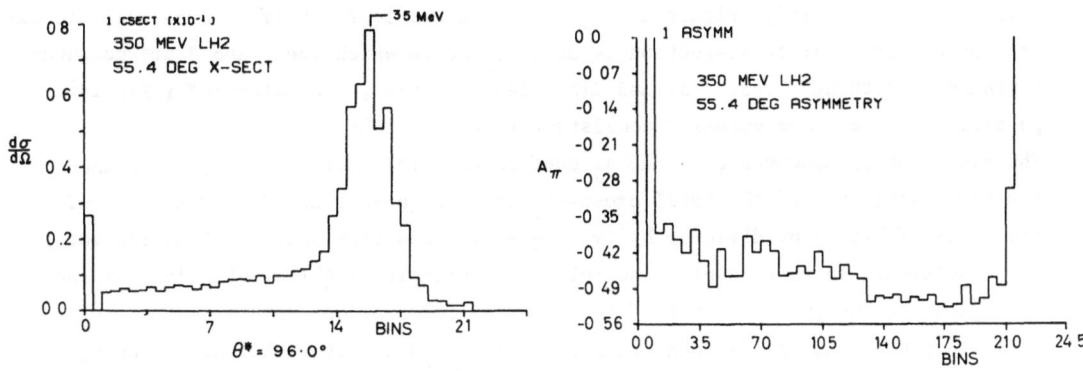

Fig. 1. $p+p \rightarrow (pn)+\pi^+$ (Double) Differential Cross-section and Analysing Power as a Function of Hodoscope Number.

The analysing power for the pions from the break-up reaction $p+p \to p+n+\pi^+$ is seen to be strikingly similar to that for the two-body reaction. Extraction of the λ_j coefficients from the analysing power data require knowledge of the unpolarized differential cross section ($\gamma_0 + \gamma_2 \cos^2\theta$ in this energy range). Because of the limited angular range accessible to our spectrograph (particularly at the lower proton energies), we fitted all the recent data[6,7,8,9] for the shape of the differential cross-sections to functions of the form: $\frac{\gamma_0}{\gamma_2} = X + (X + \frac{1}{3}) \frac{\alpha}{\beta} \frac{1}{\eta^2}$.

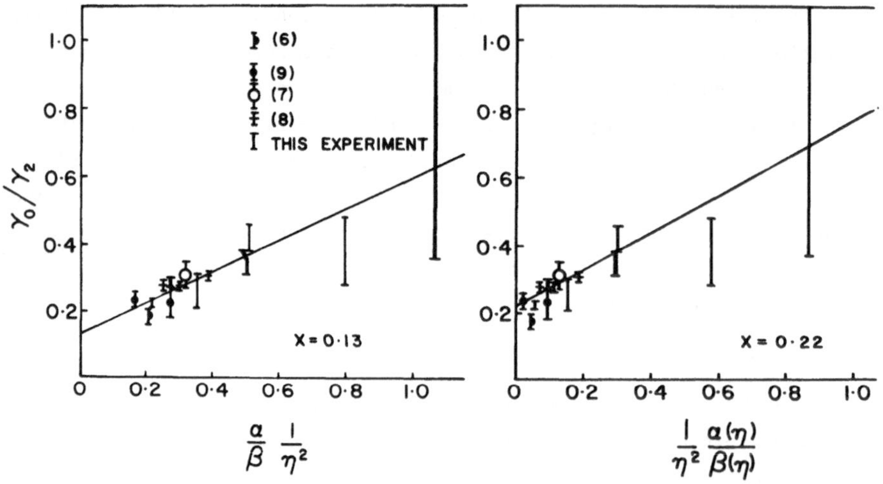

Fig. 2. $p+p \to {}^2H+\pi^+$ Isotropy Coefficient (γ_0/γ_2) as a Function of Pion Momentum (cms).

Figure 2(a) shows such a fit with α and β taken to be 0.188 and 0.90 respectively as determined from a momentum-dependent expansion of the total cross-section near threshold[10]: $\sigma_t = \alpha\eta + \beta\eta^3$. Figure 2(b) shows the equivalent fit if a more realistic expansion of the total cross-section is used[11], one in which the α and β are momentum dependent. Although Figs. 2(a) and 2(b) clearly differ, the values of γ_0/γ_2 interpolated either way are mutually consistent to within 10%.

The gross energy dependence of the λ_j coefficients (basically a (3,3) resonance behaviour like that of the total cross-section) has been removed for the sake of the plots of Fig. 3 by dividing by $(\gamma_0 + \frac{1}{3}\gamma_2)$, a quantity approximating the total cross-section. In this respect, we follow the example of Akimov et al[12]. As seen in Fig. 3 the energy dependence of the λ_j extracted from our data are completely consistent with the trends found in earlier measurements at rather higher energy[12,13]. It is clear from the non-zero values of the λ_1 and λ_2 of Fig. 3 that d-wave pion production is playing a significant role for pion cms momenta as low as $\eta = 0.5$. It is an open question whether λ_3 is non-zero or not in this energy range. The data do not extend over a large enough angular range to permit determining four unknown parameters. However, when the composite data (our data together with that of Dolnick)[6] at $\eta = 1$ was fitted in this way, a small negative value of λ_3 was obtained although with a statistical uncertainty too large to be conclusive: $\lambda_3(\gamma_0 + \frac{1}{3}\gamma_2)^{-1} = -.075 \pm 0.11$. The values for the coefficients of the odd powers of cosθ are strongly correlated. Thus the accuracy of the values for $\lambda_1(\eta)$ obtained by earlier workers[12,13] who rather arbitrarily assumed $\lambda_3 = 0$ in order to obtain reasonable statistical accuracy for the remaining three parameters, is open to question.

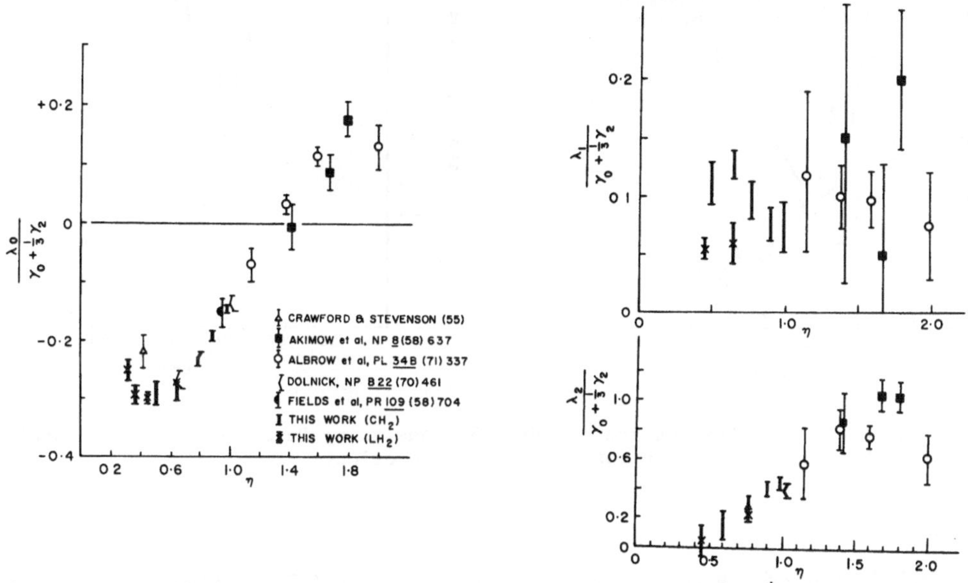

Fig. 3. λ Coefficients versus Pion Momentum (cms) for $p+p\to{}^2H+\pi^+$.

RESULTS AND DISCUSSION: $p+p\to{}^2H+\pi^+$

To date we have obtained data at two incident proton energies, 305 and 330 MeV. In Fig. 4 the differential cross-section is illustrated while the analysing power for the reaction is displayed in Fig. 5. The absolute value of the cross-section was obtained by normalizing to the known[10] cross section for the $p+p\to{}^2H+\pi^+$ reaction. The solid line in Fig. 4 is the theoretical prediction of Fearing[4] based on an impulse approximation. The theoretical prediction clearly reproduces the trend of the experimental data in the forward hemisphere, but deviates markedly at large angles. In addition, there is as yet no calculation of the analysing power expected on the basis of such a theory. The resemblance of the analysing power for this reaction to that characterizing the primary $p+p\to{}^2H+\pi^+$ is indeed striking and serves to reinforce the point of view that the $p+{}^2H\to{}^3H+\pi^+$ reaction is dominated by the elemental pion production reaction associated with the proton-proton interaction within the three-body system.

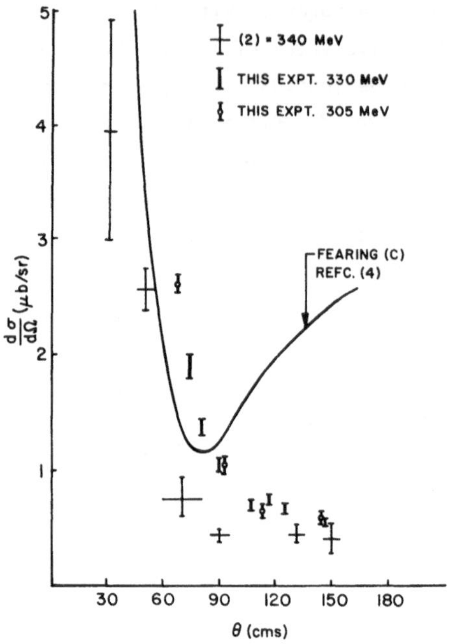

Fig. 4. Differential Cross-section for the $p+{}^2H \rightarrow {}^3H+\pi^+$ Reaction.

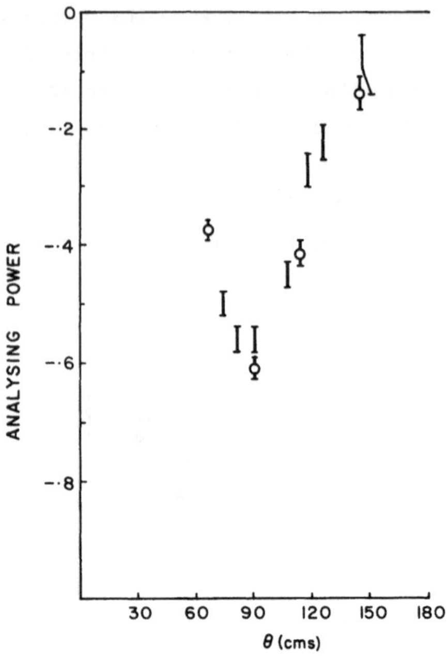

Fig. 5. Analysing Power for the $p+{}^2H \rightarrow {}^3H+\pi^+$ Reaction.

REFERENCES

1. M. Gell-Mann and K. Watson, Ann. Rev. of Nucl. Sci. 4 (1954) 219.
2. W.J. Frank, K.C. Bandtel, R. Madey and B.J. Moyer, Phys. Rev. 94 (1954) 1716.
3. K.R. Chapman, J.D. Jafar, G. Martelli, T.J. MacMahon, H.B. Van Der Raay, D.H. Reading, R. Rubinstein, K. Ruddick, D.G. Ryan, W. Galbraith and D. Sharp, Nucl. Phys. 57 (1964) 499.
4. H.W. Fearing, Phys. Rev. C11 (1975) 1210.
5. "Polarization Phenomena in Nuclear Reactions", Proc. 3rd International Symposium, Madison 1970, Ed. H.H. Barschall and W. Haeberli, U. of Wisconsin Press, Madison (1971).
6. C.L. Dolnick, Nucl. Phys. B22 (1970) 461.
7. D. Axen, G. Duesdieker, L. Felawka, Q. Ingram, R. Johnson, G. Jones, D. Lepatourel, M. Salomon and W. Westlund, Nucl. Phys. A256 (1976) 387.
8. B.M. Preedom, C.W. Darden, R.D. Edge, T. Marks, M.J. Saltmarsh, K. Gabathuler, E.E. Gross, C.A. Ludemann, P.Y. Bertin, M. Blecher, K. Gotow, J. Alster, R.L. Burman, J.P. Perroud, R.P. Redwine, B. Goplen, W.R. Gibbs and E.L. Lomon, Phys. Lett. 65B (1976) 31.
9. D. Aebischer, B. Favier, L.G. Greeniaus, R. Hess, A. Junod, C. Lechanoine, J.-C. Niklès, D. Rapin and D.W. Werren, Nucl. Phys. B108 (1976) 214.
10. C. Richard-Serre, W. Hirt, D.F. Measday, E.G. Michaelis, M.H.J. Saltmarsh and P. Skarek, Nucl. Phys. B20 (1970) 413.
11. J. Spuller and D.F. Measday, Phys. Rev. D12 (1975) 3550.
12. Yu. K. Akimov, O.V. Savchenko and L.M. Soroko, Nucl. Phys. 8 (1958) 637.
13. M.G. Albrow, S. Andersson-Almehed, B. Bosnjakovic, F.C. Erne, Y. Kimura, J.P. Lagnaux, J.C. Sens and F. Udo, Phys. Lett. 34B (1971) 337.

QUASI-ELASTIC SCATTERING OF POLARIZED PROTONS*

P. Kitching, L. Antonuk, D.A. Hutcheon, W.J. McDonald
C.A. Miller, G.C. Neilson, W.C. Olsen, and G.M. Stinson
Physics Department, University of Alberta, Edmonton

E.D. Earle
Chalk River Nuclear Laboratories, Chalk River, Ontario

A.W. Stetz
Oregon State University, Corvallis, Oregon, U.S.A.

Quasi elastic scattering, which has proved so useful in the investigation of hole states in nuclei, has not yet been used as a tool for studying the effects of the nuclear environment on nucleon-nucleon scattering. Although recently suggested techniques [1] for observing off-energy shell effects in (p,2p) reactions seek to avoid some of the theoretical uncertainties inherent in the DWIA, these techniques still depend on the validity of the reaction model. One way of studying the basic assumptions underlying the DWIA is to measure the asymmetries produced in $(\vec{p},2p)$ scattering from $\ell = 0$ states in nuclei. These asymmetries must be the same as those obtained by scattering on unpolarized free protons unless spin orbit distortion effects or off-energy shell effects are important.

We have, therefore, made measurements in a kinematically complete experiment of the asymmetries produced in the ^{40}Ca $(\vec{p},2p)$ reaction, using a 200 MeV polarized proton beam from the TRIUMF cyclotron. The energies of the outgoing protons were measured in NaI(Tℓ) crystals. Missing energy spectra for events with recoil momenta $\sim$ 0 MeV/c and $\sim$100 MeV/c are shown in Figures I(a) and I(b) respectively. In the first case, in which states with $\ell = 0$ are enhanced, one can see a peak arising from protons scattered from hydrogen contamination in the target at $E_{miss}=0$ and a peak arising from knockout of $2s_{1/2}$ protons in ^{40}Ca. In Figure I(b) states with $\ell \neq 0$ are enhanced and peaks corresponding to knockout of $1d_{3/2}$ and $1d_{5/2}$ protons are seen:

From data such as those shown in Figure 1(a), the asymmetry for the $2s_{1/2}$ state was obtained and compared with the free p-p elastic scattering asymmetry (calculated with kineamtics appropriate to the initial state) in Figure II(a) and (b). When DWIA calculations which include the effects of spin orbit distortion become available, it will be interesting to compare the data with off shell calculations of the p-p matrix elements and to extend the measurements to large values of recoil momenta and to s states with different separation energies.

Reference

1 A.A. Ionnides and D.F. Jackson, preprint.

*Work supported in part by the National Research Council of Canada.

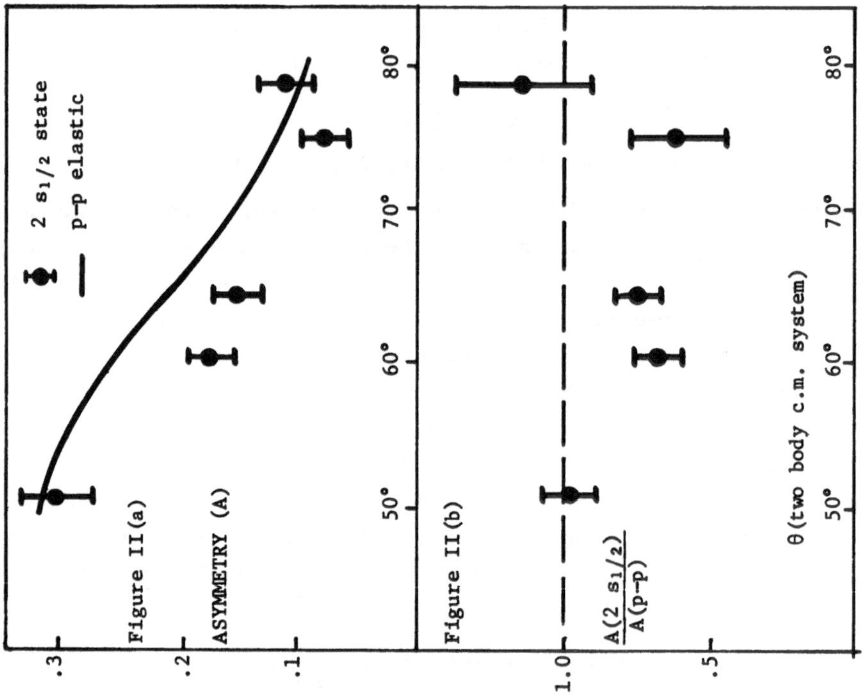

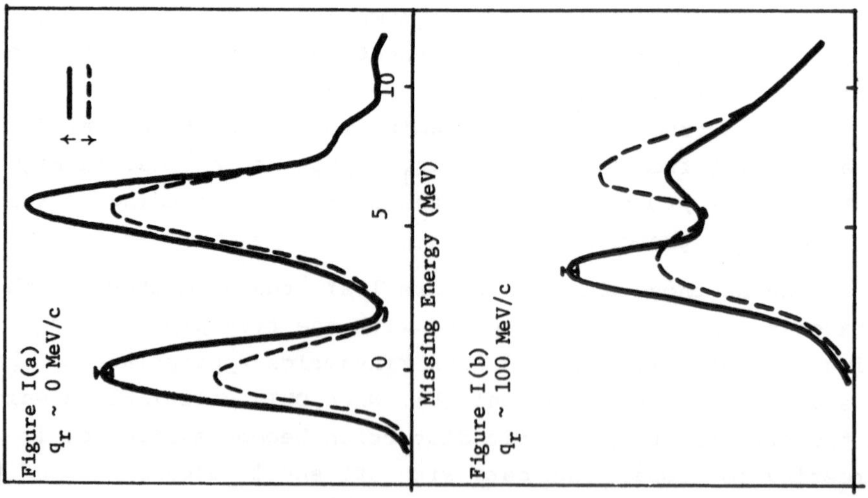

THE THREE-NUCLEON BOUND STATE COMPUTED WITH EXPLICIT Δ(1236) COMPONENTS

Ch. Hajduk and P. U. Sauer
Theoretical Physics, Technical University
3000 Hannover, Germany

Some properties of the three-nucleon bound state are computed in the framework of Faddeev equations allowing the wavefunction to contain baryon-resonance components. As is common practice, the two-nucleon interaction is taken to act in the $^3S_1$-$^3D_1$ and ^1S_o partial waves only. Assuming that at most one nucleon be excited to a Δ resonance, only the ^1S_o two-nucleon channel couples to the ^5D_o nucleon-Δ channel. The Δ resonance is treated as a stable elementary particle of rest mass m_Δ=1236 MeV, spin $\frac{3}{2}$ and isospin $\frac{3}{2}$, to which nonrelativistic quantum mechanics can be applied. It is antisymmetrized with the nucleons.

The $^3S_1$-$^3D_1$ partial wave of the two-nucleon interaction is taken from the Reid soft-core potential. The interaction in the coupled ^1S_o-^5D_o partial wave is based on a nonsingular version of the transition potential of Ref. 1. This potential is now outdated[2] and not really appropriate for an ambitious three-nucleon calculation. However, our specific aim is to isolate the effect of Δ components in the wave function on the three-nucleon ground-state properties by comparing results of a coupled ^1S_o-^5D_o potential and a standard ^1S_o potential without coupling to the inelastic ^5D_o channel. The present calculation therefore pays careful attention to the <u>equivalence</u> between the two potentials used for comparison. A hermitian, energy-independent single-channel equivalent of the coupled ^1S_o-^5D_o potential is derived according to the inverse scattering theory[3] for the transition matrix and made identical to the coupled-channel potential (i) with respect to all on-shell scattering data in the direct ^1S_o nucleon-nucleon channel and (ii) with respect to the symmetric part of the half-shell transition matrix in the direct ^1S_o nucleon-nucleon channel. Since the latter part (ii) determines the uncoupled ^1S_o off-shell transition matrix in full, the equivalence between the two potentials cannot be pushed any further. Thus, all differences, which show up in the three-nucleon results with the single-channel and the two-channel ^1S_o force models respectively, arise solely from the explicit treatment or neglect of the inelastic nucleon-Δ channel in the coupled ^1S_o-^5D_o or uncoupled ^1S_o potentials. All other sources, which could have created a difference in the results and could have disturbed their meaningful comparison, are eliminated. Furthermore, in order to stay also computationally equivalent, even the coupled ^1S_o-^5D_o potential is constructed with the help of the inverse-scattering theory[3] using the ^1S_o phase shifts of the Reid soft-core potential and the potential of Ref. 1 just as choice for the necessary off-shell extension.

The results for the three-nucleon bound states, given in obvious notation, are collected in the Table. Though the weight of Δ configurations is with 1% rather small, their explicit presence yields a large repulsion of 0.6 MeV, which has two sources: First, the nucleon-Δ channel changes the ^1S_o nucleon-nucleon transition matrix off-shell as compared to the single-channel potential, which - since energy-independent - cannot adjust the process of Fig. 1 to the nuclear medium. This dispersive effect[4], here 0.5 MeV repulsion, is well-known from nuclear matter[5]. Second, and this is the surprise, the three-body force contribution ΔE_3 of Fig. 2 also turns out to be repulsive in the present calculation with 0.1 MeV. This unexpected result is independently confirmed by a crude primitive perturbative estimate

$$\Delta E_3 = 6 \sum_\alpha \int p^2 dp q^2 dq \frac{\langle\psi| v_{(NN),(N\Delta)} P_{123} |pq\alpha\rangle\langle pq\alpha| v_{(N\Delta),(NN)} |\psi\rangle}{E - \frac{p^2}{2M_\alpha} - \frac{q^2}{2\mu_\alpha} - m_\Delta c^2}$$

In the Equ. $|\psi\rangle$ and E denote the three-nucleon wave function and binding energy obtained in a standard calculation without Δ configurations. The potential $v_{(NN),(N\Delta)}$ couples two-nucleon to nucleon-Δ channels. P_{123} is the cyclic permutation operator, and the $|p\,q\,\alpha\rangle$ are three-nucleon plane-wave basis states which contain a single Δ resonance and in which the two particles described by the coordinate of momentum p are antisymmetrized. When using, however, - in contrast to the confirmation of the 0.1 MeV repulsion of the three-body calculation - just the pure π-exchange of the transition potential $v_{(NN),(N\Delta)}$, but also P and D isospintriplet partial waves we obtain a massive attraction of -1.6 MeV (the π-exchange contribution of ^1S_o-^5D_o is just -0.1 MeV). This result is consistent with the estimate of Ref. 6, but gets substantially reduced to -0.8 MeV, when a transition potential, better defined at intermediate ranges (we use Equ. (2.14) of Ref. 5), is employed. We therefore have to conclude: When baryon-resonance configurations are included in the three-nucleon wave function, the traditional belief that the higher partial waves beyond $^3S_1$-$^3D_1$ and ^1S_o are unimportant for the ground-state properties of the three-nucleon system has to be revised. Furthermore, especially the inclusion of the ρ-exchange can significantly correct previous three-body force estimates[6] based on the π-exchange only.

The charge form factor of ^3He, shown in Fig. 3, is computed with the nucleonic charge form factors of Ref. 7. The included contribution of the Δ is the process of Fig. 4 with the prescription of Ref. 8 for the Δ

Table

			^1S_o-^5D_o	^1S_o
^3H :	E	[MeV]	-6.5	-7.1
	P_S	[%]	89.15	89.85
	$P_{S'}$	[%]	1.98	1.87
	P_D	[%]	7.96	8.28
	P_Δ	[%]	0.91	0.00
	R_{ch}	[fm]	1.83	1.78
	$q^2_{min}(F_{ch})$	[fm^{-2}]	14.8	15.6
^3He:	R_{ch}	[fm]	2.04	1.96
	$q^2_{min}(F_{ch})$	[fm^{-2}]	14.0	14.4

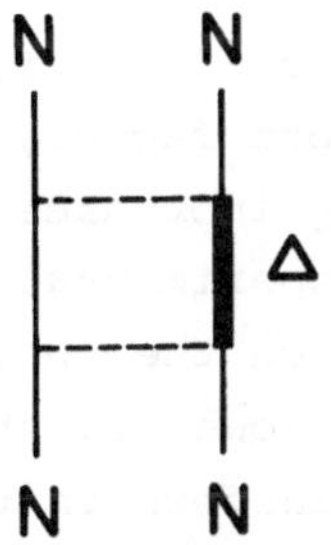

Fig. 1

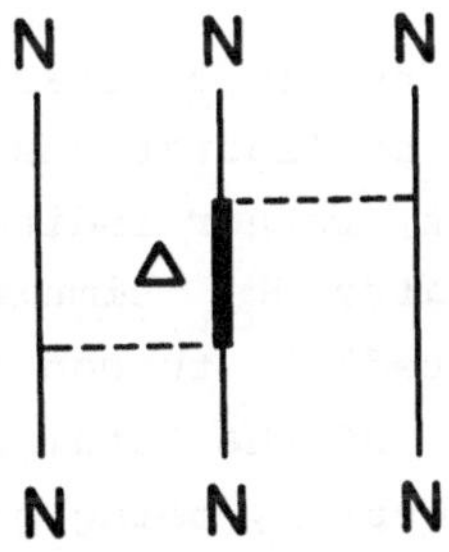

Fig. 2

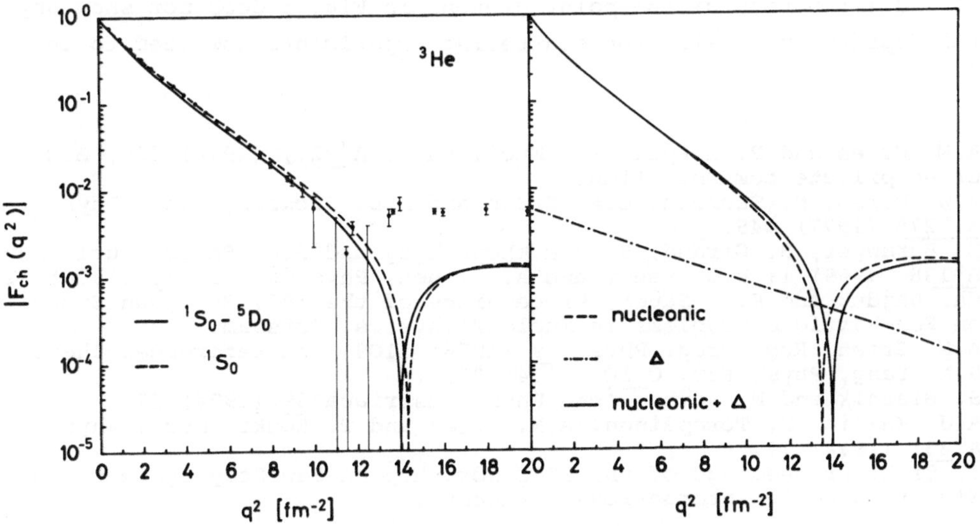

Fig. 3. ^3He charge form factor calculated from the equivalent single-channel and two-channel potentials (left side). In the two-channel case the nucleonic and Δ contributions are disentangled (right side).

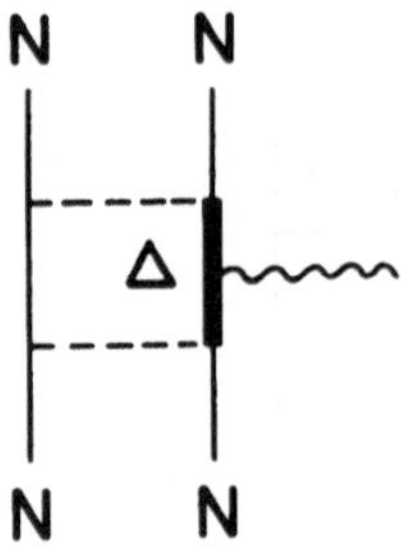

Fig. 4

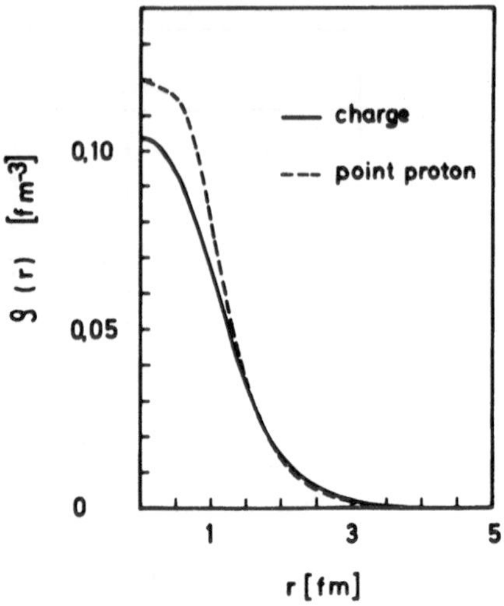

Fig. 5. Charge and point-proton density of ^{3}He.

charge form factors. In contrast to Ref. 8, there is an overall beneficial effect of the Δ configurations on the dip of the charge form factor, which is, however, rather indirect: The decrease in binding increases the r.m.s. charge radius R_{ch} simultaneously (in view of the experimental value even unwantedly), though the direct effect of the Δ on the charge radius is with 3% of the total 0.08 fm increase very weak. However, the direct Δ contribution, being positive for the momentum transfers in the region of the minimum, partly balances this shift of the form factor minimum. The distribution of the point protons in Fig. 5 does not show any central depression as electron scattering experiments now seem to indicate[9].

1. A.M. Green and P. Haapakoski, Nucl. Phys. A 221, (1974) 429; A.M. Green private communication.
2. J.W. Durso, M. Saarela, G.E. Brown and A.D. Jackson, Nucl. Phys. A. 278 (1977) 445.
3. M. Baranger, B. Giraud, S.K. Mukhopadhyay and P.U. Sauer, Nucl. Phys. A 138 (1969) 1; P.U. Sauer and A. Sevgen, Phys. Rev. C 13 (1976) 720.
4. Ch. Hajduk and P.U. Sauer, Proceedings of the 1977 European Symposium on Few-Particle Problems in Nuclear Physics, Potsdam.
5. A.M. Green, Rep. Prog. Phys. 39 (1976) 1109, and references there.
6. S.N. Yang, Phys. Rev. C 10 (1974) 2067.
7. S. Blatnik and N. Zovko, Acta Phys. Austriaca 39 (1974) 62.
8. A.J. Kallio, P. Toropainen, A.M. Green and T. Kouki, Nucl. Phys. A 231 (1974) 77.
9. I. Sick, Proceedings of the 1978 Workshop on Few-Body Systems and Electromagnetic Interactions, Frascati.

The authors acknowledge the help of R.A. Brandenburg, S.A. Coon and A. Kelemen in setting up numerical procedures for using the inverse scattering theory of Ref. 3.

SOLUTION OF THE BOUND TRINUCLEON INCLUDING THE EFFECTS OF THE ISOBAR Δ(1236)

E.P. Harper†
Dept. of Physics, The George Washington University
Washington, D.C. 20052, USA

and

Y.E. Kim+ and A.Tubis[x]
Department of Physics, Purdue University
West Lafayette, Indiana 47907

I. Introduction

We would like to report a preliminary solution of the bound trinucleon system in which the effects of virtual isobars Δ(1236) are taken into account in an approximate but realistic manner. We use the equations developed by us and reported in an earlier communication to this Conference [1]. In order to represent the two-body interaction we use the One-Boson-Exchange model of Holinde and Machleidt, hereafter referred to as HM [2]. The particular potential that we use is that referred to by HM as HM1 + Δ and though realistic insofar as it fits the NN phase shifts it fairly substantially underbinds the deuteron at 2.2089 MeV. In addition, only the π-contribution to the $N \leftrightarrow \Delta(1236)$ transition is included though it is now well known that the ρ-contribution is very important [3]. We hope to remedy this and other drawbacks (adumbrated below) in future calculations. Holinde and Machleidt [2] have reported a number of other interactions in which the meson-baryon vertex function is represented by an eikonal form factor derived from an approximate representation of the vertex contributions and crossed ladder diagrams of soft neutral vector mesons [4]. We do not consider such interactions in view of the serious difficulty involved in continuing their explicit energy dependence below the 2M threshold [5].

The great advantage of our approach i.e. that of using the equations presented previously and using the HM interaction to represent the two-body force is that it treats the nucleon and isobar in a unified manner from the start and various uncertainties [6] associated with the more ad hoc introduction of transition potentials into already existing NN-NN potentials, such as the Reid Soft Core Potential [7], are avoided. In addition, of course, the effect of the Δ(1236) on the trinucleon binding energy emerges (to a good approximation) in a natural fashion from the implementation of our procedure.

II. Two-Body Equations and Results

The multichannel Lippmann-Schwinger equation for the two-body t-matrix is

$$t_{ij} = v_{ij} + \sum_k v_{ik} G_{o,k} t_{kj} \tag{1}$$

where the subscripts take the values 0,1,2 corresponding respectively to NN, NΔ, and ΔΔ.

This three-channel equation can be approximated by the one-channel equation

$$\begin{aligned} t_{00} &= \tilde{V}_{00} + \tilde{V}_{00} G_{0'0} t_{00} \qquad \text{for } j = 0 \\ t_{10} &= V_{10} + V_{10} G_{0'0} t_{00} \\ t_{20} &= V_{20} + V_{20} G_{0'0} t_{00} \end{aligned} \tag{2}$$

where $\tilde{V}_{00} = V_{00} + V_{01} G_{0'1} V_{10} + V_{02} G_{0'2} V_{20}$ is an effective NN-NN energy dependent interaction. The matrix elements of the various V_{i0} in the L-S basis are calculated from the helicity state matrix elements given by HM. In Table I we list the potential matrix elements in the L-S basis which are used in calculating the purely nucleonic t-matrix t_{00}. As pointed out in our previous paper [1] we only consider transition t-matrices for which L = 0 for the nucleonic channel. As a partial check on our potential matrix elements we solved the deuteron problem obtaining precisely the HM result for the binding energy and D-state probability $P_D = 5.73$. This is the "unrenormalized" D-state probability. When we include the ΔΔ components we find for the deuteron $P_S^{NN} = 93.90$ $P_D^{NN} = 5.71$ $P_{01}^{\Delta\Delta} = 0.060$ $P_{21}^{\Delta\Delta} = 0.004$ $P_{23}^{\Delta\Delta} = 0.334$ $P_{43}^{\Delta\Delta} = 0.02$ giving a total isobar probability of 0.418 - rather smaller than is usually obtained. Using these probabilities and Arenhovel's estimate of the Δ(1236) magnetic moment we find for the deuteron magnetic moment the value

$$\begin{aligned} \mu_D = {} & 2\tfrac{1}{2}(\mu_p + \mu_n)(P_S^{NN} - P_D^{NN}/2) + 3/4\, P_D^{NN} \; + \\ & 2\tfrac{1}{2}\,\mu_p (P_{01}^{\Delta\Delta} - \tfrac{1}{2} P_{21}^{\Delta\Delta} + 2P_{23}^{\Delta\Delta} - \tfrac{3}{2} P_{43}^{\Delta\Delta}) \; + \\ & \frac{M}{M^*}(\tfrac{3}{4} P_{21}^{\Delta\Delta} - \tfrac{1}{2} P_{23}^{\Delta\Delta} + \tfrac{5}{4} P_{43}^{\Delta\Delta}) \end{aligned}$$

$$= 0.8009 + 0.0428 + 0.0195 - 0.0059 = 0.8573\ \mu_N$$

a rather gratifyingly close concurrence of theory and experiment [8]. This agreement may be fortuitous in view of our neglect of other isobaric configurations, our approximation of the original L-S equations, our ansatz for the isobar isoscalar magnetic moment and the neglect of relativistic effects and N-$\bar{N}$ virtual states [9] in the deuteron. It will be very interesting to see whether or not this good agreement with experiment persists when the transition potential matrix elements include contributions from vector meson exchange.

III. Three-Body Results

Using as input the solutions to the set of equations (2) we have solved the Faddeev equations in the form

$$\begin{bmatrix} \Psi_0 \\ \Psi_1 \\ \Psi_{1'} \end{bmatrix} = 2 \begin{bmatrix} G_0 \; t_{00} & G_0 \; t_{01} & G_0 \; t_{01} \\ 0 & 0 & G_1 \; t_{11'} \\ G_1 \; t_{1'0} & 0 & 0 \end{bmatrix} \begin{bmatrix} \Psi_0 \\ \Psi_1 \\ \Psi_{1'} \end{bmatrix} \quad (3)$$

to obtain the eight all-nucleon configuration wave function components, one (NN,Δ) component $\Psi_1(1)$ and two (NΔ,N) components $\Psi_{1'}(1,2)$. We obtain a binding energy of 6.76 MeV. We regard this result as tentative because an exhaustive study of the accuracy of our numercal methods has not yet been made. This is necessary in view of the fact that the configurations containing isobars have stronger high momentum components than the purely nucleonic configurations and this necessitates an increase in our cut-off parameter. We are using a cut-off of $Cq = 4.0\ fm^{-1}$ instead of the usual $Cq = 3\ fm^{-1}$.

The low value for the H^3 binding energy is somewhat surprising in view of Yang's [10] perturbation theory result for the contribution of the Δ(1236) three-body force diagram to the trinucleon binding which amounted to a net <u>increase</u> of 1.98 MeV. Our result seems to indicate that the single Δ(1236) contribution results in a net <u>decrease</u> in binding energy since previous calculations with monopole regularized OBE potentials gave binding energies E_3 in the range 6.73 to 7.38 MeV [11]. It is very possible that the present result ensues from the approximate treatment of the two-body Lippmann-Schwinger equation which reduces the multi-channel equation to a single channel plus simple quadratures. The effective NN interaction thus generated causes the NN t-matrix to become, perhaps, excessively weak at energies below the 2M threshold. In order to eliminate this possibility we are solving the two-body problem including the full transition potentials and without making any truncations. The resulting NN t-matrix will not then give as good a fit to the NN phase parameters as the t-matrix obtained by the truncation procedure (this method was used by HM in fitting to the above threshold NN and deuteron data). We think that it should be good enough to reveal whether or not the effective potential is a faithful description significantly below the

2M threshold by a direct comparison with the t-matrix already obtained using the effective potential method.

Using the solutions Ψ_0, Ψ_1 and Ψ_1' for the all-nucleon and single isobar configurations we can now obtain the two-isobar and three-isobar configurations by a single quadrature from the equations

$$\begin{aligned} \Psi_2 &= 2\, G_2 \quad t_{20} \quad \Psi_o \\ \Psi_{2'} &= 2\, G_2 \quad t_{2'1'} \; \Psi_{1'} \\ \Psi_3 &= 2\, G_3 \quad t_{31'} \quad \Psi_{1'} \end{aligned} \tag{4}$$

We intend to use these wave functions (after complete antisymmetrization) to calculate the trinucleon magnetic moments, electromagnetic form factors and Gamow-Teller matrix element in the decay $H^3 \rightarrow He^3 + e^- + \nu_e$.

IV. Discussion

We have presented the results of a calculation of the trinucleon bound state using a procedure which we believe is capable of yielding configuration wave function components which are a close approximation to the exact non-relativistic wave functions. The enhancement of the high momentum components arising from the isobar containing configurations is evident (see Fig.1). The binding energy (6.76 MeV) that we obtain is disappointingly small and may be due to our truncation of the two-body Lippmann-Schwinger equations. On the other hand the effect may be a real one and, if it is, we have at present no idea why it is so since Yang [10] (and others) have found that the contribution to the three-body force arising from the diagram

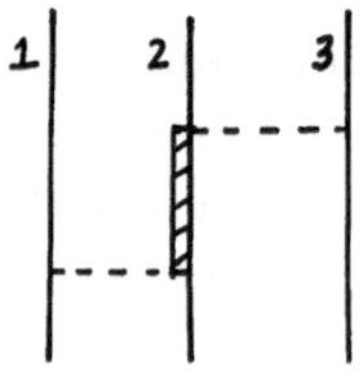

always gives a substantial increase in the total trinucleon binding. Notice that our procedure includes this diagram as a three-body correlation contribution to the trinucleon binding energy [12]. Apart from numerical refinements we intend to check this result by a) using a transition interaction which contains single vector meson exchange in addition to one-pion-exchange and b) doing the calculation without approximating the two-body Lippmann-Schwinger equations. Our approximation of the Faddeev equations is unavoidable if we are to have a manageable set of equations but at any rate we strongly feel that the set of equations that we use give a solution which is a close approximation to the solution of the exact equations.

$\langle 00|V^{01}_{NN,NN}|00\rangle$

$\langle 01|V^{10}_{NN,NN}|01\rangle, \langle 21|V^{10}_{NN,NN}|01\rangle, \langle 01|V^{10}_{NN,NN}|21\rangle, \langle 21|V^{10}_{NN,NN}|21\rangle$

$\langle 00|V^{01}_{NN,N\Delta}|22\rangle$

$\langle 00|V^{01}_{NN,\Delta\Delta}|00\rangle, \langle 00|V^{01}_{NN,\Delta\Delta}|22\rangle$

$\langle 01|V^{10}_{NN,\Delta\Delta}|01\rangle, \langle 01|V^{10}_{NN,\Delta\Delta}|21\rangle, \langle 01|V^{10}_{NN,\Delta\Delta}|23\rangle, \langle 01|V^{10}_{NN,\Delta\Delta}|43\rangle$

$\langle 21|V^{10}_{NN,\Delta\Delta}|01\rangle, \langle 21|V^{10}_{NN,\Delta\Delta}|21\rangle, \langle 21|V^{10}_{NN,\Delta\Delta}|23\rangle, \langle 21|V^{10}_{NN,\Delta\Delta}|43\rangle$.

Table 1. Two-body potential matrix elements used in the calculation of the two-body t-matrices.

References

(1) E.P. Harper, Y.E. Kim and A. Tubis "A Formalism For Incorporating Nucleon Isobars In Bound Trinucleon Systems" This conference.

(2) K. Holinde and R. Machleidt, Nucl. Phys. A280, 429 (1977).

(3) P. Haapakoski, Phys. Lett. 60B, 405 (1974)
A.M. Green and P. Haapakoski, Nucl. Phys. A221 429 (1974).

(4) R. Woloshyn and A. Jackson, Nucl. Phys. A185, 131 (1972).

(5) E.P. Harper "The Trinucleon System Bound By A One-Boson Exchange Force" in Proc. Of The VII International Conference on Few-Body Problems In Nuclear And Particle Physics, New Delhi (1977), Edited by A. Mitra et al. (North-Holland, 1977).

(6) A.M. Green and T.H. Schucan, Nuc. Phys. A188, 289 (1972).
K. Ohta and M. Wakamatsu, Prog. Theor. Phys. 55, 131 (1976).

(7) R.V. Reid,Ann. Of Phys. 50, 411 (1968).

(8) See, for example, A.M. Green, Reports On Progress In Physics, 39, 77 (1976)

(9) F. Gross, "Relativistic Effects In Few-Body Systems", Proc. Of The VII International Conference On Few-Body Problems In Nuclear and Particle Physics, New Delhi (1976), (North-Holland, 1977).

(10) S.N. Yang, Phys. Rev. C10, 2067 (1974).

(11) E.P. Harper, Phys. Rev. Lett. 34, 677 (1975) and unpublished results.

(12) B.H.J. McKellar, "Three-Body Forces In Nuclei", Proc. VII International Conference In Nuclear And Particle Physics, New Delhi (1976), (North-Holland, 1977).

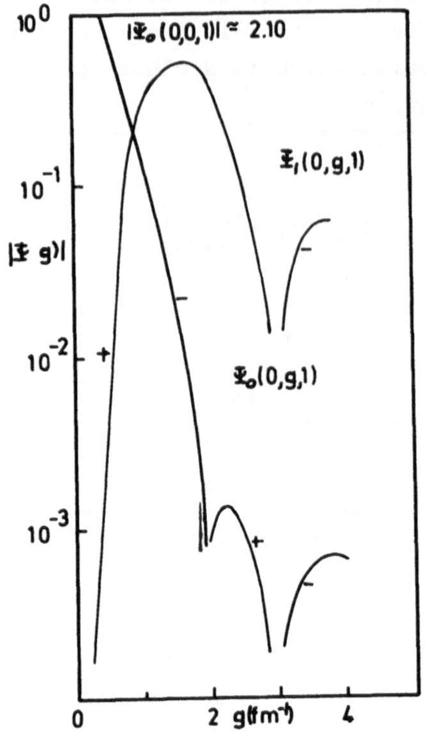

+ Presently visiting the Institut für Kernphysik der Universität Mainz, Mainz, West Germany. Work supported by U.S. National Science Foundation.

* Work supported by U.S. Department of Energy.

† Work supported by the George Washington University Committee on Research.

Figure 1. Wave functions for an all-nucleon and two-nucleon plus one $\Delta(1236)$ components of the bound trinucleon system.

PERTURBATION APPROACH TO THE ^{3}He BOUND STATE WITH THE REID SOFT-CORE POTENTIAL

T. Sasakawa and T. Sawada[+]

Department of Physics, Tohoku University, Sendai 980, Japan

[+] Department of Applied Mathematics, Faculty of Engineering Science

Osaka University, Toyonaka 560, Japan

The exact solution of the ^{3}He bound state is calculated with the perturbational approach to the Faddeev equation in the coordinate space[1,2] using the Reid soft-core potential without the separable approximation. The three-body states included are those with the two-body pair in the $^1S_0$ and $^3S_1+^3D_1$ states and the third particle in the $\ell=0$ state. In Fig.1 we show typical examples of convergence. The quantity D plotted is essentially the difference between the left side and the right side of Eq.(23) of reference 2. It is sensitive to the choice of $|E|$, and converges within five iterations. The value E = 6.70 MeV at which D converges to zero provides the ^{3}He binding energy.

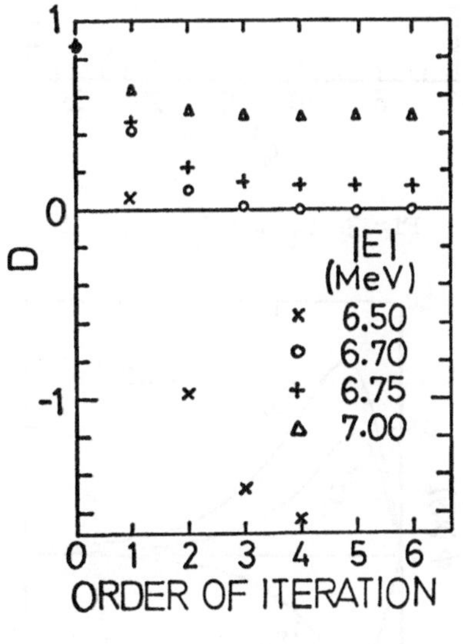

Fig.1 Examples of Convergence

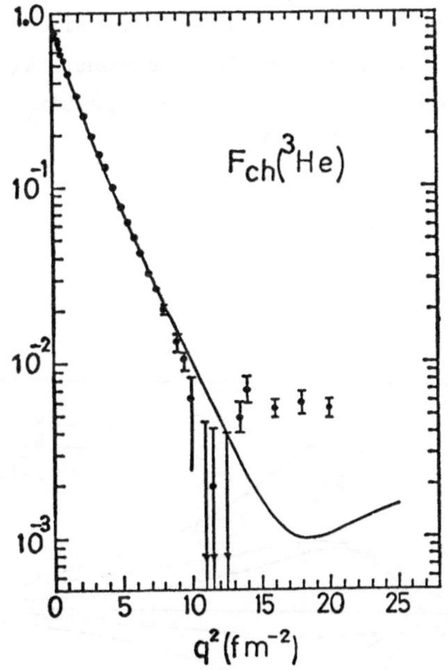

Fig.2 ^{3}He charge form factor

The charge form factor $F_{ch}(^3He)$ is compared with the experimental data[3] in Fig.2. Characteristically, it stays positive and has a broad minimum at $q^2 \simeq 18 fm^{-2}$. It is given by

$$2F_{ch}(^3He) = (3F_a + F_b)\cdot F^p_{ch} + 2F_b\cdot F^n_{ch} \qquad (1)$$

where F^p_{ch} (F^n_{ch}) is the proton (neutron) charge form factor[4], and

$$F_a = F_1^S + \frac{1}{2} F_2^S - 2 F_3^{SS} + \frac{1}{2} F_4^{SS} + \frac{3}{2} (F_2^A - F_4^{AA}) + \sqrt{3} (\varepsilon \cdot F_3^{SA} + \frac{1}{2} F_4^{SA}) \qquad (2)$$

with ε=1. For F_b, a similar expression holds with ε= -1 and S and A interchanged, where S(A) indicates component with the $^3S_1$+ $^3D_1$ ($^1S_0$) two-body state. F_{ch} (^{3}He) is dominated by F_1 and F_2 which are shown in Fig.3 (before the over-all normalization). Unlike F_1 and F_2, F_3 and F_4 change signs, but they are smaller than F_1 by an order of magnitude at large q^2, and are at most of the same magnitudes as $^1S_0$ at small q^2. The ^{3}He wave function $\Phi(x, y)$ has a node in x (=the two-body separation) at about 0.5 fm for the $^3S_1$ and $^1S_0$ components (Fig.4). No node is seen in the $^3D_1$ component. $\Phi(x, y)$ extends quite far in y (=the spectator coordinate) (Fig.5).

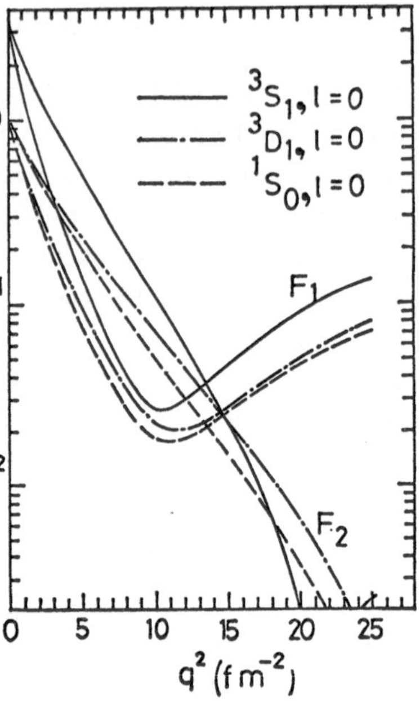

Fig.3 F_1 and F_2 of Eq.(2)

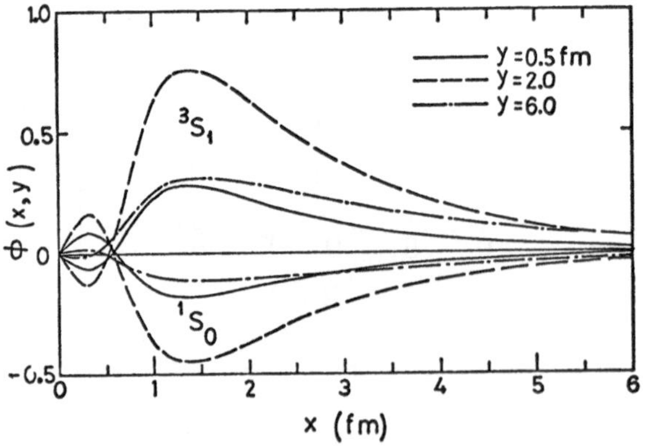

Fig.4 x-dependence of $\Phi(x, y)$

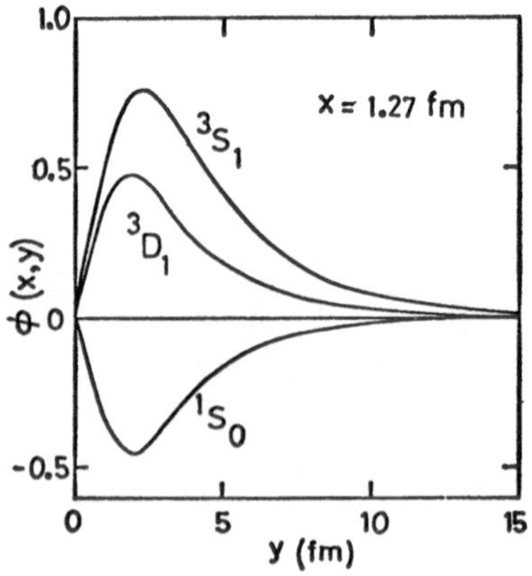

Fig.5 y-dependence of $\Phi(x, y)$

1. T. Sasakawa and T. Sawada, Suppl. Prog. Theor. Phys. 61 (1977) 61.
2. T. Sasakawa, T. Sawada, and S. Shioyama, Proc. Int. Conf. Nuclear Structure, Tokyo, 1977, VII-D-c ; Suppl . J. Phys. Soc. Japan 44 (1978) 298.
3. J.S. McCarthy et al., Phys. Rev. Letters, 95 (1970) 884.
4. T. Janssens et al., Phys. Rev. 142 (1966) 922.

Three nucleon observables with meson theoretical interactions

R. Offermann and W. Glöckle
Institut für Theoretische Physik
Ruhr-Universität Bochum
4630 Bochum, Germany

The quantitative understanding of the binding energy and charge form factor of three nucleons is still lacking. Besides possible contributions of three-body forces and relativistic effects even different models for the nucleon-nucleon force may give different results for the three-body observables and one may use them to favor or object some of the force parameters. Recently we used[1] 4 versions of the OBEP of the Bonn group[2]. The potentials HM3 A-C differ essentially in the form of the pion-nucleon vertex formfactor (cut off mass Λ) $F(k^2) = \Lambda^2/(\Lambda^2 - k^2)$ and the coupling constants of the σ- and δ-mesons, the potential HM1 uses different form factors for all mesons. Except for the version HM3-C the fits to the phase shifts and the deuteron are satisfying. To solve the Faddeev equations we used our method to expand the Faddeev-component into a suitable set of basis functions[3]. In table 1 we give the resulting triton binding energies E_t using 3 partial waves (spectator orbital angular momentum $\lambda = 0$ only), 5 partial waves ($\lambda = 2$) and the additional change of E_t due to the inclusion of p and d-waves. We also give from [2] the pion cut off mass Λ, the coupling constants g_σ and g_δ, the deuteron d-state probability P_D and its quadrupole moment Q.

Table 1

	HM1	HM3-A	HM3-B	HM3-C	Exp.
(MeV)	completely	1530	1265	1000	?
g_σ	different	8.20	8.67	9.202	?
g_δ		4.99	2.88	0.704	?
P_D	5.75	5.18	4.70	3.63	?
$Q(fm^2)$	0.284	0.281	0.277	0.263	0.2875 0.0020
$E_t(\lambda=0)$	-7.25	-7.37	-7.49	-7.84	-8.49
$E_t(\lambda=0,2)$	-7.83	-7.94	-8.00	-8.26	
ΔE_t(p-waves)	+0.04	+0.03	+0.01	+0.01	
ΔE_t(d-waves)	-0.07	-0.06	-0.05		

The increase in the binding energy E_t going from HM1 to HM3-C can be well understood from the decrease of P_D. Including $\lambda = 0$ and $\lambda = 2$ waves but excluding p- and d-wave forces we calculated the charge form factor of ^3He and ^3H. In table 2 we show the calculated values for the charge radius $\sqrt{\langle r^2 \rangle}$, the positions of the form factor dip Q_o^2 and the maximum Q_m^2 and the value of the form factor $M = 10^3 \times f(Q_m^2)$.

Table 2

		HM1	HM3-A	HM3-B	HM3-C	Exp.
^3He	$\sqrt{\langle r^2 \rangle}$	1.91 (1.93)	1.90 (1.92)	1.89 (1.91)	1.86 (1.88)	1.88 ± .05
	Q_o^2	16.9 (17.1)	16.7 (16.9)	16.9 (17.1)	18.6 (18.8)	11.6
	Q_m^2	22.	21.5	21.5	23.	17.
	M	.71	.75	.70	.43	6.3
^3H	$\sqrt{\langle r^2 \rangle}$	1.79 (1.76)	1.78 (1.75)	1.77 (1.74)	1.75 (1.72)	1.70 ± .05
	Q_o^2	18.3 (17.7)	18.1 (17.5)	18.2 (17.6)	19.9 (19.3)	?

We used the proton and neutron form factors of Iachello et al. [4] as well as those of Janssens et al. [5] (numbers in paranthesis).

To draw conclusions we want to exclude HM3-C from further consideration because it does not fit the $^3D_1$ phase shift, which forbids pion cut off masses below 1000 MeV in this potential model. For all potentials the p and d waves cancel nearly to zero in the 3 body system. One can gain slightly more binding energy by introducing a weaker pion-nucleon vertex without making the charge formfactor still worse than it is compared to experiment.

1) R. Offermann and W. Glöckle, to be published
2) K. Erkelenz, Phys. Rep. 13C, (1974), 191
   K. Holinde and R. Machleidt, Nucl. Phys. A256, (1976), 479
   K. Holinde, preprint and private communication.
3) W. Glöckle and R. Offermann, Phys. Rev. C16, (1977), 2039
4) F. Iachello, A. Jackson and A. Landé, Phys. Letters 43B, (1973), 191
5) T. Janssens et al., Phys. Rev. 142, (1966), 922

Charge-dependent Three-body Force with $N^*(1236)$ in the Intermediate States

S. N. Yang
Dept. of Physics, National Taiwan University, Taipei, Taiwan

The charge-dependent three-body force due to the excitation of one nucleon into resonance state $N^*(1236)$, as shown in Fig. 1, is considered. Very detailed analysis[1] including magnetic force, momentum-dependent electromagnetic force, p-n mass difference and exchange current corrections can best give the energy difference of ^{3}He-^{3}H E_C= 683±29 KeV, which is 81±29 KeV less than the experimental value. This work is carried out in the hope of understanding this discrepancy.

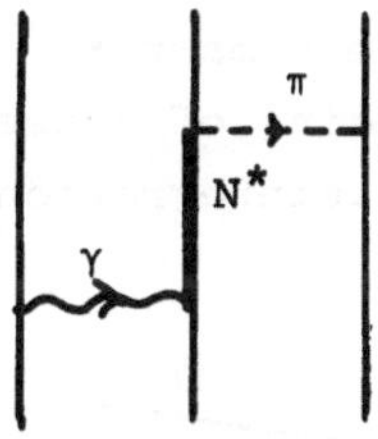

Fig. 1

The CD three-body potential that arises from the mechanism as depicted in Fig. 1 is derived by employing a chiral-invariant Lagrangian[2] which gives favorable comparison with experiment for single pion photoproduction at threshold. The contribution of this CD three-body potential is then estimated by calculating its expectation value in a variational triton wavefunction obtained with HJ potential[3].

We have previously investigated[4] the CD three-body forces which arise from the Born terms of the pion photoproduction for the internal process with one photon and one pion exchanged. It gives a contribution of -7 KeV to E_C. However, contribution of N^* to E_C might be larger due to its strong tensor character. This has been shown to occur in the case of nuclear three-body force. Further numerical results will be presented.

References

1. Brandenburg, R.A., S.A. Coon and P.U. Sauer, Nucl. Phys. A294 (1978) 305.
2. Peccei, R.D., Phys. Rev. 181 (1969) 1902.
3. Delves, L.M. and J. M. Blatt, Nucl. Phys. A98 (1967) 503.
4. Yang, S.N., to be published.

THE N-D AND N-D* MOMENTUM DISTRIBUTIONS+

E.P. Harper, D.R. Lehman and F. Prats
Department of Physics
The George Washington University
Washington, D.C. 20052, USA

I. Introduction

The vertex functions corresponding to the virtual dissociation of the trinucleon into a deuteron plus a nucleon or into an (np) pair in a $^1S_0$ state plus an uncorrelated nucleon are of basic importance for the prediction of inelastic cross-sections involving strong, weak and electromagnetic projectiles on He^3 or H^3[1]. In particular the S-matrix approach of Shapiro [2] relies heavily on an understanding of these vertex functions. Thus, one might calculate the photodisintegration of the trinucleon from the diagrams [3]

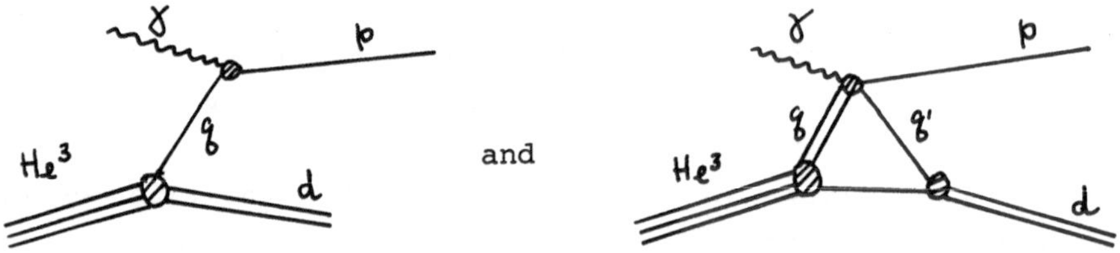

both of which require a knowledge of the momentum distributions $\Phi^t(q)$ and $\Phi^o(q,k)$ where q is the momentum of the struck nucleon(s) and k is related to the relative energy of the correlated (np) pair. The techniques implied by the preceding diagrams can be applied to photo- and electrodisintegration of trinuclei at intermediate energies since the Born approximation is valid in these cases. It has also been applied with some success to (p,pp') and (p,pd) interactions on He^3 at intermediate inergies [1].

In this communication we present the results of a calculation of these momentum distributions using a One-Boson-Exchange potential [4] to represent the N-N interaction. This potential is an unpublished version of the Holinde-Machleidt [5] monopole regularized interaction and gives a H^3 binding energy of 7.38 Mev while binding the deuteron at the experimental value of 2.225 Mev. We present the results for the n-d and n-(pn)* momentum distributions, the n-d-H coupling constant $|C_t|^2$, and the probability for the n-d configuration in H^3, P_{dn}.

II. Formulation of the Problem

Our expressions for the vertex functions are defined by the formulas

$$f^{t}(\vec{q}) = \frac{1}{\sqrt{3}} \sum_{ij} i \langle \psi_d \; q\beta | V_j \; \bar{\delta}_{ij} | \Psi \; \alpha \rangle \tag{1A}$$

$$f^{s}(q) = \frac{1}{\sqrt{3}} \sum_{ij} i \langle \psi_{\vec{k}} \; \vec{q}\beta | V_j \; \bar{\delta}_{ij} | \Psi \; \alpha \rangle \tag{1B}$$

Formulas (1A) and (1B) correspond respectively to the triplet and singlet two-body states. The symbol β denotes spin-isospin quantum numbers and α specifies the quantum numbers of the trinucleon system. In the present case, we restrict our attention to the $L = \ell = 0$ contribution to the spatially symmetric part of the trinucleon wave function. In general, we adhere to the formalism used in ref.[11].

After making an angular momentum reduction, equations (1A) and (1B) assume the explicit form

$$\phi^{t}(q) = \sqrt{\frac{3}{8\pi}} \int_0^\infty dp \; p^2 \; \psi_d \; (p) \; \Psi^{S} \; (p,q) \tag{2A}$$

$$R \; \phi^{s}(q,k) = \sqrt{\frac{1}{8\pi}} \left[\Psi^{S}(k,q)\{1+ \tfrac{\pi}{2}mk\beta(k,k)\} + ⨍_0^\infty dp \; p^2 \; \frac{\alpha(p,k)}{k^2-p^2} \Psi^{S}(p,q) \right]$$

$$I \; \phi^{s}(q,k) = \sqrt{\frac{-1}{8\pi}} \left[-\tfrac{\pi}{2}mk\alpha(k,k)\Psi^{S}(k,q) + m ⨍_0^\infty dp \; p^2 \; \frac{\beta(p,k)}{k^2-p^2} \; \Psi^{S}(p,q) \right] \tag{2B}$$

where $\Psi^{S}(p,q)$ is the spatially symmetric trinucleon wave function, $\alpha(p,k)$ and $\beta(p,k)$ are the real and imaginary parts of the two-nucleon half-shell t-matrix, $t_{oo}^{ol}(p,k;k^2+i\varepsilon)$. The symbols R and I mean real and imaginary, m is the nucleon mass and the bars on the integral sign indicate a principal value.

The momentum distribution for the triplet case is related to the vertex function defined by equation (1A) thus

$$\phi^{t}(q) = \frac{-f^{t}(q)}{(2\pi)^{3/2} m(E_3-E_2)} \tag{3}$$

where E_3 and E_2 correspond respectively to the three- and two-nucleon binding energies.

The two-body t-matrix is obtained by solving the integral equation for the K-matrix

$$K(p,q) = V(p,q) + m ⨍_0^\infty dk \; \frac{k^2}{q^2-k^2} \; V(p,k)K(k,q) \tag{4}$$

and

$$\alpha(p,q) = K(p,q)/(1 + \tfrac{\pi}{2}mqK(q,q))^2$$

$$\beta(p,q) = \tfrac{\pi}{2}mqK(q,q)\ \alpha(p,q) \tag{5}$$

The singular integral equation (4) was solved by means of the method of Padé approximants using the OBE potential of Holinde and Machleidt to represent the ^1S_o N-N interaction. This potential gives the values a_s = -23.83 fm. and r_s = 2.703 fm. for the singlet scattering length and effective range, respectively.

The momentum distributions obtained from equations (2A) and (2B) are shown in figures 1 and 2, respectively. In figure 1, we also show the constant vertex and Tabakin potential [6] results for comparison.

From $\phi^t(q)$ we can easily extract the probability for the d-n configuration in H^3 via the expression

$$P_{dn} = 4\pi \int_0^\infty dq\, q^2\, |\phi^t(q)|^2 \quad . \tag{6}$$

Evaluation of this integral produces (for the OBE result) the value

$$P_{dn} = 0.727$$

which is to be compared with the empirical value $P_{dn} = 0.68 \pm 0.09$ obtained by Kitching et al. [7].

Another important parameter which is fairly readily available from our triplet spectral function is the coupling constant $|C_t|^2$ for the H^3-d-n vertex. As stressed by Kim and Tubis [8] the prediction of this number provides a discriminating index of the quality of the underlying trinucleon wave function and potential used in calculating that wave function. A number of methods exist for the extraction of C_t such as the empirical method of Bolsterli and Hale [9] and the conformal mapping technique used by Kisslinger [10], the Laurents series representation of Kim and Tubis [11] and the integral representation of Gibson and Lehman [12]. Using the method of Kim and Tubis and retaining only two terms in the expansion (they keep five, but their two term result is within 1% of their converged result) we obtain the equation:

$$\pi\sqrt{\frac{2}{x_t}}\ \phi^t\ (q) = \frac{C_t}{q^2 + x_t^2} + \alpha \tag{7}$$

where $x_t = m(E_3 - E_2)$ and α is a constant to be determined. The linear relation implied by equation (7) gives the value

$$|C_t|^2 = 2.60.$$

III. Discussion

As can be seen from fig.1 the result for $\phi^t(q)$ using the OBE (normalized to the Tabakin value at q=0) agrees with the Tabakin result for the same function at low q and this common behavior agrees with the constant vertex value whose three-body binding energy is set equal to the Tabakin value of 9.33 Mev. This agreement remains good up to about $q = 0.5\ fm^{-1}$. After this point the high momentum components of the present $\phi^t(q)$ are sharply attenuated relative to those produced by the Tabakin interaction (as mentioned, this potential binds the triton at 9.33 Mev and is separable). In order to use the OBE function in a calculation of ^3He (e,e'p)d cross-sections we intend to complement the results already obtained by calculating in addition the contributions arising from the deuteron L = 2 and He^3 L $\neq$ 0 states as has been done by Dieperink et al. [13]. These latter contributions become very important at high values of the momentum transfer involved in the e,e' experiment.

The probability P_{dn} for the d-n configuration in H^3 is well within the experimental limits and to our knowledge this is the first time this number has been predicted.

In fig. 2 we show the curves obtained for the spectral function S(q,w) using our method and the OBE interaction. The complete function is defined by

$$S(q,w) = \sum_{S=0}^{1} \int d\hat{k}\ \frac{mk}{2}\ |\phi^S(\vec{q},\vec{k})|^2$$

where $w = k^2/m$ and m is the nucleon mass. We have not included the triplet contributions in our curves, but this can be done without difficulty. We intend to use these results to calculate the e,e' cross-section for electrodisintegration of He^3 for energies below pion-production threshold [17].

Another extremely interesting parameter that we hope to extract from $\phi^S(q,k)$ is the d^*-n-H coupling constant $|C_s|^2$ and to this end we are adapting the integral representation of Gibson and Lehman [12]. There has been a recent experimental determination of the $He^3 \rightarrow p+d^*$ asymptotic normalization by Plattner et al. [14] and their value is not in agreement with values calculated by a number of authors (see, however, ref.[16]) using various potentials [15]. Apart from being difficult to determine experimentally it appears that $|C_s|^2$ is extremely sensitive to the N-N interaction and we feel that a calculation using our methods would be very interesting.

References

[1] D.R. Lehman, Phys.Rev. C3, 1827 (1971).
D.R. Lehman, Phys.Rev. C6, 2023 (1972).
[2] I.S. Shapiro, "Dispersion Theory of Direct Nuclear Reactions" in "Selected Topics in Nuclear Physics" International Atomic Energy Agency, Vienna (1962).
[3] F. Prats, "Photodisintegration of He^3" to be published.
[4] K. Erkelenz, Phys. Letts. C13, 193 (1974).
[5] K. Holinde, (Private communication).
[6] F. Tabakin, Phys.Rev. B75, 137 (1965).
[7] P. Kitching et al., Phys.Rev. C6, 769 (1972).
[8] Y.E. Kim and A. Tubis, Ann.Rev.Nucl. Science, Vol.24, 69 (1974).
[9] M. Bolsterli and G. Hale, Phys.Rev.Lett.28, 1285 (1972).
[10] L.S. Kisslinger, Phys.Rev.Lett. 29, 505 (1972).
[11] Y.E. Kim and A. Tubis, Phys.Rev. Lett. 29, 1017 (1972).
[12] D.R. Lehman and B.F. Gibson, Phys.Rev. C13, 35 (1976).
[13] A.E.L. Dieperink, T. De Forest, I. Sick and R.A. Brandenburg, Phys.Lett. B63, 261 (1976).
[14] G.R. Plattner, M. Bornand and R.P. Viollier, Phys.Rev.Lett. 39, 127 (1977).
[15] A.G. Baryshnikov, L.D. Blokhintsev and I.M. Nerod, Skii Zh. ETP. 19, 608 (1974) (Translation; JETP Lett. 19, 315 (1974)); V.B. Belyaev, B.F. Irgaziev and Yu. V. Orlov, Yad. Fiz. 24, 44 (1976) (Translation; Sov. J. Nucl. Phys. 24, 22 (1976)).
[16] V.S. Bhasin and A.N. Mitra, Phys.Rev.Lett.40, 1130 (1978).
[17] J. Mougey, Bull. Am. Phys. Soc. 23, 535 (1978).

+Work supported by the George Washington University Committee on Research

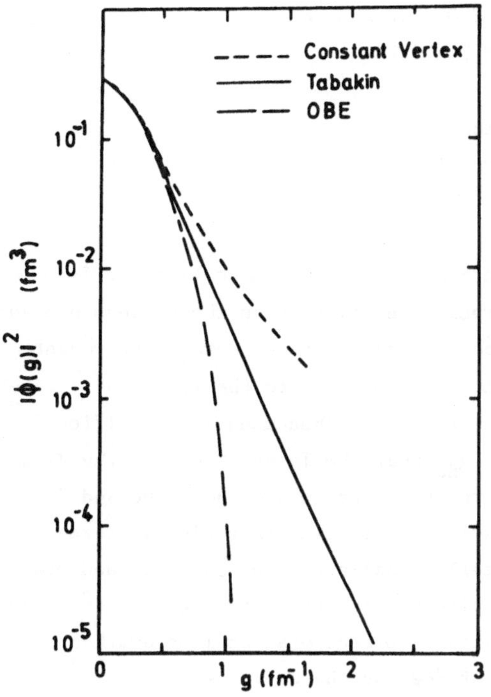

Figure 1

The n-d momentum distribution in H^3 derived using the OBE of Holinde-Machleidt. The OBE result is normalized to the Tabakin result at q = 0. The OBE gives $|\phi(0)|^2 = 3.58$ while the Tabakin gives $|\phi(0)|^2 = 2.8\ fm^3$.

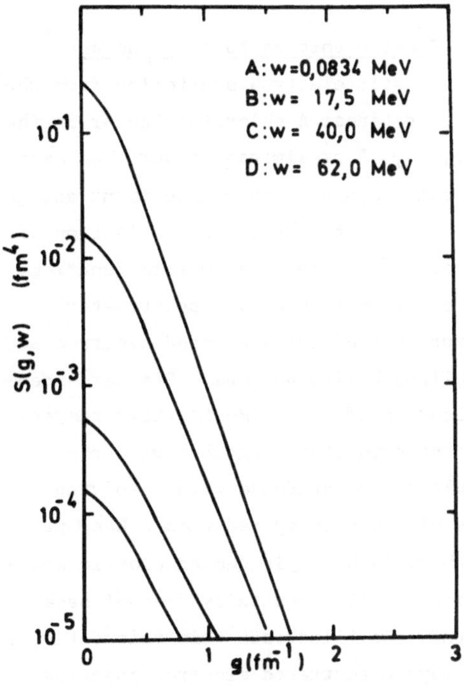

Figure 2

The ^1S_o n-d^* spectal distribution derived from the OBE of Holinde-Machleidt.

ELECTROMAGNETIC STRUCTURE OF ^{3}HE AND ^{4}HE AT LARGE MOMENTUM TRANSFER

B.T. Chertok
American University
Washington, D.C. 20016

Measurements of electromagnetic form factors at large momentum transfer from the few-nucleon systems provide direct and crucial information on the nucleon-nucleon and few-body interactions. Three principal points are presented here. The recent determination of elastic form factors for both ^{3}He and ^{4}He into the region $q^2 > 1$ GeV^2 has given theory a significant impetus; very interesting phenomena and specifically quark-constituent behavior may be visible in $F_{^3He}$ near the large-q^2 end of the data; and the new results for νW_2, the inelastic structure functions of both ^{3}He and ^{4}He provide independent and complementary information to elastic scattering in terms of the approach to scaling, the Fermi momentum tail, final-state interactions and pion electroproduction. The paper is divided as follows- I. $F_{^3He}$ and $F_{^4He}$ for $q^2 \geq 0.8$ GeV^2, II. Comparison with the Extended Version of Dimensional Scaling Quark Model, and III. Results for Electrodisintegration of ^{3}He and ^{4}He at Threshold for $q^2 \geq 0.8$ GeV^2.

Ia. Elastic Form Factors of ^{3}He and ^{4}He

Elastic electron scattering from the nuclei ^{3}He and ^{4}He was measured at the Stanford Linear Accelerator Center in the q^2-ranges 0.7 to 4 and 0.8 to 2.4 GeV^2 respectively. A preliminary report has been published(1) so that I will discuss one salient feature of this experiment and go on to a comparison with theory.

As in our previous investigation of the elastic deuteron structure function, the experiment is a two spectrometer measurement of the scattered electron and recoiling Helium nucleus. Standard identification of e^- and He in their respective spectrometers together with the double arm coincidence with resolving time of ~ 4 ns compared with a beam pulse width of 1500 ns yielded an elastic scattering signal essentially free of background. To illustrate this point, Fig. 1 displays a scattered electron spectrum from ^{3}He at q^2=2.0 GeV^2(E_o = 10.32 GeV and θ_e=8°). The open circles represent single

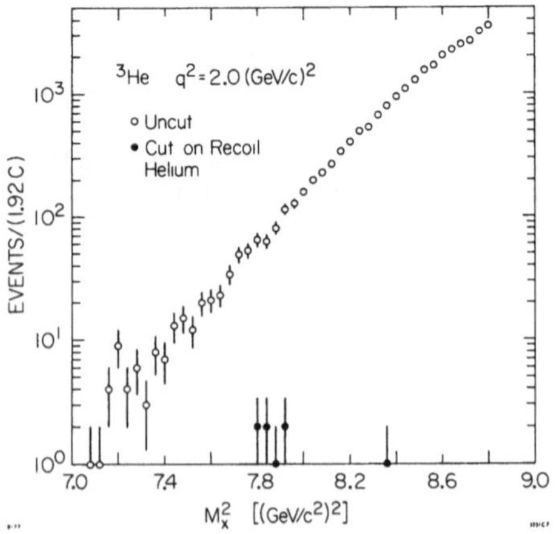

Fig. 1 Missing Mass (Squared) Spectrum

arm data uncorrected for background, while the closed circles are the double arm result which clearly displays the elastic scattering peak with a single event in the radiation tail. This experiment is capable of cross-section determinations less than 10^{-39} cm^2/sr. with errors which are dominated by event-statistics.

Three features of the new data are notable.

1. Both F_{3He} and F_{4He} decrease approximately exponentially for $0.8 < q^2 < 2$ GeV^2 with $F_3 = .03e^{-2.7q^2}$ and $F_4 = .15e^{-4.0q^2}$. The reduced chi-squares are 1.03 and 1.15 respectively, over the entire range of the new data.

2. Near q^2_{max} other interpretations are not ruled out including diffractive phenomena in both ^3He and ^4He and the preasymptotic falloff from quark counting in ^3He.

3. The two form factors are quite similar throughout the q^2-range. The cross-over occurs beyond $q^2 = 1.2$ GeV^2 so that $F_3 > F_4$; this could be directly related to the reverse situation near $q^2 \sim .7$ GeV^2 or signal the appearance of other large q^2-differences in the Helium isotopes.

Ib. Nuclear Theories for F_{3He}

The comparison of the data with two specific nuclear theoretical approaches is presented in Fig. 2 assuming that $F_{ch} = F_{exp}$ for the new data. (Q^2 in Fig. 2 is used interchangeably with q^2 in the text.) Other comparisons are given in ref. 1. The lower dashed curve is a Faddeev calculation(2) with the following input-wave function determined by OBE potential of Holinde - Machleidt with $P_D = 5.18\%$, monopole vertex function, 3.0 F^{-1} momentum cutoff for the spectator, and nucleon form factors of Janssens et al. This calculation gives a ^3H binding energy of 7.38 MeV and rms charge radius of 1.89 F. The solid upper curve is a published prediction(3) based on a variational calculation of ^3He with the Δ-resonance included as part of the ^3He wave function; this one-body term has been augmented by two-body electromagnetic interactions in the form of meson exchange currents. The nucleon form factors used are from Iachello et al.

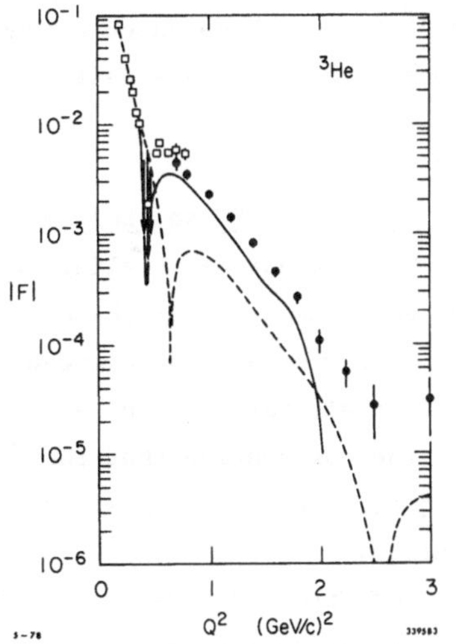

Fig. 2 ^3He Form Factor.

Among the interesting features is the relatively good agreement out to $q^2 \sim 1.8$ GeV^2 of the variational plus MEC calculation and the poor agreement of the Faddeev calculation with experiment. The one-body part of the solid curve, which is the variational calculation, accounts for 2/3 of F_{ch} for $q^2 \geq 1.2$ GeV^2 and this contribution alone is 2.5 times larger than the Faddeev prediction in Fig. 2.

Future work should include the following: I. Reconciliation of the large and long-standing differences in the two calculational approaches to ^3He. II. Calculation of relativistic corrections to the impulse approximation. (Important work on $A(q^2)$ for the deuteron[4] has indicated large corrections--roughly 60% at q^2=1 increasing to 250% at q^2 = 2 GeV^2.) III. Measurement of F_{mag} for ^3He well beyond the q^2 = 0.7 GeV^2 boundary to provide further constraints on the ^3He system.

Ic. Large q^2 Calculations for F_{He}

The He elastic form factors are added to the series of curves which represent the prediction of the Dimensional Scaling Quark Model, $(q^2)^{n-1} F_n \to$ const. as $q^2 \to \infty$.[5] The ansatz that n is an integer given by the minimum number of quantum fields is extended to include atomic nuclei where n = 3A. It is clear from the results that neither $(q^2)^8 F_{3He}$ nor $(q^2)^{11} F_{4He}$ is asymptotic; the comparison with an extended version of DSQM will be examined in the next section. In order to approach the asymptotic region, it is estimated that $q_i = q/n$ be of order of the mean transverse momentum of a constituent, $\langle k_\perp^2 \rangle^{1/2} \sim 0.3$ GeV. Thus $q \sim k_\perp 3A \sim A$ GeV, and $q^2_{3He} \sim 9$ GeV^2.[6,7]

Other theoretical approaches to nuclear form factors for $q^2 > 1$ GeV^2 have been made[8,9] and these calculations have received impetus for refinements and elaboration as a result of the new Helium data.

II. Extended DSQM and Helium Form Factors

The Dimensional Scaling Quark Model has been generally successful in describing high momentum transfer phenomena when s, t, u >> M^2. Two different extended versions of the DSQM have been advanced to explain nuclear form factors and reactions for intermediate rather than asymptotic conditions, i.e. $-t > 1$ GeV^2[6,8]. Here we will make the observation that the ^3He form factor may be exhibiting quark-like behavior for $q^2 \gtrsim 2$ GeV^2. The basis for this statement and the consequences will be briefly examined.

The extended version of DSQM explicitly partitions q among the nucleons and/or quark constitutents and retains the mass corrections on the struck quark of the target.[6] Quark interchange (hereafter QIM or CIM) is a natural mechanism to transfer q/A to each nucleon. Brodsky has referred to this as the "new synthesis" between N-N physics and quark-gluon asymptotic behavior. Representative diagrams for the leading term for F_{3He} at large q^2 are presented in Fig. 3. The momentum of one of the target lines in Fig. 3a or 3b is $p_i = x_i p + \kappa_i$ with $x_i = m_i/M = 1/9$ and κ_i is transverse to p. Evaluating the off-shell propagator on the struck fermion line in Fig. 3a, $((q + p/9 + \kappa_i)^2 - m_i^2 + i\varepsilon)^{-1}$, yields $(8\, q^2/9 + \beta^2)^{-1}$ as the contribution to F_{3He}. Spacelike q^2 is positive here. The mass corrections are $\beta^2 = \kappa_i^2 + \overline{2q \cdot \kappa_i}$, which are related to the mean square quark momentum in the nucleon's rest frame and are set at β^2=.235 GeV^2 from a study of the pion and nucleon elastic form factors. Partitioning q as in Fig. 3a yields $F_{^3He} \sim F_p(q^2/9)\ F_D(\frac{4q^2}{9})\ [1 + q^2/\frac{9}{8}\beta^2]^{-1}$. Calculation of the q^2-dependence of the leading term for the underlying 3-nucleon structure yields the prediction,

$$F_{^3He} \sim F_p^3\left(\frac{q^2}{9}\right)\ \left[1 + \frac{q^2}{\frac{9}{8}\beta^2}\right]^{-1}\ \left[1 + \frac{q^2}{\frac{27}{10}\beta^2}\right]^{-1} \tag{1}$$

with two quark interchanges of q/3. Partitioning q in the chain model, Fig. 3b, gives approximately

$$F_{^3He} \sim \left(1 + \frac{q^2}{9\beta^2}\right)^{-8} \tag{2}$$

The normalization of the above equations was uncertain since it depended on many other lower q^2-features. Accordingly the q^2-dependence was normalized to plausible form factor curves at $q^2 \sim 1$ GeV^2 (Section IV.D of ref. 6). These predictions obey dimensional counting as $q^2 \to \infty$, i.e. $F \sim (q^2)^{-8}$. Similarly $F_{^4He}$ can be described as a skeletal 4-nucleon structure, N-^{3}He, D-D or chain of 12 quarks.

The comparison of the predictions of Eqs. 1 and 2, and the diagram in Fig. 3a with the ^{3}He data indicate a change in slope for $q^2 \gtrsim 2$ GeV^2. The ratio in Fig. 4 examines all three nuclei compared to the quark interchange model which is Eq. 1 for ^{3}He. Within large errors the dashed lines in Fig. 4 provide the evidence for scale invariance in ^{3}He for $q^2 \gtrsim 2.2$ GeV^2. One can generalize on this conjecture that the mass-scale for the threshold for scaling in nuclei is given approximately by the denominator of Eq. 2, i.e. $n\beta^2$ = .7, 1.4, 2.1 and 2.8 GeV^2 for A=1 to 4. This threshold would be

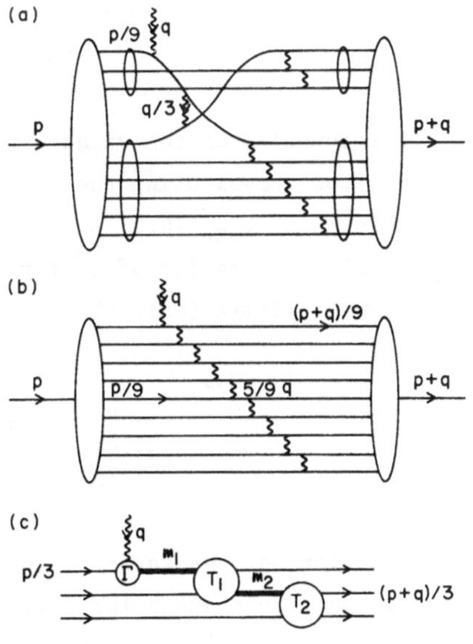

Fig. 3 Diagrams for $F_{^3He}$

$$q_o^2 \sim 0.71A \text{ GeV}^2 \tag{3}$$

which is to be contrasted with the fully asymptotic behavior of the DSQM where $q_o^2 \sim A^2$ may be the proper mass scale. The ^{4}He ratio in Fig. 4 is steeply falling out to the boundary at 1.8 GeV^2 which is less than the 2.8 GeV^2 threshold for flattening out from Eq. 3. An important consequence of this conjecture on the scaling of F_{3He} is to permit an estimate of the relativistic bound state wave function $\psi_{3He}(x,y)$ at x_μ and $y_\mu=0$. Using the diagrams in Fig. 3c for the reactions $\gamma_V + p + (n+p) \to p' + (n'+p')$ the form factor can be evaluated in terms of elastic scattering for $np \to n'p'$, and $pp \to p'p'$, using the techniques developed previously.(6) The leading amplitude from Fig. 3c is

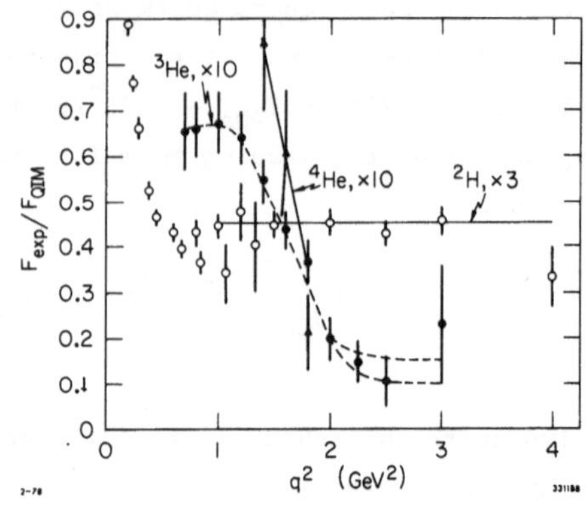

Fig. 4 Nuclear Elastic Form Factors Compared to QIM

$$F_{3He} \sim \frac{6\Gamma\Delta(m_1^2)\, T_1\, \Delta(m_2^2) T_2\, \psi^2_{3He}(0,0)}{F_N(q'^2)} \tag{4}$$

Dimensional counting on the right hand side of this equation gives $(q^2)^{-8}$ as required. The result from substituting the off-shell values of T and the estimated asymptote of F_{3He} into Eq. 4 is $\psi_{3He}{}^2(0,0) \sim 3 \times 10^{-10}$ GeV^4 within characteristic sizes Δx, $\Delta y \sim 1/q_{MAX} = 0.1F$. The ratios of ψ^2 for the deuteron and ^3He are more certain. In the non-relativistic limit (NR), the ratio is

$$\left.\frac{\psi_D^2(0)}{\psi^2_{^3He}(0,0)}\right|_{NR} \simeq 5300\ GeV^{-3} = 41\ F^3. \tag{5}$$

Examinations of the ratio in Eq. 5 as 3A-quark states will require explicit use of quark flavor, spin and color degrees of freedom.(10) Taking a plausible constant-density model at the nucleon level with internucleon distances, $r,\rho \sim 2/q_{MAX} \sim 0.2$ to 0.3F, yields a ratio of probabilities from Eq. 5 of

$$\frac{P(0)}{P(0,0)} \sim 10^3. \tag{6}$$

The large repulsion in adding a sincle nucleon to the overlapped N-N system presumably reflects the strong underlying N-N repulsion at short distances. At the quark-constituent level this repulsion could be attributable to the increased number of fermion degrees of freedom. Equations 5 and 6 could have significance in astrophysics.

III. νW_2 of ^3He and ^4He

Preliminary results of the inelastic structure functions $\nu W_2\ (q^2,\nu)$ for ^3He are presented in Figs. 5 and 6. The ^4He results appear similar. These e,e' measurements at θ_e=8° were made simultaneously with the elastic scattering measurements and cover the excitation region in ^3He and in ^4He from threshold to ~ 180 MeV. For the nucleon and deuteron, it had been observed that for q^2, $\nu \to \infty$, νW_2 becomes an approximate function of the scaling variable $\omega' = 1 + W^2/q^2$ with $W^2 = M^2 + 2M\nu - q^2$, the square of the missing nucleon mass. In the region forbidden for a nucleon, $\omega' < 1$, it was further observed that νW_2 scales for the deuteron for $q^2 \geq$ GeV^2.(11) As νW_2 for ^3He approaches the limit $\omega' \to M/M_{^3He}$ in Fig. 5, one observes the approach to scaling for $q^2 \gtrsim 3$ GeV^2. In Fig. 6 where $E'_{e\ell}$ = 7.75, 9.96 and 12.15 GeV respectively, a less than exponential increase of νW_2 is observed with excitation energy in ^3He. A finite step at threshold is not visible as it is in ^2H and ^4He. The electro-excitation through the region of pion threshold is unremarkable in ^3He as well as ^2H and ^4He.

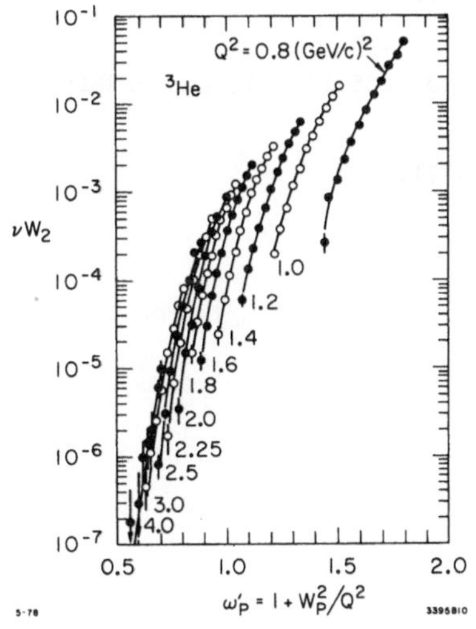

Fig. 5 Inelastic Structure Function

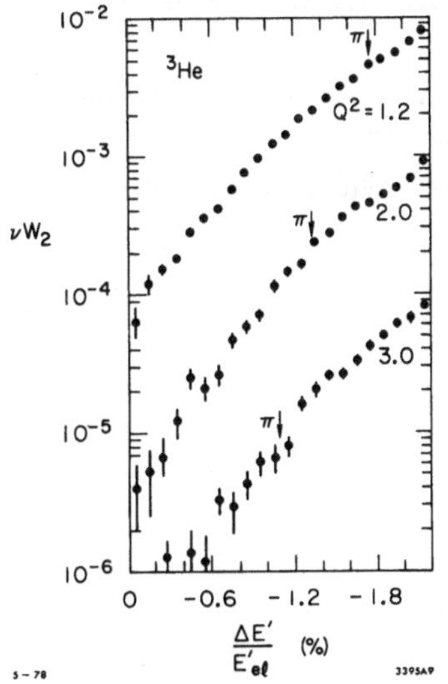

Fig. 6 Inelastic ^3He Spectra

Acknowledgements

The experimental work presented is by the collaboration of ref. 1. In addition, it is a pleasure to acknowledge Professor Stanley J. Brodsky for helpful discussions on Section II, Dr. S. Rock for Section III, and the National Science Foundation for partial support through Grant PHY75-15986.

References

1. R.G. Arnold, B.T. Chertok, S. Rock, W.P. Schütz, Z.M. Szalata, D. Day, J.S. McCarthy, F. Martin, B.A. Mecking, I. Sick and G. Tamas, Phys. Rev. Lett. 40, 1429 (1978).
2. E.P. Harper (private communication).
3. E. Hadjimichael, Nucl. Phys. A294, 513 (1978).
4. R.G. Arnold, C. Carlson and F. Gross, Phys. Rev. Lett. 38, 1516 (1977).
5. S.J. Brodsky and G. Farrar, Phys. Rev. Lett. 31, 1153 (1973); V.A. Matveev, R.M. Muradyan and A.N. Tavhelidze, Lett. al Nuov. Cim 7, 719 (1973).
6. S.J. Brodsky and B.T. Chertok, Phys. Rev. Lett. 37, 269 (1976), and Phys. Rev. D14, 3003 (1976).
7. L.L. Frankfurt and M.I. Strikman, LINP Preprints 349 and 329 (1977).
8. I.A. Schmidt and R. Blankenbecler, Phys. Rev. D15, 3321 (1977); I.A. Schmidt, SLAC - Report - 203 (1977).
9. S. Gurvitz, Y. Alexander and A. Rinat, Ann. Phys. 98, 346 (1976); and R.D. Amado and R.M. Woloshyn, Phys. Lett. 62B, 253 (1976).
10. V.A. Matveev and P. Sorba, Lett. al. Nuov. Cim. 20, 435 (1977); and FERMILAB Pub. 77/56.
11. W.P. Schütz et al., Phys. Rev. Lett. 38, 259 (1977).

S-MATRIX POLE TRAJECTORIES IN THE ^{3}N-SYSTEM

K. Möller
Zentralinstitut für Kernforschung Rossendorf/Dresden, GDR

Three-particle systems are suitable for the investigation of resonance phenomena since on one hand they can be treated mathematically in a rigorous way and on the other hand they have a multichannel decay structure. By studying such systems one hopes to get results which can be generalized to the description of resonances in systems with more than three particles. In a recent paper /1/ we investigated the possibility of the existence of resonances in a system of three neutrons (quantum numbers: $(T,S,L^{\pi})=(3/2,1/2,1^-)$) interacting by s-wave Yamaguchi potential (range parameter $\beta=1.165$ fm^{-1}). Resonances were defined as poles of the S-matrix in the plane of the total cms-energy z_3 of the system or alternatively by the condition $\lambda_n(z_{3,n}^{res})=1$ with λ_n being the eigenvalues of the Faddeev equation kernel. More generally one investigates the pole trajectories $z_{3,n}^{res}$ (α), i.e. the position of the poles in dependence on the two-particle interaction strength α. In paper /1/ the pole trajectory of the first eigenvalue was presented. Fig.3 from paper /1/ is reprinted here (Fig.1). In the preliminary interpretation of this curve there were problems as to the behaviour of the trajectory for small α. To solve these problems we calculated

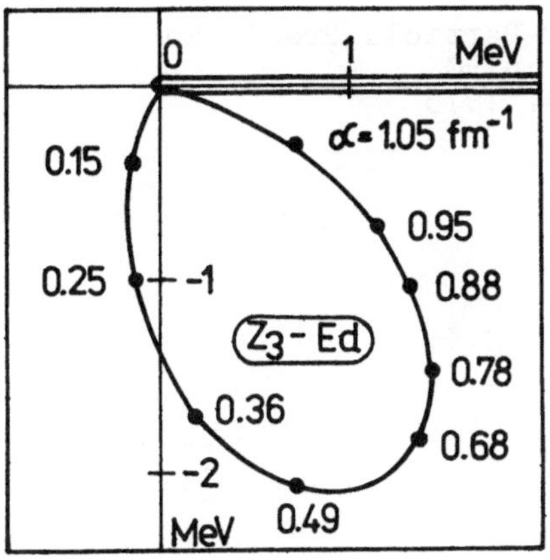

Fig.1 Pole trajectory of the first eigenvalue (Trajectory of the conjugate pole omitted)

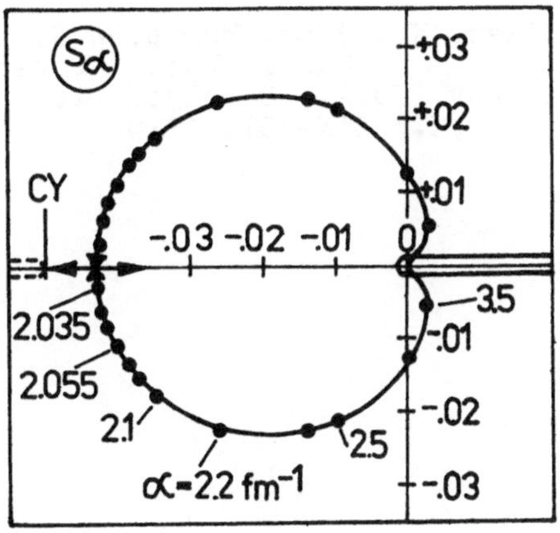

Fig.2 Pole trajectory of the second eigenvalue (s_α defined by: $z_3=-E_d\ (1-s_\alpha)$)

the trajectories for $n>1$. The resulting trajectory for $n=2$ is shown in Fig.2. For $n>2$ qualitatively one gets the same picture as for $n=2$. The behaviour of the trajectories can be summarized as follows: Let (α decreasing) the bound state of order n approach the two-particle threshold. Then a conjugate pole on the negative real axis of the unphysical sheet (virtual state) also approaches the threshold. After forming a double pole at threshold the pole pair describes the loops shown in Fig.2. At some point on the negative real axis the poles coincide again. Then one pole moves to the left toward the logarithmic branching point CY. There it passes to another unphysical sheet. The other pole moves to the right to become a virtual state of order n-1 at the two-particle threshold. From this result together with the behaviour of the eigenvalues near the two-particle threshold which is known analytically /2/ one can conclude that for $\alpha=0$ the first trajectory (Fig.1) reaches the point $z_3=0$ and no bound state is formed. For $\alpha<0$ one would expect to find a continuation of the first trajectory on the unphysical sheet adjacent to the physical sheet above three-particle threshold, but no pole has been found there /1/.
So we conclude that there are no resonances in this model for the ^3n-system.
We mention that there is a complete different situation in a system of three identical spinless bosons with total orbital momentum L=0 /3/.

References

1) Möller, K.: Proc. European Symp. on Few-Particle Probl. in Nucl. Phys. , Potsdam 1977, p.90
2) Simonov, Yu.A., Badalyan, A.M.: ITEP-89, 1973
3) Möller,K.: ZfK-357, 1978

Is there a three neutron resonance?

W. Glöckle
Institut für Theoretische Physik
Ruhr-Universität Bochum
D-4630 Bochum, Germany

A three neutron model with a pure $^1S_0$ interaction is studied [1)] within the framework of the Faddeev equations. We enlarge artificially the potential strength until a three neutron bound state (the first one) occurs. Decreasing the strength again we find the pole-trajectory on which one has to expect the resonance closest to $E = 0$.

We use the fact that at the position of the pole the eigenvalue of the Faddeev kernel is unity and establish an analytical continuation of the eigenvalue equation of the Faddeev kernel into nonphysical energy sheets. To simplify the formal and numerical work the separable Yamaguchi force is used.

We find that in agreement with the shell model the state with spectator angular momentum $l = 1$ is energetically favoured over $l = o$. The enhancement factor for the $^1S_0$ force to get a three body bound state is $\lambda = 4.2$. Decreasing λ the bound state pole overtakes the two body threshold and moves into the lower half plane, bends up again, contrary to what would occur in potential scattering, and goes towards zero together with the two body binding energy. This happens at $\lambda = 1.06$. A careful numerical investigation rules out that the trajectory enters into the lower half plane below the positive real axis. Thus the very pattern of the pole trajectory forbids a low energy resonance in that model.

This investigation has to be supplemented by including p-wave forces. As a first orientation of the effect of p-wave forces we used [2)] the Reid potential in our three body bound state code. We searched for the enhancement factors for the $^1S_0$ and the $^3P_0$ force to find a three neutron bound state near zero energy. Thereby a drastic increase of the actual $^3P_0$ force was necessary. Thus for a binding energy of around -0.5 MeV one is forced to choose $\lambda_{^1S_0} = 1.25$ and $\lambda_{^3P_0} = 5.8$, where both forces support two body bound states around - 0.5 MeV. This refers to the $1/2^-$ state. The inclusion of the $^3P_2$-$^3F_2$ forces has negligeable effect, since they occur together with $l = 2$. The $3/2^-$ state and the positive parity states, proposed by Mitra[3)], are energetically less favoured and need larger enhancement factors for the same binding energy. It is interesting to note that the stronger binding energy of the $3/2^-$ state familiar from the shell model does not show up for T = 1 forces, where the $^3S_1$-$^3D_1$ force is absent.

Within the described theoretical framework there is hardly a possibility for a low energy three neutron resonance left. A final study of the pole trajectory including p-wave forces is underway.

1) W. Glöckle, to be published in Phys. Rev.
2) W. Glöckle, R. Offermann, to be published
3) A.N. Mitra, Phys. Rev. 150 (1966) 839
Phys. Rev. Lett. 16 (1966) 523

THREE-BODY REACTIONS WITH CHARGED PARTICLES

E.O. Alt
Institut für Physik, Universität Mainz, D 65 Mainz, West-Germany

1. Introduction

Three-body processes can be treated reliably by means of three-body integral equations provided that at least two of the participating particles are uncharged. This restriction imposes severe limitations on the applicability of the theory, since in most nuclear and atomic collisions processes two or even all three particles carry charge. In these cases besides any short-ranged interaction between two of the bodies, the Coulomb potential acts in addition. But it is already well known from two-particle scattering that the infinite range of the latter introduces difficulties which render inapplicable conventional (short-range) scattering theory (a short discussion of some aspects relevant in this context and some important references can be found in Ref.1). Of course, there exists a variety of procedures which yield the correct definition of the two-particle scattering amplitude in the presence of the Coulomb force, and a recipe for its calculation. But only recently it has been shown that methods which are well known in the two-particle case can be used with advantage also in three-body reactions. In fact, in this way it was possible to rigorously define two-fragment scattering amplitudes for systems consisting of one neutral and two charged[1)], and even of three charged particles[2)]. The break-up amplitudes leading to three free particles have been investigated up to now[1)] only for situations in which not more than two charged bodies emerge. One decisive aspect of the above mentioned approaches is that at the same time integral equations are provided which render feasible practical calculations of Coulomb corrections in nuclear reactions and of atomic scattering processes. They have already been used for numerical investigations of elastic proton-deuteron scattering[3)]. This reaction is particularly interesting since, e.g., the comparison with sophisticated neutron-deuteron calculations can provide insight into the problem of charge symmetry breaking.

In Refs. 1 and 2 the starting point are the well-known three-body integral equations for short-ranged potentials which then are reformulated to accommodate also Coulomb potentials. However, an alternative approach has been proposed recently which is based directly on the resolvent equations and the fundamental operator relations[4)]. Due to this fact it can be written down easily for an arbitrary number of charged particles, leading to a correct definition of two-fragment scattering amplitudes. Of course, when the latter are specialized to the three-charged particle case they coincide with those introduced in Ref. 2.

2. Two-Fragment Processes of Three Charged Bodies

Consider three particles with charges e_1,e_2,e_3, interacting via pairwise short-ranged plus Coulomb potentials. The transitions we are interested in lead from an incoming state characterized by a channel index α designating the free particle, by the quantum numbers of the bound state of β and γ collectively denoted by "m", and by the relative momentum $\vec{q}_\alpha$ between α and the center-of-mass of $(\beta+\gamma)$, to an outgoing configuration $(\beta,n,\vec{q}_\beta')$. In the case of short-ranged potentials this process can be described by the amplitude $T_{\beta n,\alpha m}(\vec{q}_\beta',\vec{q}_\alpha) = \langle\vec{q}_\beta'|T_{\beta n,\alpha m}(E+io)|\vec{q}_\alpha\rangle$ where the effective two-body transition operators $T_{\beta n,\alpha m}(z)$ fulfill multichannel LS-type equations. With V and G_0 denoting the effective potential and free Green's function they read explicitly[5)]

$$T_{\beta n,\alpha m}(z) = V_{\beta n,\alpha m}(z) + \sum_{\gamma,rs} V_{\beta n,\gamma r}(z) G_{0;\gamma,rs}(z) T_{\gamma s,\alpha m}(z). \tag{1}$$

But in the presence of Coulomb potentials the on-shell scattering amplitudes can no longer be obtained in this way. This situation is completely analogous to what is encountered in two-charged particle scattering.

The remedy is, as in the two-body case, to decompose the transition amplitude, by application of the two-potential formalism, into two parts, one of which is calculable by integral equations with a shorter-ranged potential, the other one being expressible in terms of explicitly known functions. This is made possible by the intuitively expected but rigorously derived fact that the effective potential V occurring in Eq.(1) admits a separation into an infinite-ranged part and a remainder of shorter range,

$$V_{\beta n,\alpha m}(\vec{q}_\beta',\vec{q}_\alpha) = \delta_{\beta\alpha}\delta_{nm}\frac{e_\alpha \bar{e}_\alpha}{(\vec{q}_\alpha'-\vec{q}_\alpha)^2} + V'_{\beta n,\alpha m}(\vec{q}_\beta',\vec{q}_\alpha). \tag{2}$$

The former can be interpreted as a genuine two-body potential giving rise to Coulomb scattering of a particle with charge e_α off another (structureless) particle with charge $\bar{e}_\alpha = e_\beta + e_\gamma$, the so-called "Coulomb scattering of particle α off the center of mass of particles β and γ". (In conventional approaches such a type of interaction is introduced by hand).

Due to the essentially two-body nature of the pure Coulomb contribution in (2) the methods established for two-charged elementary particle scattering can be taken over with minor modifications, resulting in the proof of the following representation for the on-shell arrangement amplitudes

$$\begin{aligned} T_{\beta n,\alpha m}(\vec{q}_\beta',\vec{q}_\alpha) = {} & \delta_{\beta\alpha}\delta_{nm}\bar{\delta}(\bar{e}_\alpha,0)\, T_c(\vec{q}_\alpha',\vec{q}_\alpha) \\ & + \bar{\delta}(\bar{e}_\beta,0)\bar{\delta}(\bar{e}_\alpha,0)\langle\vec{q}^{\,\prime(-)}_{\beta,c}|t^{sc}_{\beta n,\alpha m}(E+io)|\vec{q}^{\,(+)}_{\alpha,c}\rangle + \delta(\bar{e}_\beta,0)\bar{\delta}(\bar{e}_\alpha,0)\langle\vec{q}_\beta'|t^{sc}_{\beta n,\alpha m}(E+io)|\vec{q}^{\,(+)}_{\alpha,c}\rangle \\ & + \bar{\delta}(\bar{e}_\beta,0)\delta(\bar{e}_\alpha,0)\langle\vec{q}^{\,\prime(-)}_{\beta,c}|t^{sc}_{\beta n,\alpha m}(E+io)|\vec{q}_\alpha\rangle + \delta(\bar{e}_\beta,0)\delta(\bar{e}_\alpha,0)\langle\vec{q}_\beta'|t^{sc}_{\beta n,\alpha m}(E+io)|\vec{q}_\alpha\rangle. \end{aligned} \tag{3}$$

Here we have defined $\delta(\bar{e}_\alpha,0)$ being equal to 1 for $\bar{e}_\alpha = (e_\beta + e_\gamma) = 0$ and equal to 0 for $\bar{e}_\alpha \neq 0$, and $\bar{\delta}(\bar{e}_\alpha,0) = 1 - \delta(\bar{e}_\alpha,0)$. The first term on the r.h.s. is the explicitly known two-body Coulomb amplitude describing the scattering of particle α off the center of mass of particles β and γ. The remaining terms, called Coulomb-modified short-range amplitude, have been written as matrix elements of an operator t^{sc} between two-body Coulomb scattering states $|\vec{q}^{\,(\pm)}_{\alpha,c}\rangle$ (for $\bar{e}_\alpha \neq 0$) which take care of the long-ranged distortion of the motion of the two charged fragments in the initial and/or final state. It can be calculated from an integral equation of the type (1) with the shorter-ranged potential V'.

Two special cases are worth mentioning. First consider the situation that the bound states of the two charged particles both in the initial and the final state are neutral. Then the last form in (3) for the Coulomb-modified short-range amplitude applies, implying that conventional short-range scattering theory is valid here. Secondly, all formulae remain valid when only two of the three particles are charged. We have only to set one of the charges equal to zero.

3. Break-up processes

Here we discuss only break-up reactions for three-body systems composed of one neutral and two charged particles, a prominent example being the deuteron break-up by proton impact. Without the Coulomb force the scattering amplitude leading from an incoming configuration $(\alpha, m, \vec{q}_\alpha)$ to an outgoing three-free particle state $|\vec{p}_3{}', \vec{q}_3{}'\rangle$ with $\vec{p}_3{}'$ being the relative momentum between particles 1 and 2, is given by $T_{0,\alpha m}(\vec{p}_3{}', \vec{q}_3{}'; \vec{q}_\alpha) = \langle \vec{p}_3{}', \vec{q}_3{}' | T_{0,\alpha m}(E+io) | \vec{q}_\alpha \rangle$. The effective transition operators $T_{0,\alpha m}(z)$ fulfill multichannel LS-type equations analogous to (1),

$$T_{0,\alpha m}(z) = V_{0,\alpha m}(z) + \sum_{\gamma, rs} T_{0,\gamma r}(z) G_{0;\gamma,rs}(z) V_{\gamma s,\alpha m}(z). \tag{4}$$

But if long-ranged Coulomb potentials are present in addition, the on-shell break-up amplitudes can not be calculated according to this prescription.

As can be seen from Eq.(4) not only does the infinite-ranged part of the arrangement potentials $V_{\beta n,\alpha m}$, which is explicitly displayed in Eq.(2), occur but also the one arising from the break-up potentials $V_{0,\alpha m}$. However, the latter has been shown in Ref.1 to effect only a long-ranged distortion of the movement of the emerging two charged bodies. Thus, after the separation of both $V_{\beta n,\alpha m}$ and $V_{0,\alpha m}$ into one part of long and one of shorter range has been performed the following representation for the on-shell break-up amplitude can be derived (we choose particle 3 to be the neutral one)

$$T_{0,\alpha m}(\vec{p}_3{}', \vec{q}_3{}'; \vec{q}_\alpha) = \bar{\delta}_{\alpha 3} \langle \vec{q}_3{}' | \langle \vec{p}_{3,c}'^{(-)} | \bar{\chi}_{\alpha m} \rangle | \vec{q}_{\alpha,c}^{\,(+)} \rangle$$
$$+ \bar{\delta}_{\alpha 3} \langle \vec{q}_3{}' | \langle \vec{p}_{3,c}'^{(-)} | t^{sc}_{0,\alpha m}(E+io) | \vec{q}_{\alpha,c}^{\,(+)} \rangle + \delta_{\alpha 3} \langle \vec{q}_3{}' | \langle \vec{p}_{3,c}'^{(-)} | t^{sc}_{0,3m} | \vec{q}_3 \rangle . \tag{5}$$

The first term on the right-hand side is the pure Coulomb break-up amplitude which can be calculated by quadrature from analytically known two-body Coulomb scattering wave functions ($|\bar{\chi}_{\alpha m}\rangle$ is related to the bound state wave function of the pair $\beta+\gamma$). The other terms of Eq.(5) are called Coulomb-modified short-range break-up amplitude and are written here already as matrix elements of $t^{sc}_{0,\alpha m}$ between two-body Coulomb scattering states $\langle \vec{p}_{3,c}'^{(-)}|$ for the outgoing charged pair (1+2), and $|\vec{q}_{\alpha,c}^{\,(+)}\rangle$ (for $\alpha \neq 3$) for the incoming two charged fragments. The operator $t^{sc}_{0,\alpha m}$ can be obtained as solution of an integral equation with the shorter-ranged parts of $V_{\beta n,\alpha m}$ and $V_{0,\alpha m}$, or by quadrature from the corresponding non-break-up operator $t^{sc}_{\beta n,\alpha m}$.

References

1. E.O. Alt, W. Sandhas and H. Ziegelmann, to be published in Phys. Rev. C.

2. E.O. Alt, Invited Talk given at the Workshop on Few Body Problems in Nuclear Physics, Trieste, 1978.

3. E.O. Alt, in Few Body Dynamics, edited by A.N. Mitra et al. (North Holland, Amsterdam, 1976); E.O. Alt, W. Sandhas, H. Zankel and H. Ziegelmann, Phys. Rev. Lett. 37, 1537 (1976); H. Ziegelmann, Invited Contribution to this Conference.

4. E.O. Alt and W. Sandhas, Contribution to this Conference, and to be published.

5. E.O. Alt, P. Grassberger and W. Sandhas, Nucl. Phys. B2, 167 (1967).

UNEXPECTED BEHAVIOR OF THE CCA TRANSITION OPERATORS UNDER PARTICLE INTERCHANGE

F. S. Levin*
Physics Department, Brown University
Providence, Rhode Island 02912/USA

and

H. Krüger
Fachbereich Physik, Universität Kaiserslautern
6750 Kaiserslautern, FRG

When the kernel of the integral equations defining a set of n-particle transition operators is "label transforming", then effects of particle identity can be included in a straightforward way, as discussed recently by Bencze and Redish [1]. The kernel of the CCA class of transition operator equations obtained from a channel permuting array (CPA) [2] is, however, not label transforming, so that the Bencze-Redish method is inapplicable. We establish this here for a three-body example using one of the simplest procedures one might adopt to include exchange effects, that involving a projection operator. Although this method does succeed for the CCA equations obtained using a Faddeev-Lovelace (F-L) array [2], it unexpectedly fails for the CPA equations.

Consider a three-particle system interacting via pair potentials V_i, $i = 1,2,3$ being a pair index for particles (j,k). Denoting the kinetic energy operator by H_o, the $i^{\underline{th}}$ channel Hamiltonian H_i is $H_i = H_o + V_i$, and the $i^{\underline{th}}$ channel interaction is $V^i = H - H_i = V_j + V_k$, where H is the full Hamiltonian. We work with wave function components [3] rather than with transition operators for which, however, our results are also valid. In the present instance, particles 2 and 3 are assumed to be identical. That is, if $P = P_{23}$ interchanges the labels 2 and 3, with $P^2 = 1$, then $PH_oP = H_o$, $PH_2P = H_3$, $PH_1P = H_1$ and $PHP = H$. There are two equivalent sets of CPA equations (of which we consider one) and one set of F-L equations, each involving matrix Hamiltonian operators [3]: $\mathcal{H}^{CPA}\underset{\sim}{\psi} = E\underset{\sim}{\psi}$ and $\mathcal{H}^{F-L}\underset{\sim}{\chi} = E\underset{\sim}{\chi}$, where $\underset{\sim}{\psi}$ and $\underset{\sim}{\chi}$ are column vectors of wave function components. The operators are [3]:

$$\mathcal{H}^{CPA} = \begin{pmatrix} H_1 & V^2 & 0 \\ 0 & H_2 & V^3 \\ V^1 & 0 & H_3 \end{pmatrix}, \quad \mathcal{H}^{F-L} = \begin{pmatrix} H_1 & V_1 & V_1 \\ V_2 & H_2 & V_2 \\ V_3 & V_3 & H_3 \end{pmatrix}.$$

It is straightforward to show that $R\mathcal{H}^{F-L}R = \mathcal{H}^{F-L}$, where R is a 3 x 3 matrix with all elements equal to zero except for $R_{11} = R_{23} = R_{32} = P$. Hence the projection operator matrix $\Lambda_\pm = (1 \pm R)/2$ will effectively decouple the F-L equations into two pairs of equations, one pair for the symmetric (+) and one pair for the antisymmetric (-) combinations, $(1 \pm P)\chi_1$ and $\chi_3 \pm P\chi_2$. Thus, inclusion of the effects of identity under particle interchange is achieved.

An analogous procedure fails for $\mathcal{H}^{CPA}$, as we now demonstrate. The only possibility for success is first to find a matrix S such that $S\mathcal{V}S = \mathcal{V}$, where $\mathcal{V}_{jk} = \mathcal{H}_{jk}^{CPA} - H_j\delta_{jk}$, and then next to show that S leaves the diagonal elements of $\mathcal{H}^{CPA}$ unchanged. Let S_{jk} be the nine elements of such an assumed S. It is a simple matter of matrix multiplication to show that the only S satisfying $S\mathcal{V}S = \mathcal{V}$ is an arbitrary multiple of the unit matrix, which clearly demonstrates that no projection operator analogous to $\Lambda_{\pm}$ can be found for $\mathcal{H}^{CPA}$. Since $\Lambda_{\pm}$ involves the symmetrizer or antisymmetrizer for this system, the inapplicability of the Bencze-Redish method is made evident. Furthermore, the procedure of Goldflam and Tobocman [4], which reduces the CPA equations when all particles are identical, also does not apply to this special case wherein only two of the three particles are identical.

The implication of this result, viz., that the n-particle CPA equations cannot be reduced when fewer than all n particles are identical, has recently been shown by Levin and Bencze [5] to be false for precisely the present example: a reduction can be effected, but only by using an entirely different approach. Methods to extend this result to more general cases are being studied. The reduction of the CPA equations when identical particles are present would seem to be one of the major problems of n-particle scattering theory, particularly in light of the successful applications of these equations to atomic and molecular structure [6].

REFERENCES

*Work supported in part by the United States Department of Energy

[1] Gy. Benzce and E. F. Redish, J. Math. Phys. (in press).

[2] D. J. Kouri and F. S. Levin, Nucl. Phys. A253, 295 (1975).

[3] D. J. Kouri, H. Krüger and F. S. Levin, Phys. Rev. D15, 1156 (1977).

[4] R. Goldflam and W. Tobocman, Phys. Rev. C (in press).

[5] F. S. Levin and Gy. Bencze, to be submitted for publication.

[6] F. S. Levin, Application of Few-Body Methods to Atomic and Molecular Structure, invited paper, this Conference.

MINIMAL RELATIVISTIC DYNAMICS OF THE 3π AND ππN SYSTEMS

I. J. R. Aitchison

Department of Theoretical Physics,
Oxford University, Oxford OX1 3NP, U.K.

At the Delhi Conference I described an approach to the relativistic three body problem based on two "minimal" ingredients: two-body unitarity, and analyticity. The amplitude for producing a three hadron final state is written as a sum of three terms, one for each subenergy channel, each involving certain angular factors and the two-body denominator function for the ith pair $D_i(s_i)$. A simple "isobar model" of this kind can be made to satisfy two-body unitarity by introducing correction functions $\phi_i(s_i,m^2)$ multiplying the D_i^{-1} factors, where m is the total cms energy. The imposition of unitarity and analyticity in s_i leads to an integral equation for ϕ_i of the form

$$\phi_i(s_i,m^2) = \phi_i^o(s_i,m^2) + \int_{-\infty}^{s_{jmax}} K_1(s_i,s_j,m^2)\frac{1}{D_j(s_j)}\phi_j(s_j,m^2)ds_j + \int_{LH} K_2(s_i,s_j,m^2)\frac{1}{D_j(s_j)}\phi_j(s_j,m^2)ds_j \qquad (1)$$

where ϕ_i^o has no threshold branch point in s_i. The K_1 integral includes a part which runs over the physical region, while the K_2 integral does not - the latter is confined to the unphysical ("left hand") region. This integral equation has the same structure as those derived in approaches to the three body problem formulated (via potential theory analogues) more directly in the *three body* channel. Thus the dynamical theory produced so as to embody, in a "minimal" fashion, unitarity corrections to the isobar model in the *subenergy* channels, may also serve to investigate some features of dynamics in *three hadron* channels. Since Delhi, a number of detailed calculations using this approach have been done by myself, and collaborators. I shall discuss work on the 3π and ππN systems.

1. 3π Systems (a) The ω Channel.

First, I shall review work done by Golding and myself on the ω channel[1]. We included only the p-wave πρ channel, and found the following results:

(i) For a choice of the ππ D function corresponding to the physical ρ, it was possible to generate an ω, whose mass could be adjusted by choice of a cut-off parameter "c", but whose width was then a prediction of the model; we obtained an ω *width of 17 MeV.* That a three body resonance could actually be generated at all in such a simple model is worth remarking on and the width is encouragingly realistic. But how significant is the result, really? The actual physics input (unitarity and analyticity) is believable as regards that part of the s_i - s_j plane which corresponds to the physical region, or its vicinity. In terms of eqn. (1) this means that the physics of the kernel K_1 is well-founded, and in fact K_1 is just the appropriate projection of the one-pion exchange potential in $\pi\rho \to \rho\pi$. However, by itself this part of the kernel is insufficient to generate an ω, for physical ρ parameters. Rather, it is the more distant unphysical part of the total kernel $K_1 + K_2$ which provides the additional "force" to generate the ω. The parameter c controls the asymptotic behaviour of the ρ D function, and thus allows us to vary the extent to which the unphysical part of the s_j integration contributes. I do not believe that the precise from of this contribution is physics: I interpret the "far left" contribution in $K_1 + K_2$ as representing short range forces in the ω channel, which the model is flexible enough to incorporate, but which are only phenomenologically controllable via the parameter c. To get anything more fundamental out, one will surely have to put something more realistic in (explicit short range contributions, quark structures, etc.). Nevertheless, given some kind of short range contribution strong enough to produce a three body resonance, I think the result for the *width* is significant: it has to do with the physical decays, of course, so that here the physical region input should be determinative.

(ii) We also experimented with a wide variety of "ρ" parameters (mass and width), and different c's. In the three body sector, we found we could generate (by choice of c, for physical ρ) *an effect reminiscent of the A1 situation* (though of course in the wrong J^P state) - namely, a broad peak, with a phase rising barely to 90° and then dropping. In all cases we found that the rescattering effects were mainly dependent on the ρ pole parameters (mass and width), and were largely independent of the asymptotic behaviour controlled by c. *Three body resonance effects (sensitive to the short range pieces in $K_1 + K_2$) could be clearly separated from isobar model corrections in the subenergy variables (controlled by the physical region part of the kernel).* We also found that we needed less "short range" piece if m_ρ were less, or Γ_ρ more (both manifestations of a stronger effective $\pi - \pi$ forces,presumably).

B. <u>Other 3π Channels</u>.

It is obviously important to know whether these conclusions carry over into other physically realistic cases. Another simple 3π system is the A2 ($I^G J^P = 1^- 2^+$). Together with a student at Oxford, Mr. K. R. Parker, I have set up the integral equation for this case ($\pi\rho$ in a d-wave). Once again, the equation involves a single cut-off c. It turns out that, for a physical ρ, one has to allow even more distant unphysical regions of the kernel to enter, before one comes near to generating an A2. Possibly this corresponds to the necessity of having a larger short range attraction, to compensate for the larger centrifugal barrier in the d-wave case. Numerical problems have been encountered for the very large cut-off masses necessary to get an A2 for the physical ρ case. But the same qualitative features are clear in the A2 channel as appeared for the ω: for stronger $\pi\pi$ forces (smaller m_ρ or larger Γ_ρ) the cut-off mass is not so large and an A2 is generated without difficulty. For a physical ρ, a reasonable A2 width could probably be obtained in the model. Independent of these calculations, the rescattering corrections are, as before, dominated by the physical region parameters, and are in fact very smoothly varying in subenergy. Parker's calculation definitely shows that *the isobar model should be perfectly reliable for extracting the A2 phase*, which is just as well in view of the beautiful Breit Wigner that has been found experimentally!

A case of more pressing urgency is the A1. A vast amount of new data will very soon be appearing from the ACCMOR collaboration. Experience gained in the previous calculations indicates that, as far as the rescattering corrections are concerned, the *first iteration* to the integral equations gives an excellent approximation. The equations for the A1 (retaining $\pi\rho$ in s- and d-waves, and $\pi\varepsilon$ in the p-wave) have been obtained,(2) and Parker is engaged in evaluating the first iteration. We will then feed the correction functions to those doing the partial wave analysis at Oxford, and see how the extracted A1 phase is affected. Personally, I don't anticipate a dramatic result. Certainly there may be quite large corrections due to the ρ feeding into the ε channel (see the $\pi\pi N$ case below), but the ε phase is not a crucial benchmark. The ρ phase is far more critical, but the $\varepsilon \to \rho$ feeding is likely to be insignificant (see below), while any effect from $\rho \to \rho$ will be concentrated near threshold, and hence de-emphasised by the ρ D-function.

II <u>Meson- meson- nucleon systems</u>.

Another important area for the application of these ideas is of course the 2 meson + 1 baryon one - specifically, the $\pi\pi N$ sector, for which an elaborate isobar model analysis was made a few years ago. The main technical impediment in the way of a straightforward implementation of our two requirements of unitarity and analyticity is, in this case, the nucleon spin. This turns out to lead to awkward square root factors in the unitarity relations, which have to be eliminated before analyticity can be used. John Brehm (at University of Massachusetts at Amherst) and I have recently solved this problem,(3) and have shown how to extract the kinematical singularities associated with the nucleon spin. We also, incidentally, treat the case of three different masses (e.g. $K\pi N$), which as usual presents some non-trivial complications in a dispersion-theoretic approach. The way is therefore clear to proceed with applications. As a start, I have been looking at $\pi N \to \pi\pi N$ in the P11 wave, retaining the $\pi\Delta$ and εN isobar channels, at cms energies in the range 1300-1600 MeV. This is a wave in which Aaron et al.(4) found large unitarity corrections to the isobar model: these authors, however, did not impose analyticity, and consequently part of the rapid variation they encountered was unphysical.(5) I have calculated the first iteration correction to the isobar model for this case, and find that, although there is some theoretical uncertainty about the absolute magnitude of the corrections (due to uncertainties in the "left hand" part of the kernels, again), the shape of the corrections is, as in the 3π cases, determined by the long-range physical region parts (π exchange in $\varepsilon N \leftrightarrow \pi\Delta$, N exchange in

$\pi\Delta \leftrightarrow \Delta\pi$). There is a striking variation (Fig. 1) in the first iteration correction to ϕ_ε, in the vicinity of the $\pi\pi$ threshold, due to the Δ feeding the ε. But there is no sign of the other pronounced bump at higher $\pi\pi$ mass found by Aaron et al.,[4] and the latter is therefore, as expected,[5] spurious. The peak near threshold is associated with the square root singularity in ϕ_ε, accentuated by a (non-spurious) logarithmic singularity which moves away from the physical region as m increases past about 1600 MeV. Similar calculations show that the correction in the $\pi\Delta$ channel from εN has no interesting subenergy variation; calculations for $\pi\Delta \to \pi\Delta$ have not yet been completed. The results obtained so far tend to confirm those of early (non-relativistic and spinless) calculations of mine[6]: that the most significant corrections are likely to be those (c.f. $\Delta \to \varepsilon$) in which a pronounced resonance quite near threshold feeds into a strong, *but non-resonant*, channel. The sub-energy variation of the amplitude in the non-resonant channel near threshold is likely to be markedly affected by the corrections: this may well influence the extraction of low energy $\pi\pi$ (or πK) parameters from single pion production at low energies.

(1) I. J. R. Aitchison and R. J. A. Golding, J.Phys.G. 4 43 (1978).

(2) I. J. R. Aitchison, J.Phys.G. 3 121 (1977).

(3) I. J. R. Aitchison and J. J. Brehm, to appear in Phys. Rev.

(4) R. Aaron et al., Phys. Rev. D12 1984 (1975).

(5) I. J. R. Aitchison and R. J. A. Golding, Phys. Lett. 59B 288 (1975).

(6) I. J. R. Aitchison, Nuovo Cim. 51A 249, 272 (1967).

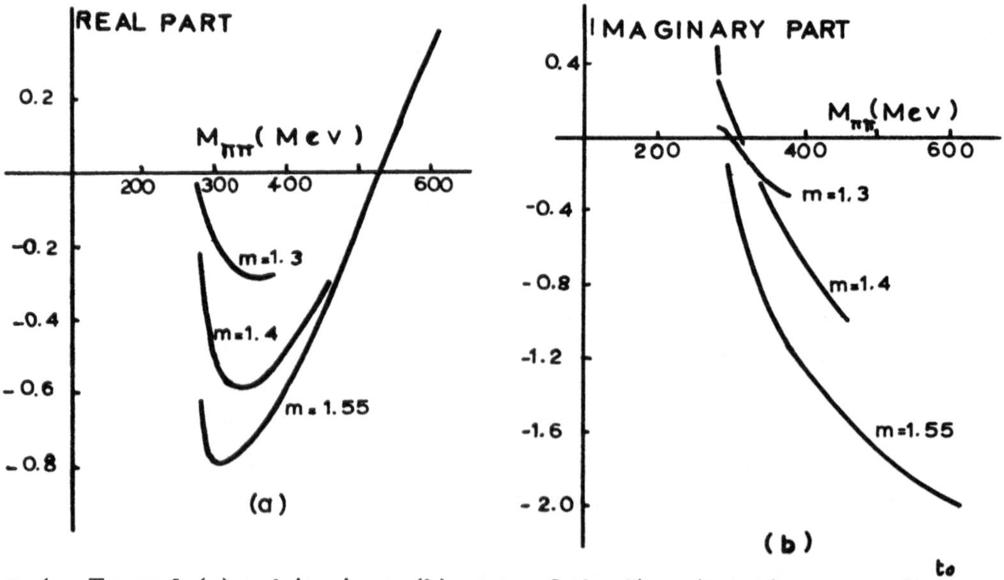

Fig. 1. The real (a) and imaginary (b) parts of the first iteration correction to ϕ_ε for various $\pi\pi N$ cms energies m (in GeV), in the P11 wave (Δ feeding ε). The normalisation is arbitrary.

(n,d) ELASTIC DIFFERENTIAL CROSS SECTION AT 2.48 AND 3.28 MeV AND RELATED PHASE SHIFT ANALYSES*

P. Chatelain, Y. Onel, R. Viennet and J. Weber
Institut de Physique de l'Université, Rue A.-L. Breguet 1,
CH - 2000 Neuchâtel, Switzerland

1. INTRODUCTION

In the case of (n,d) scattering, at energies of a few MeV, the elastic differential scattering cross sections are not well determined, especially beyond cos $\theta_{CM} = -0.8$ and the number of data points is rather scarce [1]. This lack of experimental data causes great difficulty to perform a phase shift analysis even in the simple case when partial waves up to $\ell_{max} = 2$ are included as free parameters in the parametrization of the scattering amplitude. We have therefore, in the framework of our study of the (n,d) scattering [2,3], measured this cross section at 2.48 and 3.28 MeV with special emphasize on large scattering angles.

2. EXPERIMENTAL METHOD

We have used the recoil energy spectrum method [4]. The neutrons were produced by the $D(d,n)^3He$ reaction. The incident deuteron beam was produced by our 3 MeV Van de Graaff accelerator. In order to reduce the background as much as possible, we have discriminated between neutrons and gammas and used the method of the associated particle by detecting the ^3He recoil nuclei. The experimental set-up as well as the method which we have used to measure the resolution of our detector and its light output function is fully described elsewhere [5].

3. DATA REDUCTION

In order to solve the problems due to the finite resolution of the recoil detector, we have used a method described in ref. [4] where the resolution function and the light yield function were determined independently of the cross section measurement itself [5]. The end products of this procedure were two sets of coefficients a_j's for the development of the cross section at 2.48 and 3.28 MeV :

$$\frac{d\sigma}{d\Omega} = \sum_{j=0}^{j_{max}} a_j \cos^j \theta_{CM}$$

(j was determined by the statistical F_{1,ν_2} test (at 5%)).

*Supported in part by the Swiss National Science Foundation.

Two sets of discrete data were extracted from these two sets of coefficients to be used as input data to our standard phase shifts analysis code [6].

4. RESULTS

In Fig. 1 and 2 the solid line represents our results for $d\sigma/d\Omega$ and the dotted line is the curve given in ref. [1]. A more detailed description of these results will be presented at the conference as well as the results of our phase shifts analyses. Our results show that partial waves up to $\ell = 3$ are statistically significant and can therefore be determined without using a model dependent calculation.

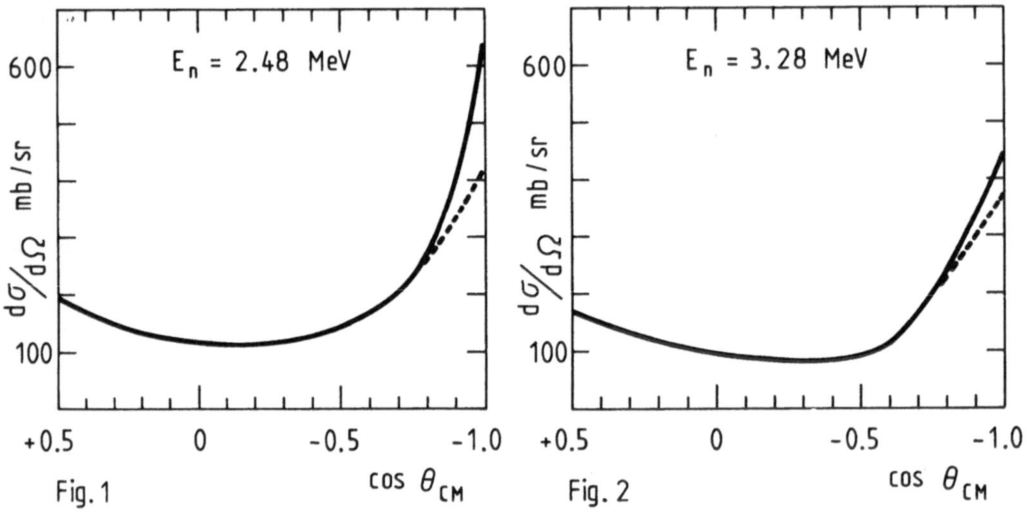

Fig. 1 Fig. 2

[1] M.D. Goldberg, V.M. May and J.R. Stehn, BNL-400, Vol. 1, (1962).
[2] D. Bovet, P. Chatelain, S. Jaccard, Y. Onel, R. Viennet and J. Weber, Proceedings of the ICINN Conference, Lowell 1976, p. 1357 (Conf-760715-p2).
[3] D. Bovet, P. Chatelain, R. Viennet and J. Weber, J. Phys. G, August 1978.
[4] V. Morgan and R.L. Walter, Phys. Rev. 168, (1968), 1114.
[5] P. Chatelain, Y. Onel and J. Weber, accepted for publication in Nucl. Inst. Meth.
[6] R. Viennet, Thesis, Nucl. Phys. A189, (1972), 424.

Backward Elastic nd-Scattering in the Energy Range 350 - 550 MeV +)

Th. Fischer, G. Hammel, W. Hürster, K. Kern, R. Kettle, M. Kleinschmidt,
L. Lehmann, E. Rössle, H. Schmitt

Fakultät für Physik der Universität Freiburg
D - 7800 Freiburg i.Br.

The steep backward rise of the elastic nd cross section stimulated a theoretical discussion leading to several different models. It was felt that angular distributions at closely spaced bombarding energies would be useful to test the different models as of Craigie and Wilkin[1] and of Gurwitz et al.[2]
The neutron beam of the "Schweizerische Institut für Nuklearforschung" (SIN) provides an excellent facility for this purpose. The beam has a wide energy range and a high flux. The neutrons impinge on a conventional liquid deuterium target with a diameter of 8 cm. The scattered deuterons are analyzed in a magnetic spectrometer equipped with drift chambers. The angular resolution is better than $.4^\circ$, the momentum resolution is about 3 %. The restmass of the detected particles was determined from its momentum and its time of flight through the spectrometer. Above $E_n = 430$ MeV it is possible to separate the elastically scattered deuterons from the deuterons due to the reaction $nd \rightarrow dN\pi$. The angular distributions are displayed in fig. 1 - 5. The error bars comprise all errors arising from counting statistics, background subtraction and target empty correction. When reading the absolute scale of the cross sections one has to add an error of 20 %. We hope to reduce this error to 10 % in the near future after we have completely evaluated our monitor reaction $np \rightarrow d\pi^0$.

If one tries to understand the backward scattering as one particle exchange in the u channel, one expects the cross section to vary like

$$\frac{d\sigma}{d\Omega} = \alpha \exp(\beta(u_{max} - u)). \qquad (1)$$

One observes that equation (1) is a fair representation of the data. Only the points close to u_{max} ($\vartheta > 177^\circ$) do not lie on the straight line. The slopes β have been fitted using only those points that are on the straight line. The results are shown in fig. 6. A comparison of our results to the models of Craigie and Wilkin and of Gurwitz et al. is currently in progress.

+) Work supported by Bundesministerium für Forschung und Technologie

[1] N. S. Craigie and C. Wilkin, Nucl.Phys. B14 (1969) 477
G. W. Barry, Ann.of Phys. 73 (1972) 482
[2] S. A. Gurwitz et al. Ann.of Phys. 93 (1975) 152, 98 (1976) 346

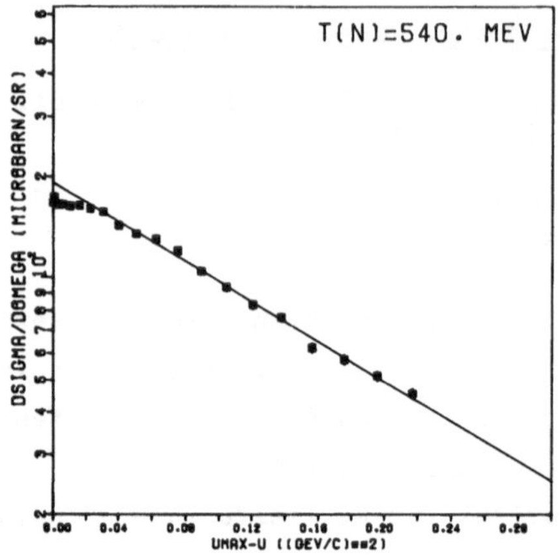

Fig. 1

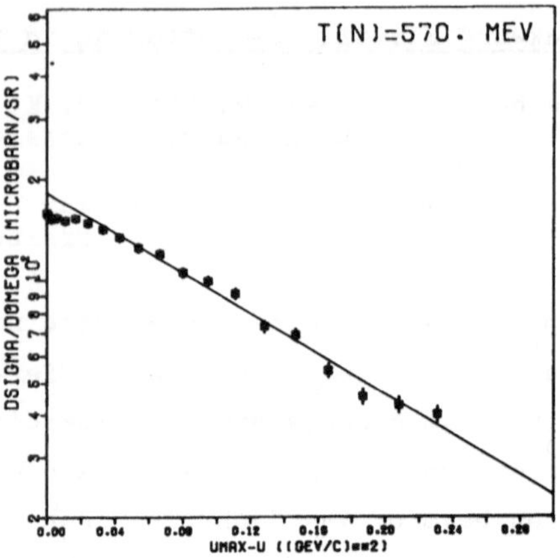

Fig.2

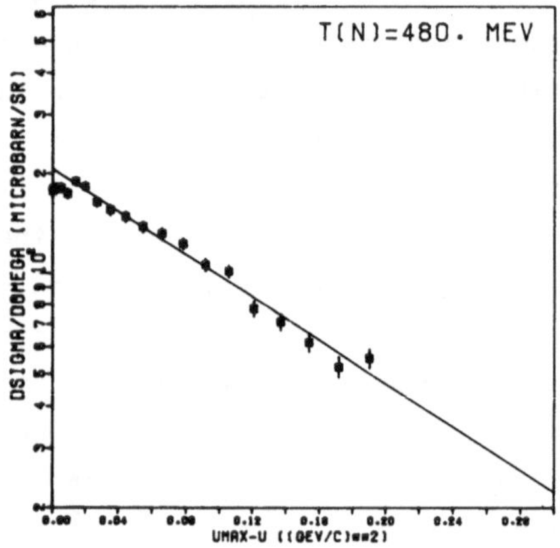

Fig. 3

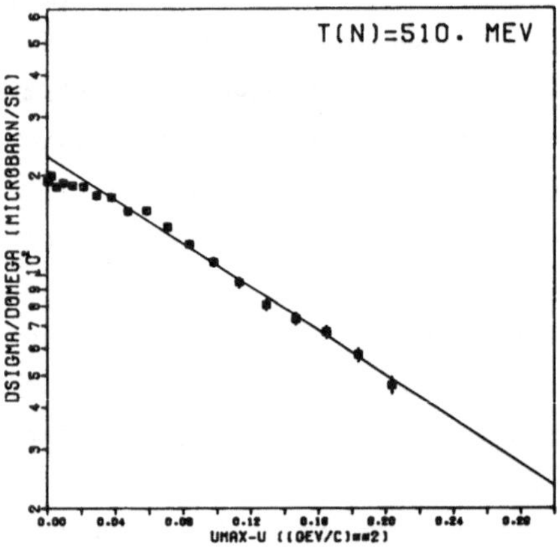

Fig. 4

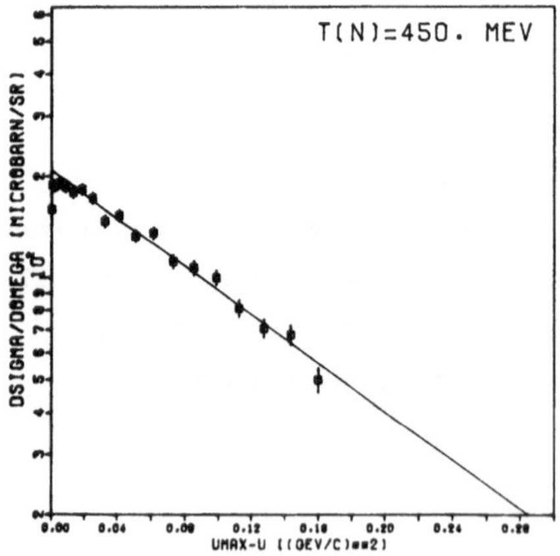

Fig. 5

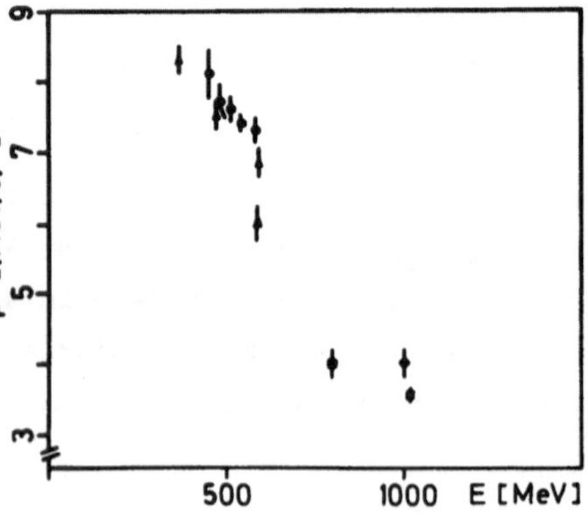

Fig. 6 β Parameter as function of incident energy. (o this work).

nd BACKWARD SCATTERING FROM 200 TO 800 MeV*

B. E. Bonner
Los Alamos Scientific Laboratory, University of California
Los Alamos, New Mexico 87545

G. Glass
Texas A & M University, College Station, Texas 77843

C. L. Hollas, C. R. Newsom, and P. J. Riley
University of Texas, Austin, Texas 78712

Backward nucleon-deuteron (Nd) elastic scattering, being a high-momentum transfer process, in principle probes the deuteron wave function at short distances. The failure of one-nucleon exchange calculations[1] to fit the 1-GeV pd data[2] led to the introduction of nucleon isobar components in the deuteron wave function and myriad searches for experimental manifestations of such components in nuclei.[3] The evidence for such exotic admixtures at the magnitude originally proposed is still far from compelling.

All previous measurements of Nd scattering above 150 MeV have utilized either proton or deuteron beams with the exception of a single 800-MeV measurement.[4] We have recently completed the analysis of an experiment using a continuum neutron beam incident on a deuterium target in which recoil deuterons were detected at laboratory angles from 0° to 20°, corresponding to neutron center-of-mass angles from 140° to 180°. Preliminary results for the extreme back angle ($\theta_n^* > 175°$) measurements have been reported[5] and details of the experimental setup are given therein. We report here on the energy and angular variation of the nd elastic scattering cross section for angles 140° to 180° c.m. and energies 200 to 800 MeV.

The data consist of 27 angular distributions spanning the energy range. Each distribution is parameterized in terms of a simple exponential in the momentum transferred. The choice of variable that is most appropriate has been the subject of some controversy.[1,6,7] Although we have studied the effect of using different variables, we report here on the variation with energy of the parameters determined using the relativistic momentum transfer variable[6] $[(p\cdot d')^2/M_d^2 - M_p^2]^{1/2}$. This is the momentum of the exchanged nucleon in the rest frame of either deuteron.

As reported[5] previously the excitation function of the 180° c.m. cross section falls exponentially with increasing energy up to about 250 MeV and then remains almost constant up to about 550 MeV after which an even more precipitous decrease is observed. It has also been noted[5] that such behavior is not consistent with nucleon isobar exchange although calculations[8] using triangle diagrams in addition to one-nucleon exchange do yield the general features of the observed excitation function.

When the 27 angular distributions are plotted as a function of $q = Q(\theta_n^*) - Q(180°)$ then it is observed that the simple exponential $d\sigma/d\Omega^* = A \exp(-Bq)$ provides a good fit to the data below 350 MeV and above about 650 MeV. For the region in between however, the very backward points ($\theta_n^* \gtrsim 165°$) fall below the exponential fit. This effect has not been noted before since most pd measurements do not extend into this region. The slope parameter B extracted from fits to each of the distributions decreases from 0.013 $(\text{MeV/c})^{-1}$ at 200 MeV to 0.0085 at 300 MeV, after which it remains constant up to about 600 MeV. Above 600 MeV, it begins to decrease, reaching 0.007 at 800 MeV.

REFERENCES

*Work supported by the U. S. Department of Energy.

1. A. K. Kerman and L. S. Kisslinger, Phys. Rev. 180 (1969) 1483.
2. G. W. Bennett et al., Phys. Rev. Letters 19 (1967) 387.
3. A review of the status of isobars in nuclei is given in the article by H. J. Weber, Proc. Conf. on Meson-Nuclear Physics, AIP, New York, 1976.
4. B. E. Bonner et al., Phys. Rev. C 17 (1978) 671.
5. B. E. Bonner et al., Phys. Rev. Letters 39 (1977) 1253; Nucleon-Nucleon Interactions-1977, AIP Conf. Proc. No. 41, New York, 1977, p. 387.
6. J. V. Noble and H. J. Weber, Phys. Letters 50B (1974) 233.
7. L. Dubal and C. F. Perdrisat, Lett. Nuovo Cimento 11 (1974) 265.
8. V. M. Kolybasov and N. Ya. Smorodinskaya, Yad. Fiz. 17 (1973) 1211 [Sov. J. Nucl. Phys. 17 (1973) 630].

QUASI FREE SCATTERING IN THE $^2H(n, 2n)p$ REACTION AT $E_n = 21.5$ MeV

J. M. Cameron, H. W. Fielding, S. T. Lam, G. C. Neilson and J. Soukup
University of Alberta
Edmonton T6G2N5, Canada

Recent advances in our understanding of the 3-body problem suggest that one can, with some confidence, extract information on on-shell nucleon-nucleon scattering parameters from quasi free scattering experiments (QFS). In particuler it has been shown that effects due to differences in the off-shell potential [1] or to higher partial waves [2] are small.

We have thus measured n-n QFS at $E_n = 21.5$ MeV in a complete experiment for several sets of co-planar symmetric angles. NE 230 liquid scintillators were used as targets and to provide a start pulse for neutron time of flight measurements over at 1.3 m flight path Scattered neutrons were detector in two NE 213 scintillators. The product of target thickness times neutron flux was monitored by simultaneous recording of events from $n + {}^2H$ elastic scattering. The spectra measured projected along the kinematic locus, at $\theta_1 = \theta_2 = 30, 35$ deg. are shown in the figure.

Also shown are curves calculated using the Ebenhöh [3] with the Yamaguchi form factor for various scattering parameters. A comparison of experimental and theoretical cross sections, average for Yamaguchi and Exponential form factor yields a best fit with $r_{nn} = 2.80 \pm 0.4$ fm when $a_{nn} = -16.4$ fm is assumed. The quoted error includes contributions from statistical and scale errors only.

References

1) D. D. Brayshaw, Phys. Rev. Lett. 32 (1974) 382.
M. I. Haftel et al. Phys. Rev. C14 (1976) 419.
2) J. Bruinsma and R. Van Wageningen, Nucl. Phys. A282(1977) 1.
3) W. Ebenhöh, Nucl. Phys. A191(1972) 97.

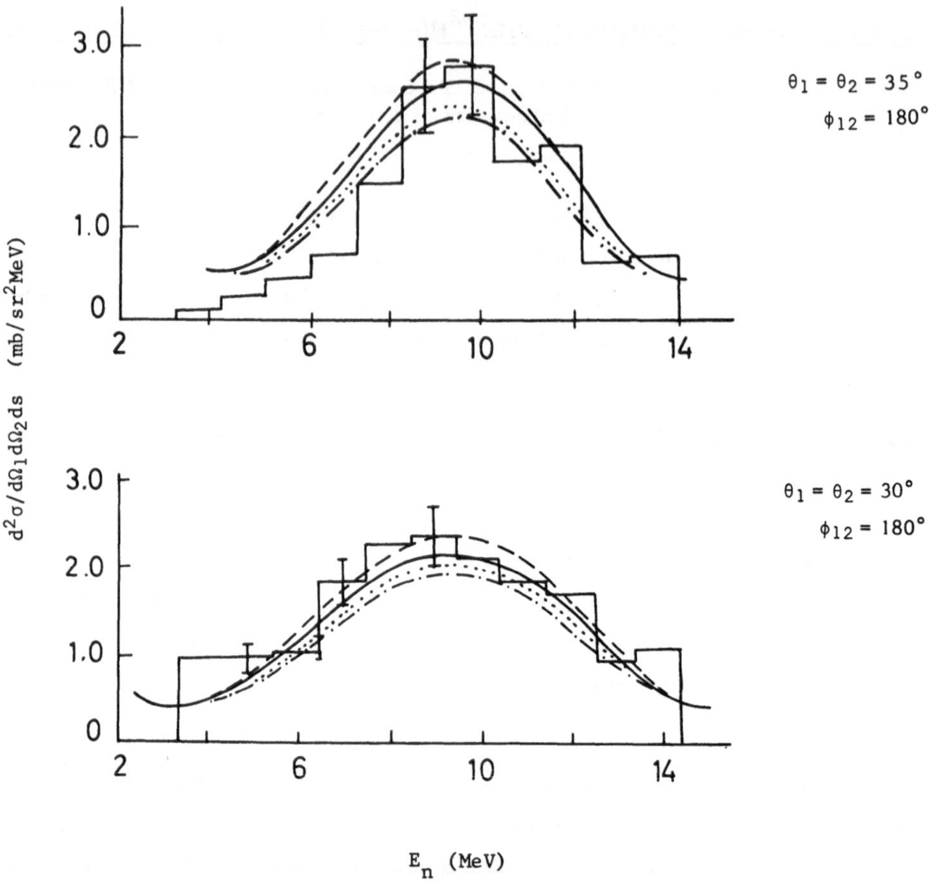

Projection of breakup events on the kinematic locus.
Calculated cross sections shown ware obtained with Ebenhöh code and Yamaguchi form factor. The two body scattering parameters used were :
a_{nn} = - 16. 4 fm ; r_{nn} = 2. 86 fm for solid line : a_{nn} = -23. 7 fm , r_{nn} = 2. 86 fm for dashed line ; a_{nn} = -23. 7 fm , r_{nn} = 3. 4 fm for dotted line and a_{nn} = -16. 4 fm r_{nn} = 3. 4 fm for dashed-dotted line

A FOUR-DIMENSIONAL APPROACH TO THE ALMOST 4π H(d,2p)n data

G.J.F. Blommestijn, Y. Haitsma*, R. Mooy, R. van Dantzig,
Instituut voor Kernphysisch Onderzoek (IKO), Amsterdam,
The Netherlands

and

Ivo Šlaus**
"Rudjer Bošković" Institute, Zagreb, Yugoslavia

The H(d,2p)n data at E_d=26.5 MeV were obtained using the multidetector system BOL[1] covering about 50% of the total phase space. The representation of the data in terms of measured laboratory angles and energies does not reveal the symmetries present and thus, the following 4 independent kinematical variables are introduced: T_{pp}-the relative energy of two protons, Θ_n-the polar angle of the neutron in the overall c.m. and the polar angle, Θ_R, and the asimuthal angle, ϕ_R, of the p-p relative momentum in the recoil c.m. (RCM is the right-handed system with its origin in the p-p c.m., positive Z_{RCM} in the direction of p-p c.m. and Y_{RCM} parallel to Y_{cm}).

The BOL system offers the possibility to obtain a global view and to zoom on any particular detail - apart from limitations due to detection geometry and statistical accuracy. Thus, the four dimensional (4D) cross section is displayed in 2D arrays of 2D arrays. Every

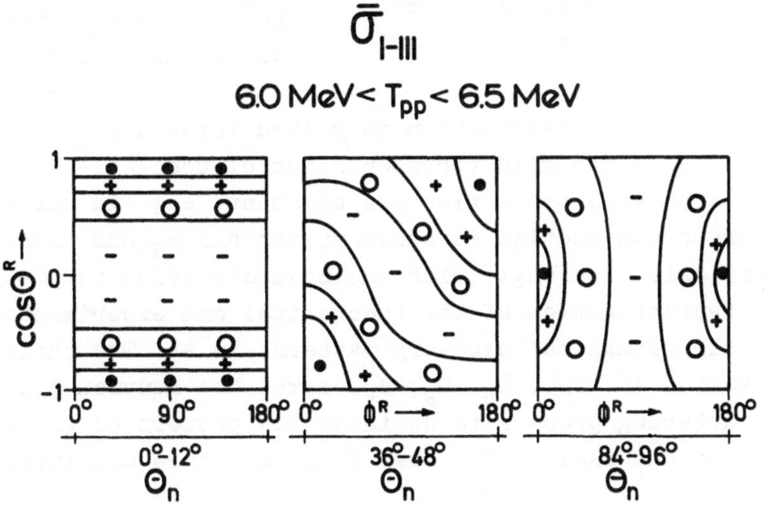

Fig.1. Faddeev cross section $\bar{\sigma}$(I-III) presented in a part of the 4D phase space $6 \leq T_{pp} \leq 6.5$ MeV and $0^o \leq \Theta_n \leq 96^o$. The symbols denote: .0.03 mb, -0.06 mb, o 0.16 mb, + 0.4 mb and • 1.3 mb.

---

* Present address: Physics Dept., Free University, Amsterdam

** Supported by PL480 grant no. F6F005 and SRH-SIZ-I grant 2.5.10.

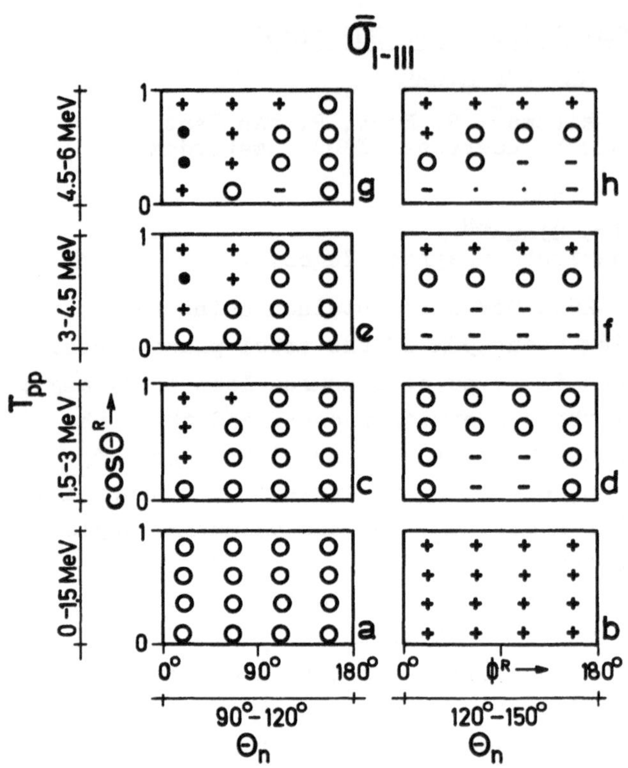

Fig.2. Cross section $\bar{\sigma}$(I-III). The meaning of the symbols as in Fig.1.

point in the T_{pp}- Θ_n space is a 2D space of Θ_R, ϕ_R. The identity of two protons and the mirror symmetry lead to $\sigma(\Theta_R\phi_R)=\sigma(\pi-\Theta_R,\phi_R+\pi)=\sigma(\Theta_R,-\phi_R)$.

The symmetries in the cross section are shown in Fig.1. by presenting the Faddeev cross section calculated for the MT I-III potential[2] and averaged over the instrumental resolution and over the grid interval. The portion $6\leq T_{pp}\leq 6.5$ MeV corresponds to the almost highest T_{pp} (T_{pp}^{max}=6.6 MeV) and thus, the neutron is almost at rest in the c.m. system. (The momentum triangle is reduced to a straight line[3].) The Θ_n-intervals differ only in the rotation angle around the Y axis. The "max T_{pp}" angular distribution is peaked forward (Θ_R=0 and $=\pi$).

Fig.2. is a coarse grid representation of the $\bar{\sigma}$(I-III) cross section wherefrom some symmetries are apparent; e.g. in the grid interval (f) the cross section is constant for all ϕ_R and thus, Z_{RCM} is a symmetry axis. Its angle with the beam direction is $\pi-\Theta_n=45^0$. A fine grid representation of the theoretical and experimental cross section reveals rotational symmetry patterns in 4D. The choice of these four variables: T_{pp}, Θ_n, Θ_R, ϕ_R, makes the neutron-proton quasifree scattering process to dominate the pattern of the cross section even in the regions far away from quasifree conditions.

1) L.A.Ch. Coerts et al., Nucl. Instr. and Methods 92 (1971) 15
2) R.A. Malfliet and J.A. Tjon, Ann. Phys. 61 (1970) 425
3) J.M. Lambert et al., Phys. Rev. C13 (1976) 43

THE D(p,2p)n REACTION AT 50 MeV

G.J.F. Blommestijn, R. Mooy, R. van Dantzig, Instituut voor Kernphysisch Onderzoek, Amsterdam, The Netherlands

and

Ivo Šlaus*, "Rudjer Bošković" Institute, Zagreb, Yugoslavia

The reaction D(p,2p)n is studied at 50 MeV. The experimental and

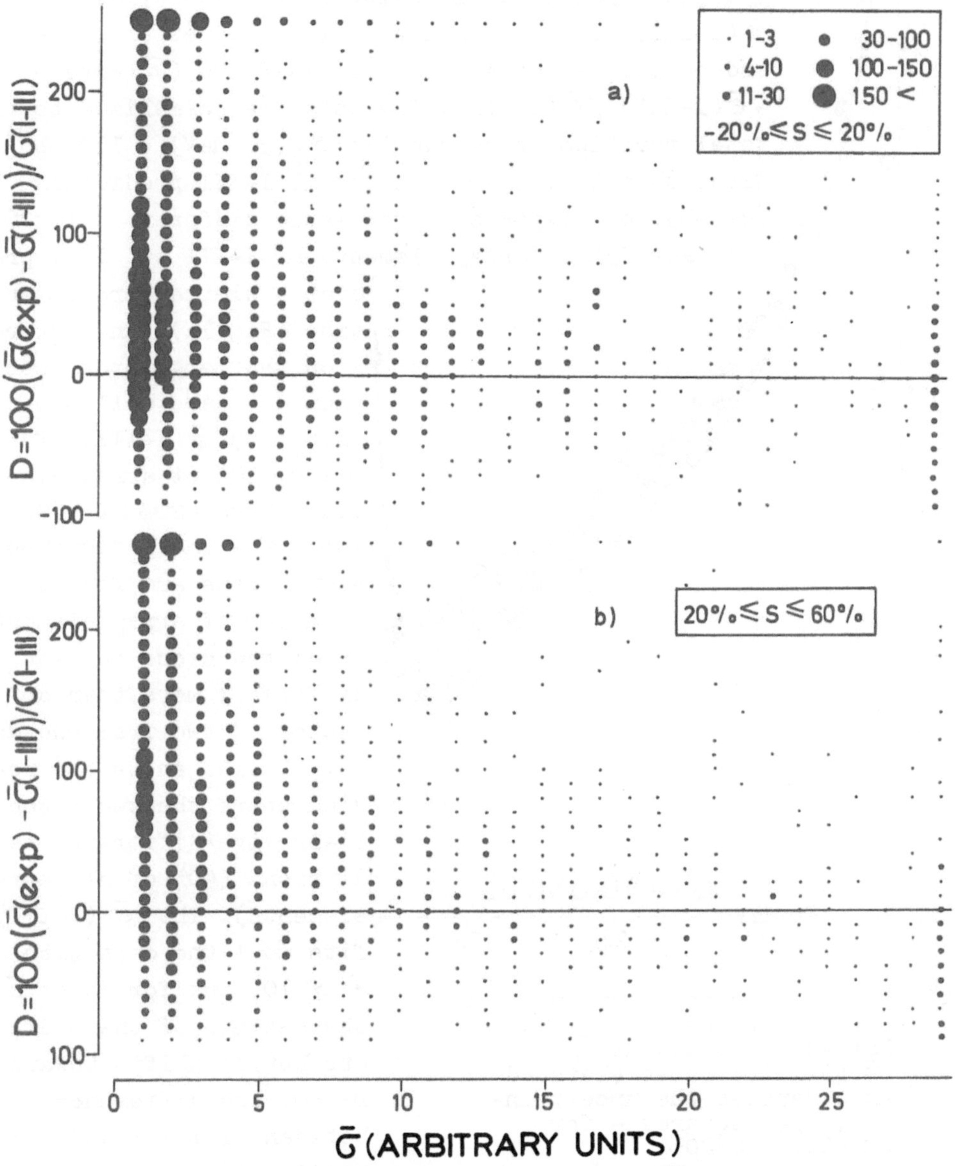

Fig.1. Number of array elements vs D and $\bar{\sigma}$. $\bar{\sigma}_{max}$ is defined as $\bar{\sigma}$(n-p QFS). D_{max} includes D > 260.

* Supported by PL480 grant no. F6F005 and SRH-SIZ-I grant 2.5.10.

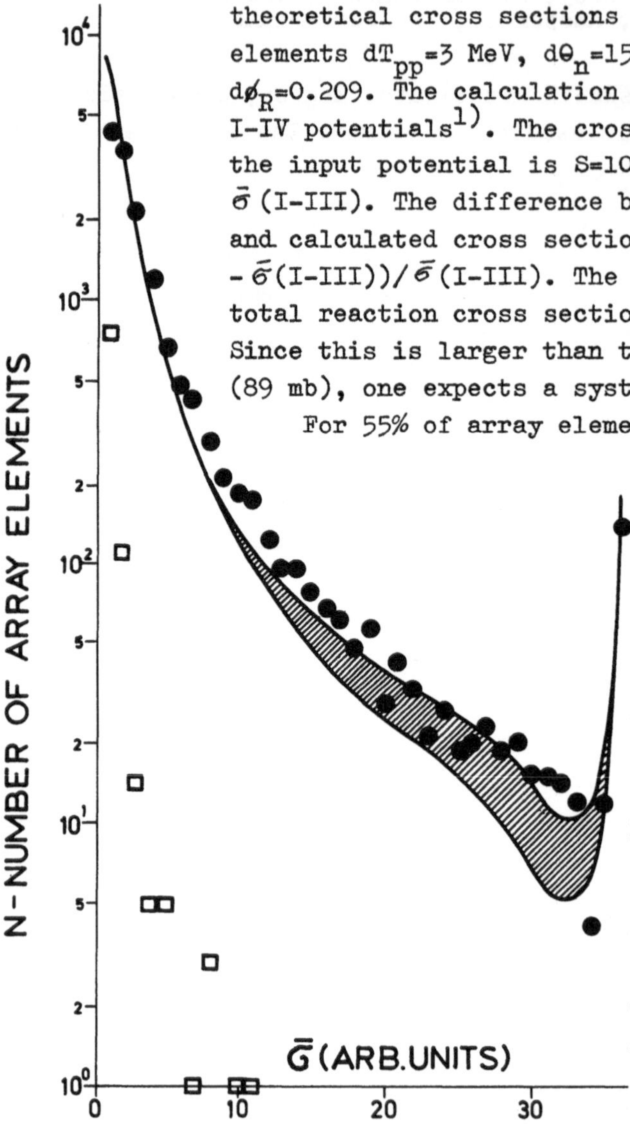

Fig.2. Number of array elements vs $\bar{\sigma}$ (same units as in Fig.1). Data-dots. Calculation $\bar{\sigma}$ (I-III): curve (hatched area denotes the uncertainty); open squares - for S>60%, S<-20%.

theoretical cross sections are averaged over array elements dT_{pp}=3 MeV, $d\Theta_n$=15°, $d\cos\Theta_R$=0.067 and $d\phi_R$=0.209. The calculation is done for MT I-III and I-IV potentials[1]. The cross section sensitivity to the input potential is S=100 ($\bar{\sigma}$(I-IV)- $\bar{\sigma}$(I-III))/ $\bar{\sigma}$ (I-III). The difference between the experimental and calculated cross sections is D=100 ($\bar{\sigma}$(exp) - - $\bar{\sigma}$(I-III))/$\bar{\sigma}$(I-III). The data are normalized to the total reaction cross section σ_R(50 MeV) = 113 mb[2]. Since this is larger than the MT I-III prediction (89 mb), one expects a systematic difference of ~20%.

For 55% of array elements MT I-III and I-IV predict a similar cross section (Fig.1a). In that region the large cross section data are well predicted by $\bar{\sigma}$(I-III): for $\bar{\sigma} \geq 9$, 60% of array elements have $-20\% \leq D \leq 40\%$. For $\bar{\sigma} < 9$ the D distribution widens and for more than 20% of array elements predicted cross section differs by more than a factor of two from the data. Fig.1b. shows the region where the two potentials give different predictions (40% of all array elements). The $\bar{\sigma}$ (I-III) fits well the data with $\bar{\sigma} \geq 10$, but for lower $\bar{\sigma}$ the maximum of the D distribution shifts toward D~80. The difference between MT I-III and I-IV predictions is larger than 60% in 5% of array elements. Most of phase space has a low calculated and a low measured cross section (Fig.2).

1) R.A. Malfliet and J.A. Tjon, Ann. Phys. 61 (1970) 425
2) R.F. Carlson et al., Lett. Nuovo Cimento 8 (1973) 319

WHAT COLLINEARITY EFFECT?

J. Birchall, M.S. de Jong, M.S.A.L. Al-Ghazi, J.S.C. McKee,
W.D. Ramsay and N. Videla
Cyclotron Laboratory, Department of Physics,
University of Manitoba, Winnipeg, Manitoba, R3T 2N2, Canada

R.E. Warner[1)], while studying charge exchange effects in the ^{3}He + d, three-body breakup reaction, observed an apparent enhancement in the cross section for single particle production at a laboratory energy where the detected particle had a velocity equal to that of the center of mass of the three-body system. This observation could however have been instrumental in origin. Berovic[2)] then searched for, and indeed found evidence for an enhancement of the cross section for the ^{2}H(d,dp)n reaction at the kinematic condition of collinearity in the centre of mass. The data obtained were later interpreted by Reitan[3)] as being largely due to rescattering effects, and a function therefore of the kinematics rather than evidence for a more fundamental physical phenomenon.

Lambert[4)], on the other hand, studied the ^{2}H(p,2p)n reaction at 23 MeV under the condition of collinearity, and produced some evidence for a collinearity effect without publishing experimental data. The ratio of the experimental cross sections to the predictions of two specific nucleon-nucleon potentials was examined and interpreted as generating evidence for an enhancement in the observed cross section. The peak in the ratio was several MeV wide for one potential, and rather less well identified for the other. The authors indicated that the collinearity effect disappeared, as the kinematic condition was relaxed. Fujiwara[5)] then performed a similar measurement using the same reaction, at 156 MeV, and published experimental data which suggested a slight peaking of the cross section in the vicinity of the kinematic condition of interest. The enhancement was seen to be a less than two standard deviation fluctuation in the mean cross section at that point. This brief review of the situation raises several important questions

1) Does the collinearity effect exist, or does it not?

2) If it does exist, is it a function of kinematics or direct evidence of some more fundamental physical effect?

3) If the latter, what fundamental physics is likely to be involved?

Lambert[4)] has suggested following the work of Yang[6)] that the collinearity condition may be unique with respect to studies of the three-body force. It should be remembered however that the effect, whatever its nature, was first suggested by experiments that did not involve three nucleons in the breakup channel. Also, because 'exact' three body codes do not yet include Coulomb forces, the phenomenon could in principle if it exists, be Coulombic in origin.

It was with all these facts in mind that a study of the ^{2}H(p,pp)n reaction at

28.5 MeV was undertaken using the University of Manitoba spiral ridge cyclotron facility. In the present experiment protons from the $^2H(p,2p)n$ reaction were detected in coincidence by two detector telescopes set at laboratory angles of $\pm 58.5^\circ$. Each telescope comprised silicon surface barrier passing and stopping detectors of depletion depths 0.5 mm and 2 mm respectively. A self-explanatory electronics diagram is shown in Fig. 1. Data were accumulated online using a PDP15-20 computer and a multi-parameter data acquisition program. Proton events were selected from a display of $\Delta E+E$ against ΔE. Prompt plus random and random coincidence events were separately identified from a spectrum of ΔT versus ΔE_L+E_L. With a singles count rate of $\sim$ 1 kHz in the passing counters the true to random ratio along the $^2H(p,2p)n$ kinematic locus was $\sim$ 60:1. No significant background in the kinematic region of interest was seen from events arising from the target windows, or from air contamination of the target.

The data are shown in Figure 2 in the form of a projection along the arc of the kinematic locus. The collinearity point lies between channels 9 and 10. The cut-off at either end of the plot is instrumental. An 'exact' three-body calculation is in progress. Preliminary analysis indicates that the rise at either end of Fig. 2 might be attributable to a final-state interaction between the neutron and one of the detected protons. Phase space varies only slightly with energy over this region. Our thanks are due to Dr. Pal Doleschall for several stimulating discussions and to Lidia Hak for assistance in collecting the data.

References

1. R.E. Warner Private Communication 1971.
2. N. Berovic, C. Blyth, R. Maughan, J.S.C. McKee and C. Pope in Few Particle Problems in the Nuclear Interaction ed. Slaus et al. 605, (1972).
3. A. Reitan Nucl. Phys. A268, 358 (1976).
4. J.M. Lambert, P.A. Treado, R.G. Allas, L.A. Beach, R.O. Bondelid and E.M. Diener Phys. Rev. C13, 43 (1976).
5. N. Fujiwara, E. Hourany, H. Naramura-Yokota, F. Reide and T. Yuasa, Phys. Rev. C15, 4 (1977).
6. S.N. Yang, Phys. Rev. C10, 2067 (1974).

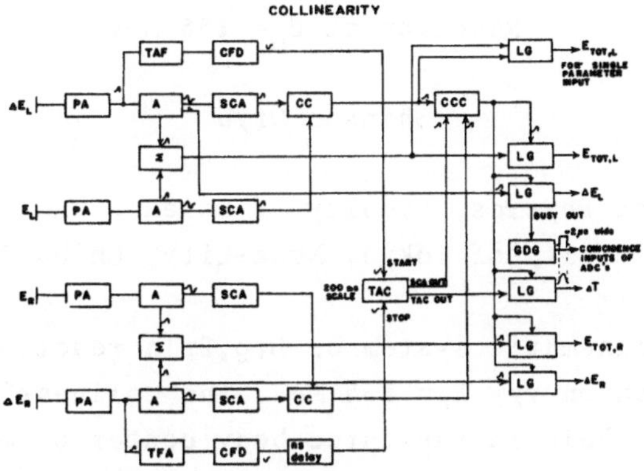

Figure 1

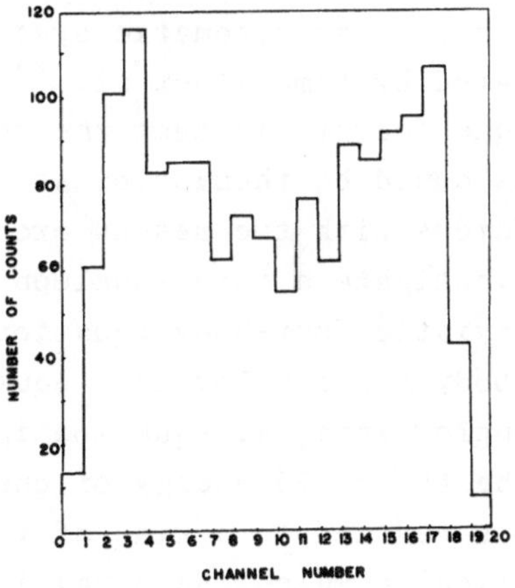

Figure 2

A Test on p_1-n- p_2 Colinearity Resonance of $D(p_0,p_1p_2)n$ Reaction at E_0= 156 MeV

Shinsho Oryu

Department of Physics, Faculty of Science and Technology, Science University of Tokyo, Noda-City, Chiba 278, Japan

In a special kinematic system of D(p,2p)n reaction, $\Theta_1=-\Theta_2= 58.3°$ at incident proton energy E_0= 156 MeV, the colinearity of three outgoing nucleons is held in the three-body center of mass system. Yuasa et al. suggest a narrow resonance near at outgoing neutron energy E_n= 0 (CM) or outgoing proton energy E_1= 68 MeV(Lab.).[1] The phenomenon has been given attention by reason of which it might be originated in the three-body force or a relativistic effect.

Before Yuasa et al., a similar kinematic system at lower energy E_0= 23 MeV was investigated by Lambert et al..[2] This particular geometry also has a unique feature in that the colinearity of the three outgoing particles could be thought of as leading to a shielding effects of three-body forces with two mesons exchanges.[2]

In this paper, we investigate a three-nucleon + one pion system which satisfies a relativistic four-body equation. A formal extension of the three-body Faddeev-Lovelace equation to four-body system gives a seven-coupled integral equation in the pion-nucleon isobar model. Below the threshold energy of one pion production, the four-body equation is reduced to a kind of three-nucleon Faddeev -Lovelace integral equation in which the usual inhomogeneous terms and kernels are modified by a few terms of subjoined diagrams with a virtual pion exchange. One of the subjoined diagrams is illustrated as in Fig. 1. This diagram contains the Fujita-Miyazawa type three-body force as Fig. 2 in it by the methodology of the Faddeev formalism.

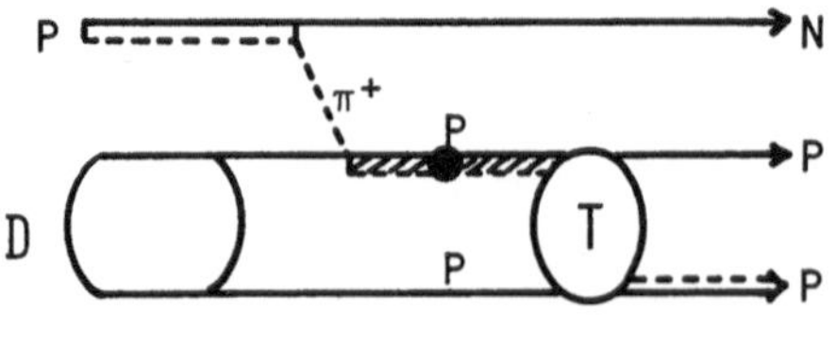

FIG. 1

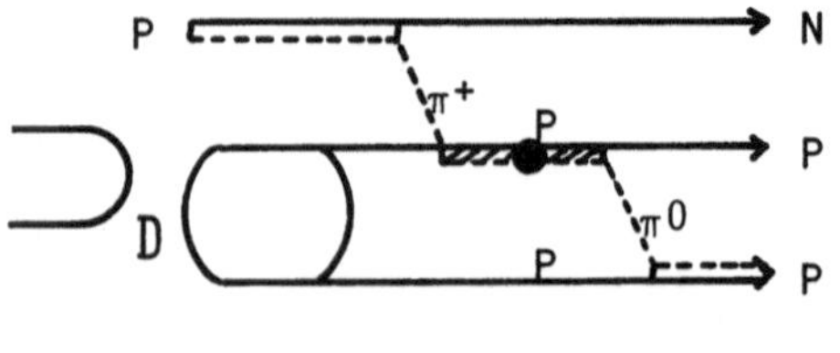

FIG. 2

The calculation of the diagram as Fig. 1 is accomplished by using a pion nucleon vertex function, a deuteron wave function, a pion propagator, a Baryon propagator and two-nucleon off-shell amplitude. Here, we can safely replace the intermediate energy integrals of the deuteron wave function, the Baryon propagator and two-nucleon off-energy-shell amplitude by off-energy-shell $T(\pi^+ + D \to p + p)$ amplitude. Our present interest is near at $E_n = 0$ (CM) or $E_1 = 68$ MeV (Lab.). As E_n is very small and a smooth function of E_1 around $E_1 = 68$ MeV, the pion-nucleon vertex function and the pion propagator are constants or smooth functions of E_1 variable. Therefore, the amplitude given by Fig. 1 is proportional to $T(\pi^+ + D \to p + p)$ amplitude in which the pion comes into a head-on collision with the deuteron. Moreover, $T(\pi^+ + D \to p + p)$ amplitude is S- and P-waves dominant and the angular dependence is not so sensitive to E_1 around 68 MeV. Consequently, those diagrams contribute to the absolute values of the cross section but less to the narrow resonance.

Our final theoretical results are given by using Padé approximant with some fitting parameters but a different kinematic set of angles $\Theta_1 = 45.0°$, $\Theta_2 = -57.0°$ at the same energy 156 MeV (Fig. 3). For the colinearity case of p-n-p : $\Theta_1 = -\Theta_2 = 58.3°$ is investigated in the consistent theoretical status with the case Fig. 3 without any additional parameters. Although our results (solid line in Fig. 4) could not express the fine structure of the experimental data by Yuasa et al., it seems that our curve shows a good fitting to the avarage values of those experimental data.

1) T. Yuasa, H. Nakamura-Yokota and N. Fujiwara, Suppl. Prog. Theor. Phys. (Kyoto) No.61 (1977) 161.
2) J.M. Lambert, P.A. Treado, R.G. Allas, L.A. Beach, R.O. Bondelid and E.M. Diener, Phys. Rev. C13 (1976) 43.

FIG. 3

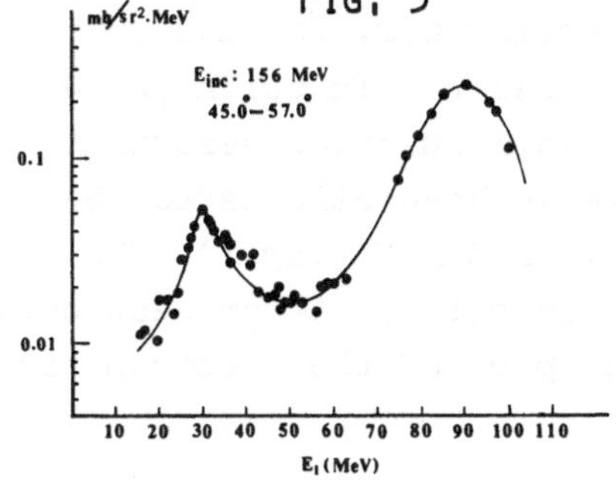

FIG. 4

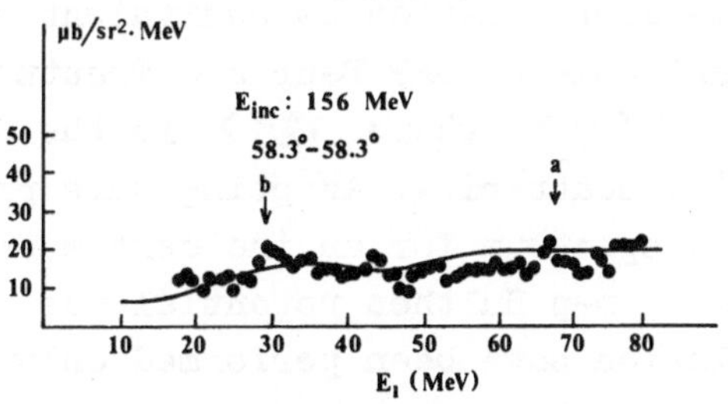

a: p-n-p colinearity resonance $E_n = 0$

b: N-N final state interaction

COULOMB BREAK-UP OF DEUTERON BY MUON AND n-p OFF-ENERGY-SHELL EFFECTS

Ž. Bajzer*

"Rudjer Bošković" Institute, Zagreb, Yugoslavia

The break-up process $\mu^+ + d \rightarrow \mu^+ + n + p$ has so far (to the author's knowledge) been considered neither theoretically nor experimentally. It seems, however, that this process might be interesting in several aspects. In the present paper the attention is paid to the break-up at low incident muon energy (up to 5 MeV). In this case, the weak interaction can be completely neglected and the (μ^+,n,p) system represents a very simple three-body system with only two dominant pairwise interactions: electromagnetic (μ^+p) and strong (np). The process could be described in terms of nonrelativistic quantum scattering theory using the Coulomb potential $(V_{\mu p})$ for the electromagnetic and the n-p potential (V_{np}) for the strong interaction. The n-p interaction appears off the energy shell and this makes possible the investigation of n-p off-energy-shell effects. The obvious advantage of considered process is that there are no effects of 3-nucleon forces appearing in 3-nucleon systems.

The break-up of the deuteron in a Coulomb field of a heavy nucleus has been studied by Baur and Trautmann[1)]. We follow their formalism and write the break-up amplitude for $\mu^+ + d \rightarrow \mu^+ + n + p$ as

$$T_{fi} = \langle \chi_f^- \varphi_f | U_{fi}(E_i + i0) | \chi_i^+ \varphi_i \rangle \qquad (1)$$

Where $U_{fi} = V_f[1+(z-H)^{-1}V_i]$, $H = H_o + V_{np} + V_{\mu p} = H_i + V_i = H_f + V_f$. H_o is the total kinetic energy operator, V_i is the "Coulomb polarizing potential" (CPP) (i.e. $V_i = V_{\mu d} - V_{\mu p}$) and $V_f = V_{np}$. $|\chi_f^- \varphi_f\rangle$ and $|\chi_i^+ \varphi_i\rangle$ are eigen functions of H_f and H_i respectively. In DWBA approximation the break-up amplitude becomes

$$T_{fi}^{DWBA} = \langle \chi_f^- \varphi_f | V_f | \chi_i^+ \varphi_i \rangle \qquad (2)$$

This approximation is equivalent to neglecting CPP. To take into account somehow CPP Baur and Trautmann replaced the function $|\chi_i^+ \varphi_i\rangle$ by $|\chi^+ \varphi_i\rangle$ where $|\chi^+\rangle$ is the Coulomb wave function describing μ^+-p scattering. Adopting this procedure we have calculated the muon spectrum for an incident muon energy of 5 MeV (Fig.1). The simple n-p Hulthèn potential was used as in ref.1). At present calculation have been performed only for the part of the spectrum since

---

* Supported by PL480 grant no. F6F005 and SRH-SIZ-I grant 1.3.7.3.

the argument of hypergeometric functions appearing in the break-up amplitude[1] have a singularity at a certain outgoing muon energy. This problem could be solved by appropriate techniques of analytic continuation.

To investigate the off-energy-shell effects a class of phase equivalent potentials has been introduced using the unitary transformation[2]:

$U=1-2|h\rangle\langle h|,\ \langle h|h\rangle=1,\ \langle\vec{r}|h\rangle=a^{5/2}\left[\pi(a^2-3ba+3b^2)\right]^{-1/2}\cdot(1-br)\exp(-ar).$

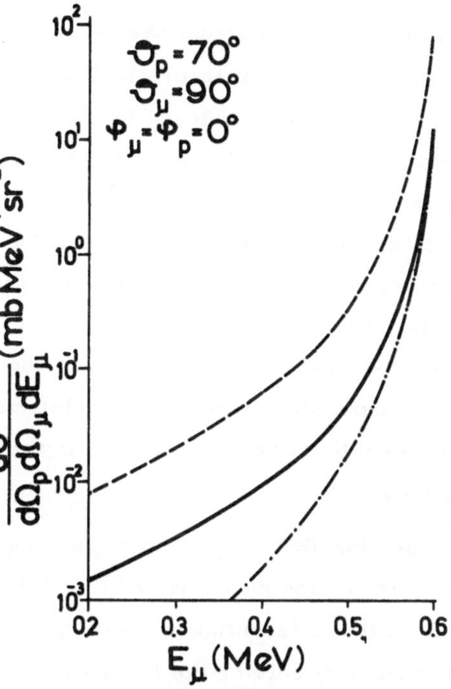

Fig.1. Muon spectrum in $\mu^+ + d \rightarrow \mu^+ + n + p$.
— Hulthèn potential,
--- a=2f, b=f;
-.-.- a=6f, b=5f;
$f=5\cdot10^{-3}fm^{-1}$.

The corresponding muon spectrum for two of such potentials is shown in Fig.1. The off-shell effects observed are rather strong. However, this is only an indication, since the calculation is approximate. More exact calculation should be performed using for example recently developed formalism for 3-particle scattering including the Coulomb interaction[3]. It is interesting to note that T_{fi}^{DWBA} (eq. (2)) coincides with the "pure Coulomb break-up amplitude" given by Alt et al.[3].

The author is grateful to Dr. M. Furić and Dr. Đ. Miljanić for helpful discussions.

1) G. Baur, R. Trautmann, Nucl. Phys. A191 (1972) 321; Phys. Rep. 25C (1976) 294
2) M.I. Haftel, F. Tabakin, Phys. Rev. C3 (1971) 921
3) E.O. Alt, N. Sandhas, H. Ziegelmann, preprint MZ-TH 77/7; to be published in Phys. Rev. C;
P. Sauer, Three Body Scattering with two Equally Charged Particles, preprint, 1978

DETAILED CALCULATIONS OF π - d ELASTIC SCATTERING AT MEDIUM ENERGIES

N. Giraud, Y. Avishai*, C. Fayard and G. H. Lamot
Institut de Physique Nucléaire de Lyon and IN2P3
Université Claude Bernard Lyon-I
43, Bd du 11 Novembre 1918, 69621 Villeurbanne, France

The elastic scattering of medium energy pions off deuterium has recently been given a considerable attention. In this report we present theoretical calculations of elastic cross-section and polarization observables induced by elastic π-d scattering at pion laboratory kinetic energy $T_\pi = 142$ MeV. Despite the large progress in this field, there have been some important effects which have so far been neglected in an exact three-body calculation. These effects are 1) The inclusion of all S and P π-N partial waves and not only the P_{33}. 2) Retaining the non diagonal terms $\ell \neq \ell' = J \pm 1$ in the partial wave elastic π-d amplitude $T^J_{\ell \ell'}$. 3) An adequate description of the two nucleon subsystem. This cannot be achieved by using Yamaguchi form factors. For example, in the "fully relativistic" (FR) treatment of Rinat and Thomas[1], these effects are not included in the calculations, due mainly to numerical limitations.

We use relativistic kinematics only for the pion (RPK approach of Thomas[2]) which should be adequate at $T_\pi = 142$ MeV (where for the nucleons $v/c \sim 0.1$), and concentrate on the influence of points 1-3 on the differential cross-section $d\sigma/d\Omega$, and polarization observables. This is done by solving the Faddeev equations in an exact way, with all effects taken to all order.

The two-body input we have used is as follows. For the N-N $^3S_1$-$^3D_1$ channel, we choose either the Yamaguchi parametrizations with D-state percentage values $P_D = 0\%$, 4% and 7% (Y0, Y4, Y7), or the Pieper rank 1 and 2 (P1, P2), which give the same deuteron wave function as Reid soft core ($P_D = 6.5\%$). For the π-N channels we adopt the parametrizations used by Thomas[2], and compare results obtained with inclusion of all S, P partial waves (the SP scheme) with those obtained by using the P_{33} alone. More details about the calculations described hereafter can be found in Ref. 3.

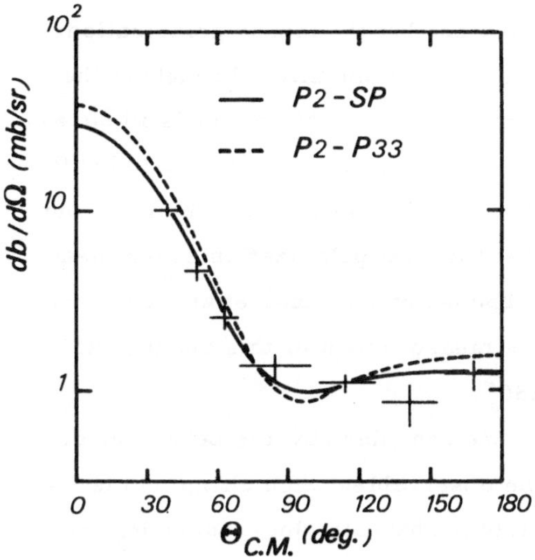

Figure 1

First, we investigate the effect of interference of other than P_{33} π-N channels by introducing in an exact way all S, P π-N channels. If we compare the P2-SP and P2-P33 results (Fig. 1), we see that the differential cross-section is considerably lower in the SP scheme than in the P33 scheme (except in the minimum region). Referring to the experimental data[4], the improvement of P2-SP calculation in comparison with P2-P33 throughout the angular range is obvious. The effect of the SP scheme on the vector polarization it_{11} is very strong as shown in Fig. 2, while it has been found to be moderate on the tensor polarizations.

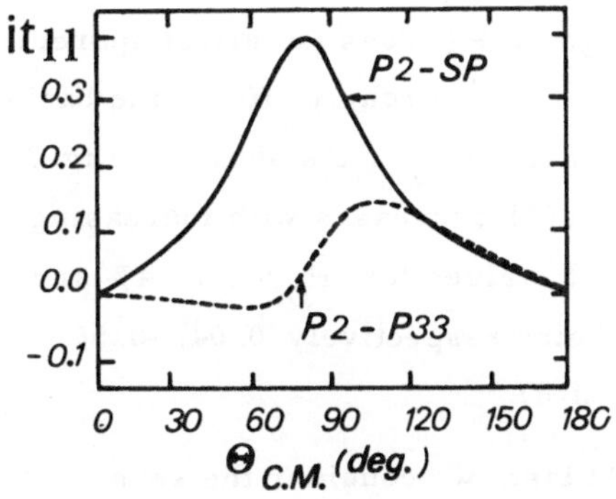

Figure 2

In the next step, we investigate the effect of the ℓ, ℓ' coupling. We found that the diagonal scattering amplitudes with $\ell = \ell' = J \pm 1$ are very close to those obtained without coupling, while the non diagonal terms $T^J_{J+1,J-1} = T^J_{J-1,J+1}$ are not negligible but of same order of magnitude as the $T^J_{J+1,J+1}$. The effect on $d\sigma/d\Omega$ is small and is apparent only in the backward part (at 180° the curve with coupling is 7% higher than the curve without coupling). The quantity it_{11} is slightly affected, but the tensor polarizations are strongly modified throughout the angular range. This effect is illustrated in Fig. 3a for t_{20} calculated with the P2-SP interactions. We point out that the first polarization experiment proposed by Grüebler et al.[5] concerns the measurement of this quantity at 180°.

We consider now the behaviour of the observables when changing the D-state probability. Increasing P_D from 4% to 7% lower slightly $d\sigma/d\Omega$ throughout the angular range as well in the Y-P33 as in the Y-SP calculations, but this effect is small (~ 4 % at 180°) in comparison with the effect found by Rinat and Thomas in the FR approach (~ 25 % at 180°). While a change in P_D does not affect appreciably it_{11}, it strongly affects the backward part of t_{20}. As shown in Fig. 3b, t_{20} (180°) decreases with increasing P_D, its values for P_D = 0%, 4% and 7% being respectively 0.04, -0.56 and -0.67.

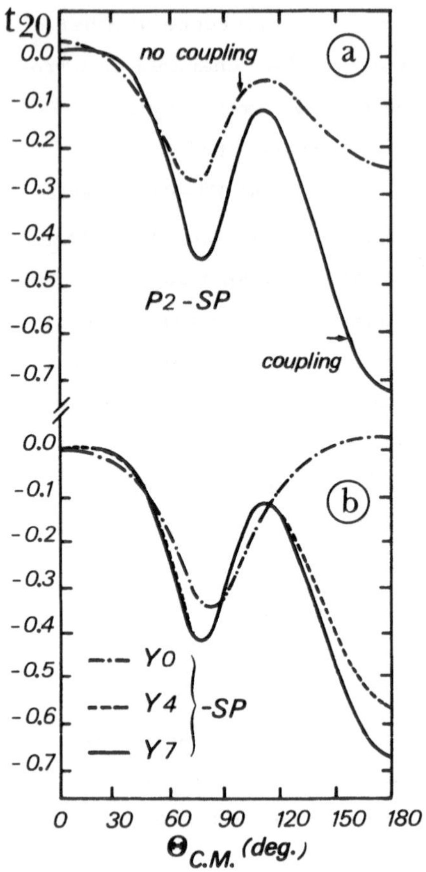

Figure 3

At last, we consider the sensitivity of the observables to the description of the deuteron wave function by comparing the P1-SP, P2-SP and Y7-SP calculations. The reasons why we use the Y7, P1 and P2 tensor forces are the following : these parametrizations have similar P_D, P1 and P2 have a more realistic deuteron wave function than Y7, and P2 gives a better description of the phase shifts than P1 and Y7 (in particular, $\delta(^3D_1)$ has the good sign). The P1-SP and P2-SP observables have been found to be practically identical (the differences are at most ~ 4 %), while the Y7-SP results are appreciably different, mainly in the backward part of the angular distributions ($d\sigma/d\Omega$ (180°) takes the values 1.60 mb/sr with Y7 and 1.22 mb/sr with P2, and for t_{20} (180°) we get -0.67 with Y7 and -0.73 with P2). Therefore, the model dependence of the observables is directly related to the description of the deuteron wave function rather than to the details of the N-N phase shifts.

We point out that similar conclusions hold at 180 MeV.

References

1. A.S. Rinat and A.W. Thomas, Nucl. Phys. A 282 (1977) 365.
2. A.W. Thomas, Nucl. Phys. A 258 (1976) 417.
3. N. Giraud, C. Fayard and G.H. Lamot, Phys. Rev. Lett. 40 (1978) 438.
N. Giraud, Thèse de Troisième Cycle, Université Lyon-I (unpublished), report n° LYCEN/7805.
N. Giraud, Y. Avishai, C. Fayard and G.H. Lamot, Phys. Lett. (to appear).
4. E.G. Pewitt, T.H. Fields, G.B. Yodh, J.G. Fetkovitch, and H. Derrick, Phys. Rev. 131 (1963) 1826.
5. W. Grüebler, SIN proposal and private communication.

* On leave of absence from Ben Gurion University, Beer-Sheva, Israel.

PRACTICAL SCHEME FOR LOW ENERGY π-d SCATTERING

Y. Avishai,* N. Giraud, C. Fayard and G.H. Lamot
Institut de Physique Nucléaire de Lyon and IN2P3
Université Claude Bernard Lyon-I
43, Bd du 11 Novembre 1918, 69621 Villeurbanne, France

Recently, it became clear that the solution of the π-d scattering problem in the presence of pion absorption rests outside the Faddeev theory. The most one can expect from this theory is the N-N' model of Afnan-Thomas[1], in which the Pauli principle is violated. In the present work, we impose the exclusion principle on the Afnan-Thomas model as an ad-hoc assumption, and get a modified set of equations in which the two nucleons are identical through all intermediate states, and non-Faddeev terms with two successive pion emissions are included (but states of more than one pion are eliminated).

We also make the approximation (tested and justified to within 2 %[2]) that the N-N interaction in three-body states is allowed strictly in the $^3S_1 - {}^3D_1$ quantum numbers. At the expense of imposing " external " assumption on a self consistent model we arrive at a compact set of equations, which couples the amplitudes for the reactions $\pi + d \rightarrow \pi + d$, $N + \Delta \rightarrow \pi + d$ and $N + N \rightarrow \pi + d$.

The fact that the two nucleon amplitudes in T = 1 states are generated (as by-products) do not lead to bootstrap since these states are not used in advance, and there is no input - output overlap. In addition, these amplitudes are not used as input for the production process, since all the unknown amplitudes are generated by multiple scattering. Therefore, our equations are free of overcounting.

From a numerical point of view, our equations are relatively simple and no extra work is needed beyond the use of three-body codes. Exact solution has recently been reported[3]. Here we give the following results : 1) Singlet scattering length $a_{\pi d}$ and triplet scattering volume $A_{\pi d}$. 2) The effect on the real part of the singlet πd scattering amplitude (at threshold) of including the coupling to the N-N channel, denoted by $\Delta\, a_{\pi d}$. 3) Production coefficients at threshold, α, β defined by $\sigma(N+N \rightarrow \pi + d)_{k \rightarrow 0} \approx \alpha k + \beta k^3$ ($[k] = m_\pi^{-1}$) (β has contribution from the partial wave amplitudes T_{11}^J, J = 0, 1, 2.)

L = 0

	Theor.	Exp.
$a_{\pi d}(m_\pi^{-1})$	-0.042	$-0.052 \pm \binom{0.022}{0.017}$ [4]
$\Delta a_{\pi d}(m_\pi^{-1})$	-0.0057	
α (μb)	258	$200 < \alpha < 300$ [5]

L = 1

	J = 0	J=1	J=2	Exp.
$A_{\pi d}(m_\pi^{-3})$	-0.29	0.11	0.48	none
β (mb)	small	0.	1.20	1.(?) [5]

References

1. I.R. Afnan and A.W. Thomas, Phys. Rev. C10, 109 (1974)
2. N. Giraud, Thèse de troisième cycle, Université Lyon-I (unpublished)
3. Y. Avishai, N. Giraud, C. Fayard and G.H. Lamot, Lyon preprint (1978)
4. J. Baily et al., Phys. Lett. B50, 403 (1974)
5. J. Spuller and D.F. Measday, Phys. Rev. D12, 3550 (1975).

* On leave of absence from Ben Gurion University, Beer-Sheva, Israel.

AN IMPORTANT CONTRIBUTION TO πD SCATTERING IN THE RESONANCE REGION

F. Myhrer
NORDITA, Copenhagen Ø, Denmark

and

A.W. Thomas
TRIUMF, Vancouver, Canada

In the resonance region the Faddeev calculations of Rinat and Thomas [1] seem to show a systematic deviation from the measured πd elastic differential cross section [2]. The measured $d\sigma/d\Omega$ shows a dip around an angle of 100° which becomes deeper with increasing pion energy, whereas the calculated πd $d\sigma/d\Omega$ does not give this deep minimum. There are at least two contributions that have not been included in ref.[1], which could produce this dip in $d\sigma/d\Omega$. One is higher nucleon-nucleon (NN) partial waves (only intermediate S-wave NN scattering is included in [1], and the other is effects from including π absorption in πd scattering calculations. The latter we will discuss at the end and first we will present results of a calculation of a triple scattering ("TS") term in the Faddeev πd series. The "TS" term involves two πN scatterings with an intermediate P-wave NN scattering.

We know that the "TS" term with $^3S_1$ NN scattering, "TS" ($^3S_1$), gives a 10% - 20% correction to $d\sigma/d\Omega$ for $T_\pi \simeq 200$ MeV [3,4]. Further, the P-wave NN phase shifts are larger than the S-wave ones at the relevant NN energies, and therefore we should expect the "TS" terms with intermediate P-wave NN interaction to give a significant contribution to πd elastic $d\sigma/d\Omega$ (backwards). However, the three ^3P_J ($J = 0,1,2$) NN phase shifts tend to cancel each other at the relevant NN energies. And the $^1P_1$ NN scattering requires nucleon spin-flip in the πN interactions before and after to contribute to elastic πd scattering. The πN (3/2, 3/2) amplitude has the following spin structure.

$$t_{\pi N} \sim 2\vec{k}\cdot\vec{k}' + i\vec{\sigma}\cdot(\vec{k} \times \vec{k}') \qquad (1)$$

where $\vec{k}$ and $\vec{k}'$ are relative πN initial and final momenta respectively. Only the last term in eq.(1) will contribute to the "TS" term with intermediate $^1P_1$ NN scattering. So although the $^1P_1$ phase shifts are much larger than the S-wave ones, the angular momentum factors in eq.(1) will reduce its contribution. But from these on-shell arguments we do expect

"TS" ($^1P_1$) to give a contribution to $d\sigma/d\Omega$ of the same size as do "TS" ($^3S_1$).

We have calculated the "TS" ($^1P_1$) term with different parametrizations for the πN and NN t-matrices to test for off-shell sensitivities. As $\pi N(3/2,3/2)$ amplitudes we have used the ones of refs.[3] and [5]. Both reproduce the πN (3/2,3/2) phase shifts well when the pion is treated relativistically. The NN t-matrix is either the Mongan II one or the one of Tabakin. We find that the value of the "TS" ($^1P_1$) contribution varies with different two-body inputs, i.e., the off-shell structure of the πN and NN t-matrices matters. In fig.1 we show the elastic πd $d\sigma/d\Omega$ for T_π(lab) = 234 MeV, the crosses are from the Faddeev calculation of ref.[1] without the "TS" ($^1P_1$) and the solid line the results with "TS" ($^1P_1$) included. Fig.1 shows the results when "TS" ($^1P_1$) is calculated with the πN model of ref.[5] and Mongan II's NN forces. For all other two-body input the effects are smaller.

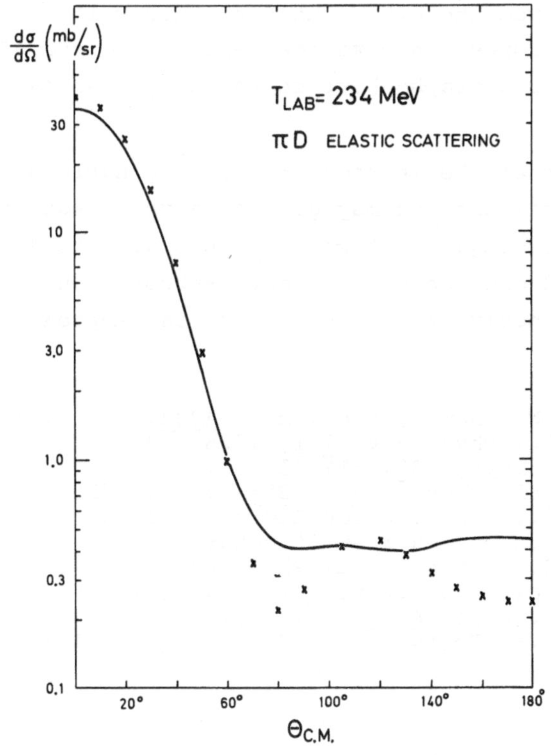

Fig.1. The elastic πd $d\sigma/d\Omega$ at T_π(lab) = 234 MeV with and without the "TS" ($^1P_1$) term in the Faddeev series.

Because the "TS" ($^1P_1$) term is sensitive to the off-shell two-body t-matrices, we examined the πN amplitudes more carefully. Our two πN amplitudes differ mainly in the treatment of the static Chew-Low pole at the pion energy $\omega = (k^2 + m_\pi^2)^{1/2} = 0$. This pole in the static model comes from the πN u-channel. In the πN amplitude of ref.[3] this pole is a pole in the energy variable. Ref.[5] use the fact that left-hand singularities in the t-matrix can be simulated by potential poles. The range of this πN potential will then correspond to the distance to the left-hand singularities, which in this case is m_π^{-1}. This means the static Chew-Low pole below πN threshold makes the Δ-resonance appear with a soft form factor. We find that the different treatment of the pole at $\omega = 0$ matters when we calculate the correction term "TS" ($^1P_1$). But to calculate the effect of this pole consistently we must remember its origin, it is the crossed graph of πN → N → πN. Therefore we are lead to conclude that a consistent treatment of the off-shell πN-matrix forces us to consider the effects of π absorption on πd scattering as well as crossed πNN graphs, e.g. the pion is absorbed on one nucleon after it is emitted from the other nucleon. This will introduce effective three-body forces into πd scattering, and the effects of these crossing symmetric graphs have so far only been calculated in πd at threshold [6].

To conclude we find that the introduction of $^1P_1$ nucleon-nucleon forces into the πd Faddeev calculation may give a large effect on elastic πd $d\sigma/d\Omega$ in the resonance region. It will not produce the dip at 100° in $d\sigma/d\Omega$ as measured and one has to consider π-absorption which probably cannot be handled properly in a purely potential approach to πd.

References

[1] A.S. Rinat and A.W. Thomas, Nucl.Phys. A282, 365 (1977).
[2] E.G. Pewitt et al., Phys. Rev. 131, 1826 (1963).
J.H. Norem, Nucl. Phys. B33, 512 (1971).
K. Gabathuler et al., Nucl. Phys. B55, 397 (1973).
R. Minehart et al., Contribution E10 to International Conf. on High Energy Physics and Nuclear Structure, Zürich, Sept. 1977.
[3] F. Myhrer and D.S. Koltun, Nucl. Phys. B86, 441 (1975).
[4] R.M. Woloshyn et al., Phys.Rev. C12, 909 (1975).
[5] A.W. Thomas, Nucl. Phys. A258, 417 (1976).
[6] T. Mizutani and D.S. Koltun, Ann. Phys. 109, 1 (1977).

THE NNπ SYSTEM--AN EFFECTIVE 3-BODY PROBLEM WITH UNUSUAL DISCONNECTEDNESS STRUCTURE

M. Stingl*)
University of Washington, Seattle, WA 98195 USA
and University of Münster, 4400 Münster, W. Germany

and

A.T. Stelbovics
South Australian Institute of Technology
Ingle Farm, South Australia 5098

To date, the most detailed descriptions of the π-d system--a key problem of intermediate-energy pion physics--use relativistic extensions of Amado-Lovelace three-particle theory, which implies the assumption of strict particle-number conservation[1)]. Recently, two formulations have appeared[2),3)] which account for the non-conserved nature of the pion. Using formally quite different methods, they both reduce the problem to coupled-channel equations between channels of two-particle (N+N) and three-particle (N+N+π) type, the latter in turn being represented in ref. 2 by quasi-two-body configurations such as d+π or Δ+N. Effects of sectors N+N+kπ, with k>2, are implicit in the effective interactions and vertices of this coupled-channel problem. These formulations naturally describe processes such as d+π ↔ N+N and their coupling to elastic d+π scattering.

In cases where the focus is on d+π scattering and three-body breakup, it is natural to go one step further by trying to completely reduce the problem to an effective 3-body problem in the N+N+π sector. Our motive for reporting here on such a formulation[4),5)] is that we have found this route to lead directly into a novel, interesting, and nontrivially solvable problem in scattering theory. The model we studied is a simple, semi-relativistic N-π field theory with $\vec{\sigma}\cdot\vec{\nabla}$ interaction, and with a Hilbert space restricted to the sectors N+N+π ("P space") and N+N, N+N+2π ("Q space"), sectors with k≥3 being neglected. While somewhat schematic (the only NN interaction arising is one-pion exchange), the model has the rare and pleasant feature that Q space can be eliminated exactly while still producing, in the P space, a highly nontrivial three-body problem whose energy-dependent and complex interactions can be given analytically. This problem takes the form

$$\{[G_0(E+i0)]^{-1} + \sum_{n=1}^{5} V_n(E+i0)\}|\Psi_P^+(E)\rangle = 0, \tag{1}$$

*) Stiftung Volkswagenwerk Fellow 1977/78.

where G_0 is a (renormalized) propagator for three free particles, $V_{1,2}$ are effective N-π forces (the π is labeled no. 3) comprising the usual direct and crossed Born graphs, V_3 is an (absorptive and retarded) nucleon-nucleon force of OPE type, and $|\Psi_P\rangle$ is the P-space projection of the total scattering state at energy E. The interesting new feature is the presence of effective forces $V_{4,5}$, which describe absorption of the π on one nucleon and its re-emission from the other. E.g., V_4 looks like

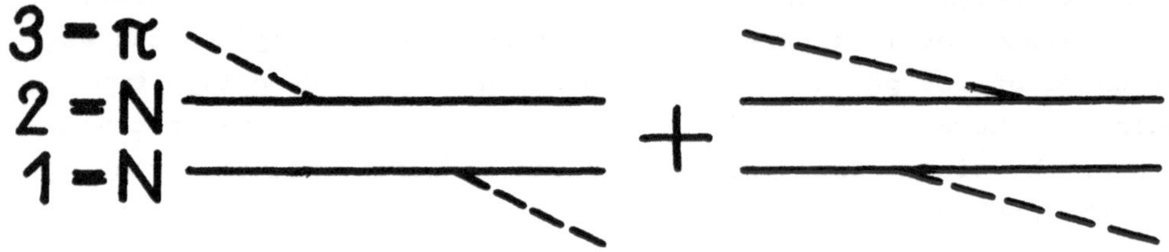

While the first three forces have, in momentum space, the usual factors $\delta_n = \delta^3(\vec{q}_n{}'-\vec{q}_n)$ with $\vec{q}_n$ = CMS momentum of particle n and n=1,2,3, the new forces are easily seen to have δ functions $\delta_4 = \delta^3(\vec{q}_1{}'+\vec{q}_2)$ and $\delta_5 = \delta^3(\vec{q}_2{}'+\vec{q}_1)$, that is, a new type of disconnectedness. This singular behavior makes them quite different from connected three-body forces. While the isolated V_4 and V_5 are kinematically forbidden on the energy shell, they produce, on iteration of the total interaction $V = \sum_n V_n$, two distinct kinds of effects: (i) mass and vertex renormalizations both in other forces and in themselves, and (ii) generation of connected 3-body graphs not obtainable from the other forces. The complicated interplay between the five disconnectedness types is best described by a "multiplication table" of semigroup structure, which for any two operators A_i, A_j of disconnectedness types δ_i, δ_j specifies the type of the product $A_i\ G_0\ A_j$, and which for the present case reads

$$\delta_1\cdot\delta_1 \to \delta_1,\quad \delta_2\cdot\delta_2 \to \delta_2,\quad \delta_3\cdot\delta_3 \to \delta_3, \tag{2a}$$

$$\delta_1\cdot\delta_4 \to \delta_4,\quad \delta_4\cdot\delta_2 \to \delta_4,\quad \delta_2\cdot\delta_5 \to \delta_5,\quad \delta_5\cdot\delta_1 \to \delta_5, \tag{2b}$$

$$\delta_4\cdot\delta_5 \to \delta_1,\qquad \delta_5\cdot\delta_4 \to \delta_2. \tag{2c}$$

All products not listed explicitly are understood to fall into the class c of fully connected terms; in particular,

$$\delta_4\cdot\delta_4 \to c,\quad \delta_5\cdot\delta_5 \to c. \tag{3}$$

The first line of (2) is a compact description of the usual, particle-number conserving 3-body problem, as covered by Fadde'ev theory. (The disconnectedness semigroup becomes a trivial direct product of three identical factors.) Lines (2b,c), which render the multiplication law nondiagonal and non-abelian, substantially complicate the task of summing disconnectedness classes, which can no longer be performed by separate ladder summation in each of the 3 subsystems (except for δ_3 which obviously is still "decoupled"). Instead it requires solving a system of four coupled equations[4], whose physical content turns out to be both ladder summation and extraction of the renormalization effects (i) above. It must be emphasized that even if one wishes (as most workers seem to do) to discard all renormalization effects beyond the introduction of physical masses and coupling constants, this may be done only after having isolated and summed them in this way--otherwise they will always "contaminate" the solution with disconnected terms. With the summed quantities w_i^R (i=1....5) known, the "genuine" scattering effects (ii) above are then generated by the new fourth and fifth lines and columns in the generalized Fadde'ev equations[5]

$$\bar{T}_i = w_i^R + \sum_{j=1}^{5} K_{ij} G_0 \bar{T}_j, \quad \sum_i \bar{T}_i = T \quad , \tag{4}$$

where the kernel matrix,

$$(K_{ij}) = \begin{pmatrix} 0 & w_1^R & w_1^R & 0 & w_1^R \\ w_2^R & 0 & w_2^R & w_2^R & 0 \\ w_3^R & w_3^R & 0 & w_3^R & w_3^R \\ w_4^R & 0 & w_4^R & w_4^R & 0 \\ 0 & w_5^R & w_5^R & 0 & w_5^R \end{pmatrix} \quad , \tag{5}$$

is seen to have zeros in exactly those places where the "multiplication table" has nontrivial entries, and where therefore the kernel becomes connected after one iteration.

On the basis of eq. (5) it is possible to define new effective transition operators between quasi-two-body channels, analogous to and nontrivially different from those of Alt, Grassberger, and Sandhas[6]. Assuming separability of the Nπ t matrices $w_{1,2}^R$ alone (a nonseparable remainder in the NN t matrix can be taken into account exactly), the equations for these reduce to modified Amado-Lovelace equations, where driving terms involving an $I = \frac{1}{2}$ N-π quasiparticle are defined by an auxiliary equation in one vector variable[4].

We believe this model may be of interest not only as an example of a fully understood three-body problem with non-conserved particle number, but also for the field theorist, since the subtle interplay between renormalization and disconnectedness can seldom be studied and solved as explicitly as here.

References:

1) A.W. Thomas, in: Proceedings of the 7th International Conference on High-Energy Physics and Nuclear Structure, (Zürich 1977) (M.P. Locher, ed.), p. 109, and literature quoted there.

2) A.S. Rinat, Nucl. Phys. A287 (1977) 399

3) T. Mizutani and D.S. Koltun, Ann. Phys. (N.Y.) 109 (1977) 1

4) M. Stingl and A.T. Stelbovics, J. Phys. G (in print)

5) A.T. Stelbovics and M. Stingl Nucl. Phys. A294 (1978) 391

6) E.O. Alt, P. Grassberger, and W. Sandhas, Nucl. Phys. B2 (1967) 167

PION-DEUTERON OPTICAL POTENTIAL

I.R. Afnan and A.T. Stelbovics
School of Physical Sciences, Flinders University of South Australia
Bedford Park, S.A. 5042, Australia.

Recently[1] it has been shown that the Faddeev equation with relativistic kinematics gives a good description of π-d scattering and below the (3,3) resonance. It would then be interesting to compare the exact solution of the Faddeev equation with approximations commonly used for π-nucleus scattering. In the present report we will compare the exact solution with (i) the optical potential approach as suggested by the multiple scattering series (MSS) of Watson[2] and Kerman *et al*[3] (KMT). (ii) A possible truncation of the Faddeev equation.

Following the procedure of Tandy *et al*[4] the scattering amplitude for π-d system can be written in terms of a hierarchy of integral equations of the form

$$T_e = U_e + U_e \Gamma_e T_e$$
$$U_e = \sum_{i=1}^{A} U_e^{\,i} \text{ and } U_e^{\,i} = \tau_e^{\,i} + \tau_e^{\,i}(G_e - \Gamma_e)\sum_{j\neq i} U_e^{\,j} \tag{1}$$
$$\tau_e^{\,i} = \hat{t}_e^{\,i}\,(1 - \Gamma_e\tau_e^{\,i}) \text{ with } \hat{t}_e^{\,i} = t_p^{\,i} + t_p^{\,i}\, G_o\, t_e\, G_o\, \hat{t}_e^{\,i}$$

where t_p and t_e are the π-N and N-N amplitudes, G_o is the free Green's function for the π-d system, while G_e is the Green's function for the free pion and two interacting nucleons. Γ_e is to be chosen to get the two different approximations. The lowest order optical model involves taking $U_e^{\,i} \cong \tau_e^{\,i}$. With this approximation and the fact that the target particles are identical, we can write the Eq. for T_e in KMT form

$$T_e = A\hat{t}_e + (A-1)\tau_e\Gamma_e T_e, \tag{2}$$

or the Watson MSS form

$$T_e = A\tau_e(1 + \Gamma_e T_e). \tag{3}$$

In general the determination of τ_e or $\hat{t}_e$ involves the solution of a three body problem[4]. The usual optical potential[4] for π-A scattering involves taking $\Gamma_e = PG_e$, where P is the projection operator for the ground state of the target. In this case, if we solve the equations for τ_e or $\hat{t}_e$ then T_e includes the effect of break-up of the target[4]. However, in all pion optical model calculations, the approximation $\hat{t}_p \simeq t_p$ is made. In the present report we will give results with $\tau_p \cong t_p$ in Eq.(3) and $\hat{t}_e \cong t_p$ in Eq.(2). These will be referred to as approximation (I) and (II) respectively.

An alternative approximation suggested by the Faddeev equation is to take $\Gamma_e = G_o t_e G_o$. In this way the intermediate state deuteron is represented by the full N-N T-matrix, which not only includes the deuteron pole but also the square root branch point corresponding to break-up of the deuteron. This approximation is identical to the solution of the Faddeev equation with the restriction that each π-N interaction is followed by an N-N interaction. Furthermore, this choice of Γ_e gives us the result that $\tau_e = t_p$. This with Eq.(3) we refer to as approximation (III).

To compare the three different approximations to the exact result, we have taken for the N-N interaction a $^3S_1$ Yamuguchi type potential, while for the π-N interaction we take a one term separable potential in the P_{33} channel[5]. Since π-d scattering is dominated by the $J^\pi = 2^+$ channel we have restricted our results in Fig.1 to this channel. By comparing the total cross section for the different approximation to the exact result as a function of three-body energy, we see that none give a good approximation to the exact answer. In particular, (III), which includes the effect of deutron breakup, gives the poorest result. Furthermore, we see that approximations (I) and (II) shift the peak in the cross section in different ways. We are now in the process of solving the three-body equation for $\hat{\tau}_e$ to see if that will improve the agreement with the exact result.

The authors would like to thank Dr. P.C. Tandy for many enlightening discussions.

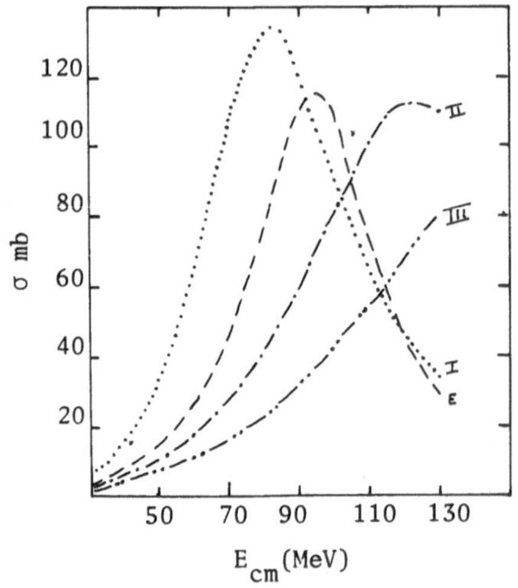

Fig 1. Comparison of Exact (E) and approximate (I, II, and III) total cross section for π-d scattering in $J^\pi = 2^+$ channel.

REFERENCES

1. N. Giraud, G.H. Lancot and D. Fayard, Phys. Ref. Letters 40, 438 (1978)
A.W. Thomas, Nucl. Phys. A258, 417 (1976).
2. K.M. Watson, Phys. Ref. 89, 575 (1953).
3. A.K. Kerman, H. McManus, and R.M. Thaler, Ann. Phys. (N.Y.) 8, 551 (1959).
4. P.C. Tandy, E.F. Radish, and D. Bollé, Phys. Rev. C16, 1924 (1977).
5. I.R. Afnan and A.W. Thomas, Phys. Rev. C10, 109 (1974).

DIBARYON RESONANCES IN πNN AND $\pi\pi$NN DYNAMICS

T. Ueda

Faculty of Engineering Science, Osaka University, Toyonaka, Osaka

Recently experimental evidences of dibaryon resonances have been reported in the states of (I=1, $J^P=3^-$, M=2.26 GeV) [1], (I=1, $J^P=2^+$, 2.16 GeV) [2] and (I=0, $J^P=3^+$, 2.38 GeV) [3]. In πNN and $\pi\pi$NN systems a strong attractive force is produced between the two nucleons (2N) by the mechanism in which the 2N keep in common one or two π in the Δ state. The sum of this attractive force and the attractive nuclear force originally existing in the $^3S_1$ and $^3P_2$ states is strong enough to make resonances in the πNN and $\pi\pi$NN systems. The author has shown that the (I=1, $J^P=3^-$) and (I=1, $J^P=2^+$) resonances are explained by this mechanism in the πNN system where the 2N are in the $^3P_2$ and $^3S_1$ states respectively [4]. In this note we show that the same mechanism in $\pi\pi$NN system explains the (I=0, $J^P=3^+$) resonance and also predict existence of several other dibaryon resonances produced by the mechanism.

In the $^3S_1$, $^3P_2$, $^1S_0$, $^1D_2$, $^3D_2$ and $^3P_0$ NN states strong attractive nuclear forces exist. Among the eigenstates of πNN systems in which the 2N are in those states and the π has the orbital angular momentum 1 around one of the 2N, we note that the πNN eigenstates with $J^P=2^+$ and 3^- have the πN subsystems with purely j=3/2, that is, these can be purely in the Δ state, while the other πNN eigenstates have admixtures of j=3/2 and j=1/2. Therefore in those two eigenstates the mechanism described above can work most efficiently to produce the attractive force. Similarly the $\pi\pi$NN eigenstates of $J^P=3^+$ and 4^-, associated with $^3S_1$ and $^3P_2$ NN states respectively, have the favorable condition for producing the attractive force. (Table 1)

As for isospins, the πNN resonance with $J^P=2^+$ should have only I=1, while the one with $J^P=3^-$ has both I=1 and I=2. In the $\pi\pi$NN resonances, the $\pi\pi$ subsystem has dominantly $I_{\pi\pi}=0$ because the $\pi\pi$ mass is below 500 MeV and the attraction between the two π is strongest in the $I_{\pi\pi}=0$ state. Therefore the $\pi\pi$NN states with $J^P=3^+$ and 4^-, associated with the $^3S_1$ and $^3P_2$ NN states, should have I=0 and I=1 respectively.

We proceed in analogy with Heitler-London formalism for H_2^+ and H_2. Then effective potentials between the 2N produced by the proposed mechanism are given by for the πNN and $\pi\pi$NN systems respectively,

$$V_1(R) = V + \int d\underset{\sim}{r}_1 X_1^*(k_1 + v_{1A} + v_{1B})X_1 \,, \tag{1}$$

$$V_2(R) = V + \int d\underset{\sim}{r}_1 d\underset{\sim}{r}_2 X_2^*\Big(\sum_{i=1}^{2}\{k_i + v_{iA} + v_{iB}\} + v_{12}\Big)X_2 \,, \tag{2}$$

where

$$X_1 = u_A(\underset{\sim}{r}_1) \pm u_B(\underset{\sim}{r}_1) \,, \tag{3}$$

$$X_2 = u_A(\underset{\sim}{r}_1)u_B(\underset{\sim}{r}_2) + u_A(\underset{\sim}{r}_2)u_B(\underset{\sim}{r}_1) \ . \tag{4}$$

In these eqs. R and V represent the relative cooridinate and the potential (Ueda-Green I [5]) of the 2N respectively ; $\underset{\sim}{r}_i$ and k_i denote the coordinate and the kinetic energy of the ith π ; v_{iA} is the potential between the ith π and the nucleon A which produces the Δ resonance ; v_{12} denotes the $\pi\pi$ potential. u_A is given by the solution of the following eq.

$$\{ -\nabla^2 - p^2 + 2\omega v_{1A}(r)\} \ u(\underset{\sim}{r}) = 0 \ , \tag{5}$$

where p and ω are the π momentum and energy in the Δ state. We normalize X_1 and X_2 by putting as $u_A = u(\underset{\sim}{r})$ for $r \leq 1.4$fm and $u_A = 0$ for $r > 1.4$fm.

Using the effective potentials of eqs. (1) and (2) which are now functions of R only, we solve the schrödinger eq. for the 2N. Thus we find dibaryon resonances as summarized in Table 2. In similar treatment of the $^1S_0$ NN state we find also two πNN resonances.

References

[1] I.P. Auer et al., Phys. Lett. 70B (1977) 475 ; N. Hoshizaki, Prog. Theor. Phys. 58 (1977) 716.
[2] R. A. Arndt, Phys. Rev. 165 (1968) 1834 ; H. Suzuki, Prog. Theor. Phys. 54 (1975) 143.
[3] T. Kamae et al., Phys. Rev. Lett. 38 (1977) 468, 471.
[4] T. Ueda, Phys. Lett. 74B (1978) 123.
[5] T. Ueda and A.E.S. Green, Phys. Rev. 174 (1968) 1304.

Table 1. Correspondence of the 2N eigenstates with the πNN and $\pi\pi$NN eigenstates.

2N	$^3S_1$	$^3P_2$	$^1S_0$	$^3D_2$	$^3P_0$
πNN	$0^+ \ 1^+ \ 2^+$	$1^- \ 2^- \ 3^-$	1^+	$1^+ \ 2^+ \ 3^+$	1^-
$\pi\pi$NN	$0^+ \ 1^+ \ 2^+ \ 3^+$	$0^- \ 1^- \ 2^- \ 3^- \ 4^-$	$0^+ \ 1^+ \ 2^+$	$0^+ \ 1^+ \ 2^+ \ 3^+ \ 4^+$	$0^- \ 1^- \ 2^-$

Table 2. The predicted dibaryon resonances and comparison with the experimental evidences.

system	I	J^P	2N subsystem	binding (-)/resonance (+) energy	Width (MeV)	Mass (GeV)	Mass of evidence
πNN	1, 2	3^-	$^3P_2$	+ 40 MeV	200	2.21-2.25	2.26
	1	2^+	$^3S_1$	- 50 MeV	100	2.12-2.16	2.17
	1, 2	1^+	$^1S_0$	- 33 MeV	100	2.14-2.18	
$\pi\pi$NN	0	3^+	$^3S_1$	- 99 MeV	200	2.37-2.44	2.38
	1	4^-	$^3P_2$	+ 5 MeV	300	2.47-2.55	

ELASTIC SCATTERING DIFFERENTIAL CROSS-SECTION OF $\pi^{\pm}$ ON DEUTERON AT 47 MeV

B. Balestri[†], P.Y. Bertin[††], B. Coupat[††], A. Gérard[†], L. Guechi[†], G. Fournier[†], E.W.A. Lingeman[†††], J. Miller[†], J. Morgenstern[†], J. Picard[†], B. Saghai[†], K.K. Seth[††††], C. Tzara[†] and P. Vernin[†]

† *DPh-N/HE, CEN Saclay, BP 2, 91190 Gif-sur-Yvette, France*

†† *IN2P3, Laboratoire de Physique Corpusculaire, Université de Clermont, BP 45, 63170 Aubière, France*

††† *IKO, Oosterringdijk, 18A Posthus 4395, Amsterdam 1006, The Netherlands*

†††† *Department of Physics and Astrophysics, Northwestern University, Evanston, Illinois 60201, U.S.A.*

For several reasons the understanding of pion-deuteron interaction is the most natural way to improve our knowledge of the interaction of pions with more complexe nuclei. At low energy, i.e. below 100 MeV, besides multiple scattering methods [8], three body calculations are possible [1,2]. But only one fairly precise experiment has been done at 47.5 MeV in π^+ [3]. On the other hand a comparison between the elastic scattering differential cross-section of π^+ and π^- might allow either to test the isoscalar character of the strong interaction [4], if the coulomb effect corrections are handled, or to improve this kind of calculations.

Using the facilities of the Linear Accelerator of Saclay [5], one has decided to do a rather complete investigation with both polarities at three energies. A 10 to 25 mm thick liquid deuteron target is used and the pions are detected in a range spectrometer whose response is very similar in π^+ and π^- [6]. The data acquisition at 24 and 47 MeV for an angular distribution between 30° and 145° is finished.

On Figs. 1 and 2 the results at 47 MeV for four angles between 60° and 90° are given and compared to the experimental data of Axen et al. [3], and the 3-body calculations of Thomas [1,7] and Giraud [2]. The complete angular distribution will hopefully be available by the conference. The final expected precision is about ± 3 %.

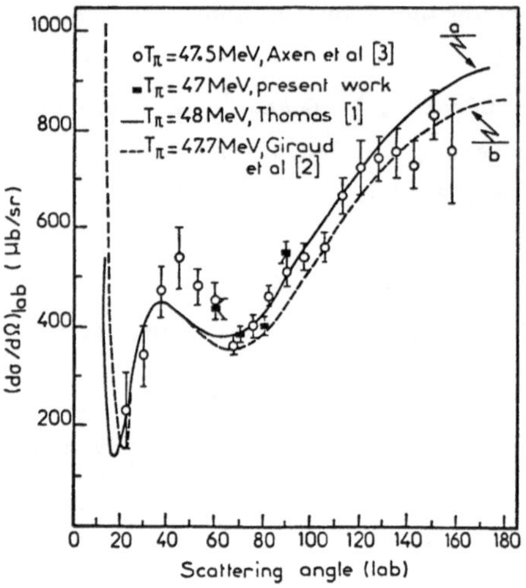

Fig. 1 - Differential cross-section elastic π^+ deuteron scattering curves (a) and (b) are Faddeev calculations assuming a deuteron wave function with 7 % D-state component.

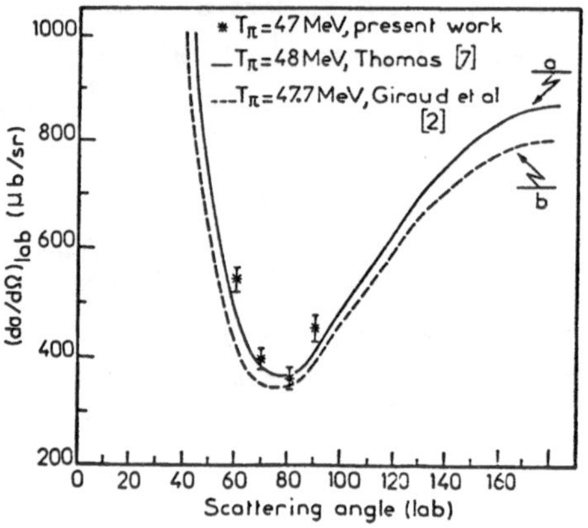

Fig. 2 - The same as Fig. 1 for π^-d.

References

[1] A.W. Thomas, Int. Conf. on few body problems in nuclear and particle physics (1974), Laval Univ. Quebec, Canada.
[2] N. Giraud, Thesis, n°683 (1978), Université Claude Bernard, Lyon 1.
N. Giraud, C. Fayard and G.H. Lamot, to be published.
[3] D. Axen et al., Nucl. Phys. A256 (1976) 387.
[4] C. Tzara, Note Interne, CEN Saclay, DPh-N/HE 73/7 (1973).
[5] P.Y. Bertin et al., Rapport Interne, CEN Saclay, DPh-N/HE 71/3 (1971).
[6] To be published by the authors of this paper.
[7] A.W. Thomas, private communication (1977).
[8] M. McMillan and R.H. Landau, Triumf TRI-74-1 (1974).
M. Van der Velde et al., Nuovo Cim. 40A (1977) 97.

APPLICATION OF FADDEEV EQUATIONS TO $K^-d \to \pi^-\Lambda p$ AT LOW ENERGY*

G. Toker and A. Gal
Racah Institute of Physics, The Hebrew University
Jerusalem, Israel

J. M. Eisenberg
Department of Physics and Astronomy, Tel-Aviv University
Tel-Aviv, Israel

The T-matrix for the $K^-d \to \pi^-\Lambda p$ breakup process is directly decomposed into the Faddeev form:

$$T = \sum_{i=1}^{3} T_i, \qquad T_i = t_i(1-\delta_{i1}) + t_i G_o (T_j + T_k) \tag{1}$$

(where the subscript 1 denotes the channels with a meson ($\bar{K}$ or π) and a baryon pair (NN, ΛN or ΣN)). S-wave Yamaguchi separable potentials are employed, with the treatment of YN and $\bar{K}$N-πY interactions in coupled channels for each class. The application of isospin formalism, neglecting all mass differences within isobaric multiplets, reduce the set (1) to a system of 13 coupled integral equations, and non-relativistic kinematics allows analytic evaluation of the kernels.

These Faddeev equations are solved by using Padé approximant technique, with a careful numerical treatment of the dynamical singularities and the two-body thresholds. These thresholds are especially troublesome in the present case due to the variety of two-body subsystems which enter here. By placing mesh points between the relevant thresholds in every channel and exploiting the logarithmic nature of the dynamical singularities we are able to reach a precision of four significant figures, or better, for numerical integration with 32 mesh points.

The calculated Λp effective mass spectrum (Fig. 1) is compared to the experimental findings of Tan [1] for $K^-d \to \pi^-\Lambda p$ at rest, and, in particular, to the enhancement at the ΣN threshold (also observed for higher incident K^- momenta [2]). The calculated peak is found exactly at the ΣN threshold for several choices of ΛN-ΣN potentials, which confirms the natural expectation that this peak is due to a two-body cusp phenomenon at the ΣN threshold. The effect of the initial $\bar{K}$N interaction is rather important in giving the right shape in the low Λp mass region. Strong final πN interactions are currently being introduced via their more realistic p-wave description.

*Supported in part by the U.S.-Israel Binational Science Foundation and by the Israel Academy of Sciences.

1. T.H. Tan, Phys. Rev. Lett. 23, 395 (1969).
2. O. Braun *et al.*, Nucl. Phys. B124, 45 (1977) and references cited here.

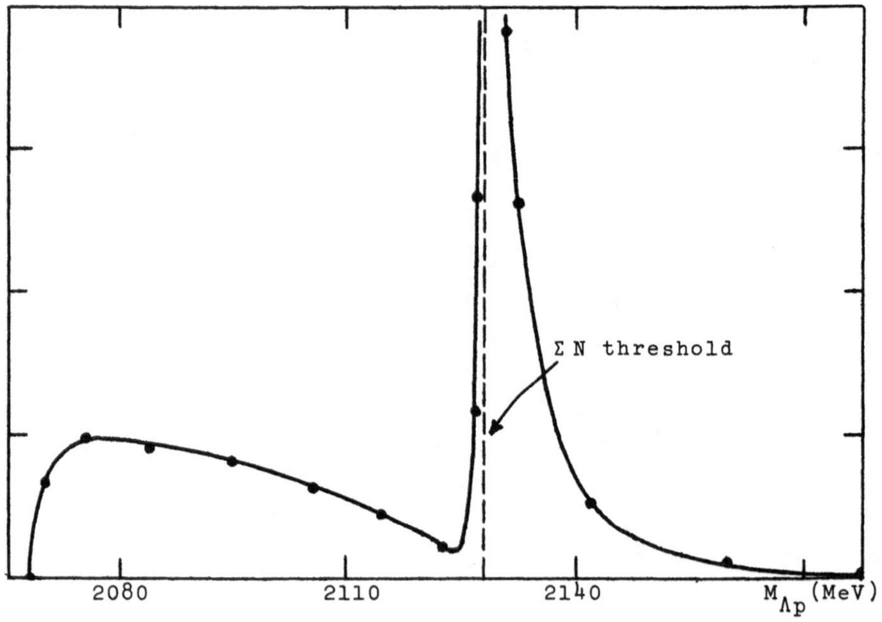

Figure 1
Calculated Λp effective mass spectrum (in arbitrary units) at very low incident K^- momenta.

EXCHANGE CURRENT EFFECTS IN THE REACTION $^3He(\gamma,\pi^+)^3H$

A.Barroso*) and A.K.Rej**)
*)Centro de Física Nuclear, Universidade de Lisboa,Lisboa 2,Portugal
**)Fysisk Institutt,Universitetet i Trondheim,7000 Trondheim ,Norway

The theoretical description of pion photoproduction in the region of the (3,3) resonance is usually made using as imput the knowledge of the photoproduction on a single nucleon. However, inside a nucleus two body processes that reveal the presence of mesonic degrees of freedom can be important. In particular for the three body system it has been shown that this is the case 1-3). For the photoproduction reaction these exchange corrections were considered previously by Lazard et al. 4) but using a wave function that only included the dominant S-state.

Following the usual S matrix prescription 1) we write down the exchange current associated with the diagram of Fig.1., i.e.

$$J_{exch} = -i(2\pi)^3 \, \delta^3(\underline{k}+\underline{p}_1+\underline{p}_2-\underline{k}'-\underline{p}_1'-\underline{p}_2')$$
$$\frac{eg}{(2\pi)^9} \; t^{\gamma\pi\pi} \; \frac{1}{(\underline{p}_2'-\underline{p}_2)^2+\mu^2} \; t^{\pi N\to N}$$

where the two pion production amplitude is given by 5)

$$t^{\gamma\pi\pi} = \frac{1}{\sqrt{4E_k E_{k'}}} \, 12\,\pi\, i\, h_{33}(E_{k'})(\underline{k}'\cdot\underline{\varepsilon}-\tfrac{1}{3}\,\underline{\sigma}_1\cdot\underline{k}'\underline{\sigma}_1\cdot\underline{\varepsilon})$$
$$\varepsilon_{3\alpha\gamma}\; \varepsilon_{3\gamma\delta}\; (\delta_{\beta\delta} - \tfrac{1}{3}\,\tau^1_\beta\, \tau^2_\delta)$$

Once we have defined J_{exch} it is straightforward to obtain the transition amplitude. This is turn when it is added to the conventional one body amplitude enables one to compute the cross section. Skipping all the details of this derivation that can be found elsewhere 6) let us concentrate on the results. Using the nucleon photoproduction amplitudes of Berends et al 7) and the Gibson 8) wave function, including S S' and D states, we obtained the results displayed on table 1. Two conclusions should be pointed out. The first one concerns the magnitude of the exchange contributions which is of the order 5-10%, i.e. smaller than the kinematics uncertanties associated with the transformation from γ-N C.m. to the γ-nucleus center of mass 6). The second conclusion is the dominance of S state which is somehow different from what happens in other exchange calculations in the 3-body system.

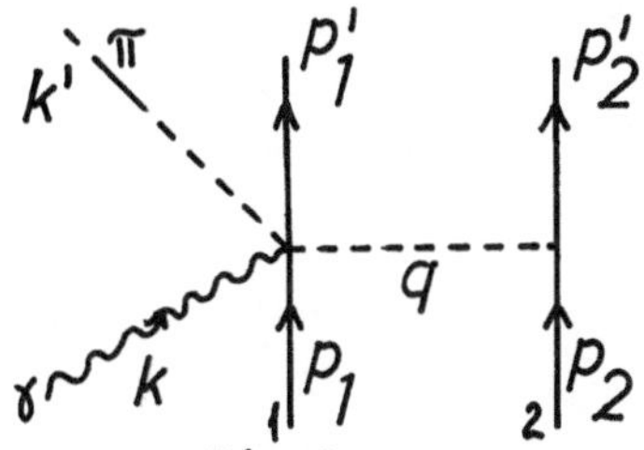

Fig.1

References

1) M.Chentob and M.Rho, Nucl. Phys. A163 (1971) 1
2) A.Barroso and E.Hadjimichael, Nucl. Phys. A238 (1975) 422
3) E.Hadjimichael, Nucl. Phys. A294 (1978) 513
4) C.Lazard et al., Lett. Nuov. Cim. 12 (1975) 405
5) P.Carruthers and H.W.Huang, Phys. Lett. B24 (1967) 464
6) A.K.Rej, University of Trondheim preprint
A.Barroso and A.K. Rej to be published
7) F.A.Berends et al., Nucl. Phys. B4 (1967) 54
8) F.B.Gibson, Nucl. Phys. B2 (1967) 501

Table 1: The differential cross-sections in the Laboratory frame (μb/sr) for $\theta_\pi^{CM} = 137^o$

Q^2 $[fm^{-2}]$	Without exch. current	With exch.current	
		S-S only	S-S+S-S'+S-D
1.4	2.404	2.373	2.382
1.8	2.307	2.259	2.276
2.2	1.756	1.697	1.715
2.7	1.396	1.332	1.349
3.9	0.798	0.744	0.754
4.9	0.436	0.406	0.410
6.0	0.167	0.158	0.159

ELECTROMAGNETIC SUM RULES FOR LIGHT NUCLEI

D. Drechsel, H. Arenhövel, W. Fabian
W. Heinze, Y.E. Kim, V. Tornow
Institut für Kernphysik
Universität Mainz
D-6500 Mainz

Electromagnetic sum rules describe gross features of the electromagnetic structure of nuclei[1]. A well known example is the Thomas-Reiche-Kuhn (TRK) sum rule, which relates the integrated total E1-absorption cross section to the ground state expectation value of the double commutator of the dipole operator D with the nuclear Hamiltonian. While the kinetic energy gives a model independent contribution, i.e., the classical sum rule $\sum_{cl}$ = 60 NZ/A MeV mb, the nuclear two-body potential gives an additional contribution in the presence of exchange and/or momentum dependent (or nonlocal) forces. In this case, the sum rule is enhanced by $1 + \kappa$, where $\kappa = M \frac{A}{NZ} \langle 0 | [D_Z, [H, D_Z]] | 0 \rangle$ is a measure of exchange effects. Because of the two-body character of the double commutator, κ depends sensitively on two-body correlations. In particular, the importance of the nuclear tensor force is reflected by a large contribution of tensor correlations[2]. While total absorption cross sections of light and medium weight nuclei ($A > 6$) integrated up to π-production threshold gave an enhancement of about 100%, theoretical evaluations were controversial. In order to clarify this situation, we have evaluated the sum rule for the lightest nuclei consistently with the Reid-soft-core potential using exact wave functions for ^2H, ^3H (^3He) and an oscillator shell model for ^4He modified by two-body correlations[3]. The results are summarized in table 1.

Table 1: Enhancement κ for the lightest nuclei

	$^1S_0$	$^3S_1 + ^3D_1$	others	total
^2H	-	0.50	-	0.50
^3H(^3He)	0.11	0.60	0.01	0.72
^4He	0.24	0.69	-	0.93

It is evident that the enhancement κ approaches about 100% for ^4He. For nuclei beyond ^4He one does not expect a further increase because the number of relevant pairs in relative S-states does not increase faster than $\frac{NZ}{A}$. Furthermore, a closer inspection of the normalized integrals over the relative two-body wave functions contributing to κ reveals that they are not very different for the three nuclei, supporting the idea of the quasi-deuteron model. The reason for this is that the operator $r^2V(r)$ tests only the intermediate range part of the relative two-body wave function which differs not so much for these nuclei. Hence one may expect to obtain a reliable estimate for heavier nuclei by simply counting the pairs in relative S-states and using the normalized integrals from the lightest nuclei which include effects of tensor correlations.

Above π-production threshold the photoabsorption cross section is dominated by the $\Delta(1232)$ resonance, and it is reasonable to expect that this will be accounted for in the sum rule by explicitly including the Δ-degrees of freedom in the nuclear wave function as isobar configurations[4)]. We obtain for the deuteron an additional contribution of 20% and for ^4He, due to its larger isobar admixture, about 120%.

More general sum rules can be studied in electron scattering by choosing different paths in the q-ω plane, where q and ω are momentum and energy transfer, respectively, for example (i) q = const., (ii) $q = x\omega$ or (iii) $q^2 - \omega^2$ = const. In the first case (q = const), by varying the momentum transfer, one can in principle determine the spatial distribution of exchange forces[5)]. Fig. 1 shows the enhancement $\Delta(q)$ of the longitudinal sum rule

$$\sum(q) = \frac{Ze^2}{2M}(1 + \Delta(q))$$

for ^2H and ^3H (^3He). The value $\Delta(0)$ can be related to the enhancement κ by $\Delta(0) = \frac{2N}{A}\kappa$. The decrease with increasing momentum transfer shows clearly the importance of the long and intermediate range of the exchange force. Unfortunately, this sum rule cannot directly be compared with experiment, because by crossing the photon line it samples also contributions from time-like virtual photons inaccessible in electron scattering.

The sum rules which are obtained by integrating along photon-like lines ($q = x\omega$) are particularly interesting since they may serve as independent tests of photonuclear sum rules ($x \to 1$)[6)]. In addition they

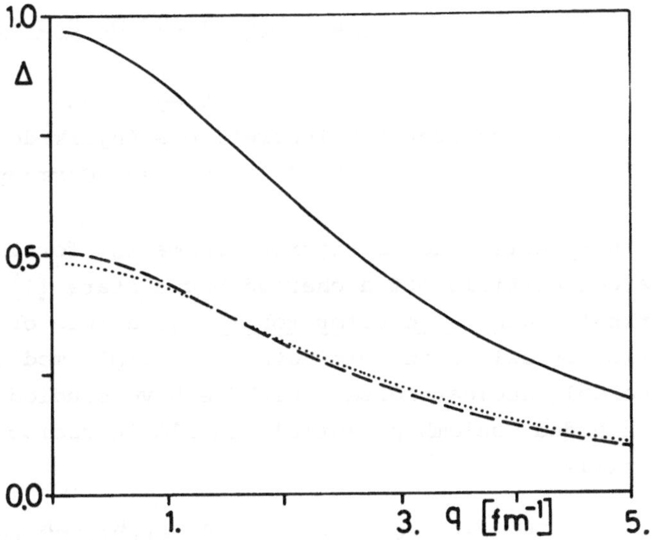

Fig. 1: Enhancement Δ of the longitudinal sum rule at constant momentum transfer q, as function of q, for ^2H (dashed curve), ^3H (full curve) and ^3He (dotted curve).

have the virtue of not crossing the photon line. One can relate these sum rules to the absorptive part of the forward scattering of a virtual photon and can derive relations similar to the sum rule of Gell-Mann, Goldberger and Thirring by exploiting dispersion relations if microcausality, i.e., analyticity holds. We have investigated this explicitly in a schematic model for a relativistic and nonrelativistic particle in a δ-like force acting only in S-states[7]. For the relativistic case analyticity holds for real and time-like photons, but not for space-like photons, and simple sum rules follow. However, for the nonrelativistic case this is not true due to the well known pathology of the nonrelativistic kinematics allowing, for example, the absorption of a real photon by a free particle. One can derive, however, a simple scaling relation, which relates the general sum rule along $q = x\omega$ to the photonuclear sum rule of a hypothetical system with scaled length ($r' = xr$) and mass ($M' = M/x^2$).

References:

1. D. Drechsel, invited paper for IVth Seminar on Electromagnetic Interactions in Nuclei, Moscow (1977)
2. A. Arima, G.E. Brown, H. Hyuga, M. Ichimura, Nucl. Phys. A205 (1973) 27
3. H. Arenhövel, W. Fabian, Nucl. Phys. A292 (1977) 429
   D. Drechsel, Y.E. Kim, Phys. Rev. Lett. 40 (1978) 531
   W. Heinze, H. Arenhövel, G. Horlacher, to be published
4. H.J. Weber, H. Arenhövel, Phys. Rep. 36C (1978) 277
5. E.V. Inopin, S.N. Roshchupkin, Sov. Journ. Nucl. Phys. 17 (1973) 526
6. D. Drechsel, Int. School on Electro- and Photonuclear Reactions Erice (1976), in Lecture Notes in Physics 62 (1977) 92
7. H. Arenhövel, D. Drechsel, H.J. Weber, to be published.

PRACTICAL CALCULATIONS OF P-D SCATTERING

H. Ziegelmann
Institut für Theoretische Physik der Universität
D-7400 Tübingen, Germany

There exists now a rigorous formalism for the scattering of one charged particle off a charged bound state [1]. Whereas this formalism works for general nuclear forces (separable or non-separable), we restrict ourselves in this note on a simple model for p-d scattering with separable nuclear forces which we have studied numerically [2]. Using a shielded Coulomb potential (shielding radius R) we get for rank-1 potentials

$$T^R_{\beta\alpha}(z) = V^R_{\beta\alpha}(z) + \sum_{\gamma=1}^{3} V^R_{\beta\gamma}(z)\tau^R_\gamma(z)T^R_{\gamma\alpha}(z) \quad . \tag{1}$$

By these Amado-type equations our original three-body problem is reduced to a two-body problem with shielded Coulomb potential. If one tries to solve eq. (1) in the no screening limit, one is faced with "one of the most famous and frustrating anomalies in scattering theory" (Taylor, ref. [3]). Anomalies show up in different ways: 1) The momentum space representation of the unshielded Coulomb potential does not exist. 2) The phase shift diverges. 3) Eq. (1) is no longer unique. 4) The t-matrix of the screened problem does not converge. 5) Angular momentum decomposition is impossible. The only quantity in which we can trust in this chaotic world is the Coulomb modified strong phase shift. Calling the phase shift of the complete solution of eq. (1) δ^R_l and the phase shift of the solution of the same equation with only the Coulomb force acting between p and d $\delta^R_{l,c}$, the Coulomb modified strong phase shift is given by

$$\delta_{l,sc} = \lim_{R\to\infty} \delta^R_{l,sc} \ , \quad \delta^R_{l,sc} = \delta^R_l - \delta^R_{l,c} \quad . \tag{2}$$

This quantity has a limit because the divergent part of δ^R_l and $\delta^R_{l,c}$ cancel:

$$\delta^R_l \xrightarrow[R\to\infty]{} \sigma_l + \delta_{l,sc} + \phi(k,R) \ , \quad \delta^R_{l,c} \xrightarrow[R\to\infty]{} \sigma_l + \phi(k,R)$$

$$\phi(k,R) = -\frac{\mu}{k}\int_{1/2k}^{\infty} v^R_c(r)dr \quad , \quad \sigma_l = \arg(\Gamma(l+1+i\eta)) \ , \tag{3}$$

The justification of the subtraction procedure, eq. (2), is given in ref. [3] for the two-particle case. It was shown in ref. [1] that similar renormalization procedures also work in the three-particle case (even for breakup amplitudes). In practical calculations one cannot, of course, go to the limit as is required by eq. (2). It is, therefore, interesting to investigate numerically if the difference in eq. (2) becomes stable and close to the exact answer for finit screening radius. This one would expect naively for $R = 10^4 F$ because the Coulomb potential of the target nucleus is, in fact, shielded by the electronic charge distribution. The momentum representation of the Coulomb potential is, however, for such a large shielding radius so strongly peaked that a stable numerical calculation is almost impossible. Avoiding this peak by contour deformation in the complex plane is also impossible because this peak is the result of two singularities lying just above and below the integration path. Consequently we have to stay on the real axis and have to try shielding radii for which the integral equation (1) can be solved with confidence. It is crucial for our method to show that the asymptotic form of eq. (3) is already valid for these smaller shielding radii.

We, therefore, performed the following test (see table 1): We calculated the phase $\delta^R_{1,c}$ for screened Coulomb potential with sharp cut off. This potential oscillates in contrast to a potential with smooth cut off, but has the advantage that one can compare with the analytically known Coulomb wave. If the numerical values of $\delta^R_{1,c}$ is good, then there should be no phase shift $\delta_{1,sc}$ between the calculated wave and the exact Coulomb wave at point R. (This is the essence of the method of Vincent and Phatak [4], which allows for finding the Coulomb modified phase shift from the phase calculated for sharp finit cut off. I am indebted to P.U. Sauer for supplying me with his Vincent and Phatak code.) As can be seen from table 1, $\delta_{1,sc}$ is very small. That means that $\delta^R_{1,c}$ has been calculated with good accuracy. Next we calculated the phase shift $\delta^R_{1,c}$ from the asymptotic formula (3) and found that for small energy the difference between this value and the exact one is about 6%, whereas for higher energies the exact value agrees with the asymptotic form. From this we conclude that with the screening value of R = 30 F we can expect to get resonable results via eq. (2) for intermediate and high energies, whereas for smaller energies the screening radius has to be increased.

In order to investigate p-d scattering in our model one has to antisymmetrize eq. (1), [5]. In our calculations we approximate the Coulomb t-matrix in $V_{\beta\alpha}$ and τ_γ by its Born term. Nevertheless, we have interaction in the p-p subsystem in all partial waves. Using a spin and charge dependent Yamaguchi force fitted to the low energy nucleon-nucleon data, we get the phase shifts of fig. 1.

One important difference between n-d and p-d scattering is that by the energy loss in the Coulomb field the p-d reaction finally takes place at some smaller energy than the p-d reaction. Therefore we expect the p-d curve shifted to the right with respect to the n-d curve. This is quantitatively expressed by the formula [7]

$$\delta_{l,sc}\,(p\text{-}d) - \delta_{l,s}\,(n\text{-}d) = \frac{2\mu e^2}{2l+1}\frac{\partial}{\partial k}\,\delta_{l,s}\,(n\text{-}d) \qquad (4)$$

The differences predicted by this formula and found by our calculation are compared in table 2. As the formula (4) was derived for elementary particles it needs further investigation why it gives so good results in our case of composite particles.

In the doublett (l = 0) case we get almost no Coulomb effects. This is again (at least qualitatively) in agreement with eq. (4).

Table 1

Phase shifts (in radians) for pure Coulomb scattering with sharp cut off (R = 30 F).

E_{Lab} [MeV]	$\delta^R_{o,c}$ (calculated)	$\delta^R_{o,c}$ (asymptotic)	$\delta_{o,sc}$
0.67	- 0.4607	- 0.4336	$-0.14\cdot10^{-4}$
24	- 0.1302	- 0.1307	$-0.42\cdot10^{-4}$
48	- 0.0999	- 0.1003	$-0.47\cdot10^{-4}$

Table 2

Differences between p-d and n-d quartett l = 0 phase shifts in radians.

E_{Lab} [MeV]	${}^4\delta_{o,sc}(p\text{-}d) - {}^4\delta_{o,s}(n\text{-}d)$	
	Eq. (2)	Eq. (4)
1	0.17	0.19
3	0.14	0.15
5	0.11	0.12
8	0.09	0.09

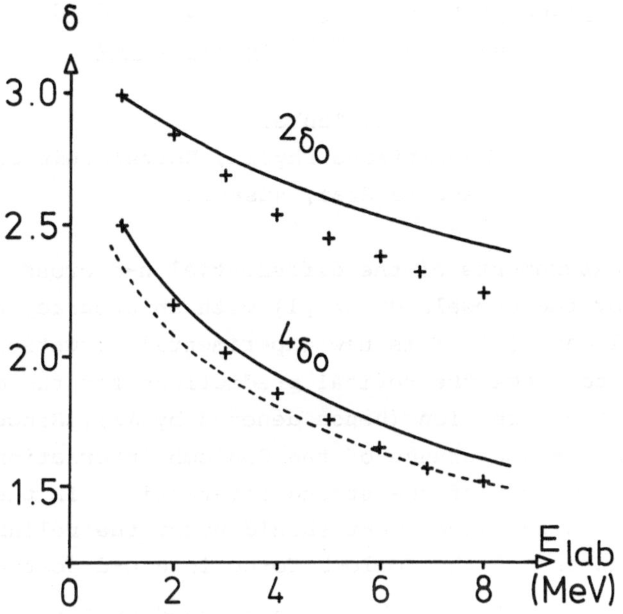

Fig. 1

The $^4\delta_0$ and $^2\delta_0$ phase shifts in radians for p-d (full line) and n-d (broken line) scattering. The results of the phase shift analysis of Arvieux [6] are shown by crosses.

References

[1] E.O. Alt, W. Sandhas, H. Ziegelmann, Phys. Rev., in press; E.O. Alt, in Few Body Dynamics, edited by A.N. Mitra et al. (North Holland, Amsterdam, 1976).

[2] E.O. Alt, W. Sandhas, H. Zankel, H. Ziegelmann, Phys. Rev. Lett. 37 (1976) 1537.

[3] J.R. Taylor, Nuovo Cim. B23 (1974) 313; M.D. Semon, J.R. Taylor, Nuovo Cim. A26 (1975) 48.

[4] C.M. Vincent, S.C. Phatak, Phys. Rev. C10 (1974) 391.

[5] E.O. Alt, to be published.

[6] J. Arvieux, Nucl. Phys. A221 (1974) 253.

[7] W. Plessas, L. Streit, H. Zingl, preprint Univ. Graz - VTP 04/74, 1974.

DIFFERENTIAL CROSS-SECTIONS FOR ELASTIC p-d AND n-d SCATTERING AT 10 MeV

H. Zankel
Institut für Theoretische Physik, Universität Graz
A-8010 Graz, Austria

Only recently measurements of the differential n-d cross section have been performed by the Uppsala-group [1] with an accuracy comparable to p-d experimental data [2]. This new experimental situation provides us the possibility to check theoretical predictions for the difference of the n-d and p-d cross section (hence denoted by $\Delta\sigma$). Since the difference is mainly due to the appearance of the Coulomb interaction (presumed there is charge symmetry of the strong interaction) in the p-d system the comparison with the experiment should proof the reliability of the theoretical treatment of the Coulomb force in a p-d scattering formalism.

Now a method is available that has been propsed by Alt et al.[3,4] which includes the Coulomb interaction exactly and which has been shown to be practical too [5]. Applying this p-d formalism one is able to evaluate the Coulomb modified strong phase shifts $\delta_{sc,\ell}$ as well as the absorption parameter $\eta_{sc,\ell}$ to the lowest order in e^2. The phases are used then to obtain the differential cross section via the formula [6]

$$\frac{d\sigma}{d\Omega} = \frac{2}{3}\left(\frac{d\sigma}{d\Omega}\right)_{quartet} + \frac{1}{3}\left(\frac{d\sigma}{d\Omega}\right)_{doublet} \tag{1}$$

with

$$\frac{d\sigma}{d\Omega} = \left| f_c(\Theta) + \sum_\ell (2\ell+1) P_\ell(\cos\Theta)\,\eta_{sc,\ell}\, e^{2i\sigma_\ell}\, \frac{e^{2i\delta_{sc,\ell}}-1}{2ip}\right|^2 . \tag{2}$$

Here f_c and σ_ℓ are the pure Coulomb amplitude respectively phase shift.

Switching off all pure Coulomb terms and using the strong phases (they can be derived easily from the p-d model by neglecting the Coulomb force and assuming the nuclear forces to be charge symmetric) in equ.(1) one gets the n-d cross section. The S-wave potentials used in our calculation are separable ones of the Yamaguchi type with the parameters fitted to the experimental low energy data. To gain some more detailed information about the mechanism of the interference of the Coulomb and the strong interaction we have calculated p-d cross sections in three ways:

i) We consider the long-ranged Coulomb distortion only which amounts to replacing in equ.(1) $\delta_{sc,\ell}$ and $\eta_{sc,\ell}$ by their pure strong analogs.

ii) In addition to the long-ranged distortion we take into account part of the shorter-ranged Coulomb corrections represented in $\eta_{sc,\ell}$ and $\delta_{sc,\ell}$, which are calculated by employing a two-nucleon Coulomb modified T-matrix in first order of e^2) in the p-p subsystem.

iii) Exact calculation including all shorter-ranged Coulomb corrections as performed in ref.[5].

The resulting cross sections are displayed in Fig.1.

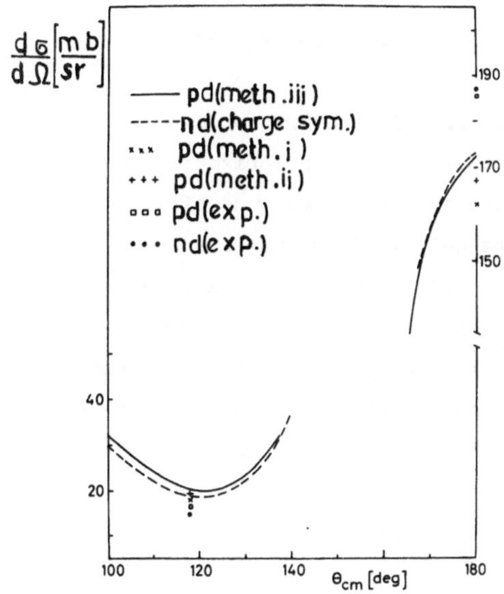

Fig.1: Differential cross section at 10 MeV.

For comparison with experimental data we concentrate on the cross section minimum around 118° and on the backward angle of 180°. In both cases it is obvious that the theoretical $\Delta\sigma$ derived with the full p-d theory is in fair agreement with the experiment. Discrepancies between the absolute values of the theoretical and the experimental points should be due to the simplicity of the nuclear potentials used in our theory. The cross section evaluated with method i) reproduces the exact result only qualitatively and one may conclude that this approximative procedure as used by Bruinsma et al. [7] is not adequate to describe the Coulomb corrections. Calculations of the cross section at 46 MeV indicate that for higher energy too the method i) fails to reproduce the experimental $\Delta\sigma$ [8] at the cross section minimum by far. The method ii) which incorporates part of the shorter-ranged Coulomb corrections appearing in the doublet channel only is rather successful at the cross section minimum, which can be understood since the contribution of the doublet scattering is more important at 118° than at 180°. Nevertheless the results in Fig.1 recommend to use the exact p-d theory even for differential cross sections at higher energies.

References

[1] L.Amten, A.Johansson and B.Sundquist, Report TLU 52/77, Uppsala 1977.

[2] D.C.Kocher and T.B. Clegg, Nucl.Phys. A132 (1969) 455.

[3] E.O.Alt, in Few Body Dynamics, ed.by A.N.Mitra et al.(North Holland, Amsterdam, 1976).

[4] E.O.Alt, W.Sandhas and H.Ziegelmann, to be published in Phys.Rev.C.

[5] E.O.Alt, W.Sandhas, H.Zankel and H.Ziegelmann, Phys.Rev.Lett. 37 (1976) 1537.

[6] W.T.H. van Oers and K.W. Brockman,Jr., Nucl.Phys. A92 (1967) 561.

[7] D.Bruinsma, W.Ebenhöh, J.H. Stuivenberg and R.van Wageningen, Nucl.Phys. A228 (1974) 52.

[8] J.L. Romero, J.A. Jungermann, F.P. Brady, W.J. Knox and Y.Ishizaki, Phys. Rev. C2 (1970) 2134.

INVESTIGATION OF THE p-d ELASTIC SCATTERING WITH POLARIZED PROTONS AND DEUTERONS

P.A. Schmelzbach, V. König, W. Grüebler, H.R. Bürgi and B. Jenny
Laboratorium für Kernphysik der ETH, Hönggerberg, CH-8093 Zürich, Switzerland

F. Seiler, G. Heidenreich and H. Roser
Institut für Physik der Universität Basel, CH-4056 Basel, Switzerland

W. Reichart
Physikinstitut der Universität Zürich, CH-8001 Zürich, Switzerland

The recent theoretical treatment of elastic scattering in the tri-nucleon system has reached a high degree of sophistication[1,2]. As expected, the polarization observables, particulary the tensor analysing powers, are seen to be very sensitive to the details of the NN-interaction introduced into the calculations. Unfortunately, the comparison between theoretical and experimental results often suffers from the poor accuracy and/or the limited range in which such data are available. Especially needed are complete sets of first order polarization data, i.e. proton (neutron) and deuteron analysing powers at the same c.m. energies. On the other hand, a better determination of the phenomenological phase shifts for N-d scattering (particularly the splitting of the phases and the mixing parameters) requires sets of very accurate data covering a large angular range at many energies.

New experimental data:

At tandem energies below $E_d = 13$ MeV an important contribution toward this goal has been made by White et al.[3].

In this paper we present a set of analysing power data for $H(\vec{d},d)H$ scattering obtained at 17, 20 and 24 MeV with the SIN-injector cyclotron together with new measurements of the $^2H(\vec{p},p)^2He$ analysing power performed at the ETH tandem laboratory.

At the cyclotron energies, the data were obtained by the technique described in ref.[4]. The statistical error was generally kept below 0.005. In the main part of the angular range the recoil protons were observed. In the very forward direction deuteron spectra were also used. The particles were detected in singles detectors: if necessary, proton and deuteron peaks were separated by a judicions choice of absorber foils and detectors, the thickness of which was monitored by the applied bias.

The absolute calibration of the beam polarization is estimated to be known to an accuracy better than 5%. The beam polarization was determined at each energy using the previously measured $^4He(\vec{d},d)^4He$ analysing power A_{yy} [5] and was continuously monitored during the runs with a polarimeter using the $^3He(\vec{d},p)^4He$ reaction at 0°.

The beam polarization was stable within 3%,therefore all data were normalized to the mean value of the beam polarization recorded during the measurements at one energy.

Due to the particular kinematical situation, the measurements in forward direction in the laboratory system are quite sensitive to errors in the setting of the angles and to instabilities like changes in the beam position during the measurements. The errors arising from such effects and from the background substraction may be estimated by checking the reproducibility of the data. This leads to an extra random error of 0.005 to be quadratically added to the statistical error. The results for the tensor component A_{xx} and A_{yy} are shown in fig. 1.

The quantity A_{xz} is currently not measurable at a cyclotron, but the present data are ideally completed at 17 and 20 MeV by very recent measurements performed at the UTTAC electrostatic accelerator by Sanada et al.[6]. It is our hope that these measurements can be extended close to 24 MeV in the near future.

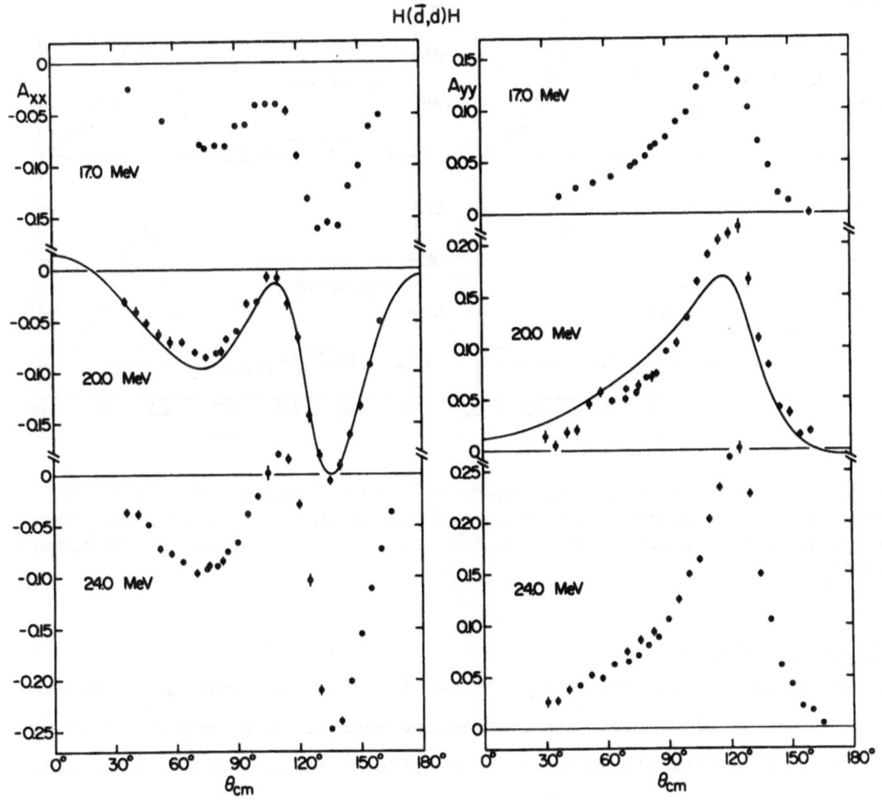

Fig. 1. The tensor analysing powers A_{xx} and A_{yy} for the $H(\vec{d},d)H$ elastic scattering between 17 and 24 MeV. Where not shown the statistical error is smaller than the dot. The curves are discussed in the text.

In order to get a complete set of new first order polarization data at the same c.m. energies, measurements of the $^2H(\vec{p},p)^2H$ analysing power were made at the ETH tandem laboratory. Here, the scale error is smaller than 2%, the statistical accuracy is better than 0.005 and the additional random error is estimated to be 0.003-0.005. These data for A_y^P at E_p = 8.5, 10 and 12 MeV are shown in fig. 2 together with the vector analysing power A_y^d for the $H(\vec{d},d)H$ scattering from the SIN experiment at the corresponding deuteron energies.

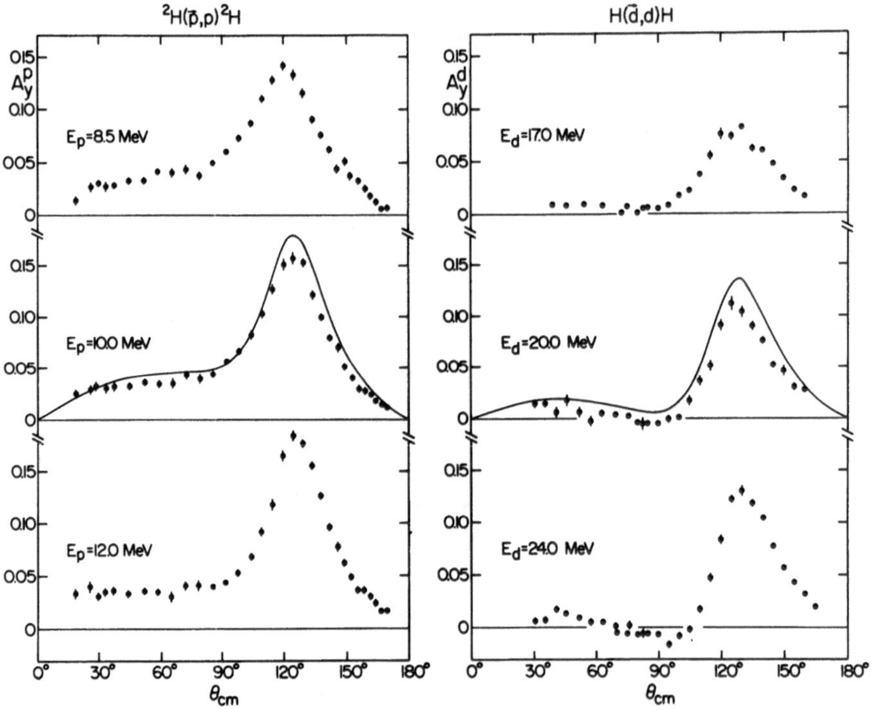

Fig. 2. The vector analysing powers A_y^P and A_y^d for the $^2H(\vec{p},p)^2H$ and $H(\vec{d},d)H$ elastic scatterings at corresponding energies. Where not shown, the statistical error is smaller than the dot. The curves are discussed in the text.

Comparison with theory:

The present data are compared at E_p = 10 MeV (E_d = 20 MeV) with a recent calculation performed by Doleshall in solving the Faddeev equation with separable NN-interactions[7]. The results are given as solid line in figs. 1 and 2. A comparison of theory and experiment for the very important quantity A_{xz} is presented by Sanada et al. in a contribution to this Conference[6].

In this calculation the NN interaction used fit the MAW[8] phase shifts at least up to 100 MeV. The two-term tensor force (2T4) reproduces the deuteron properties with the arbitrary choosen 4% d-state probability. The good agreement in shape and absolute scale shows the present state of such calculations. It will be interesting to compare the present data with calculations using other type of NN interaction[2] also.

Asymptotic normalization of the deuteron D-state:

Following a recent proposal by Amado et al.[9], we tried to use the present data at $E_d = 20$ MeV to determine the ratio of the D-state to S-state asymptotic normalization of the deuteron wave function. However, even with the good accuracy of our data, the method failed to produce a reliable value for ρ_D. The details of this investigation are discussed in ref. 10.

References

1) P. Doleschall, Proc. 4th Intern. Symp. on Polarization Phenomena in Nuclear Reactions (Zürich), eds. W. Grüebler and V. König (Birkhäuser, 1976) p. 51
2) C. Stolk and J.A. Tjon, Nucl. Phys. A295 (1978) 384
3) R. E. White et al., Contribution to this Conference
4) V. König, W. Grüebler and P.A. Schmelzbach, Proc. 4th Intern. Symp. on Polarization Phenomena in Nuclear Reactions (Zürich), eds. W. Grüebler and V. König (Birkhäuser, 1976) p. 895
5) V. König et al., Jahresbericht 1976, Laboratorium für Kernphysik der ETH, p. 64
6) J. Sanada et al., Contribution to this Conference
7) W. Grüebler et al., Phys. Lett. 74B (1978) 173
8) M.H. McGregor, R.A. Arndt and R.M. Wright, Phys. Rev. 182 (1969) 1714
9) R.D. Amado, M.P. Locher and M. Simonius, Phys. Rev. C17 (1978) 403
10) W. Grüebler et al., Contribution to this Conference

MEASUREMENT ON T_{21} FOR PROTON-DEUTERON ELASTIC SCATTERING AT E_D=20 MeV

J. Sanada, S. Seki, Y. Tagishi, W. Grüebler*, M. Sawada, Y. Nagashima, K. Furuno and L. S. Chuang**

Tandem Accelerator Center and Institute of Physics, The University of Tsukuba, Ibaraki 300-31, Japan

Here we report the results of the measurement on T_{21} for proton-deuteron elastic scattering and a comparison between the experimental data and the prediction of three nucleon Faddeev calculation made by Doleschall.

Grüebler et al.[1] have already succeeded to obtain the excellent agreement between the measured values of σ, A_y, iT_{11}, T_{20} and T_{22}, and Doleschall's prediction at $E_{c.m.}$=6.67 MeV (E_p=10 MeV or E_d=20 MeV). As for the T_{21}, however, the available data at the nearest energy is the one at E_d=16 MeV which shows less than a half values of the predicted values at E_d=20 MeV, some examples being shown in Fig. 1. Crosses are the existing data before the present work. It was thought very unlikely that the experimental values will increase by a factor of more than two from 16 MeV to 20 MeV as required by the prediction, open circles in Fig. 1 being the values predicted by Doleschall.

Polarized deuterons with the m_I=0 component from a Lamb-shift polarized ion source with spin filter were accelerated by a 12 UD Pelletron tandem accelerator vertically installed at UTTAC. The direction of the spin quantization axis at the target was regulated by the Wien-filter before entering the accelerator. To check our present procedures T_{21} for d-α elastic scattering at 17 MeV was measured. The results are summarized in Table 1, together with the data of LASL.[2] The agreement between two is quite satisfactory.

The data of T_{21} at 16 MeV, 17 MeV and 20 MeV for p-d elastic scattering were obtained in succession. The results are shown in Figs. 1 and 2 with closed circles. The errors indicated are of the statistical nature. The scale error is estimated to be $\pm$2 %. The full line in Fig. 2 is the prediction of three nucleon Faddeev calculation. The agreement is rather satisfactory up to 115°. However there remains a larger discrepancy at around 125°.

References:

1) W. Grüebler, V. König, P. A. Schmelzbach, B. Jenny, H. R. Bürgi,

P. Doleschall, G. Heidenreich, F. Seiler, H. Roser and W. Reichert, Phys. Lett. B, to be published.

2) G. G. Ohlsen, P. A. Lovoi, G. C. Salzman, U. Meyer-Berkhout, C. K. Mitchell and W. Grüebler, Phys. Rev. C8 (1973) 1262.

* On leave from Eidg. Technische Hochschule, Zürich, Switzerland.

** On leave from Chinese University of Hong Kong, Hong Kong.

Table 1. T_{21} for d-α elastic scattering at E_d=17 MeV.

$\theta_{c.m.}$	UTTAC	LASL
51.8°	0.050±0.005	0.053±0.005
58.9°	0.055±0.004	0.056±0.005
79.4°	-0.159±0.003	-0.158±0.005
85.8°	-0.173±0.002	-0.173±0.005

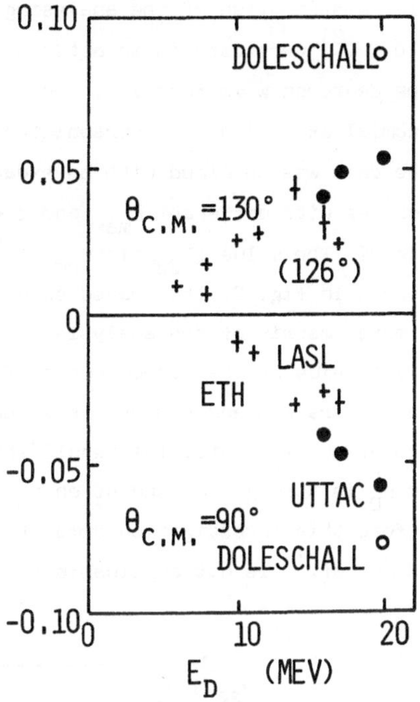

Fig. 1. Energy dependence of T_{21} for p-d elastic scattering at $\theta_{c.m.}$=90° and 130°.

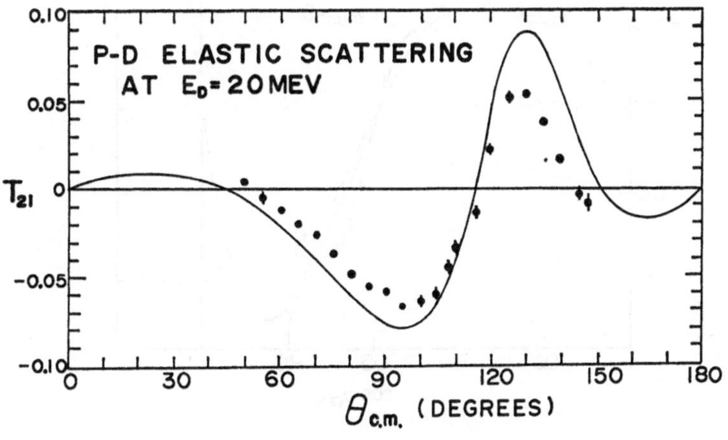

Fig. 2. Angular distribution of T_{21} for p-d elastic scattering at E_d=20 MeV. The full line is the prediction of three nucleon Faddeev calculation made by Doleschall.

ON THE ASYMPTOTIC NORMALIZATION OF THE DEUTERON D-STATE BY p-d ELASTIC SCATTERING

W. Grüebler, H.R. Bürgi, V. König, P.A. Schmelzbach and B. Jenny

Laboratorium für Kernphysik, Eidg. Technische Hochschule, CH-8093 Zürich

Following a recent proposal by Amado et al.[1] high precision measurements of the angular distribution of the analysing power T_{22} of the p-d scattering at $E_d = 20$ MeV[2] were used in an attempt to determine the asymptotic normalization ρ_D of the deuteron wave function. The quantity $T_{22}/\sin^2\theta_{cm}$ was fitted with a Legendre polynomial expansion and extrapolated to the neutron exchange pole z_p. The best fit to the data was obtained with an expansion up to $L_{max} = 9$. The experimental data, fit curves with different L_{max} and their extrapolation to the pole z_p are shown in fig. 1. The value $(T_{22}/\sin^2\theta_{cm})^{pole} = 0.2495\ E_{MeV}\cdot\rho_D$ as a function of L_{max} is presented in fig. 2. The shaded band indicates the uncertainty calculated with the full error matrix of the analysis. The probability P of the χ^2 test is shown on the right side by the dashed curves for two cases in which additional experimental random errors r=0 and r=0.002 have been added quadratically to the statistical errors of the T_{22} data. The result shows that the extrapolated value at z_p and hence ρ_D is strongly dependent on L_{max} of the Legendre polynomial expansion used. Therefore this investigation demonstrate convincingly that the simple method proposed in ref.[1] is not applicable for the calculation of a reliable value of ρ_D.

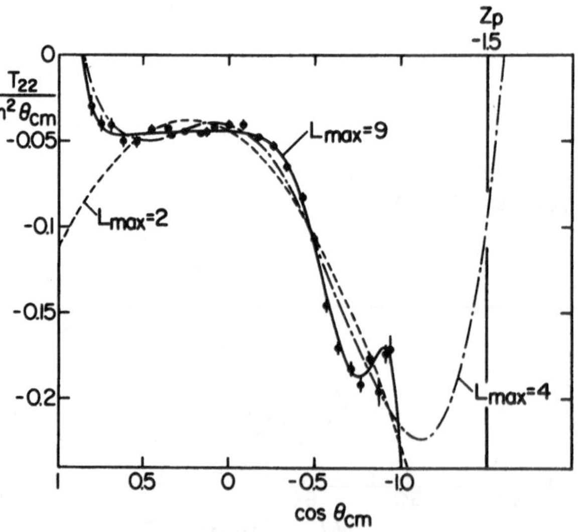

Fig. 1. The quantity $T_{22}/\sin^2\theta_{cm}$ for $E_d = 20$ MeV plotted against $\cos\theta_{cm}$. The curves are Legendre polynomial fits with different L_{max} extrapolated to $z_p = -1.5$.

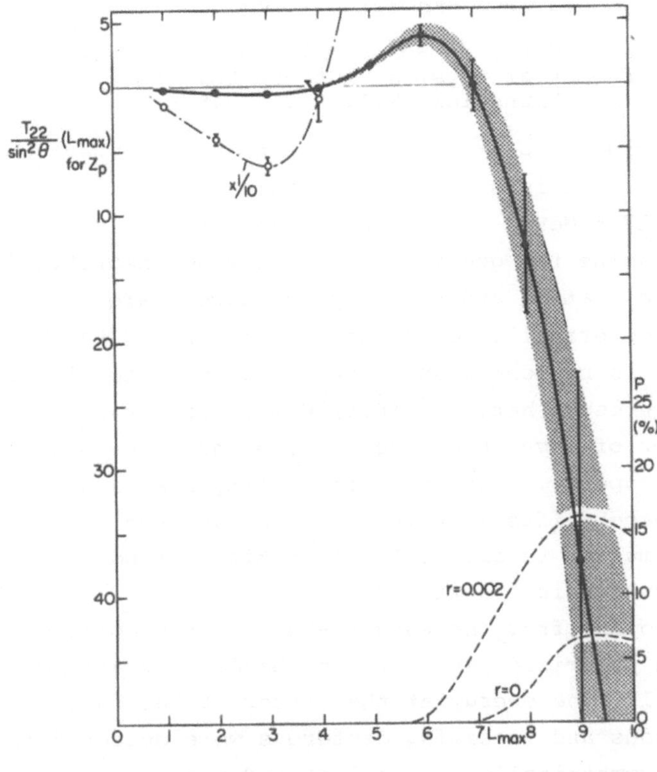

Fig. 2. $(T_{22}/\sin^2\theta_{cm})^{pole}$ as a function of L_{max}. The shaded band indicates the uncertainty. The probability P of the χ^2 test ist shown by the dashed curves for additional random errors r = 0 and r = 0.002.

References

1) R.D. Amado, M.P. Locher and M. Simonius, Phys. Rev. C17(1978)403

2) W. Grüebler et al. Phys. Lett. B74 (1978) 173

PROTON-DEUTERON ANALYZING POWER AT 14.1 MEV

J. C. Duder*, M. Sosnowski, and D. Melnik
Department of Physics, Rutgers University†
New Brunswick, N.J. 08903 U.S.A.

Despite the wide theoretical interest in nucleon-deuteron scattering at 14.1 MeV, there is only one measurement of $^2H(\vec{p},p)^2H$ between E_{lab} = 12.0 and 17.5 MeV.[1] This measurement at 14.5 MeV has been used frequently to examine the quality of 14.1 MeV nd theoretical calculations by assuming that nd and pd analyzing powers are the same. However, current theoretical calculations have improved to the point where this measurement is no longer sufficiently precise to discriminate well among them. We present here a $^2H(\vec{p},p)^2H$ measurement at 14.1 MeV which is about an order of magnitude more precise and hence has the desired discrimination property. Data of this quality also should aid in the search for pd phase shifts in this energy range, and we hope that it will help to stimulate theoretical calculations of pd analyzing powers including the Coulomb interaction.[2]

Polarized protons from the Rutgers-Bell FN tandem accelerator and atomic beam type polarized ion source bombarded a 2.8 mg/cm^2 deuterated polyethylene foil. The energy at the center of the target was (14.10 ± 0.05) MeV. Protons and recoiling deuterons were detected in ΔE-E telescopes located symmetrically to the left and right of the incident beam. The angular spread in the incident beam was limited to ±0.06° in the horizontal plane. The angular acceptance Δθ of the telescopes was never more than ±0.4°. The rms multiple scattering angle of protons or deuterons in the target material was generally smaller than this. An accurate calibration of the angular position of each detector relative to the target was carried out. There is an uncertainty in angle of ±0.05°. The incident beam polarization was continuously monitored by a helium polarimeter at the exit port of the scattering chamber. This polarimeter was calibrated by experimental comparison with the data of Ohlsen et al. at 12.03 MeV,[3] and by calculation from the phase shifts of Dodder et al.[4] at many energies. In examining the telescope spectra, the deuteron peaks were always very clean and required no background subtraction. The peaks due to protons elastically scattered from deuterons were estimated to have less than 3% background.

The new data is consistent with ref. 1 and is compared to recent theoretical calculations in Fig. 1.

We are deeply indebted to J. J. Benayoun et al. and to P. Doleschall for sending us their recent unpublished data and permitting its use in this paper. The calculations of P. Doleschall were made at the Institut für Kernphysik der Universität zu Köln. We thank N. Jarmie for calculating p-^{4}He observables for us at many energies and angles from the phase shifts of ref. 4.

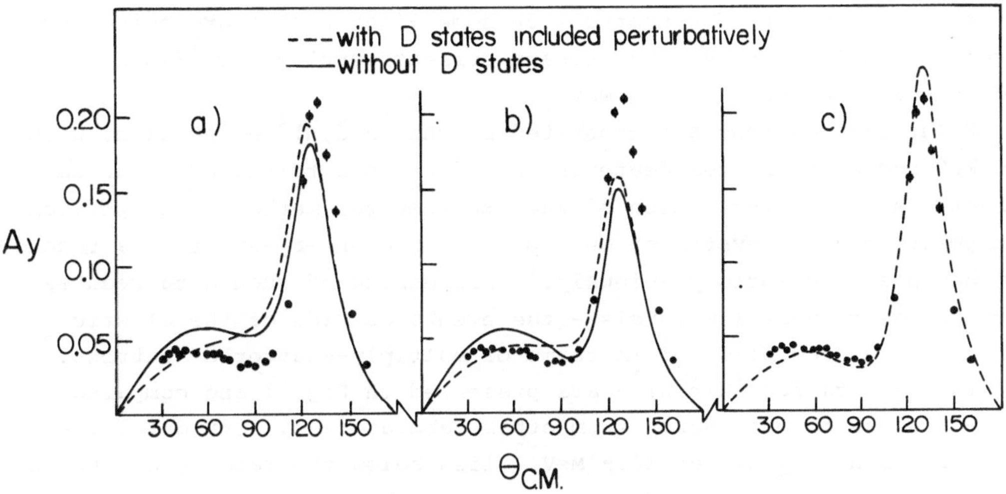

Figure 1. New $^{2}H(\vec{p},p)^{2}H$ data for E_{lab} = 14.1 MeV compared to recent nd theoretical results based on different nucleon-nucleon interactions. a) Reid soft core.[5] b) Super-soft core.[5] c) Separable interaction.[6]

References

*On leave from the University of Auckland, Auckland, New Zealand.

†Supported in part by the National Science Foundation.

1J. C. Faivre, D. Garreta, J. Jungerman, A. Papineau, J. Sura, and A. Tarrats, Nucl. Phys. A127 (1969) 169.

2E. O. Alt, W. Sandhas, H. Zankel, and H. Ziegelmann, Phys. Rev. Lett. 37 (1976) 1537.

3G. G. Ohlsen, J. L. McKibben, G. P. Lawrence, P. W. Keaton, and D. D. Armstrong, Phys. Rev. Lett. 27 (1971) 599.

4D. C. Dodder, G. M. Hale, N. Jarmie, J. H. Jett, P. W. Keaton, R. A. Nisley, and K. Witte, Phys. Rev. C15 (1977) 518.

5J. J. Benayoun, J. Chauvin, G. Gignoux, and A. Laverne, Phys. Rev. Lett. 36 (1976) 1438, and private communication.

6P. Doleschall, private communication. The result presented here used different n-n and n-p ^{1}S interactions, a two term (2T4) $^{3}S_{1}$-$^{3}D_{1}$ interaction and one term P interactions. One term D state interactions were taken into account perturbatively.

$^{1}H(\vec{n},n)^{1}H$ and $^{2}H(\vec{n},n)^{2}H$ at 14.2 MeV

J. E. Brock, A. Chisholm, J. C. Duder and R. Garrett
Department of Physics, University of Auckland†
Auckland, New Zealand

Analyzing powers for neutrons on some light nuclei are being measured in our laboratory. We present current results for $^{1}H(\vec{n},n)^{1}H$ and $^{2}H(\vec{n},n)^{2}H$ at E_{lab} = 14.2 MeV.

Polarized neutrons are generated by the $^{3}H(\vec{d},\vec{n})^{4}He$ reaction, with 150 keV vector polarized deuterons incident on a tritiated titanium target. A brief description of the experimental methods, the physical equipment, and our system of multiparameter event-by-event data recording has been presented previously.[1] Processing of such data reduces backgrounds to very low levels - the events outside of the elastic peaks can be accounted for in terms of multiply-scattered neutrons.[2]

The results for $^{1}H(\vec{n},n)^{1}H$ are presented in Fig. 1 and compared to data close to our energy from other laboratories. Our data, like those of Tornow et al. at 16.9 MeV,[3] lies below the Yale IV and LRL X phase shift predictions. However, two and three term fits to $\sigma(\theta)A_{y}(\theta)$ in order to obtain the spin-orbit phase parameters $\Delta_{LS}^{P,D,F}$ do not indicate a significant Δ_{LS}^{F}, in contrast to the finding of ref. 3. Data is still being taken, especially at back angles, in order to examine this question more closely. Results for $^{2}H(\vec{n},n)^{2}H$ are shown in Fig. 2 and compared to a recent theoretical calculation of Doleschall[6] using a separable nucleon-nucleon interaction and 14.1 MeV incident neutron energy.

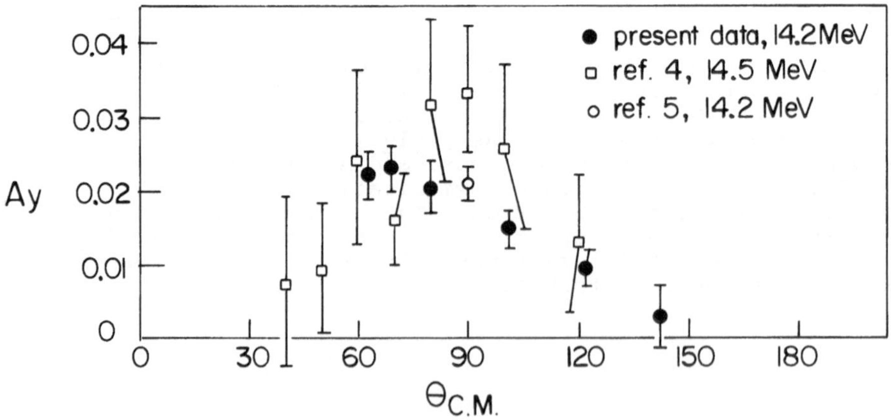

Figure 1. $^{1}H(\vec{n},n)^{1}H$ data near 14.2 MeV.

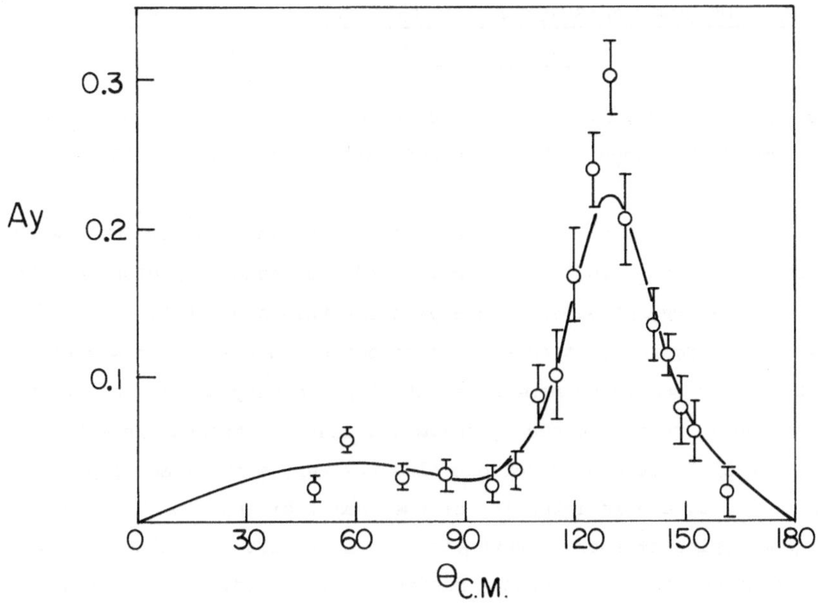

Figure 2. $^{2}H(\vec{n},n)^{2}H$ results. The curve represents a calculation by Doleschall at 14.1 MeV.

References

†Supported in part by NZ Universities Research Grants Committee.

[1]J. E. Brock, A. Chisholm, J. C. Duder, and R. Garrett, Proc. Fourth Int. Symp. on Polarization Phenomena in Nuclear Reactions (Birkhauser Verlag, 1976), W. Grüebler and V. König, eds., p. 473.

[2]J. E. Brock, A. Chisholm, J. C. Duder, and R. Garrett, Nucl. Instr. Meth. 137 (1976) 537.

[3]W. Tornow, P. W. Lisowski, R. C. Byrd, and R. L. Walter, Phys. Rev. Lett. 39 (1977) 915.

[4]R. Fischer, F. Kienle, H. O. Klages, R. Maschuw, and B. Zeitnitz, Nucl. Phys. A282 (1977) 189.

[5]B. Leeman, R. Casparis, M. Preiswerk, H. Rudin, R. Wagner, and P. Zupranski, Helv. Phys. Acta 47 (1974) 479.

[6]P. Doleschall, private communication. The result presented here used different n-n and n-p ^{1}S interactions, a two term (2T4) $^{3}S_{1}$-$^{3}D_{1}$ interaction and one term P interactions. One term D state interactions were taken into account perturbatively. The calculations were performed at Institut für Kernphysik der Universität zu Köln.

ANALYSING POWER MEASUREMENTS IN DEUTERON-PROTON SCATTERING FROM 5 TO 13 MeV

R.E. White*, W. Grüebler, B. Jenny, V. König, P.A. Schmelzbach and H.R. Bürgi
Laboratorium für Kernphysik, Eidg. Technische Hochschule, CH-8093 Zürich

Measurements[1] from 38° to 155° c.m. of the vector and tensor analysing powers in the elastic scattering of 6 to 11.5 MeV polarized deuterons by protons enabled the first detailed low energy phase-shift analysis for this process to be made[2]. Although a significant improvement on earlier attempts, this analysis was hindered by the quality of the analysing power data (the T_{kq} are very small) and its restricted angular and energy range especially below the deuteron breakup threshold where the phases are real. These considerations plus the very significant improvement in polarization stability and in intensity of the beam from the new ETH source have stimulated a new series of such measurements. In particular we wished to investigate a significant Coulomb interference minimum in T_{20} predicted at forward angles not reached in the original measurements.

Results have now been obtained using standard techniques[3] for all analysing powers at deuteron energies of 5, 7, 10 and 13 MeV over a c.m. angular range from 22.5^o to 165° by detecting both scattered deuterons and recoil protons. The much greater beam intensity now available has allowed the use of more stringent detector collimation producing a considerable improvement in the data, especially at low energies, as kinematic angular variations are very severe in this process and, in the 30° cone to which the deuterons are confined, deuteron and recoil proton energies are very similar.

The new measurements for T_{20} together with a fit to the old 10 MeV data from the existing analysis are shown in fig. 1. Statistical errors for almost all points are less than ± 0.003. The predicted minimum at forward angles is now seen clearly in the new data which agrees excellently with the earlier 10 MeV results. Measurements are in progress for extreme forward (and backward) angles at these energies to determine the magnitude of this T_{20} minimum since it provides a sensitive test of Coulomb corrected p-d Faddeev calculations, and at lower energies to provide more information in the region of real phase-shifts.

---

* On leave from the University of Auckland

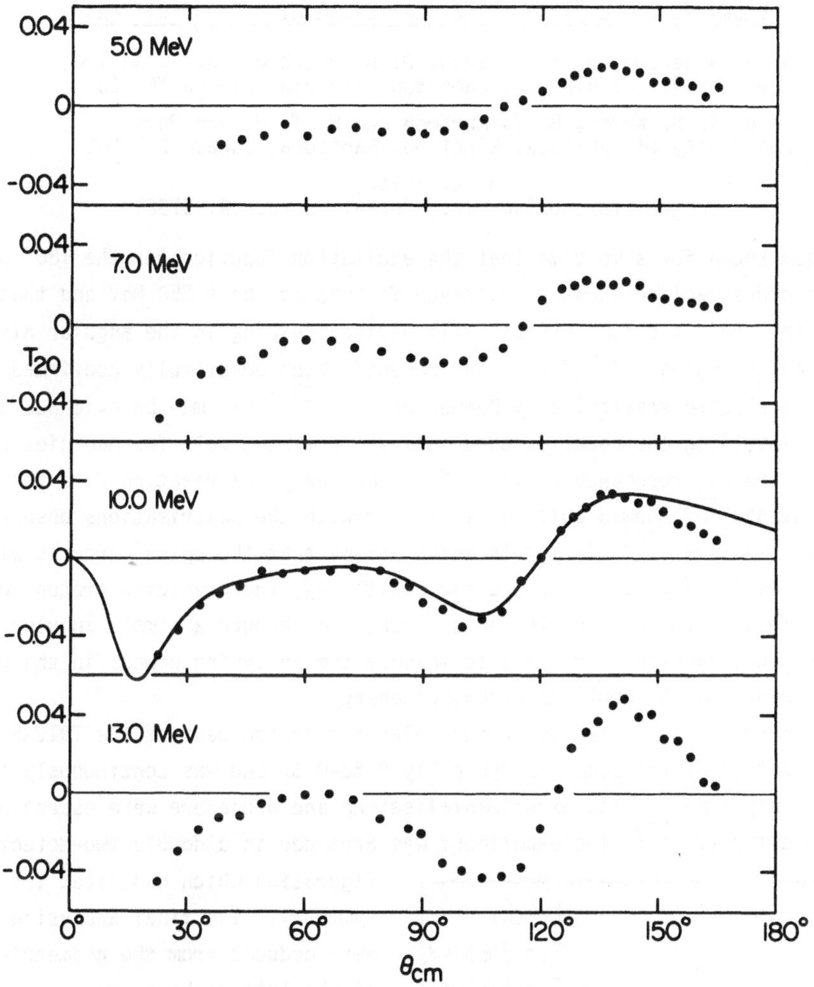

Fig. 1. The analysing power T_{20} at 5,7,10 and 13 MeV. Statistical errors are the order of the point size. The solid curve at 10 MeV is from a phase-shift analysis of earlier measurements between 38^0 and 155^0 c.m. [2].

References

1) R.E. White, W. Grüebler, V. König, R. Risler, A. Ruh, P.A. Schmelzbach and P. Marmier, Nucl. Phys. A180 (1972) 593

2) P.A. Schmelzbach, W. Grüebler, R.E. White, V. König, R. Risler and P. Marmier, Nucl. Phys. A197 (1972) 273

3) V. König, W. Grüebler and P.A. Schmelzbach, Fourth International Polarization Phenomena Proc. (1975) 893

BACKWARD ANGLE $\vec{p}$-d ELASTIC SCATTERING AT 316 AND 516 MeV*

A. N. Anderson, J. M. Cameron, D. A. Hutcheon and J. Källne,
University of Alberta, Edmonton, Alberta, Canada T6G 2J1

B. K. S. Koene, B. T. Murdoch and W. T. H. van Oers,
University of Manitoba, Winnipeg, Manitoba, Canada R3T 2N2

A. W. Stetz,
Oregon State University, Corvallis, U.S.A. 97331

It has been known for some time that the excitation function for the 180° p-d differential cross section shows a resonance feature at about 550 MeV and that the differential cross sections exhibit well defined peaking in the angular distributions at all energies.[1] The former has recently been beautifully confirmed for the case of n-d elastic scattering by Bonner et al. [2] This may be evidence that a mechanism involving the reaction pN→dπ plays a dominant role for energies corresponding to the Δ_{33} resonance region. [3] Some early polarisation data for the pd→dp reaction at 630 MeV showed quite a similarity with the polarisations observed in the pp→dπ$^+$ reaction. [4] In models which assume that the pp→dπ$^+$ process plays a dominant role in large angle pd elastic scattering, the predicted proton polarisations are related to those of the pp→dπ$^+$ reaction through a simple angular transformation. We, therefore, decided to measure the analysing powers in the pd→dp reaction below and at about the resonance energy.

The experiment was performed using the polarised proton beam of the TRIUMF cyclotron. The polarisation of the beam was typically 0.62-0.67 and was continuously measured during the experiment. The experimental set up and procedure were essentially the same as previously. [5] The experiment was arranged in a double two-detector configuration which exhibited left-right symmetry. The final analysing powers were deduced from the geometric means of the left-right asymmetries with incident proton spin up and down. The uncertainty in the beam polarisation calibration is estimated to be ± 0.03 at 316 MeV and ± 0.06 at 516 MeV.

The results are presented in Fig. 1 together with predictions deduced from the pp→dπ$^+$ reaction at the same energies. The latter analysing power values were calculated using parameterised fits to $\vec{p}$p→dπ$^+$ and $\vec{p}$p→dπ$^+$ data. [6]

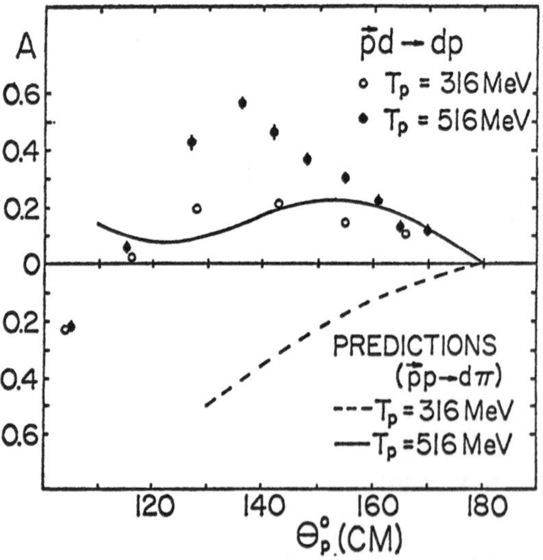

Fig. 1

*Supported in part by the National Research Council of Canada and the U.S. National Science Foundation.

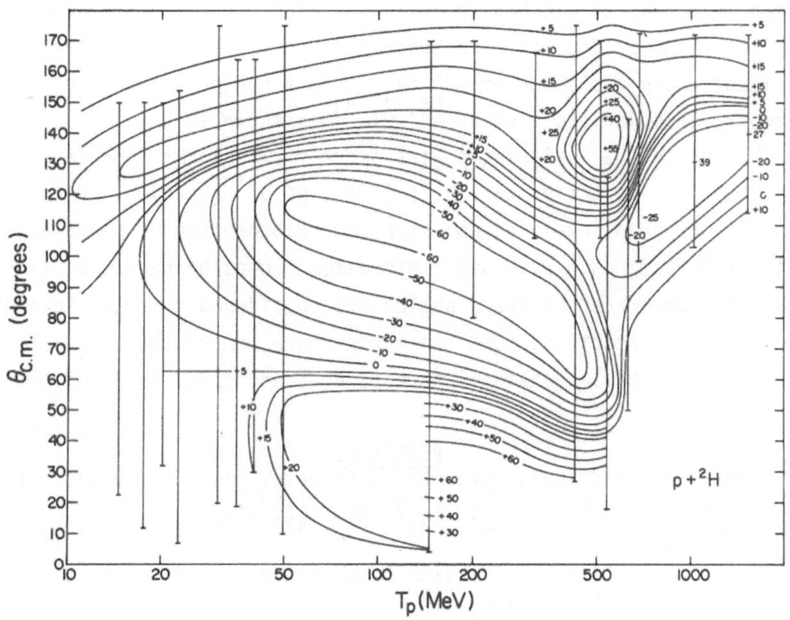

Figure 2

It can be seen that only at 516 MeV there is a reasonable measure of conformity between experiment and prediction, and apparently the agreement improves with increasing incident energy as shown at 630 MeV. [4] A contour diagram of the deuterium analysing powers for polarised protons [4, 7, 8] is shown in Fig. 2. It is seen that marked changes in the analysing powers occur for incident energies above 300 MeV which is the pion production threshold in the $pp \rightarrow d\pi^+$ reaction. A contribution of the triangular diagram involving emission and reabsorption of pions appears to be important but interference with other processes cannot be neglected since the measured analysing powers at backward angles are considerably larger than those observed in the $pp \rightarrow d\pi^+$ reaction. One can furthermore remark that the analysing power contour diagram shows additional structure around 1 GeV incident proton energy.

REFERENCES

1. L. Dubal and C. F. Perdrisat, Lett. Nuovo Cimento, 11, 265 (1974).
2. B. E. Bonner et al, Phys. Rev. Lett. 39, 1253 (1977).
3. N. S. Craigie and C. Wilkin; Nucl. Phys. B14, 477 (1969); G.W. Barry, Phys. Rev. D7, 1441 (1973).
4. L. I. Lapidus, Proceedings of the VIth International Conference on High Energy Physics and Nuclear Structure, p. 129.
5. A. W. Stetz et al, Nucl. Phys. A290, 285 (1977).
6. G. Jones, Proceedings of the IInd International Conference on the Nucleon-Nucleon Interaction, Vancouver (1977) (to be published). M. G. Albrow et al, Phys. Lett. 34B, 337 (1971).
7. H. Postma and H. Wilson, Phys. Rev. 121, 1229 (1961); R. E. Adelberger and C. N. Brown, Phys. Rev. D5, 2139 (1972); N. E. Booth et al, Phys. Rev. D4, 1261, (1971); E.T. Boschitz et al, Phys. Rev. C6, 457 (1972).
8. G. J. Igo, private communication.

NEUTRON ANALYZING POWER IN THE REACTION $\vec{n}+d \rightarrow n_1+n_2+p$

R. Fischer, H. Dobiasch, B. Haesner, H. O. Klages, R. Maschuw†,
P. Schwarz, and B. Zeitnitz

Ruhr-Universität Bochum, Germany

Recently we have measured the analyzing power in elastic and inelastic n-d scattering at 14 and 30 MeV. The results for the elastic scattering have already been published [1,2]. Here we report on the results for the analyzing power $A(\Theta_{n_1})$ for the n_2-p final state with relative energies of less than 1 MeV.

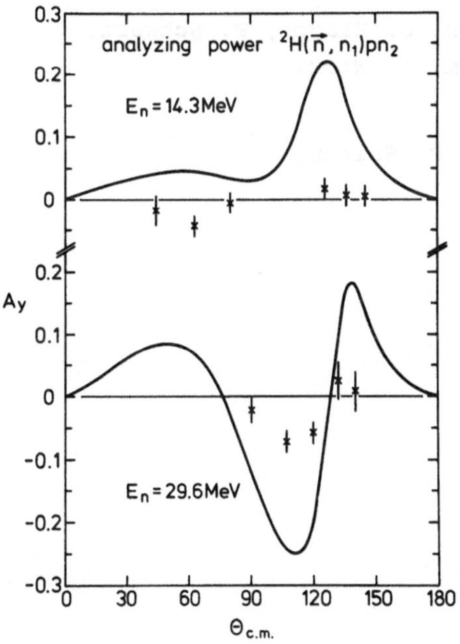

Fig.1:
Neutron analyzing power in the n-d breakup reaction. The solid curves are fits to elastic-scattering data by associated Legendre polynomials.

The relevant counting rates for the asymmetry were determined analyzing two-dimensional spectra of time of flight of the scattered neutrons versus energy of the recoil particle in the scattering target. The main problem of this analysis was to define the regions of final-state interaction in this incomplete breakup measurement. This was done by Monte Carlo calculation folding in the experimental resolutions. The size of the kinematically allowed FSI-areas was defined in a way that more than 70% of the events have relative energies $E_{pn_2} \leq 1$ MeV. The results of the neutron analyzing power at 14.3 and 29.6 MeV are shown in fig.1 together with curves representing the elastic scattering data. The error bars are purely statistical. In contradiction to a p-d measurement at 22.7 MeV by Rad et al. the similarity in the angular distributions

† II. Institut f. Experimentalphysik, Universität Hamburg

between the elastic and inelastic nucleon-deuteron scattering does not appear in our results.
One reason for this discrepancy may be that the distribution of the relative energy in the FSI-areas is quite different. The importance of the relative energy resolution is indicated by theoretical calculations of Stolk and Tjon [4].

Therefore further measurements with better energy resolution should be performed.

§ supported by Bundesministerium für Forschung und Technologie

1) R. Fischer, F. Kienle, H. O. Klages, R. Maschuw and B. Zeitnitz, Nucl. Phys. A282 (77) 189

2) H. Dobiasch, R. Fischer, B. Haesner, H. O. Klages, P. Schwarz, and B. Zeitnitz; R. Maschuw, K. Sinram, K. Wick, Phys. Lett. B (to be published)

3) F. N. Rad, H. E. Conzett, R. Roy, and F. Seiler, Phys. Rev. Lett. 35 (75) 1134

4) C. Stolk and S. A. Tjon, Phys. Rev. Lett. 39 (77) 395

VECTOR ANALYZING POWER IN FINAL STATE INTERACTION REGION FOR DEUTERON BREAKUP REACTION

J. Sanada, S. Seki, Y. Aoki, M. Sawada,
Y. Tagishi and L. S. Chuang*
Tandem Accelerator Center and Institute of
Physics, The University of Tsukuba, Ibaraki
300-31, Japan

Rad et al. reported that significant analyzing powers were found in the final state interaction region ($0<E_{np}\leq 1$ MeV) of p-d breakup reaction at 22.7 MeV and the angular distribution shows a remarkable resemblance to those of the corresponding elastic scattering.[1] We extended the same measurement to the d-d breakup reaction at 21 MeV and obtained similar results, that means, iT_{11} of d-d breakup reaction is quite close to that of elastic scattering.[2] These results are unexpected in view of the dominant contribution of the singlet n-p pairs in the final state interaction region.[3]

In the meanwhile, theoretical calculations[4,5] have been made and a considerable difference between the magnitude of the analyzing power in p-d breakup reaction near $E_{np}=0$ MeV and that at $E_{np}=1$ MeV has been predicted.

Here we report a preliminary result of the measured analyzing power in p-d breakup reaction with a better energy resolution.

Polarized protons of 21 MeV were used to bombard the dueterium gas target. Breakup protons were detected by two position sensitive solid-state detectors at the focal plane of ESP-90 spectrograph. This system is effective to a large solid angle and a better energy resolution of 200 keV.

Results of energy spectra of the spin up- and spin down-runs taken at $\theta_{LAB}=75°$ are shown in Fig. 1a. We divide the continuous spectra into several intervals with the width of 200 keV of E_{np}. The values of the analyzing power of these intervals are shown in Fig. 1b. Althogh the value near $E_{np}=0$ MeV has been still undetermined due to the difficulty of the background subtraction and the poor counting statistics, significant analyzing powers of the order of several per cent are observed from $E_{np}=1$ MeV down to $E_{np}=0.2$ MeV. Thus it becomes clear experimentally that larger analyzing power is obtained when we integrate the yields from $E_{np}=0$ MeV to 1 MeV.

Further data accumulation near $E_{np}=0$ MeV is in progress, and a similar mearsurement will be extended to the d-d breakup reaction.

References:

1) F. N. Rad, H. E. Conzett, R. Roy and F. Seiler, Phys. Rev. Lett. 35 (1975) 1134.
2) J. Sanada, S. Seki, Y. Tagishi, Y. Takeuchi, M. Sawada and K. Furuno, Proceedings of the International Conference on Nuclear Structure, Contributed papers, Tokyo, 1977, p.8.
3) H. Brückman, W. Kluge, H. Mathäy, L. Schänzler and K. Wick, Nucl. Phys. A157 (1970) 209.
4) J. Bruinsma and R. van Wageningen, Nucl. Phys. A282 (1977) 1.
5) C. Stolk and J. A. Tjon, Phys. Rev. Lett. 39 (1977) 395.

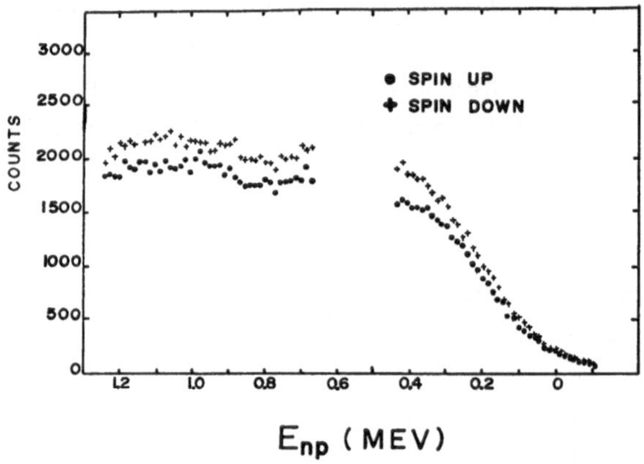

Fig. 1a.

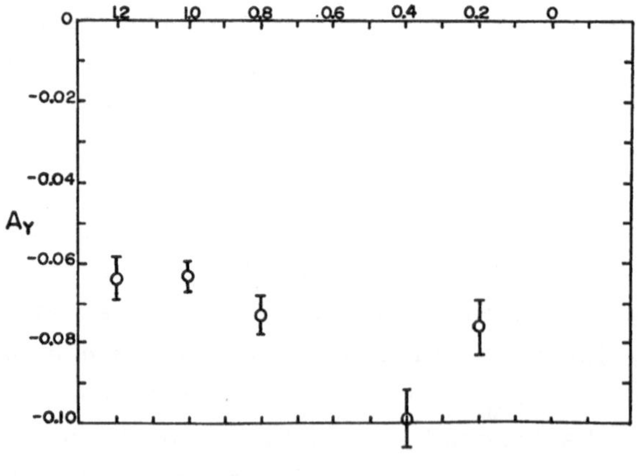

Fig. 1b.

Fig. 1. Energy spectra and analyzing power for p-d breakup reaction at E_p=21 MeV.

* On leave from Chinese University of Hong Kong, Hong Kong.

SOLUTION OF FOUR-NUCLEON INTEGRAL EQUATIONS USING THE EFFECTIVE UPA

R. Perne and W. Sandhas
Physikalisches Institut der Universität Bonn, Bonn, W.-Germany

In the three-body case it is standard to either solve the (two-dimensional) Faddeev equations directly, or to reduce them first to one-dimensional equations by means of separable approximation (expansion) of the underlying two-body interactions. The basic four-body operator identities are reduced by the latter treatment to effective three-body equations only[1]. These may be handled like their genuine three-body analoga, i.e., by directly solving them[2], or by expanding the effective interactions occurring into separable terms. Such a procedure provides us in a second step with one-dimensional integral equations for the four-body problem, too[1,3].

Recently we have proposed to perform this second step by means of a simple generalization of the three-body UPA method. Cross sections were calculated on this basis after applying additionally a K-matrix-Born approximation[4]. In the following we present n-^{3}H and p-^{3}H phase shifts and d+d→p+^{3}H cross-sections, obtained by correctly solving the final one-dimensional integral equations via Padé techniques.

Let us briefly recall the general concepts of our approach. Starting from a separable approximation

$$T_\gamma \sim |g_\gamma > t_\gamma < g_\gamma| \tag{1}$$

of the two-body transition amplitude, the three-body equations can be reduced to effective two-body equations of the matrix LS form

$$T^\tau = V^\tau + V^\tau G_o T^\tau \quad , \tag{2}$$

while we end up in the four-body case with the Faddeev type matrix relations[1]

$$U^{\sigma\rho} = \bar{\delta}_{\sigma\rho} G_o^{-1} + \sum_\tau \bar{\delta}_{\sigma\tau} T^\tau G_o U^{\tau\rho} \tag{3}$$

Here σ,ρ,τ denote the partitions (ijk,ℓ) or (ij,kℓ) of the four particles under consideration. Similarly to the genuine three-body case it is the essential aspect of Eq.(3) that its kernel is determined by the subsystem transition operators T^τ, given as solutions of Eq. (2). This suggests to approximate them by separable expressions of the form (1), too,

$$T^\tau \sim |\Gamma^\tau > t_\tau < \Gamma^\tau| \quad . \qquad (4)$$

Then, after partial wave decomposition, we end up with one-dimensional integral equations[1,3].

The construction of suitable expressions (1) is, in general, based on separable approximations of the two-body potentials. Conventionally the Hilbert-Schmidt, Bateman or UPA methods ar applied in this context. Analogously the effective potential matrices

$$V_{\beta\alpha} = \bar{\delta}_{\beta\alpha} < g_\beta | G_o(z) | g_\alpha > \quad , \qquad (5)$$

in Eq.(2), can be approximated by these techniques, providing separable amplitudes (4). Due to the complexity, in particular the energy dependence of the "potential" (5), Hilbert-Schmidt and Bateman method are rather time consuming[3,5-9]. As in the three-body UPA approach we, therefore, introduce <u>energy independent</u> form factors $|\Gamma>$ by solving the eigenvalue equation[4]

$$|\Gamma^\tau(z)> = \eta_\tau^{-1}(z)\, V^\tau(z)\; G_o(z)\, |\, \Gamma^\tau(z) > \qquad (6)$$

for a fixed adequately chosen energy z. Suitable choices in the ^{3}H-p, ^{3}H-n and d-d channels are, e.g., the binding energies of the respective bound states. Let us emphasize that proceeding along these lines we not only arrive at technically simplified equations, but, moreover, at a physically well interpretable <u>cluster model</u>.

Yamaguchi potentials with the same parameters as in Ref.4 are used as two-body input in the following numerical investigations. Since we take into account only s-wave contributions in the subamplitudes of the (3+1) channels, we are left with a system of five coupled integral equations in the spin 1, isospin 1 case, and of three equations in the other spin-isospin channels. The solution of this system was accomplished by means of the Padé method. Already the [4,4] approximant ensures sufficient accuracy in almost all cases. Numerical integration has been performed with 24 Gauss points.

For the coherent scattering length we found $A_c(n-{}^3H) = 2.93$ fm in fairly good agreement with the value 3.28 fm of Ref. 10. S-wave phase shifts δ_{oo} and δ_{o1} for spin zero and one, respectively, are given in Figs. 1-4 and are compared with the results by Tjon[7] (-·-·- ■) and Kröger and Sandhas[2] (- - - ▲). The experimental points are taken from Refs. 11,12. In view of the simplicity of the present treatment the agreement is rather satisfactory. In particular the characteristic shape of hard sphere scattering is reproduced.

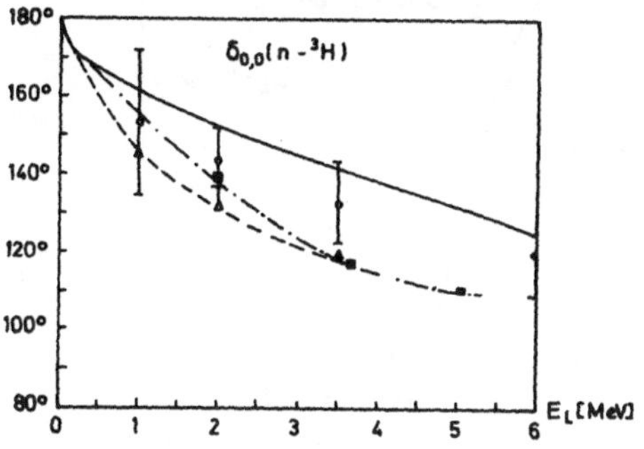

Fig. 1

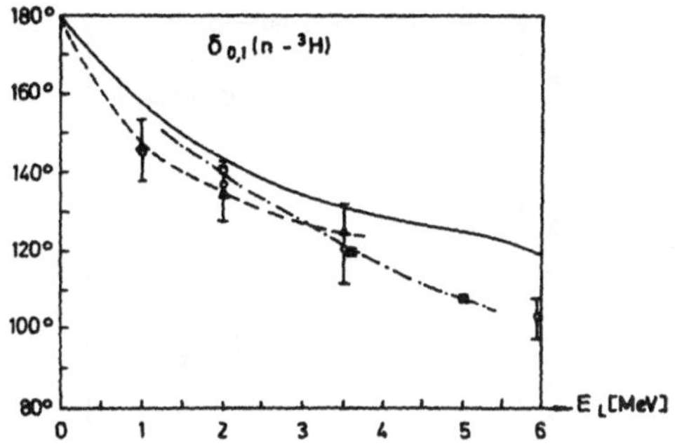

Fig. 2

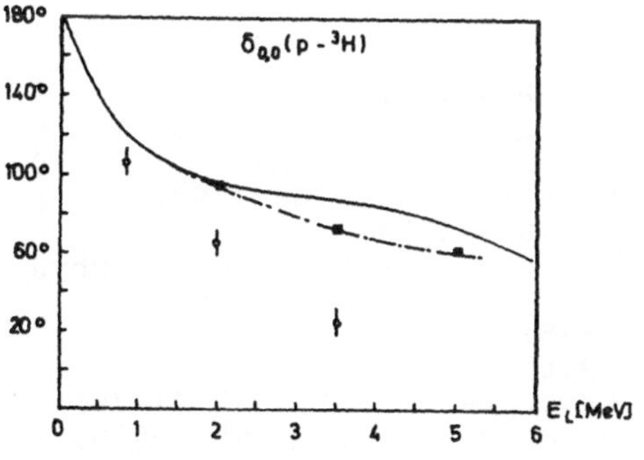

Fig. 3

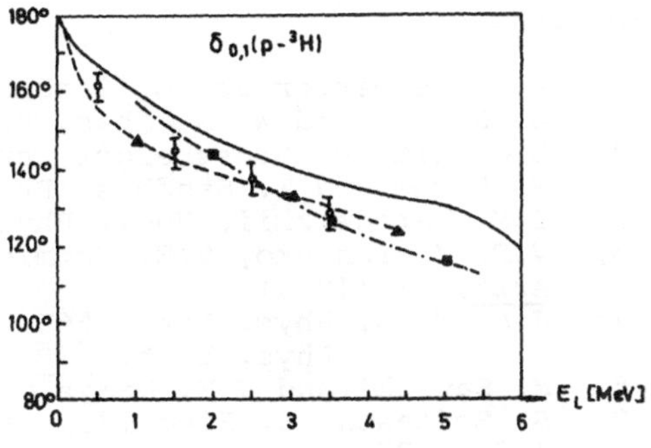

Fig. 4

The physically transparent cluster structure of our generalized UPA approach suggested the introduction of a cut-off parameter for the effective interaction, reducing its unrealistically strong attraction. In Ref. 4 it has been shown that for an optimal choice of this parameter the ^3H and ^4He binding energies and simultaneously the d+d →p+^3H cross sections, calculated in K-matrix Born approximation, could be shifted to the experimental values. Figs. 5 and 6 show the same cross sections obtained for this optimal choice by correctly solving the integral equation (E_L denotes the energy in the lab-system). Experimental points are from Refs. 13,14.

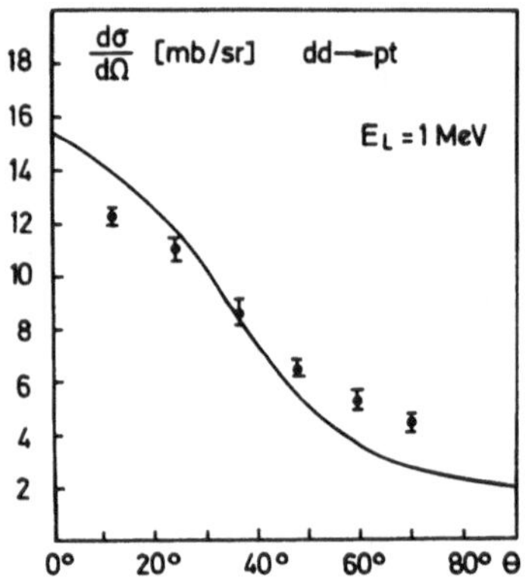

Fig. 5

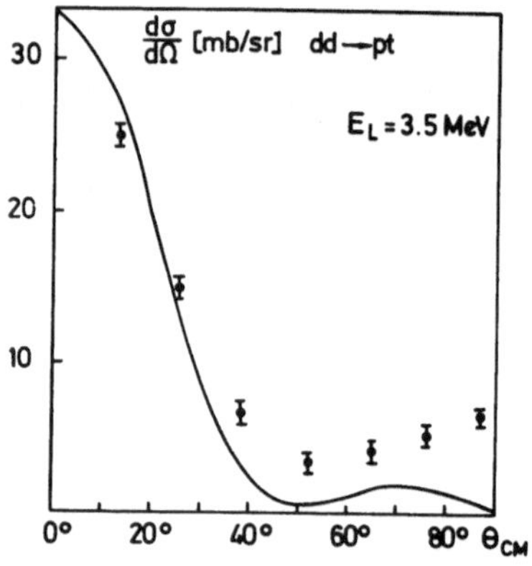

Fig. 6

References:

1) P. Grassberger and W. Sandhas, Nucl. Phys. B2, 181 (1967)
2) H. Kröger and W. Sandhas, Phys. Rev. Lett., 40, 834 (1977)
3) E.O. Alt, P. Grassberger, and W. Sandhas, Phys. Rev. C1, 85 (1970)
4) R. Perne and W. Sandhas, Phys. Rev. Lett., 39, 788 (1977)
5) I.M. Narodetzkii, Nucl. Phys. A221, 191 (1974)
6) V.F. Kharchenko, V.E. Kuzmichev, and S.A. Shadchin, Nucl. Phys. A226, 71 (1974)
7) J.A. Tjon, Phys. Lett. 56B, 217 (1975)
Phys. Lett. 63B, 391 (1976)
8) M. Sawicki and J.M. Namyslowski, Phys. Lett. 60B, 331 (1976)
9) S. Sofianos, H. Fiedeldey, and N.J. McGurk, Phys. Lett. 68B, 117 (1977)
10) V.F. Kharchenko, V.P. Levashev, Phys. Lett. 60B, 317 (1975)
11) T.A. Tombrello, Phys. Rev. 143, 772 (1966)
12) I.Y. Barit, V.A. Segreev, Sov. J. Nucl. Phys. 13, 708 (1971)
13) Blair et al., Phys. Rev. 74, 1599 (1948)
14) Schulte et al., Nucl. Phys. A192, 609 (1972)

It has been found that generally the unitary pole approximation (UPA) is an excellent one for the 3N system [1,7], indicating that it can only impose some constraints on the (singlet) deuteron wave functions. It has not even been established clearly yet, from a purely phenomenological point of view, if it is possible to construct interactions obeying all the experimental constraints, except for the ^{3}He and ^{3}H charge form factors, which can fit all the 3N data, without invoking 3N forces [3,8].

In our discussion of the 4N system we shall focus our attention on the relation between the binding energies of the triton and the α-particle, E_t and E_α. Tjon found and Becker and Sandhas [9] confirmed a remarkable linear correlation between E_t and E_α for several separable approximations to a local potential and for a number of separable NN interactions. This linear relation appeared to indicate that E_α is mainly determined by our knowledge of E_t. Hence E_α apparently would provide little independent information on the NN interaction. On the other hand the α-particle is a considerably denser system than the triton, which makes it intuitively plausible that E_α and the wave function of the α-particle will be more sensitive to the short range behaviour of the NN interaction. This conjecture has been confirmed by Sofianos et al. [10], by calculating E_α and E_t in the three- and four-boson approximations with separable potentials of rank one and two. The rank one NN interactions (averaged over the $^1S_0$ and $^3S_0$ states) were chosen to be approximately phase equivalent at low energies, up to 180 MeV for C_1 and R however, while some of them (C_1 and C_2) had identical "deuteron wave functions" $\psi_D(r)$ outside r = 1.3 fm and outside r = 2 fm (C_3) with the reference potential R. The other rank one interactions A_1-A_6 and B_1-B_6 had smoothly varying deuteron wave functions and produced values of E_t and E_α showing a strong linear correlation (Tjon line), but C_1-C_2 deviated considerably from this line. For R, C_1 and C_2 with identical deuteron wave function outside r = 1.3 fm and phase equivalent up to $E_{lab} \sim 180$ MeV, the variation in E_t amounted to less than 0.3 MeV, in agreement with many previous investigations on the sensitivity of E_t to the short range behaviour of the NN interaction, but they differed up to 6.5 MeV in E_α. This appears to confirm that the α-particle is much more sensitive to the short range behaviour of the NN interaction than the triton.

For nonlocal interactions of rank two and higher it is possible, as opposed to local potentials, to vary the high energy phase shift and the bound state wave function independently. To check and compare the validity of the UPA, three- and four-bosom approximations, for E_t and E_α, we constructed separable interactions D_1-D_6 of rank two, with a fixed deuteron wave function identical to the one produced by the reference potential and different high energy phase shifts [11].

The Tjon plot for all these interactions is given in Fig. 1. The variations in E_t and E_α, R and D_1-D_6 are respectively $\Delta E_t \approx 0.3$ MeV and $\Delta E_\alpha \simeq 6.2$ MeV clearly indicating the accuracy of the UPA to the NN interaction in the 3N system and its comparative failure in the 4N system. It also indicates the relatively high sensitivity

THE BOUND STATES OF THE TWO-, THREE- AND FOUR-NUCLEON SYSTEMS AND THE NUCLEAR FORCE

H. Fiedeldey

Physics Dept., University of South Africa, P O Box 392,
Pretoria, South Africa

Although the deuteron is the simplest, the best known and the most thoroughly investigated of the few-nucleon systems it has been rather disappointing as a constraint on the nuclear force from a phenomenological point of view, which we adopt throughout this contribution. However, recently hopes have risen that new information on the deuteron electric form factors can be obtained from tensor polarization P_e of recoil deuterons in e-d scattering experiments [1].

It has been shown by Hockert & Jackson [2] that a single measurement of P_e at $q^2 = 19.52\ fm^{-2}$ could provide a strong constraint on possible nuclear forces. This was confirmed by McGurk and Fiedeldey [3]. Hockert and Jackson suggested that such a measurement of P_e could even be used to distinguish between hard and soft interactions. A similar suggestion was put forward by Moravcsik and Ghosh [4]. In another contribution to this conference [5], however, we construct phase equivalent NN interactions by means of finite range unitary transformations out to 0.7 fm on the SSC potential of de Tourreil and Sprung, in such a manner that some of them are hard and others soft interactions, while some have nodes inside 0.7 fm. However, all these interactions are shown to produce the same fit to the deuteron electric form factor $A(q)$ and to $P_e(q)$, up to $q \simeq 8.5\ fm^{-1}$, as the SSC potential, within 4%. We therefore conclude that even a rather complete measurement of P_e out to $q \simeq 10\ fm^{-1}$ with an experimental error less than 10% would not allow us to distinguish clearly between hard and soft interactions inside 0.7 fm. However this does not imply that P_e cannot be used at all to distinguish between given interactions [3].

Measurements of P_e are also expected to yield information on the percentage D-state (P_D) of the deuteron. However it has been recently shown by Allen and Fiedeldey [6] that measurements of P_e out to $q = 4.5\ fm^{-1}$ (the region expected to be most sensitive to P_D) to within an experimental error of 10%, would not necessarily allow us to distinguish between interactions with P_D's as different as 4.5 and 7.5% until such time as there is a drastic independent improvement in our knowledge of the neutron electric form factor, or in the error on P_e and A.

We shall be very brief on the 3N system. The quantities most sensitive to the details of the nuclear force are E_t, ^2a and the 3N bound state wave function. Although the charge form factor of He^3 and ^3H have been measured, uncertainties introduced by the meson degrees of freedom are such that it is difficult to draw any firm conclusion from the disagreement of the results of most if not all NN interactions with them.

of the 4N system to the high energy phase shift, independently of its sensitivity to the deuteron wave function at short range.

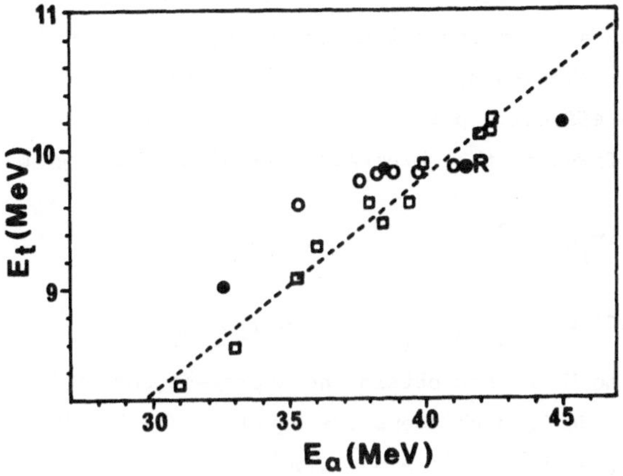

Fig. 1. The Tjon plot of the binding energies E_t and E_α of ^3H and ^4He respectively, for: (i) The interactions R, A_1-A_6 and B_1-B_6 given in table 1 of ref. [10], denoted by squares and an encircled dot for the reference interaction R. The dotted line indicates their linear correlation. (ii) The interactions C_1-C_3 of ref [10], denoted by black dots. (iii) The interactions D_1-D_6 of ref [11], denoted by circles.

Although our results are of a qualitative nature only, the qualitative conclusions drawn from them should be reliable since for instance the accuracy of the UPA in the 3N system is known to decrease for realistic interaction as compared to the three-boson approximation (and similarly for the sensitivity to the inner deuteron wave function). The presence of spin dependence and tensor forces, appears to affect a Tjon line for a given family of potentials only by shifting the line to some extent [10]. Only quantitative but no qualitative changes occur. Both the higher sensitivity to the high energy phase shift and consequently the reduced accuracy of the UPA and the increased snesitivity to the inner deuteron wave function, of the α-particle as compared to the triton, appear to be a consequence of the higher density of the α-particle, which allows it to sample the NN interaction more effectively.

On the one hand the fact that the UPA appears to be considerably less accurate in the 4N system than in the 3N system makes an already difficult computational problem even more difficult but on the other hand the rewards which can be expected from such calculations, as far as information on the nuclear force is concerned, will be greater.

The four-body AGS integral equations were reduced to coupled single variable integral equations, by expanding the NN two matrix and the (3+1) and (2+2) amplitudes in terms of separable Hilbert Schmid (HS) expansions. The disadvantage of these HS expansions is that to achieve a sufficiently high accuracy, usually 4 terms are needed in the expansion, making their extension to realistic forces quite difficult.

It would therefore be of great advantage to find an accurate pole approximation to the energy-dependent effective interactions U and Q in the (3+1) and (2+2) cases respectively. To obtain such an approximation we employ the Adhikari-Sloan Expansion (ASE) to define the three-body form factor $|\tilde{\psi}_T>$ by $|\tilde{\psi}_T> = U_T D_T|\tilde{\psi}_T>$ [12], where U_T and D_T are the effective potential and three particle propagators with their energy arguments fixed at the three-particle binding energy E_T. Then

$$U_{ASE} = U\, D_T|\tilde{\psi}_T>\Delta\dot{<}\tilde{\psi}_T|D_T U \equiv |\tilde{\psi}>\Delta<\tilde{\psi}|$$

with the propagator $\Delta^{-1} = <\tilde{\psi}_T|D_T U\, D_T|\tilde{\psi}_T> = <\tilde{\psi}|D_T|\tilde{\psi}_T>$.

With the approximation to U we then obtain the Energy-dependent Pole Approximation EDPA to the (3+1) amplitude X, which obeys the equation $X = U + UDX$, in the separable form $X = |\tilde{\psi}>H<\tilde{\psi}|$ with $H^{-1} = \Delta^{-1} - <\tilde{\psi}|D|\tilde{\psi}>$.

Similar results apply in the (2+2) case [12]. To check the accuracy of the approximation E_α was calculated for seven interactions from the two sets of interactions A_i, B_i and R of Fig. 1 by means of the EDPA and compared to the previous results obtained for the HS expansion given in Fig. 1. The error introduced by employing the EDPA was generally less than 1% and always less than 2%. This was achieved at a saving in computational time of a factor $\gtrsim 10$ depending on the number of integration points used. This hopefully will allow us to do 4N calculations with realistic interactions if a comparable accuracy can be achieved by the EDPA in such cases, without getting involved in unmanageable computational problems.

References

1. J.S. Levinger, Springer Tracts in Modern Physics, 71 (1974) 88-241.
2. J. Hockert and A.D. Jackson, Phys. Lett. 58B (1975) 387.
3. N.J. McGurk and H. Fiedeldey, Nucl. Phys. A281 (1977) 310.
4. M.J. Moravcsik and P. Ghosh, Phys. Rev. Lett. 32 (1974) 321.
5. L.J. Allen and H. Fiedeldey, contribution to this Conference.
6. L.J. Allen and H. Fiedeldey, in preparation.
7. I.R. Afnan and J.M. Read, Phys. Rev. C8 (1973) 1294.
8. I.R. Afnan and J.M. Read, Proc. of the VI Int. Conf. on Few Body Problems in Nuclear and Particle Physics, p.553, ed. R.J. Slobodrian et al., Laval University, Québec 1975.
9. J.A. Tjon, Phys. Lett. 56B (1975) 217; W. Sandhas, invited talk, Proc. of the VII Int. Congress on Few-Body Problems in Nuclear and Particle Physics, p.540, ed. A.N. Mitra et al., 1976, North-Holland.
10. S. Sofianos, H. Fiedeldey and N.J. McGurk, Phys. Lett. 68B (1977) 117
11. S. Sofianos, H. Fiedeldey and N.J. McGurk, Nucl. Phys. A294 (1978) 49
12. S. Sofianos, H. Fiedeldey and N.J. McGurk, contribution to this Conference.

ENERGY-DEPENDENT POLE APPROXIMATIONS IN THE FOUR-BODY EQUATIONS

S. Sofianos, H. Fiedeldey and N.J. McGurk
Department of Physics, University of South Africa,
P O Box 392, Pretoria, South Africa, 0001

We employ the AGS integral equations [1] for the four-body system. The formalism has been described and the explicit equations given in [2]. We restrict ourselves to separable two-body interactions: $t(p,p';z) = g(p)d(z)g(p')$. After a partial wave expansion the four-body integral equations incorporate the orresponding partial wave amplitudes X and Y of the (3+1) and (2+2) clustering respectively. (Appendix A, ref. [2]). These latter amplitudes obey the following effective multichannels LS equations.

$$X(q,q';Z_u) = U(q,q';Z_u) + \int_0^\infty q''^2 dq'' U(q,q'';Z_u)\, d(Z_{q''u})\, X(q'',q';Z_u), \tag{1}$$

with a similar equation for Y.

In operator form these can be expressed as $X = U + UDX$ and $Y = Q + QDY$. (2)

These equations are usually solved by means of a separable expansion, the most widely used being the Hilbert-Schmidt (HS) expansion [2-5].

A relatively large number of terms (more than four) are required for convergence of the four-body binding energy with the HS expansion of the X and Y amplitudes [2-5]. This is in addition to multiple solutions of the three-body system (for the (3+1) case) as well as the deuteron-deuteron system (for the (2+2) case). This procedure implies large computational times, which can only be reduced if an accurate pole approximation to the energy-dependent effective interactions U and Q can be found. To this end we define an energy-dependent pole approximation (EDPA) which reduces to the well-known unitary pole approximation (UPA) for energy-independent interactions.

Considering initially the (3+1) case, we fix the energy at the three-body binding energy E_3 and solve for the three-body form factor $|\widetilde{\psi}_T>$ which obeys the equation

$$|\widetilde{\psi}_T> = U_T D_T |\widetilde{\psi}_T> \quad , \tag{3}$$

where U_T and D_T are the corresponding quantities of eq. (2) with their energy arguments fixed at E_3.

The EDPA for U is now defined by the equation

$$U_{EDPA} = UD_T|\widetilde{\psi}_T > \Delta < \widetilde{\psi}_T|D_T U = |\widetilde{\psi} > \Delta < \widetilde{\psi}| \quad \text{with} \quad \Delta^{-1} = < \widetilde{\psi}|D_T|\widetilde{\psi}_T > . \tag{4}$$

If U were energy independent then [7] would reduce to the UPA.

Substituting U_{EDPA} for U in eq. (2) we obtain the EDPA for X

$$X_{EDPA} = |\tilde{\psi}> H <\tilde{\psi}| \quad \text{with} \quad H^{-1} = \Delta^{-1} - <\tilde{\psi}|D|\tilde{\psi}> \tag{5}$$

A similar pole approximation can be applied to the (2+2) amplitude Y. The explicit equations for ψ, H^{-1} and the corresponding quantities for the (2+2) clustering ϕ and θ will be given elsewhere [6].

Table 1. We compare E_4 calculated with the EDPA and the HS expansion (at least four terms) of ref. [5]. The three-body binding energies E_3 and the energy for the (2+2) subsystem $E_{(2+2)}$ are also given. The interaction N is the triplet interaction of ref. [3], while K is the average of the triplet and singlet interactions of ref. [3]. The interactions A_i and B_i are the same as the ones given in ref. [5], table 1. In the last column we give the percentage difference between E_4 calculated by the two methods.

Interaction	E_3 MeV	$E_{(2+2)}$ MeV	E_4 MeV		
			EDPA	HS (4 terms, (ref. [5])	% difference between HS & EDPA
N	25.53	4.45	89.0	90.3	1.44
K	12.45	0.83	53.6	54.6	1.83
A_4	9.07	0.86	35.1	35.3	0.57
A_5	8.57	0.86	32.2	32.5	0.92
A_6	8.31	0.86	30.8	31.0	0.65
B_1	10.19	0.86	41.6	42.0	0.95
B_2	9.87	0.86	39.6	40.0	1.00

In the table we compare the four-boson binding energy E_4 for seven interactions of ref. [5], calculated by means of the HS expansion, with the corresponding EDPA results. If we assume the HS result then the error introduced by the EDPA is of the order of 1% and always less than 2%. This is achieved at the saving of computational time of a factor greater than 10 depending on the number of integration points for the final matrix inversion. The small difference between the results of the two methods can to some extent be attributed to round off errors in both the HS expansion and the EDPA, as well as to incomplete convergence of the HS expansion with 4 terms.

The EDPA can be easily extended to an energy-dependent pole expansion with further separable terms. Since the EDPA already provides such an excellent approximation to E_4, it is expected that this expansion will converge considerably faster than the corresponding HS expansion.

References

1. E.A. Alt, P. Grassberger and W. Sandhas, Report E4-6688 JINR Dubna 1972.
2. S. Sofianos, H. Fiedeldey and N.J. McGurk, Nucl. Phys. A294 (1978) 49.
3. I.M. Narodetsky, Nucl. Phys. A221 (1974) 191.
4. J A Tjon, Phys. Lett. 56B (1975) 217.
5. S. Sofianos, H. Fiedeldey and N.J. McGurk, Phys. Lett. 68B (1977) 117.
6. S. Sofianos, H. Fiedeldey and N.J. McGurk, in preparation.

FOUR-BODY CALCULATION OF $^3He(p,p)^3He$ and $d(d,p)^3H$

A. C. Fonseca†
Department of Physics and Astronomy
University of Maryland, College Park, Maryland 20742 U.S.A.

The four-body equations previously described[1] have been solved in a computer for all two-to-two reactions initiated by either $p+{}^3He$ or d+d. Since our equations reduce to single variable integral equations following partial wave decomposition, and the singularity structure of the Born terms and box amplitudes is similar to that encountered in the three-body problem, the usual contour rotation method together with matrix inversion has been used. Using an 18 point integral equation mesh the CDC 7600 computer takes approximately three minutes to solve the equations for seven independent amplitudes in six partial waves (40 minutes in the IBM 370/158). The parameters of the two-body NN interaction are the coupling constant γ_ν (ν=d or ϕ) and the range parameter β_ν of the vertex function which can be easily expressed in terms of the low energy observables of the singlet and triplet nucleon-nucleon interaction. The three-body amplitudes of our model that describe Nd and Nϕ scattering in all possible spin-isospin states are characterized in each channel interaction by a coupling constant $\lambda_{\nu y}$ (y=t, t' or t") and a vertex function $g_{\nu y}(k)$ that for simplicity is taken to be energy independent ($g(k) = [(k^2+\beta_1^2)(k^2+\beta_2^2)(k^2+\beta_3^2)]^{-1}$). The parameters β_1, β_2 and β_3 are chosen by fitting $g_{\nu y}(k)$ to the appropriate s-wave three-body Sturmian function corresponding to the largest eigenvalue η at a fixed energy (three-body bound state energy for $U = V = ½$ and threshold energy $E = -\varepsilon d$ for the other spin-isospin states). In Figs. 1 through 5 we show the results of our calculations for the singlet phase shifts, inelastic parameters and differential and reaction cross sections. Curves B correspond to a different parametrization of the three-body amplitudes where a change of parameters was attempted to obtain a better fitting to the three-body observables. The s-wave phases are repulsive and as in the p.s. analyses of Trombello[2] ${}^{11}\delta_0$ are smaller then ${}^{01}\delta_0$ but do not differ greatly in magnitude. The p-wave p.s. exhibit the usual resonant structure and the d-waves are negative. Contrary to what is expected from the resonating group calculation of I. Reichstein, et al.,[3] the d-wave p.s. do not change sign. The triplet inelastic parameters are smaller than the singlet ones. Although the

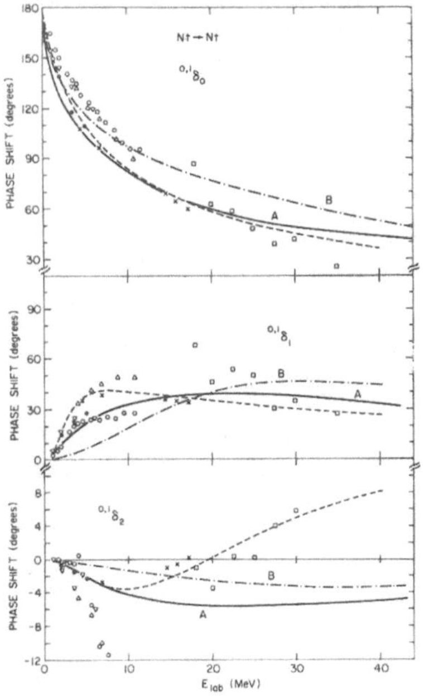

Fig. 1. Phase shifts ${}^{SI}\delta_\ell$ versus E for Nt→Nt (solid and dashed dotted line). The black dots correspond to the calculation of Tjon[4] and the dashed curve to Ref. 3. All other points are from phase shift analyses.

differential cross sections for $p\,^3He \to p\,^3He$ have the right shape and order of magnitude they do not fit the data equally well at all energies (Coulomb was added to the nucleon amplitudes). The predictions of the model for the reaction cross section seem to fall within the error bars. The $dd \to p\,^3H$ differential cross sections do not have enough structure at low energy but improve considerably at higher energies.

†Work supported by U. S. Department of Energy.

1) A. C. Fonseca, "Four-Body Model of the Four Nucleon System," contribution to this conference.
2) T. A. Trombello, Phys. Rev. 138, B40 (1965).
3) I. Reichstein, et al., Phys. Rev. C3, 2139 (1971).
4) J. A. Tjon, Phys. Lett. 63B, 391 (1976).

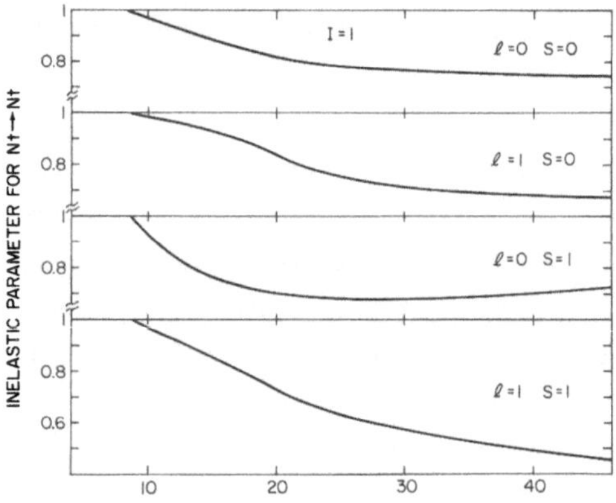

Fig. 2. Inelastic parameters versus E.

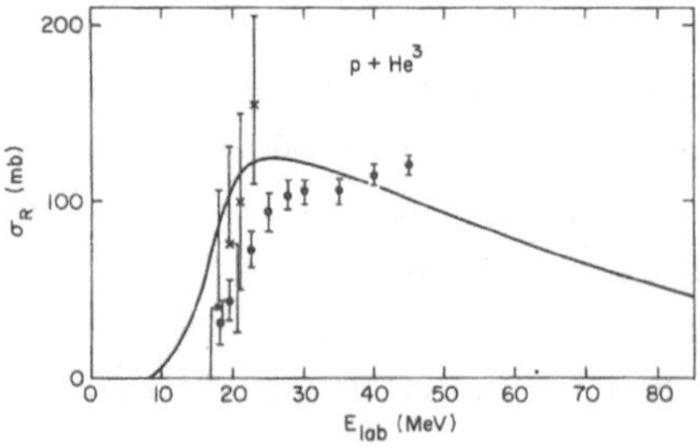

Fig. 4. Reaction cross section for $p+{}^3He$. The dots and crosses are experimental data.

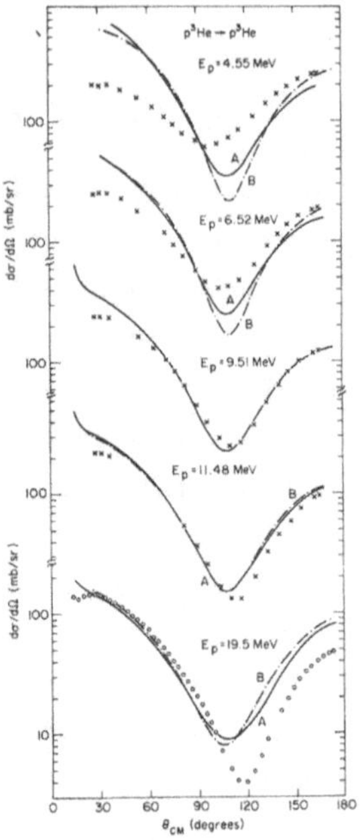

Fig. 3. Differential cross sections for $p\,^3He \to p\,^3He$. The crosses and circles are experimental data.

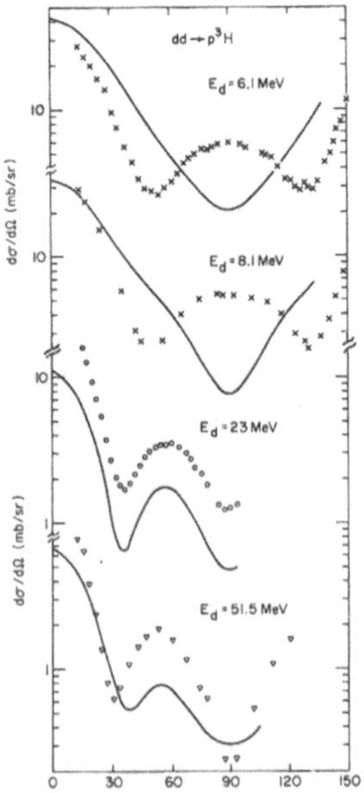

Fig. 5. Differential cross sections for $dd \to p\,^3H$. The crosses and circles are experimental points.

FOUR-BODY MODEL OF THE FOUR-NUCLEON SYSTEM

A. C. Fonseca[†]
Department of Physics and Astronomy
University of Maryland, College Park, Maryland 20742 U.S.A.

After the work of Faddeev in the three-body problem substantial progress was achieved in the formulation of connected kernel integral equations for the n-body problem ($n \geq 4$). Although the resulting equations for the four-nucleon system are, at least in principle, solvable by standard methods, their complexity is so great that few scattering solutions exist[1] for realistic two-body potentials. The main difficulty involves the need to use the off-shell 1+3 and 2+2 subamplitudes as input to the four-body equations. Even if two-body separable potentials are used between pairs, the resulting subamplitudes are nonseparable in momentum space and this leads to multivariable integral equations that are extremely time-consuming to solve numerically. Therefore we find it convenient to formulate a four-body model of the four-nucleon system that exhibits as many features of the experimental problem as possible and whose solution in the scattering region may be obtained with considerable less numerical effort than a more exact formalism would allow. In a previous work[2] we used the field theoretic method of Amado[3] to formulate a soluble model for four identical particles. Now the theory is generalized to include spin and isospin effects. As before the long range Coulomb force will be disregarded.

The elementary particle of the present model, the nucleon, is named N and has both spin and isospin ½. In the two-body sector of the model two quasiparticles, d and ϕ, are introduced with s-wave coupling to two N's. Two-body NN scattering proceeds through the d (deuteron) each time a spin triplet pair interacts and through the ϕ whenever a spin singlet pair interacts. The d is therefore a physical particle with spin $\Sigma=1$ and isospin $\Theta=0$ and the ϕ an unphysical particle with $\Sigma=0$ and $\Theta=1$. In the three-body sector, both the total spin U and the total isospin V have two possible values, ½ and $\frac{3}{2}$, and the dynamical equations that describe three-body bound states and scattering states are those of Aaron, Amado and Yam[4] (A.A.Y.). Since no tensor or spin orbit force is included, both the total three-body angular momentum ℓ and the total spin are conserved and for each ℓ there are three independent amplitudes that are characterized by their (spin, isospin) values $(\frac{1}{2},\frac{1}{2})$, $(\frac{3}{2},\frac{1}{2})$ and $(\frac{1}{2},\frac{3}{2})$.

To proceed to the four-body problem with no further approximation would lead to the numerical difficulties inherent in multivariable integral equations, so that we insert our three-body approximation at this point. Unlike most of the previous work no attempt is made to expand the exact three-body amplitudes in a complete set of separable terms. Since the number of terms needed to obtain accurate results in the four-body sector grows fast as the energy increases beyond rearrangement and breakup thresholds, such procedure would lead to a large number of coupled equations in the four-body problem and also to increasing difficulty in handling the singularities of the four-body kernel. Instead a model three-body amplitude is formulated that has the same analytical structure (poles and cuts) as the exact amplitude but contains some adjustable parameters that allow for some flexibility in fitting the on-shell three-body data. Our aim is to retain the simplicity that results whenever each independent three-body amplitude is described by a single separable term together with the ability to compensate for the suppression of the higher order terms. For that purpose we proceed as in the spinless model[2] and assume that the three-body problem of interest is dominated by the $\ell=0$ amplitudes. Three quasiparticles meant to approximate three-body scattering in each (spin, isospin) state are introduced. In the absence of a better designation they are called t, t' and t". The t (triton) is coupled to both N+d and N+ϕ and the renormalized parameters of the interactions are chosen such that equal mixtures of the N+d and N+ϕ are present in the wave function of the t. The t' and the t" are unphysical particles coupled exclusively to N+d and N+ϕ respectively. According to our approximation, Nd scattering proceeds exclusively in s-wave through the t for U=½ and through the t' for U=3/2. Since the ϕ is an unphysical particle, Nϕ scattering does not take place as an on-shell three-body process, but in the four-body sector of our model, virtual Nϕ scattering will proceed exclusively in s-wave through t or t" depending on whether the total isospin of the three-body state is ½ or 3/2. The three-body amplitudes of our model have been depicted in Fig. 1(a) and have a separable form in momentum space. The intermediate three-particle propagator is constructed by summing a series of Nd and (or) Nϕ bubbles.

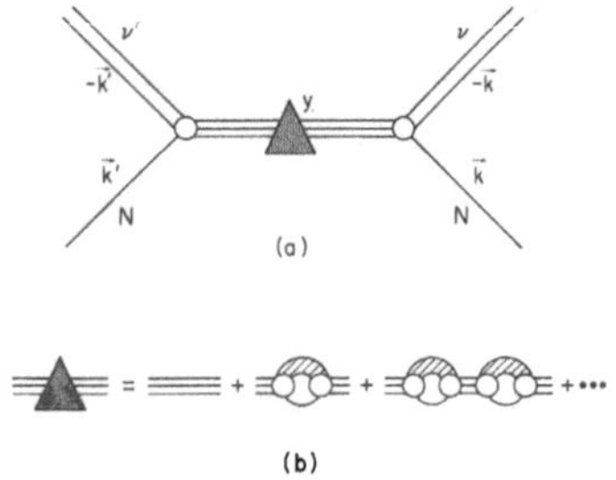

Fig. 1
Approximate N$\nu\to$Nν' amplitude. ν is either d or ϕ and y is t, t' or t".

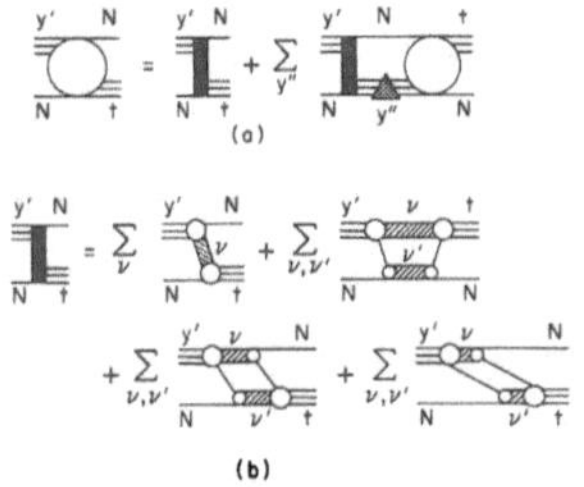

Fig. 2
Nt$\to$Ny' amplitudes (circle).

Proceeding to the four-body sector with no further approximation we obtain one vector variable integral equations for the processes Nt$\to$Nt, Nt$\to$dd and dd$\to$dd. The equation for Nt$\to$Nt is depicted in Fig. 2, where y' and y" are t, t' or t" and the driving term $I(E)$ is $I(E)=B(E)+\square(E)+\square(E)$. B(E) corresponds to the d and/or ϕ particle exchange Born term, $\square$(E) to the box amplitude depicted first in Fig. 2(b), and $\square$(E) to the sum of the last two. Both $\square$(E) and $\square$(E) may have as intermediate states the 2+2 channels dd, dϕ, and $\phi\phi$, that are treated exactly by the convolution method. Detailed numerical calculations indicate that the total cross sections obtained from unitarity for 2$\to$3 and 2$\to$4 processes are non-negative and thus we find that our three-body approximation, which involves the neglect of certain classes of graphs, leads to no gross violation of unitarity.

†Supported by the U. S. Department of Energy.

1) E. O. Alt, et al., Phys. Rev. C1, 85(1970); J. A. Tjon, Phys. Lett. 63B, 391(1976); H. Krüger and W. Sandhas, Phys. Rev. Lett. 40, 834(1978).
2) A. C. Fonseca, et al., Phys. Rev. C14, 1343(1976).
3) R. D. Amado, Phys. Rev. 132, 485(1963).
4) R. Aaron, et al., Phys. Rev. 136, B650(1964).

STUDY OF FEW-NUCLEON SYSTEMS BY MEANS OF ANALYTICAL CONTINUATION IN CORE CONSTANT

V.M. Krasnopol'sky and V.I. Kukulin
Institute of Nuclear Physics, Moscow State University;
Moscow 117234, USSR.

The available methods for calculating the few-particle systems involves serious difficulties when strong repulsion at small distances is present in two-particle interaction. At the same time, at small values of repulsive core, the developed methods can ensure a sufficient accuracy of calculations. This circumstance is used in the proposed method, namely the Schrödinger many-body equation is solved at small values of the constant λ of the repulsive part of interaction V_R

$$\{H_0 + V_A + \lambda V_R - E(\lambda)\} \Psi(\lambda) = 0. \qquad (1)$$

At such small values of λ, we calculate the values of interest to us (binding energy, wave function, etc.) and then continue them analytically using the technique of the Pade approximants (PA) of the second kind to the true values $\lambda = 1$. At certain limitations imposed on the form of the operator V_R, the eigenfunction $\Psi(\lambda)$ and the eigenvalue $E(\lambda)$ in (1) are the analytical functions of λ /1/. Equation (1) is solved for some set of values: $\{\lambda_i\}$; $\lambda_i \ll 1$; $i = 1,2,3,\ldots,k$, for which the corresponding values of energies $\{E_i\}$ are found. Then PA is constructed:

$$E^{[N,M]}(\lambda) = P_N(\lambda)/Q_M(\lambda), \qquad (2)$$

$N + M + 1 = K$, and the polynomials P_N and Q_M can be unambigousely found through the sets of initial points $\{\lambda_i\}$ and $\{E_i\}$. The value of the binding energy of the real system ($\lambda = 1$) is determined by the expression

$$E \cong E^{[N,M]}(1) = P_N(1)/Q_M(1). \qquad (3)$$

Since the function $E(\lambda)$ is analytical in the studied region of values of λ, it may be proved that $E^{[N,M]}(\lambda=1)$ at $N,M \to \infty$ converges to the true binding energy /2/. The real usefulness of such approach is characterized by the values of N and M at which a sufficient accuracy of the final result can be achieved. The rate of the convergence of PA sequence depends on the extent to which the analytical properties of the function to be continued are included. For example, the three-body energy near the two-body threshold is a two-sheet function, i.e. $E_3(\lambda) \sim (\lambda-\lambda_0)^2$. Therefore, the wave number $k_3 = [E_3(\lambda)-E_2(\lambda)]^{1/2} \sim (\lambda-\lambda_0)$ ($E_2(\lambda)$ - the moving two-body threshold) is an one-sheet function with simpler analytical structure, and the PA sequence for the function $k_3(\lambda)$ converges much more rapidly than for $E_3(\lambda)$ (see the Table).

Table

System	^3H		^4He	
Order of approximant \ Approximated function	$-E_3$ MeV	$k_3=(E_3-E_2)^{1/2}$	$-E_4$ MeV	$k_4=(E_4-E_3)^{1/2}$
1	0.602	2.727	10.6	4.404
2	1.238	2.728	16.1	4.679
3	3.573	2.728	19.2	4.735
4	6.845	2.736	21.8	4.791
accurate value	7.76	2.738	31.09	4.83

Because of the same reasons, in the four-body case we considered the function $k_4=(E_3-E_4)^{1/2}$. The table presents the results of our calculations for ^3H and ^4He with the Afnan-Tang potential S1/3/ containing a repulsive core of a 1 GeV height. It can be seen that a fairly rapid convergence takes place in case of correct inclusion of the analytical structure of continued values (the value $\lambda_{max}=0.20$ was used for initial points). The analytical continuation of the matrix elements and the wave functions are made quite similarly and are examined in /4/. The described method is convenient for calculating the near-threshold states of few-particle system both on the basis of the Schrödinger equation and on the basis of the Faddeev-Yakubovsky, etc. equations since the direct solution of such equations near the threshold (and in addition, with the interaction including the repulsive core) offer considerable calculational difficulties.

REFERENCES:

1. T. Kato, Perturbation theory of linear operators, Springer Verl. 1966
2. G.A. Baker, S.L. Gammel. The Pade' approximants in theoretical Physics, Acad. Press, N.-Y.,1970.
3. I.R. Afnan, Y.C. Tang, Phys.Rev. 175, 1337, 1968.
4. V.I. Kukulin, V.M. Krasnopol'sky, J. Phys.A: Math. Gen. 10, L33 1977, Yad. Fiz., to be published.

ANALYZING POWER MEASUREMENTS FOR THE 4-NUCLEON SYSTEM BELOW 6 MeV[†]

T. R. Donoghue, R. Detomo, Jr., L. J. Dries and H. W. Clark
The Ohio State University, Columbus, Ohio 43210

The energy level structure of the 4-nucleon system, the lightest system to have a complex level structure, reflects a tightly bound system where both strong and weak internucleon forces play important roles. The polarizations in most of the reaction channels are large indicating spin dependent interactions are important. Hackenbroich[1)] has shown that the T=0 levels of ^4He reflect a complicated balancing of odd-central and odd-tensor interactions, along with even spin-orbit interactions, whereas the T=1 levels are sensitive principally to the spin-orbit interaction. The presence of a tensor interaction is quite unusual in nuclear systems and this may well be the only one where it is so demonstrably important.

Various nuclear models have been proposed to describe the continuum level structure of ^4He, ranging from 1p-1h shell model calculations to resonating group and microscopic cluster model[2)] calculations. The importance and strengths of the various nuclear interactions are determined by comparison with the experimental level structure, such as determined by Werntz and Meyerhof[3)] in a charge independent R-matrix analysis of cross section and polarization data in the $T(p,n)^3He$ reaction. Two sets of levels were proposed (labelled WMI and WMII), corresponding to the two sets of T=1 phase shifts determined in the p-^3He phase shift analysis.[4)] The WMII set has been favored in recent compilations.[5)] However, subsequent measurements of polarization quantities in the $T(p,n)^3He$ reaction, notably the polarization transfer coefficient[6)] $K_y^y(0°)$ and the analyzing power[7)] A_y, cannot be described by either set of parameters. A comprehensive charge-independent R-matrix analysis of the A=4 system has been on-going at Los Alamos by Hale and Dodder[8)] where the goal is to describe in a consistent way all data at all energies for all open reaction channels. This analysis has been hampered in part because data on all open reaction channels, particularly at low energies where the systems are simpler, are either not available, or because existing data are not correct. For instance, Donoghue et al.[9)] recently showed that all P^y data in the T(p,n) reaction below ~ 3 MeV is incorrect. We have been using the unique polarized ion source[10)] that operates inside the high voltage terminal of our pressurized Van de Graaff accelerator to measure polarization related quantities for the various open reaction channels for energies below ~ 6 MeV for inclusion in the Los Alamos analysis. The status of some of our work is reported here.

Analyzing powers for p-^3He elastic scattering were measured for proton energies up to 4.5 MeV because of concerns over published data in this energy range. A gas scattering chamber (1/3 atm.) was used so that the low energy protons could be detected with good resolution. Our A_y angular distributions, shown in Fig. 1, are in good agreement with the 4 MeV data of Morrow and Haeberli[4)] and with the 4.46 MeV

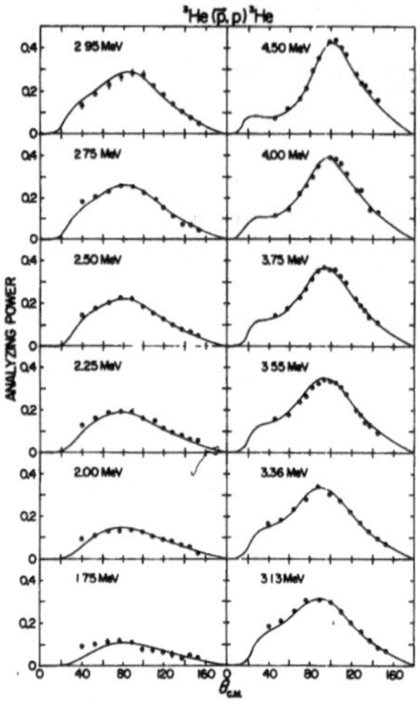

Fig. 1. p-^{3}He analyzing powers with R-matrix search fits.

data of of Drigo et al.[11] However, there are appreciable disagreements with Drigo et al. below that energy. The solid curves in the figure are the R-matrix search calculations[8] for the p-^{3}He system, where all available data up to E_p = 20 MeV were included. The characteristic energies of the R-matrix levels that produced these fits are, in order, J^π = 2^-, 1^- (S=1), 0^-, 1^- (S=0). The confirmation that the triplet 1^- level is lower than the singlet 1^- level is the WMI ordering.

$A_y(\theta)$ and $\sigma(\theta)$ were measured for p-^{3}H elastic scattering for the 1.1 to 4.7 MeV proton energy interval to explore the properties of the T=0 and T=1 states of ^{4}He. Our A_y data are plotted in Fig. 2 along with R-matrix calculations.[8] These data agree with Kankowsky et al.[12] (E_p > 4 MeV), but disagree with the double scattering data[13] (3 to 4.5 MeV). The R-matrix calculations, which are preliminary at this time as much recent accurate data on the system has yet to be included in the data base, already show surprisingly good agreement with the data. The levels included in these R-matrix calculations include the T=1 levels determined above along with the expected p-wave T=0 levels. The latter were ordered by the search calculations as J^π = 0^-, 2^- and 1^-, which is the same T=0 sequence determined[3] in the T(p,n) reaction analysis. The dashed curve here is a sketch through the recent $\vec{t}$+H A_y data.[14] An analysis of the differences in these quantities can lead to information on the 1^- mixing parameter. Hale and Dodder discuss this in a contribution to this conference.

The positioning of the 0^- level as the lowest T=0 p-wave level is indicative of a tensor term in the internucleon potential[1] and the establishment of its position can lead to a determination of the strength of that interaction. We have attempted a phase shift analysis of $\sigma(\theta)$, $A_y(\vec{p}$-T) and, when available at nearly the same energy, $A_y(\vec{t}$-H) using the Heiss formalism[15] along with a grid-gradient search routine. Absorption phases were constrained by the T(p,n)^{3}He total cross section. The search was limited to 15 phases (of 22 possible) because of earlier work.[12] Phases for the 0^- and 2^- levels that described the proton data over the whole range are shown as the solid dots in Fig. 3. When the $\vec{t}$-H data was included in the search (at 1.3 and 2.7 MeV) these phases changed to the values plotted as crosses. Additional data are clearly necessary to constrain the phases. The general trend of these phases does place the 0^- lowest, in agreement with the recent phases of Haglund et al.[14] (triangles) and the work of Barit and Sergeev[16] (dashed curve). A comparison with the phase shift predictions of Hackenbroich[1] shows the 0^- phase clearly requires

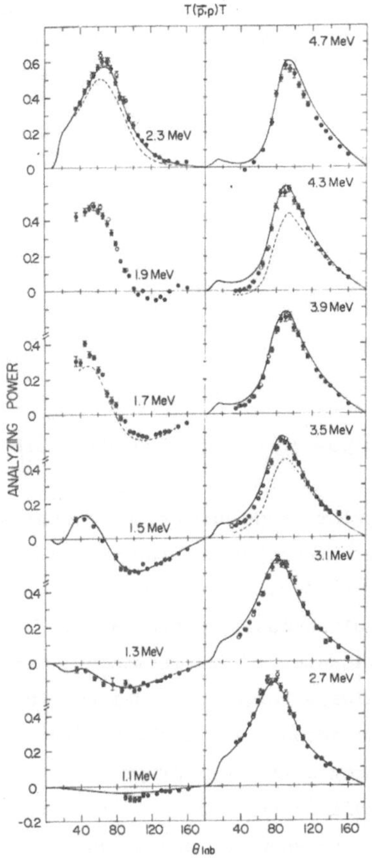

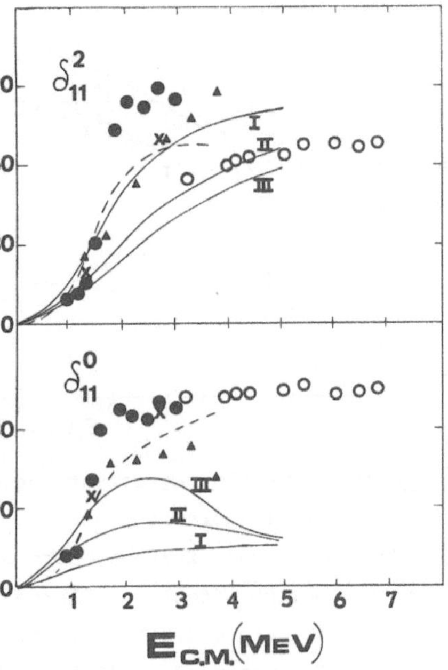

Fig. 3. The δ^{J}_{LS} phases for the 0^- and 2^- levels of ^{4}He from various p + ^{3}H analyses.

Fig. 2. p-T analyzing powers by detecting protons (•) or recoil ^{3}He (o) particles. The solid curve is an R-matrix calculation, the dashed curve is $\vec{t}$+H data of Haglund et al.

a tensor interaction (solid curve III) whereas the 2^- phase clearly does not (solid curve I). The differences in the phase shifts derived in the different analyses are clearly a strong argument in favor of the simultaneous search calculational approach of the Los Alamos group in that the additional reaction data can provide necessary constraints.

Comparisons of the D(d,n)^{3}He and D(d,p)^{3}H reactions have been used frequently to study charge symmetry effects in the nuclear interaction. Although this symmetry can be broken by the Coulomb interaction, the latter should be small in these light ion reactions. However, in a comparison of P^y data in these mirror reactions, Hardekopf et al.[17)] noted that the proton polarization was consistently larger than the neutron polarization in the 3-10 MeV energy range, but that this difference disappeared if the data sets were compared at the same exit channel energies (i.e., shifting the (d,p) data higher by ~ 1.5 MeV). This shift was proposed as an ad-hoc way to take Coulomb effects into consideration. Subsequent comparisons of vector and tensor analyzing power data for the mirror reactions show mixed results. König

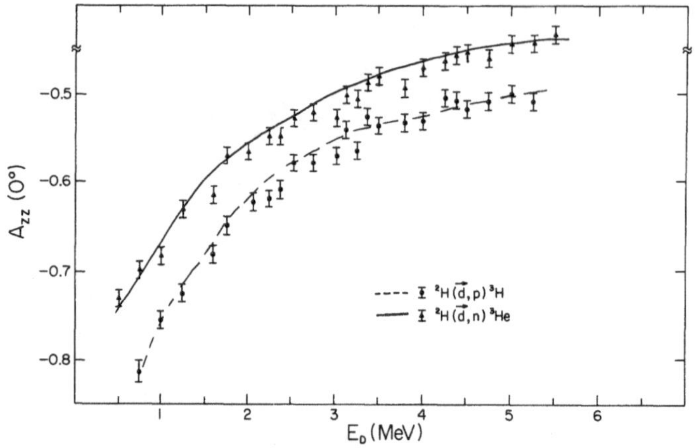

Fig. 4. $A_{zz}(0°)$ for the $^2H(\vec{d},n)^3He$ reaction, compared with data of other authors.

et al.[25] showed that agreement between some analyzing powers is improved with this shift, but others worsened.

To explore this further, both over a complete angular range and at low energies where the Coulomb effects are largest, we measured all the vector and tensor analyzing powers for the $D(\vec{d},n)^3He$ reaction to compare with the Zurich[25,26] $D(\vec{d},p)^3H$ data. We also measured $A_{zz}(0°)$ for both reactions simultaneously in such a way that differences attributed to experimental considerations could be minimized. Our results shown in Fig. 4 clearly indicate that A_{zz} is always larger in magnitude for the (d,p) reaction - and that no simple energy shift to account for Coulomb effects is possible. It is of interest to note that if an energy shift were proposed at all, it would be to shift the (d,p) data to lower energies rather than the higher energies noted by Hardekopf et al. Our $(\vec{d},n)$ measurements show serious disagreements with Lisowski et al.[20] below 3 MeV and Simmons et al.[21] at 4 and 6 MeV.

Vector and tensor analyzing power angular distributions for the D(d,n) reaction are shown in Fig. 5, along with polynomial fits (the dot-dash curve). Also shown by the dashed curves are polynomial fits to the Zurich[18,19] (d,p) data. The Hardekopf shift is easy to make here; the (d,p) curve in the left column can be compared with the (d,n) curve in the right column for a given observable. From the angular distributions that do show an energy dependence, it is clear that comparisons of the data must be made at the same projectile energy and the differences that do exist in the two reaction channels are principally in magnitude, not shape. That is, A_{xz} and A_{zz}, and to a lesser extent, $1/2(A_{xx}-A_{yy})$, change enough in shape for both channels that a shift will always worsen the comparison. It will therefore be interesting to see if the charge independent R-matrix calculations can reproduce these differences, or whether an alternate explanation must be sought. We have shown here some R-matrix predictions (solid curves) - extrapolations of the calculations from lower energies. It is interesting to note that some coefficients are described qualitatively, whereas others have serious problems. The latter data sets may therefore be of great value in directing future search calculations..

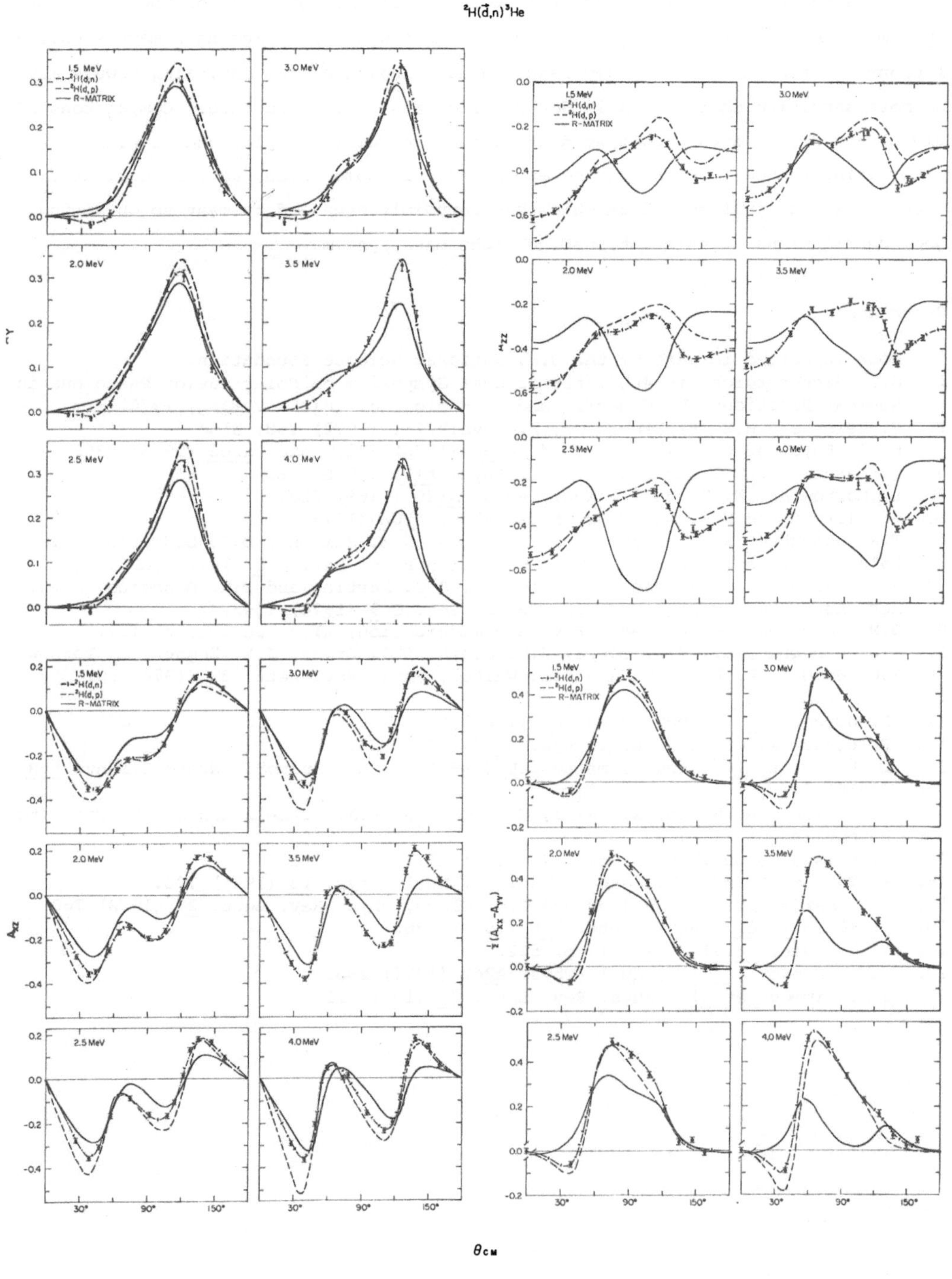

Fig. 5. Vector and tensor analyzing powers for the $^2H(\vec{d},n)^3He$ reaction. Polynomial fits to the $(\vec{d},n)$ data (-•-) and the Zurich $(\vec{d},p)$ data (---) are shown, along with preliminary R-matrix predictions that are extrapolations from lower energy.

In summary, we report a number of polarization measurements on the 4-nucleon systems, some of which show quite good agreement with the on-going R-matrix calculations. Differences in the D+d reaction data indicate some analyzing powers may be more sensitive than others in determining this level structure. Comparisons of D(d,n) and D(d,p) data show that differences are fairly complex and it will be interesting to see if these can be explained. In some cases, we find serious discrepancies with published data which has certainly been a detriment to obtaining good descriptions of this interesting light mass system.

References

† Work supported in part by the U.S. National Science Foundation.
1. H.H. Hackenbroich, in 4th International Symposium on Polarization Phenomena in Nuclear Reactions, W. Grüebler and V. König, ed. (Birk. Verlag, 1976) 133.
2. P. Heiss and H.H. Hackenbroich, Z. Physik 251 (1972) 168; also Nucl. Phys. A286 (1977) 42; Ibid. A182 (1972) 522; Ibid. A202 (1973) 335.
3. C. Werntz and W.E. Meyerhof, Nucl. Phys. A121 (1968) 38.
4. L. Morrow and W. Haeberli, Nucl. Phys. A126 (1969) 225.
5. S. Fiarman and W.E. Meyerhof, Nucl. Phys. A206 (1973) 1.
6. T.R. Donoghue, R.C. Haight, G.P. Lawrence, J.E. Simmons, D.C. Dodder and G.M. Hale, Phys. Rev. Lett. 27 (1971) 947; also Phys. Rev. C 5 (1972) 1826.
7. R.C. Haight, J.J. Jarmer, J.E. Simmons, J.C. Martin, and T.R. Donoghue, Phys. Rev. Lett. 28 (1972) 1587; also, Phys. Rev. C 9 (1974) 1292.
8. G.M. Hale and D.C. Dodder, private communication; also, ref. 1, p. 167.
9. T.R. Donoghue, Sr. M.A. Doyle, H.W. Clark, L.J. Dries, J.L. Regner, W. Tornow, R.C. Byrd, P.W. Lisowski and R.L. Walter, Phys. Rev. Lett. 37 (1976) 981.
10. T.R. Donoghue et al., ref. 1, p. 840.
11. L. Drigo et al., Nucl. Phys. 89 (1966) 632.
12. R. Kankowsky et al., Nucl. Phys. A263 (1976) 29.
13. L. Drigo et al., Nuovo Cimento 51B (1967) 43; L. Manduchi, Nuovo Cimento 55B (1968) 340.
14. R. Haglund, R. Brown, N. Jarmi, G.G. Ohlsen, P.A. Schmelzbach and D. Fick (to be published).
15. P. Heiss, Comput. Phys. Comm. 4 (1972) 371.
16. I. Ya. Barit and V.A. Sergeev, Sov. J. Nucl. Phys. 13 (1971) 708.
17. R.A. Hardekopf, R.L. Walter and T.B. Clegg, Phys. Rev. Lett. 28 (1972) 760.
18. V. König et al., Phys. Lett. 72B (1978) 436.
19. W. Grüebler et al., ref. 1, p. 522.
20. P.W. Lisowski et al., Nucl. Phys. A242 (1975) 298.
21. J.E. Simmons et al., Phys. Rev. Lett. 27 (1971) 113.

COMPARISON OF THE MIRROR REACTIONS $^2H(d,p)^3H$ and $^2H(d,n)^3He$

V. König, W. Grüebler, R.A. Hardekopf, B. Jenny, R. Risler and H.R. Bürgi
Laboratorium für Kernphysik, Eidg. Technische Hochschule, CH-8093 Zürich

In previous comparisons of the polarizations of the outgoing nucleons in the mirror reactions $^2H(d,p)^3H$ and $^2H(d,n)^3He$ it was shown[1,2] that the proton polarizations are larger than the neutron polarizations. These sizable differences can be removed by lowering the deuteron energy for the (d,n) reaction by 1.5 MeV[1,2] to give a comparison for the same energy in the exit channels. To investigate similar effects in the analysing powers the observables iT_{11}, T_{20}, T_{21} and T_{22} were measured for both reactions with high precision in steps of 1.5 MeV between 2.5 and 11.5 MeV. The angular range of comparison is restricted to backward angles mainly by the experimental technique used formeasuring the recoil ^3He for the (d,n) reaction. The experimental results have been fitted with Legendre polynomials. Fig. 1 compares the results of the (d,p) reaction with the fits of the mirror reaction at the same deuteron energy and shows discrepancies for T_{20} and T_{22}. Fig. 2 compares the results of the (d,n) reaction with the fits of the (d,p) reaction at the same energy in the exit channels and shows better agreement for T_{20} and T_{22}, but strong deviations for iT_{11} and T_{21}. For a better overall comparison, the following average deviation is given in fig. 3:

$$D_{av} = \int_{\theta_1}^{\theta_2} D_{kq}(\theta)d\theta/(\theta_2-\theta_1)$$

where: $D_{kq}=|T_{kq}(d,p)-T_{kq}(d,n)|$ and $\theta_1=100^0$ and $\theta_2=160^0$. This figure shows that when the correction for the Coulomb displacement proposed in [1,2] is applied the discrepancies persist. These almost energy independent discrepancies are observed in a region where Coulomb effects ought to be small (low charge of all particles, large energy of the emitted nucleons, backward hemisphere of scattering) and therefore a violation of charge symmetry should not be excluded from consideration. Althoug at the present time it seems too difficult to filter out the Coulomb effects in a proper way, hopefully the future will show if a combination of strong forces with charge symmetry property and electromagnetic forces can explain the present results. Model calculations of the ^4He excitations e.g. multichannel R-matrix calculations, should be helpful in the investigation of this most important question.

Results of new measurements at higher energies will be presented and disscussed.

References

1) R.A. Hardekopf, R.L. Walter und T.B. Clegg, Phys. Rev. Lett. 28, 760 (1972)
2) R.A. Hardekopf et al., Nucl. Phys. A191, 468 (1972)

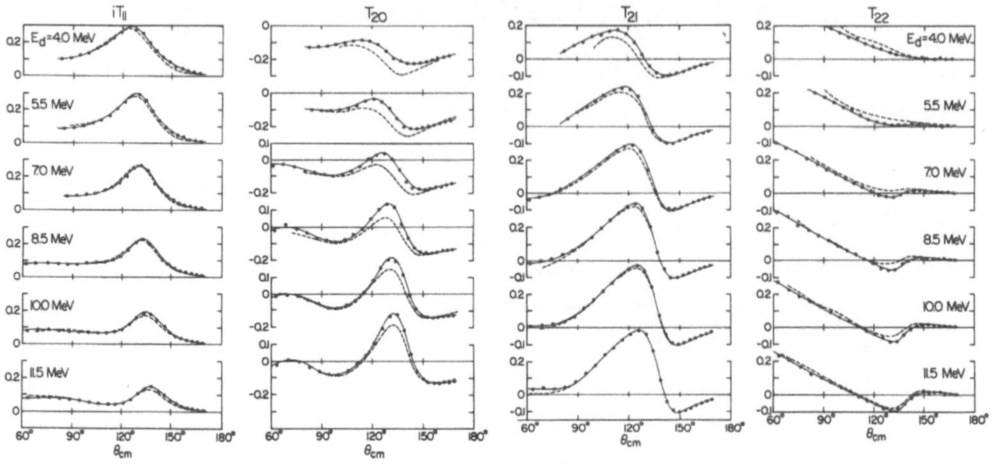

Fig. 1. Comparison of the analysing powers at the same deuteron energy. The dots are the results of the $^2H(d,p)^3H$ reaction, the solid curves fits to these data. The dashed curves are fits to the (d,n) reaction.

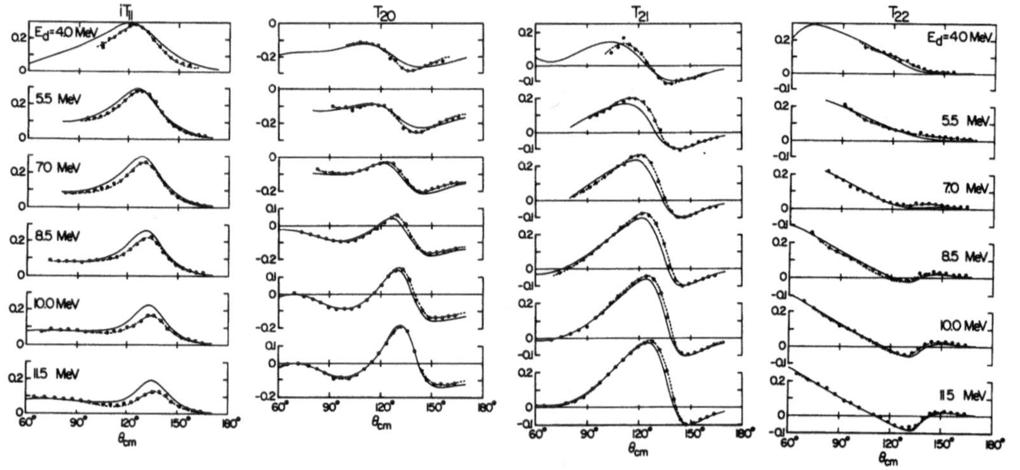

Fig. 2. Comparison of the analysing powers at the same exit energy of the nucleons. The dots are results of the $^2H(d,n)^3He$ reaction, the dashed curves fits to these data. The solid curves are fits to the (d,p) reaction, for deuteron energies lower by 1.5 MeV than the indicated energies.

Fig. 3. Average deviation D_{av} of the analysing powers for the mirror reactions as a function of energy. The dots represent comparisons at the same entrance energies, the circles at the shifted energies. The solid and dashed lines are drawn to guide the eye. The size of the dots and circles indicates the uncertainties in D_{av}.

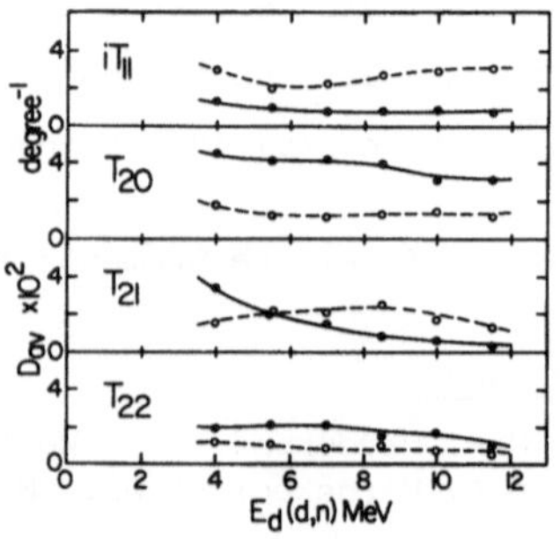

ANALYZING POWER FOR THE ELASTIC n-^{3}He-SCATTERING

H. O. Klages, H. Dobiasch, R. Fischer, B. Haesner, W. Heeringa,
P. Schwarz and B. Zeitnitz

Ruhr-Universität Bochum, Germany

The experimental situation in the A = 4 system is characterized by a lack of precise data for the neutron channels over a wide range of energy.

We measured the analyzing power A_y for the elastic scattering of neutrons from ^{3}He at neutron bombarding energies of 3.7 MeV, 10.0 MeV and 15.3 MeV. The energy values were chosen for comparison with recent data [1)] and to fill up the remaining gaps in the energy range up to 22 MeV. The experiment was performed at the Bochum neutron collimator BONCO , including a superconducting solenoid.

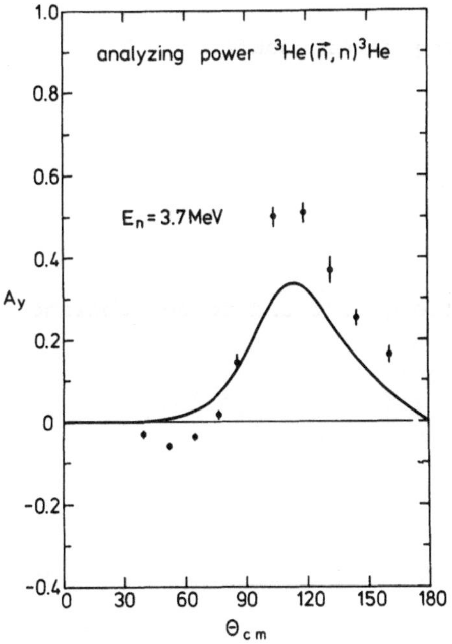

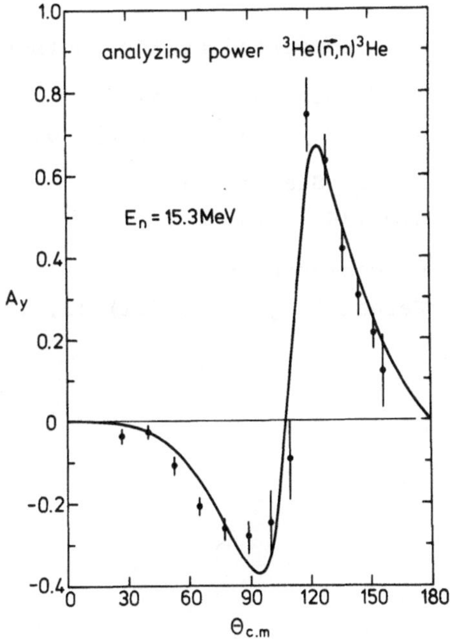

Analyzing power of ^{3}He for neutrons of 3.7 MeV and 15.3 MeV.
The curves represent calculations with extrapolated phase shift values from ref. 1.

The polarized neutrons were produced using the reactions $^{13}C(p,n)^{13}N$, $D(d,n)^{3}He$ and $^{15}N(d,n)^{16}O$, respectively. The scattering target was a liquid ^{3}He scintillating detector [2], which was also utilized as neutron beam monitor by means of the reaction $n + {}^{3}He \rightarrow p + t$. The scattered neutrons were detected at 8 different angles at the same time using scintillation detectors (NE 213) with pulse shape discrimination techniques. Details of the experimental setup and the measurement of the neutron polarization are described elsewhere [3].

Our data at the higher energies are in relatively good agreement with extrapolated values from a recent phase shift analysis [1], but there are larger discrepancies at 3.7 MeV.

We assume some of the older neutron data should be neglected for a new phase shift analysis. More precise measurements of differential cross section, analyzing power and, if possible, of spin correlation parameters etc. are necessary to get unique phase shift information.

§ Work supported by the Deutsche Forschungsgemeinschaft

1) P. W. Lisowski et al.
Nucl. Phys. A259 (1976), 61

2) R. van Staa et al.
NIM 136 (1976) 241

3) H. Dobiasch et al.
Verhandlungen DPG (VI) 13, 787, Heidelberg 1978 and to be published

THE LOW ENERGY NEUTRON SCATTERING LENGTHS OF He^3 AND T AND THEIR RELATION TO THE FOUR-BODY PROBLEM

H. Rauch
Atominstitut der Österreichischen Universitäten,
A- 1o2o Wien, Austria

Using the neutron interferometer /1/ we have started a program to measure the coherent scattering lengths of thermal neutrons for He^3 and T to get more precise data for a new evaluation of the four body system. The phase shift measured with the neutron interferometer is determined by the bound coherent scattering length b_c, which is related to the free scattering length a by the nucleus-neutron mass ratio (b=a(A+1)/A). We deal with the singlet a_s and triplet a_t scattering lengths because both He^3 and T have nuclear spin I=1/2. These individual scattering lengths determine the coherent a_c and the incoherent a_i scattering lengths and the free cross section σ_s in the form /2/

$$a_c = \frac{3}{4} a_t + \frac{1}{4} a_s$$

$$a_i = \frac{\sqrt{3}}{4} (a_t - a_s)$$

$$\sigma_s = 4\pi(| a_c|^2 + (a_i|^2)$$

Especially for He^3 we have to account for the large imaginary part of the scattering length due to the large cross section for the (n,p) reaction in the singlet state ($a_s \rightarrow a_s - i\, a_s'$ with $a_s' \simeq \sigma_a/2\lambda$). Two independent measurements are necessary to obtain values for a_s and a_t, which specify the nuclear interaction.

For He^3 a typical result of the measurement of b_c with the neutron interferometer is shown in Fig. 1/3/. The decrease of the contrast is caused by the large absorption cross section which has to be considered for the data evaluation. By this method we obtained a rather precise new value for the coherent scattering length (b_c=5,70(7).10^{-13}cm), but the error bars are now determined by the only available value for the cross section (σ_s = 3,16(20)b,/4/).The most probable data set for the individual scattering lengths is a_s = 6,7(7).10^{-13}cm and a_t = 3,5(4).10^{-13}cm. The present experimental situation for a precise determination of a_s and a_t is rather unsatisfactory as can be seen from Fig. 2, where the available experimental data are drawn with their error bars.

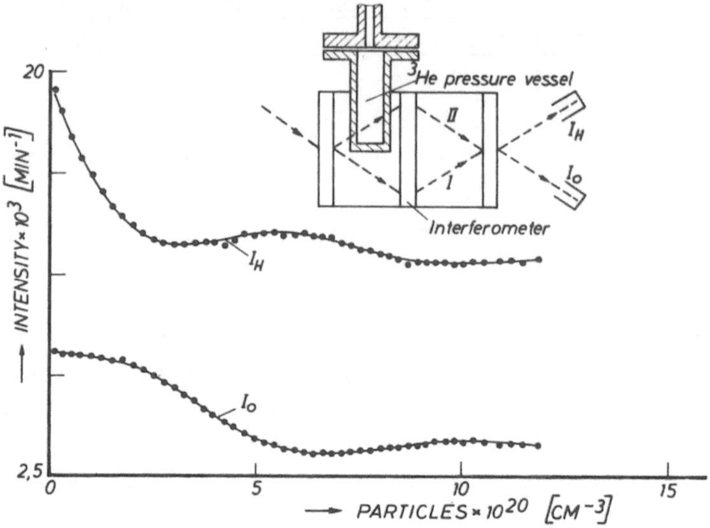

Fig.1: Experimental results for the measurement of the coherent scattering length of He^3 with the neutron interferometer./3/

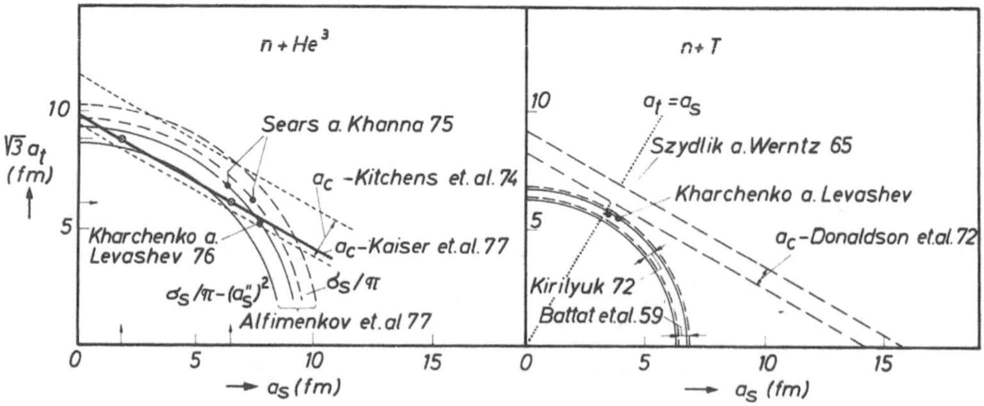

Fig. 2: Present situation for the determination of the singlet and triplet scattering lengths for the system n-He^3 and n-T from available experimental data.

The whole spin (S)- and isospin (T) set can be investigated for the four body system, if neutron and proton scattering data and the binding energy of He^4 is combined. This procedure yields a sensitive test for various theoretical models. Kharchenko and Levashev /5/ have done numerical calculations using Fadeev-Yacubovski equations to obtain the low energy neutron scattering data. With a charge independent and separable central potential with exchange character they get reasonable values for a_s and a_t, which are near to the experimental lines shown in Fig.2. According to the Pauli principle an effective repulsion exists for both interactions in T and for the triplet scattering in He^3 and therefore only little influence of the detailed form of the nuclear potential can be expected. In contrast , the singlet interaction in He^3 is attractive and therefore strongly influenced by the potential. Precise measurements on T can yield direct information on wether the repulsion is stronger in the singlet or triplet state. The present situation indicates the need for more precise experimental data and of consistent theoretical values.

Acknowledgements

The author thanks the members of the interferometer group Dortmund-Grenoble - Vienna for their cooperation and is very grateful to Dr.E. Sharapov from ECN/Netherland and to Dr. G.R. Platter from the University of Basel/Switzerland for their remark concerning the absorption term.

References

/1/ H. Rauch, W. Treimer a. U. Bonse; Phys. Lett. 47A (1974) 369
/2/ V.F. Sears a F.C. Khanna; Phys. Lett. 56B (1975) 1
/3/ H. Kaiser, H. Rauch, W. Bauspiess a. U. Bonse; Phys. Lett. 71B (1977) 321
/4/ V.P. Alfimenkov, G.G. Akopian, J. Wierzbicki, A.M. Govorov, A.M. Pikelner a. E.J. Sharapov; Sov. J. Nucl. Phys. 25 (1977) 1145
/5/ V.F. Kharachenko a. V.P. Levashev; Phys. Lett. 60B (1976) 317

$p+{}^3He$ Elastic Scattering from 18 to 48 MeV*

Ronald E. Brown,
Los Alamos Scientific Laboratory, Los Alamos, New Mexico, U.S.A. 87545
B. T. Murdoch, D. K. Hasell, A.M. Sourkes, and W. T. H. van Oers,
University of Manitoba, Winnipeg, Manitoba, Canada R3T 2N2.

The mass-4 system is a very fundamental one to the study of the nuclear interaction and the internal structure of nuclear systems. The decisive breakthrough in solving the three-body problem, using the integral equations approach following Fadeev's formulation, has given a great deal of stimulance to find adequate extensions to solving the general N-body system, e.g., the four-nucleon system. The first results of such rigorous mathematical formulations for the scattering involving four nucleons have recently appeared in the literature. [1] Data from elastic scattering of protons on ${}^3He$ in the range of incident energies from 20 to 50 MeV can contribute in several ways to this important theoretical development.

We therefore have made a series of measurements of $p+{}^3He$ elastic differential cross section angular distributions at eleven proton energies from 19.5 to 47.5 MeV (lab). These data have been combined with previously measured [2] total reaction cross sections at ten energies from 18.25 to 47.65 MeV and with analyzing power angular distributions [3] at four energies from 21.44 to 30.12 MeV. This data base of 593 points was subjected to an energy-dependent phase-shift analysis in the range 18-48 MeV. The energy parameterization used was that of the R-matrix formalism,[4] and the computer code used was developed at the Los Alamos Scientific Laboratory [5,6]. Partial waves through $\ell=7$ were allowed in the elastic channel, and absorption through $\ell=6$ was incorporated by including the channels d+pp and d*+pp as two-body channels in the (unitary) formalism. Spin (s) and total-angular-momentum (j) splittings were allowed through $\ell=5$. In the initial portion of the search there were no couplings between elastic states, but, as the search progressed, singlet-triplet couplings were allowed for $\ell=1$ and 2 and $\ell\leftrightarrow\ell+2$ couplings were allowed for $\ell=0$, 1 and 2. The initial parameters were chosen so as to yield a smooth, structureless dependence on energy for the phase shifts and to yield phase-shift values at 20 MeV approximately equal to those obtained in an R-matrix analysis[5,7] of the low-energy data. This resulted in the initial phase shifts appearing somewhat like the ${}^3S_1$, ${}^3P_0$; ${}^3D_3$, and ${}^3F_4$ phases in Fig. 1. Initially about 60 parameters were allowed to vary, with this being increased to about 120 during the middle phase of the search and being decreased to about 80 during the final phase. A χ^2 minimum was found with a χ^2 per datum of 1.08. The phase-shift solution obtained for the first four partial waves is illustrated in Fig. 1.

*Supported in part by the A.E.C.B. and N.R.C. of Canada and the U.S. Department of Energy.

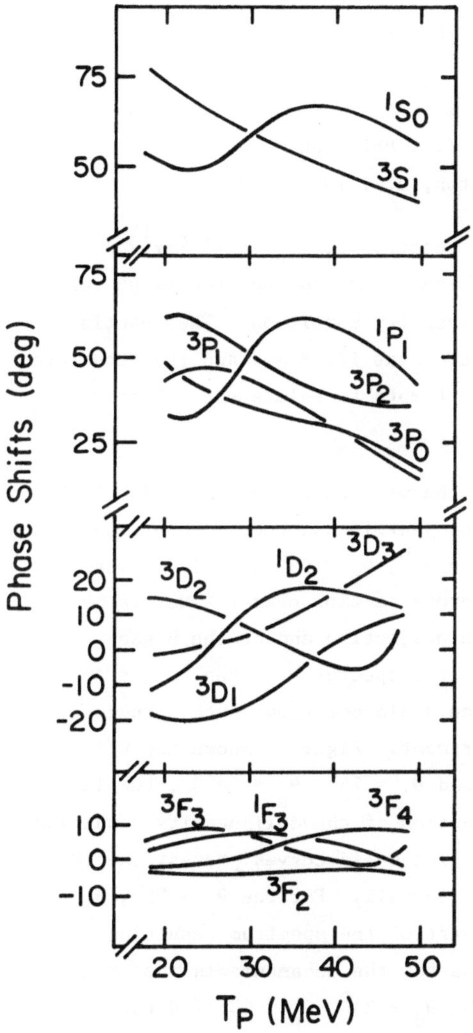

The most striking feature of the solution is the structure occurring in the singlet phases, especially in the $^1P_1$ phase. It is felt that more spin-dependent data are needed in the 20-50 MeV range to verify this structure. In the near future it is planned to initiate the more ambitious project of analyzing the data in the full energy range 0-50 MeV [5].

References

1. R. Perne and W. Sandhas, Phys. Rev. Letters 39, 788 (1978).
2. A. M. Sourkes, A. Houdayer, W.T.H. van Oers, R. F. Carlson and R.E. Brown, Phys. Rev. C 13, 451 (1976).
3. Data obtained at Lawrence Berkely Laboratory. J. W. Watson, private communication.
4. A. M. Love and R. G. Thomas, Rev. Mod. Phys. 30, 257 (1958).
5. D. C. Dodder and G. M. Hale, private communication.
6. D. C. Dodder, G. M. Hale, N. Jarmie, J. H. Jett, P. W. Keaton, Jr., R. A. Nisley and K. Witte, Phys. Rev. C 15, 518 (1977).
7. G. M. Hale, J. J. Devaney, D. C. Dodder and K. Witte. Bull. Am. Phys. Soc. 19, 506 (1974).

Fig. 1. p+^{3}He phase shifts for ℓ=0-3 vs lab proton energy T_p.

COLLINEARITY REVISITED IN D(d,pd)n AT 12.2 MeV

H.C. Rivenburg, L.T. Myers, J.M. Lambert and P.A. Treado
Georgetown University and Naval Research Laboratory
Washington, D.C. 20057 U.S.A.
Ivo Slaus
Rudjer Boskovic Institute, Zagreb, Yugoslavia and Georgetown University
R.G. Allas, R.O. Bondelid and E.L. Petersen
Naval Research Laboratory, Washington, D.C. 20057 U.S.A.

The experimental cross section of the D(p,pp)n reaction at E_{inc} = 23 MeV[1] and 156 MeV[2] shows a structure at kinematic conditions where the neutron is at rest in the overall center of mass system, the collinearity condition. The reaction D(d,pd)n also has been studied near such conditions at 12.15 MeV and the structure observed[3] led to the theoretical study which attempts to relate the enhancement to the off shell and rescattering effects.[4]

The main features of the reaction D(d,pd)n are the p-d quasifree scattering (QFS) with the spectator in both the target and the projectile and the $^3S_1$ n-p final state interaction (FSI).[5] At 12.15 MeV and $\theta_d = 25^o$, $\theta_p = 55^o$ the kinematical locus extends from 1.6 MeV to 7.8 MeV and the cross section near 1-2 MeV is dominated by the p-d QFS with the spectator in the projectile and at the higher energy region, about 7 MeV, by the p-d QFS with the spectator in the target and by the $^3S_1$ n-p FSI. Since the data of reference 3 did not show such structure, it was considered necessary to repeat the experiment. Figure 1 shows the D(d,pd)n projected cross section at $\theta_d = 25^o$, $\theta_p = 55^o$ and $\theta_d = 25^o$, $\theta_p = 68.3^o$, the latter angle pair being the precise angles for satisfaction of the collinearity condition which is indicated by the arrow in the figure. The solid curves present the PWIA predictions, normalized to fit the data reasonably well. For the $\theta_d = 25^o$, $\theta_p = 55^o$ data the normalization factor, N, for the part of the spectrum resulting from the spectator being in the target is N = 0.2 and for the enhancements when the spectator is in the projectile N = 0.2. For the $\theta_d = 25^o$, $\theta_p = 68.3^o$ data, the spectator in the target yields N = 0.2 and the spectator in the projectile yields N = 0.2 also. These normalization factors are in reasonable agreement with other PWIA results. The dashed curves of Figure 1 represent the data from reference 3 and a significant difference is apparent because the present data include the QF processes one expects to observe. Although the statistics are not too good, the structure in the $\theta_d = 25^o$, $\theta_p = 68.3^o$ data near the collinearity condition position is intriguing; it may indicate the presence of either the enhancement of off shell effects[4] or some effective three-body force manifestation.[1] Conclusive evidence requires more data (which is in progress) and a realistic four-body theory calculation.

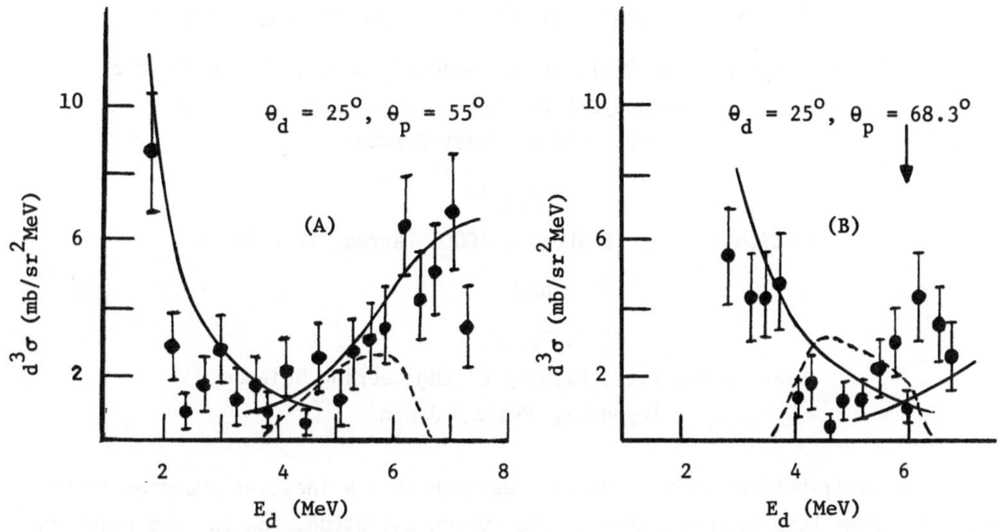

Figure 1. The D(d,pd)n projected cross section for E_i = 12.2 MeV: solid curves are PWIA predictions and dashed curves are representation of data from reference 3; (B) with $\theta_p = 67^o$.

1. J. Lambert, et al., Phys. Rev. C13 (1976) 43.
2. T. Yuasa, et al., Suppl. Progr. Theor. Phys. 61 (1977) 161.
3. N. Berovic, et al., Few Particle Problems in the Nuclear Interaction (eds. I. Slaus, et al., North Holland, 1972) 605.
4. A. Reitan, Nucl. Phys. A268 (1976) 358.
5. I. Slaus, Workshop on Few Body Problems, Trieste (1978), to be published, and references therein.

This work has been supported by PL480 (I.S.) and National Science Foundation (J.M.L. and P.A.T.) funds.

THE $^{2}H(d,N)$ REACTION: MEASUREMENT AND THEORETICAL ANALYSIS

C. Alderliesten[+], A. Djaloeis, J. Bojowald and C. Mayer-Böricke
Institut für Kernphysik der Kernforschungsanlage Jülich
D-517 Jülich, West Germany

G. Paić[++]
Institute Rudjer Bošković, 40001 Zagreb, Yugoslavia

and

T. Sawada
Osaka University, Faculty of Engineering Science,
Toyonaka, Osaka, Japan

Angular distributions of the $^{2}H(d,N)$ reaction at six incident energies in the 50-85 MeV range have been measured at the Jülich cyclotron. The charged reaction products emerging from the gas cell were detected by ΔE-E telescopes and subsequently identified.

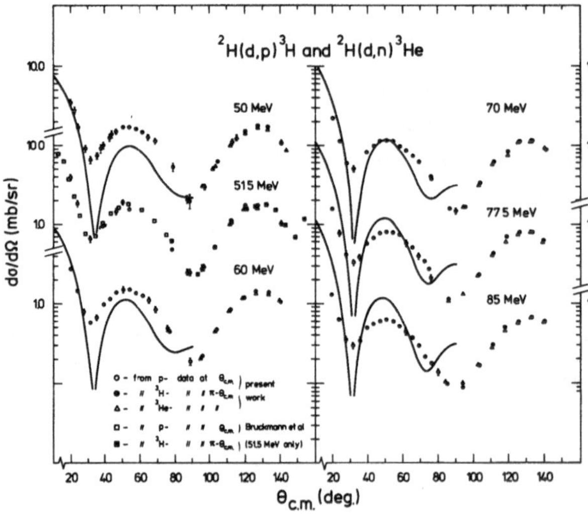

Fig. 1

The measured angular distributions are shown in fig. 1. At 51.5 MeV the results of ref. 1 are in agreement with the present data within the error bars. In the whole energy range the measured ratio of the $^{2}H(d,p)^{3}H$ to the $^{2}H(d,n)^{3}He$ cross sections at corresponding angles is consistent with unity.

The $^{2}H(d,p)^{3}H$ data have been analyzed in terms of the Plane Wave Born Approximation (PWBA), in a way similar to ref. 2. A smooth cut-off in the partial wave expansion of the reaction matrix elements has been introduced to allow the suppression of the low angular momentum components. This procedure simulates the effects of distortion and absorption in the incident and outgoing channels. Employing the Serber N-N force the reaction matrix element reduces to:

$$M_{fi}(\theta,S) = \sum_{\ell} f(\ell)\, M_{fi}\,(\theta,S,\ell)$$

where θ is the c.m. angle, S the channel spin, ℓ the relative orbital angular momentum and $f(\ell)$ is a cut-off function given by

$$f(\ell) = 1 - \left[1\text{-exp}\ \{(\ell-\ell_{c.o.})/\Delta\}\right]^{-1} .$$

The calculated angular distributions obtained with the values $\ell_{c.o.}=3$ and $\Delta=0.1$ for the whole 50-85 MeV range show reasonably good agreement with experiment (fig. 1), except at $\theta_{c.m.} \approx 90^{o}$. In this model this discrepancy is due to the contributions of even ℓ-components (here mainly $\ell=4$). This may suggest the necessity for introducing an odd-even effect.

References:

1) H. Brückmann et al., Z. Phys. 230 (1970) 383
2) T. Sawada et al., Nucl. Phys. A141 (1970) 169

+ Present address: Fysisch Laboratorium, Rijksuniversiteit, Utrecht, The Netherlands

++Guest scientist in the Institut für Kernphysik Jülich and partially supported by PL 480 Grant No. F6F005y and SIZ-I Grant 1.3.7.3.

THE ^{3}He(p,pp)d*/^{3}He(p,pp)d RATIO AT E = 136 MeV

L.T. Myers,[a] E. Karaoglan,[a] J.M. Lambert,[a] P.A. Treado,[a]
R.F. Romine,[a] D. Devins,[b] R.G. Allas,[c] and I. Slaus[a,d]

a. Georgetown University, Washington, D.C. 20057 U.S.A.; b. Indiana University, Bloomington, IN 47401 U.S.A.; c. Naval Research Laboratory, Washington, D.C. 20375 U.S.A.; d. Rudjer Boskovic Institute, Zagreb, Yugoslavia

The ratio of the ^{3}He(p,pp)d* to ^{3}He(p,pp)d cross sections at quasifree scattering (QFS) conditions, calculated in the plane wave impulse approximation (PWIA), depends on the amount of the mixed symmetry S' component (P(S')) in the ^{3}He wave function.[1,2] Measurements at 35,[3] 45,[4] 65,[5] 85,[5] 100[5] and 155[6] MeV show that the PWIA accounts for the shape of the QFS spectra, but for E_i less than 100 MeV it overestimates the absolute cross sections by factors of 2 to 10.

The objective ot this study is to measure the d*/d ratio accurately in the kinematic conditions where the PWIA provides a reasonable description of the data, i.e., it predicts the absolute cross section within 30%; thus, one argues that distortion effects should cancel or be reliably predicted for the ratio. Concentrating on the region of small momentum transfer, one can use only S and S' components in the ^{3}He wave function and S components in the ^{2}H wave function and one is assured of the dominance of the QFS processes. Experimentally, it also assures that the d and d* loci are not influenced by any unwanted events from the (p,pd) process. In order to obtain reliable and accurate d*/d ratios, it is necessary to have a system resolution good enough to separate the d and d* loci, to have a low rate of accidental events and to have an accurate determination of the E_{np} range used for determining d* events. Four-body events for large n-p relative energies must yield very few counts.

The ^{3}He + p breakup was measured for E_i = 136 MeV at IUCF with a ^{3}He gas target at 18 atmospheres. Outgoing particles were detected by Si(Li) - NaI(Tl) telescopes placed at the following angle pairs: 42.7°-42.7°, 35°-50.4° and 28°-57.7°. The Q-value spectrum depicted in Figure 1 shows a good d* vs d separation and a low yield of 4-body events. Figure 2 shows the projected energy spectra. Table I contains the experimental d*/d ratios compared with the PWIA calculations using the ^{3}He wave function obtained from a variational calculation with the Hamada-Johnson potential.[7]

The accuracy of the present measurement is such that it could yield information on P(S'), but further theoretical studies are necessary, in particular to determine whether the ^{3}He wave function used produces the measured pd and pd* coupling constants.[8] It has been pointed out[9] that the d*/d ratio contains information

Table I

Experimental		
E_{np} Range	Accidentals	Value
0-1.75 MeV	subtracted	0.173±0.009
0-1.75 MeV	not subtracted	0.197±0.008
0-2.50 MeV	subtracted	0.200±0.010

Theoretical			
E_{np} Range	P(S')	Triplet	Value
0-1.75 MeV	0%	with	0.155
0-1.75 MeV	4%	with	0.164
0-1.75 MeV	0%	no	0.150
0-1.75 MeV	4%	no	0.160
0-2.50 MeV	0%	with	0.178
0-2.5 MeV	4%	with	0.195

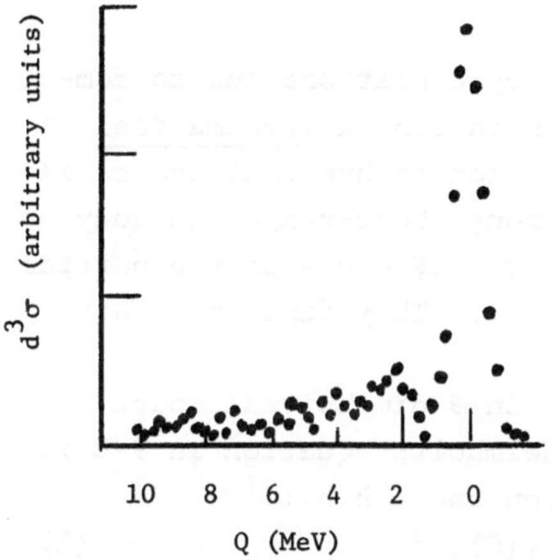

Figure 1. Binding energy data for $^3He(p,pp)d^*,d$ for 42.7° - 42.7°; d^* vs. d separation is visible.

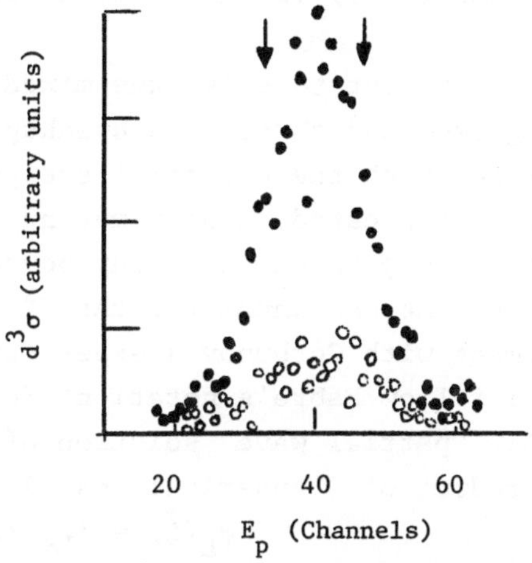

Figure 2. Locus projected data for 42.7° - 42.7°. Solid circles are d locus points and open circles are d^* locus points.

1. T. Griffy and R. Oakes, Phys. Rev. 135 (1964) 1161.
2. C.Hu, et al., Few Particle Problems (eds. I. Slaus, et al., North Holland, 1972) 648.
3. I. Slaus, et al., Phys. Rev. Lttrs. 27 (1971) 751.
4. M. Jain, et al., Lttrs. Nuovo Cim. 8 (1973) 844, and references therein.
5. A. Cowley, et al., Nucl. Phys. A220 (1974) 429.
6. R. Frascaria, et al., Nucl. Phys. A178 (1971) 307.
7. C. Hu, private communication.
8. G. Plattner, et al., Phys. Rev. Lttrs. 39 (1977) 127; V. Bhasin and A. Mitra, Phys. Rev. Lttrs. 40 (1978) 1130.
9. A. Mitra, private communication; R. Kaushal, et al., preprint (1977).

This work has been partially supported by PL480 grant no. F6F005 (I.S.) and the U.S. National Science Foundation (E.K., J.M.L, D.D and P.A.T.).

PHOTO-DISINTEGRATION OF THE ALPHA PARTICLE*

J. S. Levinger
Rensselaer Polytechnic Institute
Troy, NY 12181/USA

I make two modifications to the calculation by Fabre and Levinger[1] for electric dipole transitions in the trinucleon. First, I attack the problem of competing two-, three-, and four-body breakup with the approximate method used by Levinger and Fitzgibbon[2] (LF) for trinucleon final states of isospin $\frac{1}{2}$. Second, I expand in hyperspherical harmonics (h.h.) in the 9-dimensional space appropriate to the alpha particle.

LF attempt to solve the mixed boundary conditions due to competing two- and three-body breakup by considering a mathematical model in which the potential energy expansion in hypermultipoles is severely truncated, thus removing the strong short-range two-body force which produces two-body bound states. LF use a single partial wave for the continuum for the V^x potential. They find only fair agreement with Gorbunov's experiments.[3]

I follow Fabre's notation[4] for h.h. in 9-dimensional space. A regular "partial wave" solution of the Helmholtz equation in 9-D is the product of a spherical Bessel function and a h.h.:

$$\psi_L(\vec{\xi}) = j_{L+3}(k\xi)H_{[L]}(\Omega)\xi^{-3} \tag{1}$$

Here ξ is a vector in 9-D; the wave number k is related to the total kinetic energy E. ($E = \hbar^2k^2/M$.) The symbol [L] stands for 8 quantum numbers: the grand orbital L; L_1, and 6 conventional ℓ_i, m_i (i = 1, 2, 3). The symbol Ω stands for 8 angles. The dipole operator is proportional to $\xi H_{[1]}(\Omega)$. Orthogonality of h.h. gives the familiar triangle rule. We start with ground state grand orbital zero, and reach only grand orbital one for the final state. The total Born approximation cross section is

$$\sigma^B = (2\pi/9)\alpha(E_\gamma/k)(M/\hbar^2)(r_{if}{}^B)^2 \tag{2}$$

The Born radial overlap integral

$$r_{if}{}^B = \int_0^\infty u_o(\xi)\xi k\xi j_4(k\xi)d\xi, \tag{3}$$

With hypermultipole $V_o(\xi)$ acting in the final state, the radial function $u_1(\xi)$ is the regular solution of the differential equation

$$-d^2u_1/d_\xi{}^2 + 20\xi^{-2}u_1 + V_o(\xi)u_1 = k^2u_1 \tag{4}$$

I normalize at large hyperradius to

$$u_1(\xi) = k\xi[\cos\delta j_4(k\xi) - \sin\delta n_4(k\xi)] \tag{5}$$

I have obtained preliminary values of the total cross section using Ballot's[5] ground state wave function $u_o(\xi)$ for the Volkov spin-independent potential, together with three final state wave functions for different exchange mixtures: i) Born approximation; ii) an approximate Serber mixture; iii) Wigner exchange. The table shows two moments, the bremsstrahlung weighted σ_{-1}, and the integrated cross section σ_0. I compare with Gorbunov's experiment.

Table 1. Moments of the Photoeffect

	σ_{-1}(mb)	σ_0(MeV mb)
Born Approximation	2.6	200
Approximate Serber Exchange	2.5	130
Wigner Exchange	2.5	70
Experiment	2.5	100

All four agree on the value of σ_{-1}, as they should. The large effect of the final state interaction is shown by the factor of three between Wigner exchange and Born approximation, for the integrated cross section. (The Wigner exchange supports a bound final state with grand orbital one, that almost exhausts the Thomas-Reiche-Kuhn sum rule of 60 MeV mb.) The shape of the photoeffect curve also varies greatly with the exchange mixture. Serber exchange gives the best (but not good) agreement with Gorbunov.

*Supported in part by the National Science Foundation.

1. M. Fabre de la Ripelle and J. S. Levinger, Nuov. Cim. 25A, 555 (1975); Nuov. Cim. Letters 16, 413 (1976).
2. J. S. Levinger and R. Fitzgibbon, Phys. Rev. C (in press) 1978.
3. A. N. Gorbunov, Photonuclear and Photomesic Processes, Trudy, (Editor, D. V. Skobel'tsyn), Consultants Bureau, New York, Vol. 71, p. 1 (transl.).
4. M. Fabre de la Ripelle, Proc. Intl. School Nuclear Theory Physics, Predeal, 1969.
5. J. L. Ballot, M. Beiner, and M. Fabre de la Ripelle, Proc. Symposium Present Status and Novel Developments in Many-Body Problems, Roma 1972, Editors V. Calogero and C. Ciofi degli Atti (Bologna, 1974), p. 565.

THE BOUND STATES OF $^4_\Lambda$He - $^4_\Lambda$H

B. F. Gibson
Theoretical Division, Los Alamos Scientific Laboratory
Los Alamos, NM 87545/USA

The light hypernuclei provide us with an opportunity to study few-body bound states of baryons other than just the neutron and proton. The hypertriton, $^3_\Lambda$H, has been the subject of considerable attention.[1] That the A = 4 ground state separation energies[2]

$$B_\Lambda(^4_\Lambda\text{He}) = B(^4_\Lambda\text{He}) - B(^3\text{He}) = 2.42 \pm .04 \text{ MeV}$$
$$B_\Lambda(^4_\Lambda\text{H}) = B(^4_\Lambda\text{H}) - B(^3\text{H}) = 2.08 \pm .06 \text{ MeV} \qquad (1)$$

are not consistent with a charge symmetry hypothesis for the ΛN force (i.e., $V_{\Lambda p} \neq V_{\Lambda n}$) and that there exists a distinct excited state[3] in at least the case of $^4_\Lambda$H (with an E_γ = 1.09 MeV for the $1^+ \to 0^+$ M1 transition) makes this hypernuclear isodoublet even more interesting. In particular, one asks whether these facts are consistent with the properties of the meson-theoretic hyperon-nucleon (YN) potentials of Nagels, et al.[4]

It is known that the non-Coulomb isodoublet binding energy difference[5] (ΔB_Λ = 0.36 ± .07 MeV) is not easily explained in terms of conventional "effective 2-body" theories such as are provided by the shell model and variational approaches. Utilizing separable potential approximations to the interactions of Ref. 4 as described below, one obtains only some 0.2 MeV for the charge-symmetry-breaking contribution to ΔB_Λ even when radial compression of the trinucleon cores is included.[6] This failure is in large part due to an effect familiar to most few-nucleon physicists: For an attractive potential with negative scattering length, $|a| > |a'|$ implies that V is more attractive than V′ in both a 2-body and a 3-body binding energy calculation; however, $r > r'$ implies that V is more attractive than V′ in a 2-body calculation but <u>less</u> attractive in a 3-body calculation. Although this is an oversimplified picture, it does emphasize that potential properties which make one potential more attractive than another in a genuine 3-body or 4-body calculation may not yield the same relationship in an effective 2-body calculation for that same system.

In what follows, I wish to discuss primarily the hypernuclear 4-body problem utilizing 1) 4-body equations[7] which are equivalent to those of Yacubovsky in their separable potential formulation[8] and 2) separable potentials whose parameters were determined from the low energy scattering properties of the meson-theoretic potentials of Ref.4. The 4-body equations are easily derived using the Schrödinger formalism which is described in detail for the 4-nucleon problem in Ref. 7. One obtains coupled 2-variable integral equations for the 6 functions (A_i, B_i, C_i, i = s,t) with [3,1] symmetry and the 4 functions (D_i, F_i, i = s,t) with [2,2] symmetry.

Schematically the $J^{\pi} = 0^+$ ground state equations are

$$
\begin{aligned}
A_i &= \tau_i^A \int [X_{ij}B_j + X_{ij}D_j] \\
B_i &= \tau_i^B \int [X_{ij}^{NN}A_j + X_{ij}^{NN}D_j + X_{ij}^{N\Lambda}C_j + X_{ij}^{N\Lambda}F_j] \\
C_i &= \tau_i^C \int [X_{ij}^{\Lambda N}A_j + X_{ij}^{\Lambda N}D_j + X_{ij}^{\Lambda\Lambda}C_j + X_{ij}^{\Lambda\Lambda}F_j] \\
D_i &= \tau_i^D \int [Y_{ij}^{NN}A_j + Y_{ij}^{NN}B_j + 2Y_{ij}^{N\Lambda}C_j] \\
F_i &= \tau_i^F \int [Y_{ij}^{\Lambda N}A_j + Y_{ij}^{\Lambda N}B_j + 2Y_{ij}^{\Lambda\Lambda}C_j]
\end{aligned}
\tag{2}
$$

where the kernels X, $X^{\alpha\beta}$, $Y^{\alpha\beta}$ are themselves solutions of coupled inhomogeneous integral equations as in the 4-nucleon problem. The separable potentials underlying each of the interactions determining these kernels are assumed to be of the rank-1 Yamaguchi form.[9] Tensor components are included in all triplet potentials; however, the resulting t-matrices are truncated so that only the s-wave part appears in the integral equations,[10] an approximation which should be insignificant in a binding energy difference study.

When one includes explicit coupling of the Λ and Σ hyperons in the YN interactions, the potential combinations appearing in the 0^+ ground-state calculation are

$$V_{\Lambda N} = \frac{1}{2} V_{\Lambda N}^t + \frac{1}{2} V_{\Lambda N}^s, \quad V_{XN} = \frac{1}{2} V_{XN}^t - \frac{1}{6} V_{XN}^s, \quad V_{\Sigma N} = \frac{1}{2} V_{\Sigma N}^t + \frac{1}{2} V_{\Sigma N}^s, \tag{3}$$

where $V_{XN}^{t,s}$ are the ΛN-ΣN coupling potentials. It is clear that were not V_{XN}^s very small (see Ref. 4), one would have significant suppression of the attraction in the singlet YN channel in the 4-body binding energy calculation as compared with the free interaction. For the $J^{\pi} = 1^+$ excited states, the relevant potential combinations are

$$V_{\Lambda N} = \frac{5}{6} V_{\Lambda N}^t + \frac{1}{6} V_{\Lambda N}^s, \quad V_{XN} = \frac{1}{6} V_{XN}^t + \frac{1}{6} V_{XN}^s, \quad V_{\Sigma N} = \frac{5}{6} V_{\Sigma N}^t + \frac{1}{6} V_{\Sigma N}^s. \tag{4}$$

Since V_{XN}^t is not small, there is, in this case, sizeable "Σ suppression;" this helps separate the 0^+ and 1^+ states more than one would a priori estimate from a comparison of the free low-energy ΛN triplet and singlet scattering parameters. The possibility of similar Σ suppression in ${}^5_\Lambda He$ has been the subject of speculation for some time;[11] shell model and variational estimates of $B_\Lambda({}^5_\Lambda He)$ are of the order of 5-6 MeV compared to an experimental value of approximately 3.1 MeV, when one uses effective ΛN spin-dependent forces fitted to the binding energy of ${}^3_\Lambda H$ and the average of ${}^4_\Lambda He$ and ${}^4_\Lambda H$. Here one has a wave function of the form

$$\alpha[|{}^4\text{He},T = 0\rangle \times |\Lambda, T = 0\rangle]^{T = 0} + \beta[|{}^4\text{He}^*, T = 1\rangle \times |\Sigma, T = 1\rangle]^{T = 0} ,$$

and the even parity T = 1 states of the alpha particle have large excitation energies which should strongly suppress the Σ coupling.

Because in the A = 4 g. st. $V^s_{\Lambda N} \simeq 0$ and the coefficients of $V^t_{\Lambda N}$, V^t_{XN}, and $V^t_{\Sigma N}$ in Eq. (3) are unaltered from free scattering, it is possible to utilize effective interactions: separable potentials $V^t_{\Lambda N}$ and $V^s_{\Lambda N}$ whose parameters are determined by fitting the low energy ΛN scattering properties of the interactions in Ref. 4 (i.e., scattering lengths, effective ranges, and $^3S_1$ - $^3D_1$ mixing parameters). Thus the effects of $\Lambda\Sigma$ coupling upon the ΛN potential parameters (including charge symmetry breaking due to meson mixing, Σ mass differences, etc.) are taken into account, but there are no explicit Σ channels in the calculation. (Such a model yields $B_\Lambda({}^3_\Lambda H) \simeq 0.1$ MeV compared with an experimental value of 0.15 ± .08 MeV; the hypertriton is dominated by the singlet interaction ($V_{\Lambda N} = \frac{3}{4} V^s_{\Lambda N} + \frac{1}{4} V^t_{\Lambda N}$), so that tensor forces, $\Lambda\Sigma$ coupling, etc. are not so important.) Using the parameters indicated in the Table, where the triplet and singlet potentials correspond to

$$\begin{array}{ll} {}^4\text{He:}\ V^t_{\Lambda N} = V^t_{\Lambda p} & {}^4\text{H:}\ V^t_{\Lambda N} = V^t_{\Lambda n} \\ V^s_{\Lambda N} = \frac{1}{3} V^s_{\Lambda p} + \frac{2}{3} V^s_{\Lambda n} & V^s_{\Lambda N} = \frac{1}{3} V^s_{\Lambda n} + \frac{2}{3} V^s_{\Lambda p} \end{array} \qquad (5)$$

we obtain a value of

$$\Delta B_\Lambda \simeq 10.52 - 10.09 = 0.43 \text{ MeV} .$$

This is to be compared with the Coulomb corrected experimental value of 0.36 ± .07 MeV. This estimate is not insensitive to the ΛN scattering parameters. If $r^t_{\Lambda n}$ is decreased from 3.32 to 3.30 fm, ΔB_Λ is lowered to 0.39 MeV. (The 0.43 MeV quoted here is 0.11 smaller than the central potential result previously reported.[6] This difference is a reflection of the weaker nature of the tensor force in binding compact systems as compared with central forces, even for short range interactions.) However, the important point is not so much that this theoretical estimate is close to the experimental value but that the true 4-body result is about twice as large as the effective 2-body result for the same ΛN potential input.

Table. Potential parameterizations and low energy properties.

System	Spin	$\lambda(fm^{-3})$	β(fm)	t	γ(fm)	a(fm)	r(fm)
NN	s	0.1323	1.130			-17.0	2.84
	t	0.1430	1.241	4.495	1.948	5.40	1.73
$\Lambda N({}^4_\Lambda H)$	s	0.1093	1.260			-1.862	3.735
	t	0.1136	1.361	7.000	3.250	-1.840	3.320
$\Lambda N({}^4_\Lambda He)$	s	0.1029	1.255			-1.931	3.703
	t	0.1076	1.358	8.400	3.250	-2.060	3.180

In summary, it does seem probable that we can understand the Λ-separation energies of the s-shell hypernuclei in terms of meson-theoretic potentials and genuine few-body theory. Because ${}^{3}_{\Lambda}H$ is YN singlet dominated (where ΛΣ coupling is minimal), the conclusions of prior Faddeev type analyses[1] will not be qualitatively altered. The ground state binding energy difference in the A = 4 isodoublet can be understood from Yacubovsky type calculations. Calculation of the $1^+ \rightarrow 0^+$ transition energy will require taking into account the Σ suppression that arises in the A = 4 1^+ excited state due to the modification of the spin-isospin coefficient of the coupling potential. Such Σ suppression should be even more important in ${}^{5}_{\Lambda}He$, where the Σ couples to the even parity, T = 1 excited states of the ${}^{4}He$ core and the excitation energies are large. But even an estimate of this effect assuming that the excitation energy is infinite must await the obvious but more sophisticated computing required to treat correctly the 5-body problem.

This work was performed under the auspices of the U. S. DoE and in collaboration with Prof. D. R. Lehman of the George Washington University. The author is grateful to Prof. E. M. Henley, Prof. G. A. Miller, and other members of the Dept. of Physics at the University of Washington for their hospitality during the spring of 1978 when a part of this manuscript was drafted.

References

1. See references cited in B. F. Gibson and D. R. Lehman, Phys. Rev. C 10, 888 (1974); 14, 2346 (1976); 11, 2092 (1975) --- especially those of Dabrowski and Schick.
2. M. Juric, et al., Nucl. Phys. B52, 1 (1973).
3. A. Bamberger, et al., Nucl. Phys. B60, 1 (1973); M. Bedjidian, et al., Phys. Lett. 62B, 467 (1976).
4. M. M. Nagels, T. A. Rijken, and J. J. deSwart, Phys. Rev. D 15, 2547 (1977).
5. A Coulomb energy difference has been subtracted. This was estimated in a simple model to be

$$\Delta B_{\Lambda}^{C} = \frac{ze^2}{R + \Delta R} - \frac{ze^2}{R} \simeq - \frac{ze^2}{R}\frac{\Delta R}{R} \simeq -0.02 \text{ MeV}$$

where $ze^2/R \simeq -0.7$ MeV is the approximate Coulomb energy of ${}^{3}He$, R ≃ 1.75 fm is the radius of ${}^{3}He$, and ΔR ≃ -0.05 fm is the radial compression of the ${}^{3}He$ core in ${}^{4}_{\Lambda}He$. See J. L. Friar and B. F. Gibson, "Coulomb Energies in S-shell Nuclei and Hypernuclei" (to be published in Phys. Rev. C) for a more detailed estimate.
6. B. F. Gibson and D. R. Lehman, "The ${}^{4}_{\Lambda}He$ - ${}^{4}_{\Lambda}H$ Binding Energy Difference" (submitted to Phys. Lett.).
7. B. F. Gibson and D. R. Lehman, Phys. Rev. C 14, 685 (1976); 15, 2257 (1977).
8. O. A. Yacubovsky, Yad. Fiz. 5, 1312 (1967); Sov. J. Nucl. Phys. 5, 937 (1967); K. F. Kharchenko and V. E. Kuzmichev, Nucl. Phys. A183, 606 (1972).
9. This approximation has been found by Tjon to be quite satisfactory (priv. comm.); it leads to but a small (8%) error in our studies of the triton binding energy.
10. Y. Yamaguchi, Phys. Rev. 95, 1628 (1954); Y. Yamaguchi and Y. Yamaguchi, Phys. Rev. 95, 1635 (1954); A. C. Phillips, Nucl. Phys. A107, 209 (1968).
11. A. R. Bodmer, Phys. Rev. 141, 1387 (1966); B. F. Gibson, A. Goldberg, and M. S. Weiss, in FEW PARTICLE PROBLEMS IN THE NUCLEAR INTERACTION (ed. by I. Slaus, S. Moszkowski, R. Haddock, W. vanOers), North Holland Pub. Co. (1972), p. 188; J. Dabrowski and F. Fedorynska, Nucl. Phys. A210, 509 (1973).

Some phenomena near nuclear surface

T. Sasakawa

Department of Physics, Tohoku University
980 Sendai, Japan

If the Yakubovski equation or various version of it should be the integral equation of a N-body system with which we must deal, explanations of some nuclear phenomena must be given in a unified way on the basis of this equation. Since our number N is a finite number, the system must be confined in a region or, in general, in a number of regions in configuration space. Therefore, the system has a surface. Accordingly, the first application of our equation must be done to those phenomena which manifest itself near the surface. These phenomena are: (i) Low energy nuclear reactions; (ii) involved nuclear reactions with large angular momentum; (iii) the cluster phenomena, involving two-body clusters such as the d-d cluster of an excited state of ^{4}He, the d-t cluster of an excited state of ^{5}He,the alpha particle model of ^{8}Be; three-body clusters such as the alpha particle model of ^{6}Li, ^{9}Be, and ^{12}C; (iv) the alpha decay; (v) the asymmetric fission; and (vi) hopefully, the collective motion near the surface. So far, the topics (i)-(iii) have been studied to some extent[1-7,9]. These studies will be reviewed in the present paper.

There are two difficulties in the Yakubovski equation: (i) The number of components is formidably large. (ii) The equations for these components are not directly related to any physical wave functions such as the channel wave functions. These difficulties have been overcome by decoupling the Yakubovski components. To make clear of the idea, we will demonstrate the formulation taking a four body system.

We designate by ϕ_{ij} a component of the total wave function ψ. This component represents that a pair ij are interacting in the final state. A four-body system involves six interactions; $v_{ij}(i,j = 1,2,3,4)$. Correspondingly, ψ is expressed as a sum of six ϕ_{ij}'s,

$$\psi = \phi_{12} + \phi_{23} + \phi_{31} + \phi_{24} + \phi_{41} + \phi_{34} \quad . \tag{1}$$

If 1 and 2 are interacting in the final state, the before-final state interactions are between 1 (or 2) and 3, 1 (or 2) and 4, or between 3 and 4. Thus ϕ_{12} is expressed as

$$\phi_{12} = \phi_{12}^{123} + \phi_{12}^{124} + \phi_{12}^{12,34} \quad . \tag{2}$$

Inversely, we define a function ϕ^{123} by

$$\phi^{123} = \phi_{12}^{123} + \phi_{23}^{123} + \phi_{31}^{123} \quad . \tag{3}$$

Then, Eq.(1) is alternatively expressed as the sum of seven components[1],

$$\psi = \phi^{123} + \phi^{124} + \phi^{134} + \phi^{234} + \phi^{12,34} + \phi^{23,14} + \phi^{31,24} . \tag{4}$$

If we restrict ourselves to the two-body (open) channels, ϕ^{123} and $\phi^{12,34}$ are approximated by

$$\phi^{123} = \varphi^{(123)} f_4 \quad \text{and} \quad \phi^{12,34} = \varphi^{(12)} \varphi^{(34)} f_{12-34} , \tag{5}$$

where $\varphi^{(123)}$ denotes the wave function of the bound 123. In a phenomenological analysis, we may employ for $\varphi^{(123)}$ or $\varphi^{(12)}$ etc., some phenomenological wave function. Then we have seven coupled equations to be solved for f_4, f_{12-34} etc. Thus the number of equations we have to solve is considerably reduced from the original Yakubovski equation (eighteen for a four-body problem). A five-body system is worked out in Ref.2. The extension to a N-body system is then obvious. The manner of opening channels with increasing energy has been demonstrated in Ref.3.

Near the surface, the system is almost on the energy shell. Then, each component in Eq.(4) may be approximated by the channel wave function(5). Under this condition, the equation for ϕ^{123} and $\phi^{12,34}$ are approximated by an amenable form[4,5]

$$\phi^{123} = \varphi^{(123)} f_4 \delta_{o,i} + G_o[t_{14}(\phi^{123} + \phi^{234} + \phi^{12,34} + \phi^{31,24}) + (1\leftrightarrow 2) + (1\leftrightarrow 3)] \tag{6a}$$

and

$$\phi^{12,34} = \varphi^{(12)} \varphi^{(34)} f_{12-34} \delta_{o,i} + G_o[t_{13}(\phi^{124} + \phi^{234} + \phi^{12,34} + \phi^{14,23}) + (1\leftrightarrow 2) + (3\leftrightarrow 4) + (\substack{1\leftrightarrow 2 \\ 3\leftrightarrow 4})]. \tag{6b}$$

Here, $\delta_{o,i}$ denotes the projection operator onto the intial state. In Eq.(6), we must use Eq.(5) and its cyclic permutations of 1,2,3, and 4. The kernel of Eq.(6) is then fully connected. Equation (6) may be used for low energy reactions or for high energy reactions with large angular momentum.

From Eq.(6), we can define the optical potentials,e.g.

$$U_4 = \langle \varphi^{(123)} | \; t_{14} + t_{24} + t_{34} | \varphi^{(123)} \rangle \quad . \tag{7}$$

We can calculate the distorted wave F_4^+ satisfying

$$(E + |E_{123}| - K_4 - U_4) F_4^+ = 0 \quad , \tag{8}$$

if we employ the technique given in Ref.6. The lowest order term of the reaction amplitude obtained by Eq.(6) is DWIA (the distorted wave impulse approximation),

$$T_{12,34 \leftarrow 123,4} = \langle \varphi^{(12)} \varphi^{(34)} F^-_{12-34} | t_{14} + t_{24} | \varphi^{(123)} F_4^+ \rangle \; . \tag{9}$$

If we use Eq.(5) in Eq.(6), this is the equation of coupled channel. Since Eq.(6) is valid near the surface, the coupled channel approximation is valid only near the surface. Deep in the collision complex, the channel wave functions are collapsed under two-body interactions. It has been demonstrated in Ref.5 that once the channel wave function is collapsed, it is very difficult to revive as a channel wave function. We can interpret the collapsed wave functions as forming the compound nucleus.

The cluster phenomena in the structure problem must also be handled by Eq.(6), because the difference of the reaction and the bound state problems is only in energy. We must note that the cluster phenomena manifest itself also near the surface. Only thing which is important in the structure problem is that we must seriously take account of the antisymmetrization. This has been done for a fourbody system in Ref.7. If we require

$$P_{12}\phi_{12} = - \phi_{12} \quad \text{and} \quad P_{34}\phi_{12} = - \phi_{12} \quad , \tag{10}$$

Eq.(1) is totally antisymmetric. It has been shown that each of components

$$\phi^{123} + \phi^{124} + \phi^{134} + \phi^{234} \tag{11a}$$

and

$$\phi^{12,34} + \phi^{23,14} + \phi^{31,24} \tag{11b}$$

is totally antisymmetric. In the two-body cluster model, we approximate ϕ^{ijk} and $\phi^{ij,kl}$ by Eq.(5). The function (11a) represents (antisymmetrized) single particle model and (11b) does the (antisymmetrized) d-d cluster model. Then, Eq.(4) shows the <u>coexistence</u> of the single particle and the cluster models.

It is an important question to ask the manner that the N-body equation is approximated by the Faddeev equation of the three-body system, as in the case of ^{6}Li, ^{9}Be, and ^{12}C. In answering this question, we must bear two things in our mind: (i) The antisymmetrization of the total system is very important. Otherwise, we get a too big binding energy[8]. Therefore, we must seek for the generalized three-body Faddeev equation in which the antisymmetrization of the whole N-body system is taken into account. (ii) If the energy increases, the system will breakup into four bodies. Therefore, the kernel of the integral equation must be fully connected with respect to four particles.

To satisfy these two requirements, we define the functions[9]

$$\phi_{12,3} \equiv \phi_{13}^{123} + \phi_{23}^{123} \quad , \tau_{12,3} \equiv W_{13}^{123} + W_{23}^{123} \quad , \tau_{34} \equiv W_{34}^{12,34}. \qquad (11)$$

The interaction W_{13}^{123} has been defined in Ref.1. This is an operator which is fully connected with respect to 1,2, and 3. From equations in Ref.1, we obtain the following equations

$$\phi_{12,3} = G_o\tau_{12,3}(\phi_{12,4} + \phi_{13,4} + \phi_{23,4} + \phi_{34}^{12,34} + \phi_{24}^{31,24} + \phi_{14}^{23,14}), \qquad (12a)$$

$$\phi_{12,4} = (3 \leftrightarrow 4) \text{ of } \phi_{12,3} \quad , \qquad (12b)$$

$$\phi_{34}^{12,34} = G_o\tau_{34} (\phi_{12,3} + \phi_{12,4}). \qquad (12c)$$

On the right hand side of Eq.(12c), we have dropped without violating the antisymmetrization some terms which do not take the form of $\phi_{ij,k}$. Besides Eq.(12), we require that the amplitude of ϕ_{12} vanishes when the particles 3 and 4 are removed from the pair 12.

Eq.(12) is the generalized Faddeev equation that we wanted to have. This is seen if we keep only terms involving the pair 12 in Eq.(12).

$$\phi_{12,3} = G_o\tau_{12,3}(\phi_{12,4} + \phi_{34}^{12,34}), \qquad (13a)$$

$$\phi_{12,4} = G_o\tau_{12,4}(\phi_{12,3} + \phi_{34}^{12,34}), \qquad (13b)$$

$$\phi_{34}^{12,34} = G_o\tau_{34} (\phi_{12,3} + \phi_{12,4}) . \qquad (13c)$$

1) T.Sasakawa,Phys. Rev.C 13 (1976),1801.
2) T.Sasakawa,Prog. Theor. Phys. Suppl. 61 (1977),149.
3) T.Sasakawa,Phys. Rev.C 17 (1978) no.5, to appear.
4) T.Sasakawa,Phys. Rev.C 17 (1978) no.5, to appear.
5) T.Sasakawa, N-body theory of reactions (preprint).
6) T.Sasakawa and T.Sawada,Phys. Rev. C 17 (1978) no.6, to appear.
7) T.Sasakawa, Antisymmetrized four-body wave function (preprint).
8) R.Tamagaki and Y.Hujiwara,Prog. Theor. Phys. Suppl. 61(1977),229.
9) T.Sasakawa, Three-body model for a N-body system (preprint).

EXPLORATION OF A CONTINUUM AMBIGUITY IN THE $^3He(d,d)^3He$ PHASE-SHIFT ANALYSIS AT 0.32 MeV

B. Jenny, W. Grüebler, P.A. Schmelzbach, V. König and H.R. Bürgi
Laboratorium für Kernphysik, Eidg. Technische Hochschule, CH-8093 Zürich

Between 0.32 and 5.0 MeV deuteron energy a phase-shift analysis of $^3He(d,d)^3He$ scattering has been made[1]. Above 1.5 MeV it is based on nearly complete angular distributions of the cross section and the four deuteron analyzing powers[2], below 1.0 MeV on cross section measurements[3] only. In both energy regions serious ambiguities have been found. Even at low energies, where Coulomb effects dominate, a continuum ambiguity exists: a family of solutions gives identical fits to the elastic cross section and the total cross section of the $^3He(d,p)^4He$ reaction.

In scattering of spin zero particles, the continuum ambiguity is eliminated either by elastic unitarity or, at low energies, by Coulomb interference[4]. When both particles have spin and the partial wave series is truncated at $\ell=0$, neither Coulomb interference together with the absorption cross section, nor unitarity is sufficient to remove it.

For $\ell=0$ the differential cross section is obtained from the expression $2|a|^2 + |q|^2$ where a and q are the quartet and the doublet scattering amplitude respectively. In terms of the elements $U^{3/2}$ and $U^{1/2}$ of the scattering matrix it is given by

$$2|U^{3/2}|^2+|U^{1/2}|^2 + 2\,\mathrm{Re}\big(C(\theta)-1\big)\big(2U^{3/2}+U^{1/2}\big)$$

$C(\theta)$ is essentially the Coulomb scattering amplitude.

Considering Coulomb interference together with elastic unitarity, two discrete solutions are obtained, connected by expressions given in ref.[1]. By Coulomb interference alone only the sum $2U^{3/2} + U^{1/2}$ can be determined; by the absorption cross section the quantity $2|U^{3/2}|^2 + |U^{1/2}|^2$. One degree of freedom is left to cause the continuum ambiguity. The exploration of this ambiguity at 0.32 MeV deuteron energy assumes pure S wave scattering (significant improvement of the fit by P wave contributions was found above 0.6 MeV only). The range has been traced out by projecting the four dimensional χ^2 surface on the $(^2S_{1/2},\chi^2)$ plane, shown in fig. 1a. To do this, χ^2 had to be minimalized for fixed values of the $^2S_{1/2}$ phase-shift. There are two corresponding values of the absorption parameters, shown in fig. 1b by thin and dashed lines. The unitarity inequality confines the range to the intervals cA and Bc. The range found for the ^4S phase-shift goes from 5° to 25°.

Since in the limit of pure S wave scattering all elastic first order polarization observables vanish, a complete phenomenological determination of the $\ell=0$ matrix elements at this energy seems to be impossible as long as the $^3He(d,p)^4He$ reaction is not included explicitly in the analysis. Therefore, in order to obtain as meaningful results as possible, several physically reasonable restrictions had to be made in the $^3He(d,d)^3He$ phase-shift analysis.

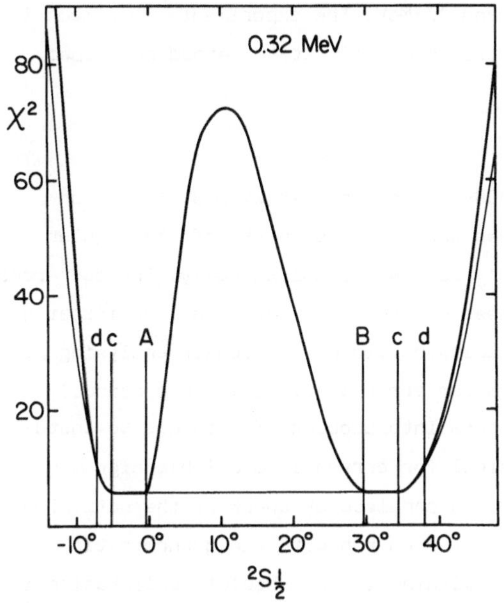

Fig. 1a. Projection of the χ^2 surface on the $(^2S_{1/2},\chi^2)$ plane. Between the verticals c the data are identically fitted. The saddle between A and B results from the bound $\eta_{2S} \leq 1$. At the verticals d η_{4S} reaches this limit.

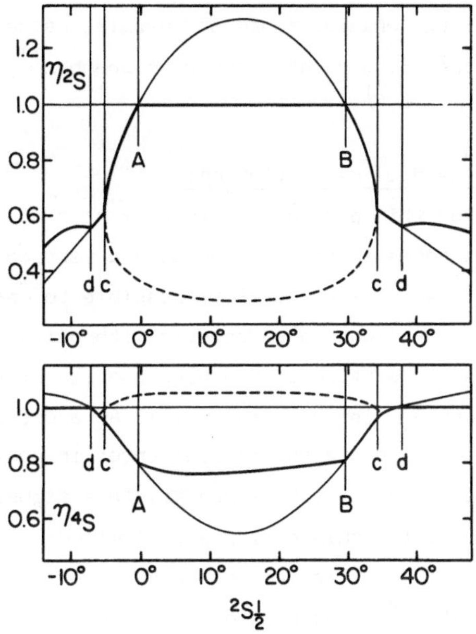

Fig. 1b. Corresponding values of the absorption parameters η_{2S} and η_{4S}.

References

1) B. Jenny, Diss. ETH Nr. 6073, ETH Zürich, 1977
2) B. Jenny et al., Proc. of the Fourth Int. Symp. on Polarization Phenomena in Nuclear Reactions, Eds. W. Grüebler et V. König (Birkhäuser Verlag Basel, 1975) p. 538 and to be published
3) L. Kraus, Thèse Nr. 681, Univ. Louis Pasteur, Strasbourg, 1971
4) D. Atkinson et al., Nucl. Phys. B77 (1974) 109

THE SCATTERING OF POLARIZED DEUTERONS ON ^{3}He BETWEEN 10 and 17 MeV *

P.A. Schmelzbach⁺, G.G. Ohlsen, N. Jarmie, R.H. Haglund, Jr. and R.E. Brown
Los Alamos Scientific Laboratory, University of California
Los Alamos, N.M. 87545

Using the LASL polarized beam facility, the cross section and vector and tensor analysing powers for the $^{3}He(\vec{d},d)^{3}He$ and $^{3}He(\vec{d},p)^{4}He$ reactions have been measured in a large angular range at energies between 10 and 17 MeV. The experiments were made in the "Supercube" scattering chamber[1)] using the three spin state method described in ref.[2)].

$^{3}He(\vec{d},d)^{3}He$ scattering:

For this part of the experiment a special 300-torr gas target with 6.3 μ-kapton windows in the horizontal and vertical planes was used. For parts of the angular range it was therefore possible to measure A_{xx} and A_{yy} simultaneously. The deuterons and helions were detected with pairs of telescopes to allow particle identification. As an example of the high quality of the results obtained, the tensor analysing power A_{xz} is shown in fig. 1. An additional random error of 0.005 is quadratically added to the statistical error in order to take into account effects due to instabilities like beam wander. This figure, typical for currently available high quality equipment, was determined from repeated consistency check of the data. The beam polarization was measured before and after each run with the quench ratio method[3)]. The beam polarization was constant within 1%. The absolute calibration is known to be better than 2%.

While the angular distribution for the cross section, A_y and A_{xx} show only very small changes in the energy range investigated, the structure observed at $\theta_{cm} \approx 60^{o}$ for A_{yy} and A_{xz} appears to be quite energy dependent.

$^{3}He(\vec{d},p)^{4}He$ reaction:

For the study of this reaction, a 2200-torr high pressure target was used. While the recoil α were detected with thin singles detectors, the high energy protons were detected by their ΔE signals in 1000 μ detectors. In order to get clean spectra and to reduce the dead time, Al absorbers were monted in front of the detectors to stop the elastically scattered deuterons. In the forward direction, coincidences with the signals from an additional 1000 μ detector were used to eliminate the neutron induced background. Measurements were made at 13, 15 and 17 MeV. The components A_y, A_{yy} and A_{xz} show only small changes as a function of energy, whereas a new strong structure develops in the backward part of the angular distribution of

A_{xx}. The present data do not show any evidence for the existence of a suggested $A_y = A_{yy} = 1$ point[4] in this energy range.

A Legendre polynomial analysis of these data is currently being made to extract the matrix elements dominating these reactions.

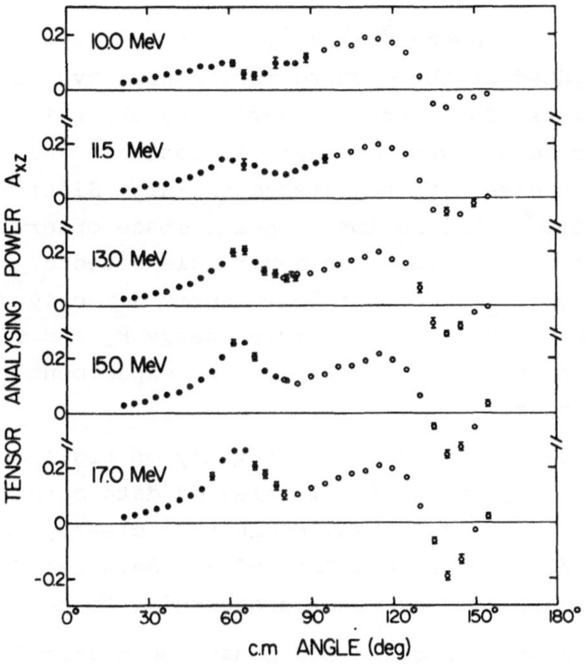

Fig. 1. The analysing power A_{xz} for $^3He(\vec{d},d)^3He$ scattering between 10 and 17 MeV. The open circles indicate that the recoil helions were detected.

* Work supported by the US Department of Energy

+ Visiting Staff Member from Eidg. Techn. Hochschule, Zürich, Switzerland

References

1) G.G. Ohlsen and P.A. Lovoi, Proc. of the 4th Int.Symp. on Pol. Phenomena in Nucl. Reactions, Zürich, eds. W. Grüebler and V. König (Birkhäuser, 1976)
2) G.G. Ohlsen and P.W. Keaton, Jr., Nucl. Instr. Meth. 109 (1973) 41
3) G.G. Ohlsen et al., Phys. Rev.Lett. 27 (1971) 599
4) F. Seiler et al., Nucl. Phys. A296 (1978) 228

The Deuteron Break-up Induced by α-Particles

B. Anders, U. Berghaus, H. Brückmann, G. Körber, P. Lara
C. Pegel, H. Salehi, K. Sinram and K. Wick

I. Institut für Experimentalphysik
Universität Hamburg, Germany

The tensor polarization of $^5Li(3/2^-)$ produced in the two-step reaction

$$\alpha + d \rightarrow n + {}^5\vec{Li}(3/2^-) \rightarrow n + \alpha + p \qquad (1)$$

has been investigated at the Hamburg isochronous cyclotron. Deuterated polyethylene targets $(CD_2)_n$ were bombarded by α-particles of 28.3 MeV energy. n-α coincidences were measured in coplanar and non-coplanar geometry. The α-particles were registered in large Si(Li)-semiconductor detectors ($60 \times 20\,mm^2$). The region of phase space observed in one experiment is shown in fig.1. Since the α-particle detector was not position-sensitive, the angle θ_α could not be measured directly, but it can be determined from the measured α-particle energy E_α and the neutron time-of-flight τ_n as can be seen from fig.1. The experiment is therefore kinematically complete.

Experimental cross-sections for the geometry of fig.1 are reproduced in figs.2a and 2b. A projection of all measured data onto the neutron time-of-flight axis is shown in fig.2a, exhibiting clearly the strong influence of the $^5Li(3/2^-)$ resonance. Part of the data measured in the neighborhood of this resonance have been projected onto the E_α-axis (fig.2b).

The full curves drawn in figs.2a and b have been calculated using the formalism of Heiss[1]). This model assumes that the break-up reaction (1) proceeds via the sequential decay process. The cross-section is expressed in terms of the even tensor moments t_{00}, t_{20}, t_{21} and t_{22} of $^5Li(3/2^-)$ produced in the first reaction step. These four parameters have been deduced from the experimental data measured at fixed neutron detection angles ($\theta_n = 53^o$ and $\theta_n = 43^o$, respectively) in coplanar and non-coplanar geometry. Only the coplanar data for $\theta_n = 53^o$ are shown in fig.2b.

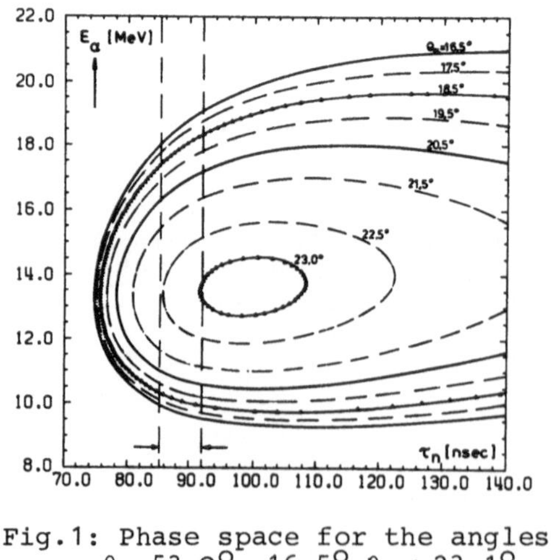

Fig.1: Phase space for the angles $\theta_n = 53.0^o$, $16.5^o < \theta_\alpha < 23.1^o$

The tensor moments t_{KQ} of the ground state ^5Li produced in the reaction $\alpha + d \rightarrow n + {}^5Li$ have been calculated independently by Schuette et al.[2]). The following table compares the results of these authors with the tensor moments determined experimentally in the present work.

Table: Experimentally and theoretically[2]) determined tensor moments t_{KQ} (divided by t_{oo}) for two production angles θ^c_{Li} of $^5Li(3/2^-)$ in reaction (1). The moments t_{KQ} are defined according to Welton[2])

θ_n	θ^c_{Li}	t_{KQ} determined by:	$\frac{t_{20}}{t_{oo}}$ (%)	$\frac{t_{21}}{t_{oo}}$ (%)	$\frac{t_{22}}{t_{oo}}$ (%)
53.0°	84°	experiment:	20.1±2.5	27±1.7	-18.4±2.2
		theory:	13.	9.	-24.6
43.0°	100°	experiment:	-2.9±2.5	23.4±1.4	10±2
		theory:	-24.	8.	3.

No quantitative agreement is found between experimental and theoretical results. However, the overall behaviour of the four tensor moments is quite well predicted by the theory.

1) P.Heiss, Z.Phys. A272 (1975)267

2) W.Schütte, H.H.Hackenbroich et al., Phys. Lett. 65B (1976)214
W.Schütte, thesis Köln 1976

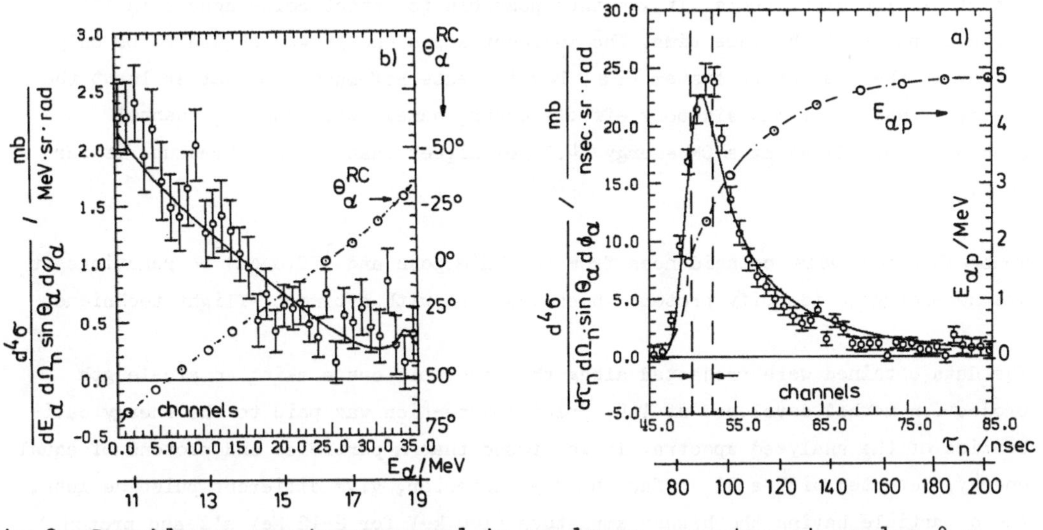

Fig.2: Concidence events measured in coplanar geometry at angles θ_n= 53.0° and 16,5°<θα < 23.1°
a) Projection of all measured data onto the time-of-flight axis τ_n.
b) Projection of the data measured near the maximum of the P3/2 resonance of ^5Li(1.85< $E_{\alpha p}$ <2.5 MeV, corresponding to the limits plotted as hatched lines in fig.1 and 2a) onto the axis E_α. θ^{RC}_α is the decay angle for the decay of $^5Li\to\alpha+p$ in the α-p subsystem.

LOW-ENERGY ALPHA-INDUCED DEUTERON BREAKUP STUDIED WITH A SIX-DETECTOR SYSTEM AND A FADDEEV ANALYSIS

L Glantz, I Koersner, A Johansson and B Sundqvist
Tandem Accelerator Laboratory, Uppsala, Sweden

An experiment and a preliminary analysis of the α-induced deuteron breakup $^2H(\alpha,\alpha p)n$ at 15 MeV laboratory energy have been carried out at the Tandem Accelerator laboratory in Uppsala (TLU).

In the past several three-body calculations based on the solution of the Faddeev equations for the α-2N system (e.g. bound states of ^6He and elastic d-α scattering) have been performed. The alpha particle has been regarded as elementary and empirical αN forces based on phase shifts have been used. One might expect that the d-α system in contrast to the three-nucleon system, would be richer in structual features because of the existance of resonances in the alpha-nucleon systems and in the three-body system (^6Li) itself. It would thus be of considerable interest to try to explore the α-d breakup reaction further.

The TLU experiment was kinematically complete and made with a (CD2)n target and a multiparameter system consisting of six Si surface barrier detectors and a CAMAC interface with eight ADC's. It was thus possible to detect coincidences in 15 detector pairs at the same time. The incident alpha energy was chosen to be only 15 MeV in the laboratory system (i e 5 MeV in center-of-mass) so that at least the energy at which obvious six-body effects occur, namely the outgoing channel consisting of $^3H+^3He$ at a CM energy 14.3 MeV higher than the α-d breakup, is far away.

To be able to handle coincidences from the $^2H(\alpha,p\alpha)n$ and $^{12}C(\alpha,\alpha\alpha)^8Be$ reactions it was necessary to identify protons which was done with a time-of-flight technique.

The data obtained were projected along the kinematic curve using an arc-length projection method developed at TLU. Special attention was paid to the energy calibration of the analyzed spectra. It was found that α particles and protons of equal energy, detected with a Si surface barrier detector, give different pulse heights, the α particle having the higher amplitude (100 keV for 2-12 MeV α's and protons). The absolute cross section were determined by counting recoiling deuterons from elastic α-d scattering in a monitor detector.

The most prominent features of the arc-length projected spectra obtained are peaks which seem to be due to the final state interactions (FSI) in the α-n and the α-p systems. In some spectra the minimum of the relative n-p energy was less than 50 keV and the kinematic condition for the observation of effects due to the n-p singlet resonance was fulfilled. However, no significant structure in the arclength projected spectra, which might have been caused by this resonance was seen.

Faddeev calculations on the $^{2}H(\alpha,\alpha p)n$ reaction have recently been published by Y Koike including comparisons with experimental data in the laboratory energy region of α's of 15-42 MeV. Calculations on the α-d breakup using Faddeev formalism are also in progress in Uppsala using the Doleschall code and some preliminary results have already been obtained. One conclusion which can be drawn at the present state is that already the correct treatment of the dynamics of the reaction gives the gross features of the spectra both in shape and magnitude but that there are many details which remain to be accounted for.

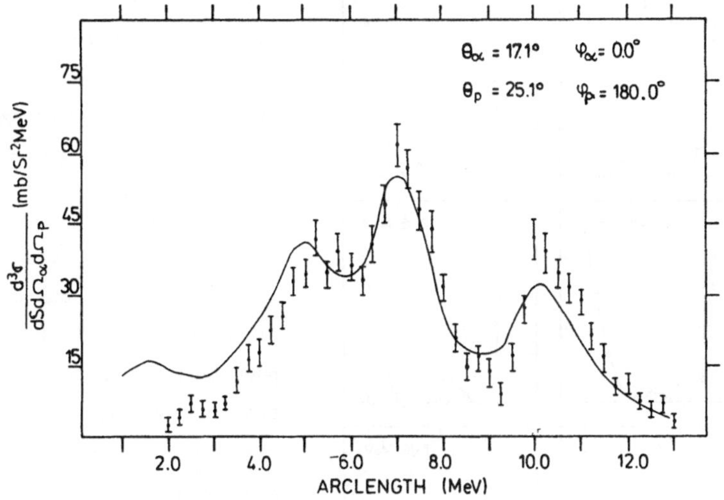

Fig 1 Arclength projected spectrum. The solid line shows the result of the Koike calculation.

References

1 P E Shanley, Phys Rev 187 (1969) 1328
2 M S Shah and A N Mitra, Phys Rev C1 (1970) 35
3 A Ghouhvanlou and D R Lehamn, Phys Rev C9 (1974) 1730
4 B Charnomordic, C Fayard and G H Lamot, Phys Rev C15 (1977) 864
5 Y Koike, Prog Theor Phys 59 (1978) 89
6 Y Koike, Nucl Phys (to be published)

INVESTIGATION OF RESONANCES IN d-α ELASTIC SCATTERING

W. Grüebler, B. Jenny, V. König, P.A. Schmelzbach and H.R. Bürgi
Laboratorium für Kernphysik, Eidg. Technische Hochschule, CH-8093 Zürich

Because of its simplicity in theoretical and experimental respect, the d-α elastic scattering has been since a long time a favoured testing ground for few nucleon problems. The T = 0, $J^{\pi} = 3^{+}$, 2^{+} and 1^{+} resonance in ^{6}Li are well established[1]. Recently complete angular distributions of the differential cross section and all four analysing powers for d-α scattering has been measured with high precision[2] in order to make a detailed study of the energy dependence of the s-wave to d-wave mixing parameter through the 1^{+} resonance. This investigation gives an indication of the tensor force contributing to the interaction and shows for the first time the necessity for a complex mixing parameter in a phase-shift analysis.

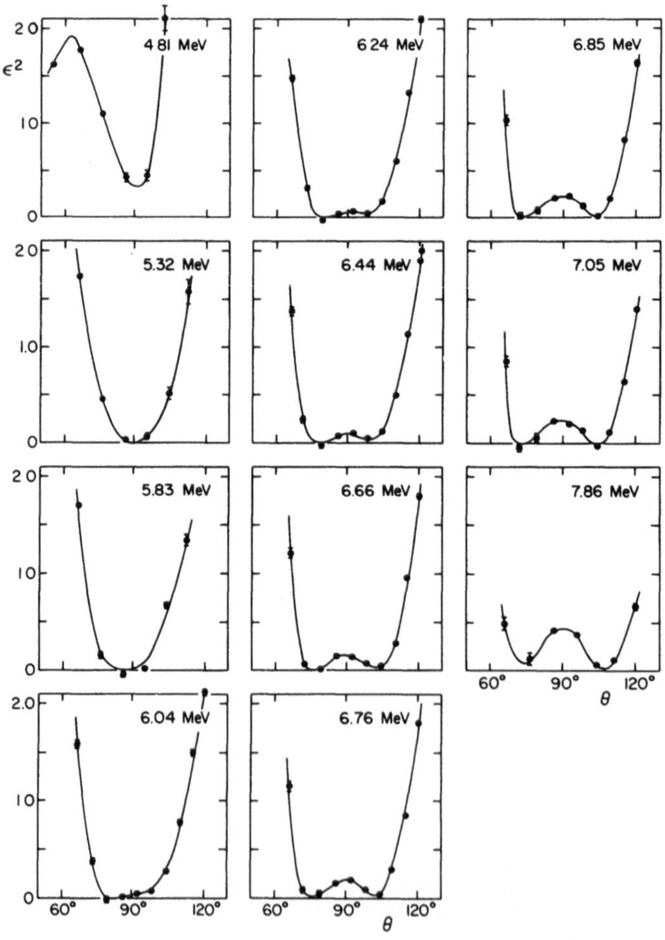

Fig. 1. The quantity ε^2 as a function of c.m. angle for the deuteron lab.energy range between 4.8 and 6.8 MeV.

As at many resonances in light nuclei states of maximum polarization have been observed in the corresponding reactions, the question was investigated if such a situation exists for this 1^{+} resonance too. While extrema in the analysing powers can be seen directly in the measured quantities, the detection of polarization maxima of the most general type requires a closer inspection of the data. In principle all states of maximum polarization correspond to points on the surface of a cone which is given by the relation[3]

$$\varepsilon^2 = (T_{20} + \sqrt{2})^2 - 3(T_{10})^2 - 6(T_{22})^2 = 0 \; .$$

The new data of ref.[2] together with older ones at lower and higher energies have been used to study the behaviour of ε^2 as a function of energy in the interesting angular range. The result is shown in fig. 1. The condition $\varepsilon^2 = 0$ seems to be fulfilled within the statistical errors over a larger energy range. Seiler et al.[3] have recently shown that the exact angle and energy can be found uniquely by investigating a quadratic condition of the relevant M-matrix elements in the complex plane. Following these lines the location of this maximum has been determined. The results are represented in fig. 2 in a contour plot of the quantity ε^2. As shown in the fig. 2 a state of maximum polarization ($\varepsilon^2 = 0$) is found near E_d = 6 MeV, $\theta_{cm} = 80^0$. In ref.[2] the location of the 1^+ resonance was established by an R-matrix fit at a deuteron energy of 6.26 MeV.

Additional measurements for the d-α scattering have been performed recently between 10 and 13 MeV. The analysis of these data should demonstrate if the presence of the $A_{yy} = 1$ point at about 12 MeV might be correlated to a possible new resonance around this energy.

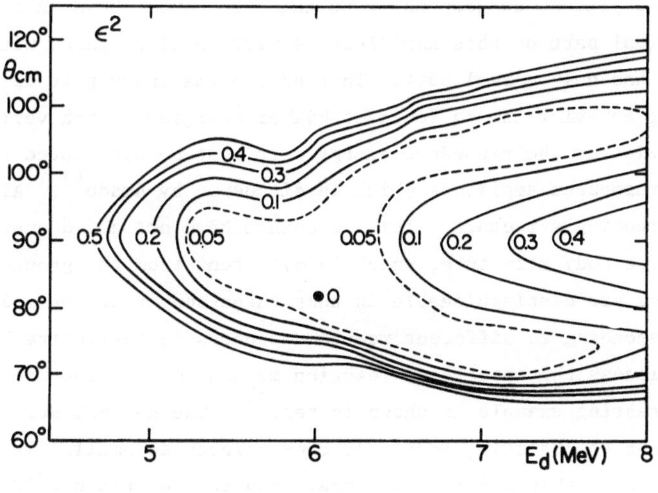

Fig.2. Contour plot of the quantity ε^2. The condition for a point on the cone is $\varepsilon^2 = 0$.

References
1) P.A. Schmelzbach, W. Grüebler, V. König and P. Marmier, Nucl. Phys. A184 (1972) 193
2) R.A. Hardekopf, W. Grüebler, B. Jenny, V. König, R. Risler, H.R. Bürgi and J. Nurzynski, Nucl. Phys. A287 (1977) 237
3) F. Seiler and H.W. Roser, Phys. Lett. 69B (1977) 411

THREE-BODY MODEL OF THE DEUTERON BREAKUP REACTION BY AN α-PARTICLE

Yasuro Koike
Department of Nuclear Engineering, Kyoto University, Kyoto, Japan

Recent development of method for solving the three-body equations enables us to calculate not only the three-nucleon system but also more complicated three-body systems such as the d-α system. Three-body calculations of the d-α elastic scattering have been succesfully done by Charnomordic et al.[1] and by the present author[2].

It is much more interesting and important to apply this theory to the deuteron breakup reaction by an α-particle. The calculation has already been done by the present author[3]. The theory is applied to the $^2H(\alpha,\alpha p)n$ reaction at incident α-particle energies of 15-42 MeV with fairly good results. In this paper we discuss some interesting features of this reaction, which suggest, we hope, further research on this and related subjects.

First we discuss the 1^+ resonance of ^6Li. This resonance exists above the deuteron breakup threshold, and hence this is a *three-body* resonance. We have pointed out in ref.[2] that *no* peaks exist in the total breakup cross section at the energy corresponding to this resonance, although the contribution of this resonance to the breakup process is very large. The resonance affects the total breakup cross section even at higher energies.

The breakup amplitude corresponding to this resonance shows an interesting behavior[3]. The real part of this amplitude is very small compared with the first Born term which has only a real part. Instead the imaginary part is extremely large. This interesting behavior can be found at higher energies, which verifies the fact that no peaks exist at the resonance energy. It must be also noted that this is one example of the resonance amplitude which is discussed by Amado[4]. All these facts suggest an interesting structure. This resonance have *not* a d-α custer structure but an α-p-n three-body structure, which is different from the ground state of ^6Li.

All particles are distinguishable in this three-body reaction. Several kinds of amplitude corresponding to different reaction process in the entire breakup amplitude. We can discuss the three-body reaction mechanism from the role of each amplitude. One interesting example is shown in ref.[3]. The n-p FSI strongly reduces the QFS peaks when the relative energy of two-nucleons is small. It is also pointed out that the range of this n-p FSI is longer than that of the n-α FSI.

Further complicated reaction mechanisms can also be discussed from the discrepancy between the three-body calculation and the experiment. Fig. 1 shows one example. The three-body calculation reproduces the recent experiment which has been done by Sagara et al[5] very well. However, it does *not* reproduce the small peak under the kinematical condition of the zero final energy between two nucleons.

Assuming the isospin conservation, we have only included the triplet s-wave in the n-p interaction. Although the n-p FSI with triplet s-wave plays an important role, its contribution to the differential cross section is rather smooth. We can conclude that this peak is due to the n-p FSI with the singulet s-wave corresponding to the virtual state of the deuteron. That is, a violence of the isospin conservation is found in this reaction, which may due to the Coulomb interaction which we have neglected in solving the three-body equations.

It is expected that informations about n-p and n-α interactions are derived from this reaction. Since the 1^+ resonance of ^{6}Li strongly affects this reaction, the differential cross sections are expected to be sensitive to two-body potentials in the three-body calculation. Polarizations may be also useful for deriving the informations about these interactions. The calculation of polarizations and calculations with different kind of two-body potentials are now in progress.

The author would like to thank Dr. K. Sagara for stimulating discussions.

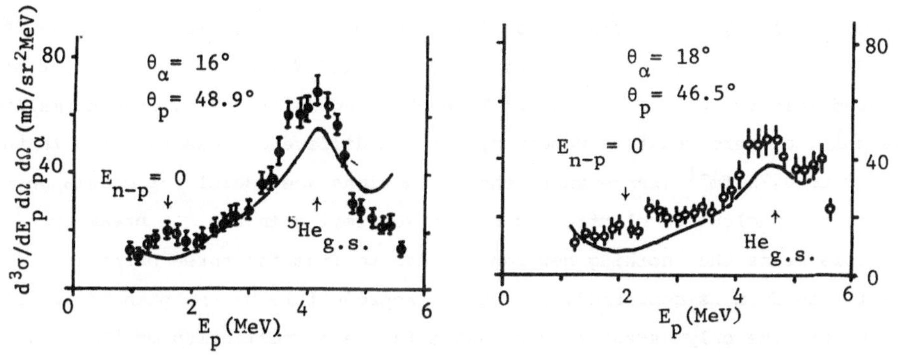

Fig. 1 Differential cross section of ^{2}H(α,αp)n reaction at 18 MeV (from ref.[5])

References

1) B. Charnomordic, C. Fayard and G. H. Lamot, Phys. Rev. C15 (1977) 864

2) Y. Koike, Prog. Theor. Phys. 59 (1978) 87

3) Y. Koike, to be published in Nuclear Physics

4) R. D. Amado, Modern three-hadoron physics, ed. A. W. Thomas (Springer-Verlag, Berlin, 1977) p. 85

5) K. Sagara et al., to be published in Nuclear Physics

THREE MODEL APPROACHES TO THE α-d RELATIVE WAVE FUNCTION IN THE ^{6}Li NUCLEUS*

F. Foroughi
Institut de Physique de l'Université, Rue A.-L. Breguet 1,
CH - 2000 Neuchâtel, Switzerland

The α-d relative wave function in the ^{6}Li nucleus has been calculated by using three different models to describe the ^{6}Li, namely phenomenological potential, phonomenological cluster wave functions and Faddeev equations. For the potential approach, we have used those given in the reference [1]. In the phenomenological case, we have used various wave functions which had the correct asymptotic behaviour (Whittaker function $\frac{1}{r} W_{-\eta,\frac{1}{2}}(2\kappa r) Y_0^{\,0}$, where η is the Coulomb parameter and $\hbar^2\kappa^2/2\mu$ is the absolute value of the α and d binding energy in the ^{6}Li). The various parameters were adjusted by fitting the electric form factor of the ^{6}Li. In the approximation of this nuclei, as a three particles system (α - n - p), we have used the Faddeev equations which were adapted for this case by Charnomordic and al [3]. Form factors of Mongan [4] and CFL [3] were used to describe respectively the N-N S wave and N-α $p^{3/2}$ wave interaction. The α-d squared relative wave functions we have obtained for the ^{6}Li, when normalized to the same point at zero relative momentum, shows a difference less than 10% in their form between 0 to 0,75 fm^{-1} (range where these functions are useful for models describing either the nucleus or electron quasifree scattering in the ^{6}Li break-up reactions). This means that nothing new can be learned from the three particles approach as far as form is concerned, this being reproducible by the phenomenological models. Therefore, the only essential observable is the normalization of the relative wave function. This can be done in two ways :

(i) by the normalization to the α-d percentage in the ^{6}Li,

(ii) by the normalization to the asymptotic behaviour which contains information on the residue of the α-d bound state in ^{6}Li [5].

We have chosen the later, because this residue is accurately determined [5] and the choice of the correct asymptotic behaviour makes this normalization reliable. It was then possible to calculate the corresponding α-d percentage in ^{6}Li. For the best phenomenological functions we have found 30%, the others giving much less.

[1] J.W. Watson, H.G. Pugh, P.G. Roos, D.A. Goldberg, R.A. Riddle and D.I. Bonbright, Nucl. Phys. A172, (1971), 513.
[2] L.R. Suelzle, M.R.Y. Yearian and Hall Crannell, Phys. Rev. 162, (1967), 992.
[3] B. Charnomordic, C. Fayard and G.H. Lamot, Phys. Rev. C15, (1976), 864.
[4] T.R. Mongan, Phys. Rev. 178, (1969), 1597.
[5] G.R. Plattner, M. Bornand and K. Alder, Phys. Lett. 61B, (1976), 21.

*Supported in part by the Swiss National Science Foundation.

α+n+P→α+n+P ELASTIC SCATTERING AT VERY LOW ENERGIES

K.K. Fang
Linear Accelerator Laboratory
University of Saskatchewan
Saskatoon, Canada S7N OWO

In our earlier calculation [1], we applied the hyperspherical method to the study of (α,n,P) three-body model of ^{6}Li. The results found there indicate that the one-term expansion, the grand orbital zero term, of the wavefunctions can provide good account of bound states of the system. We then extend the calculation to (α,n,P) continuum states and study α+n+P→α+n+P, the three-body to three-body elastic scattering [2], in pattern on our previous work for the three-nucleon elastic scattering [3].

In ref. 2 , we have investigated the three-body scattering in the energy region 0.02 mev $\leq$ E $\leq$ 450.0 MeV where E is the total kinetic energy of three free incident particles in the center of mass frame. In this work, we give numerical results of three-body to three-body phase shift δ and total cross section of α+n+P→α+n+P elastic scattering at very low energies 0.2 kev $\leq$ E $\leq$ 5.0 kev, and only the scattering channel, $J^\pi = 0^+$ (total angular momentum zero, positive parity) and T = 1 (isospin one), will be considered. We give below the equation of motion in the variable r, the hyperradius, for the scattering channel $J^\pi = 0^+$ and T = 1, (for derivation see ref.[1,2])

$$(\frac{-1}{2})[\frac{d^2}{dr^2} + \frac{5}{r}\frac{d}{dr}]\Psi_o^{(1)}(r) + \frac{(4\pi)}{\beta^3}[\frac{1}{\pi\sqrt{\pi}}] F_o'(r)\Psi_o^{(1)}(r)$$

$$+ \frac{(4\pi)}{\alpha^3}[\frac{-2}{\pi\sqrt{\pi}}] F_o(r)\Psi_o^{(1)}(r) = E\Psi_o^{(1)}(r) \quad . \tag{1}$$

The phase shift δ can be found from the solution $\Psi_o^{(1)}$ of eq.1 in the asymptotic region of r. The total cross section for three free particles scattering can be written in terms δ (see ref.[2,3]) and is

$$\sigma_T = [\frac{128\pi^2}{K^5}] \sin^2\delta . \tag{2}$$

For numerical computation, we use the potential set C defined in ref. 1 for the (α,n,P) system. Phase shift and total cross section are given in Table A. A log-log plot of δ vs k is made in Figure A. It can be seen that the phase shift δ is linearly dependent on k and this shows that the total cross section σ_T of eq.2 goes to infinity as E goes to zero. In fact, σ_T has the following form when E approaches zero

$$\sigma_T \sim \frac{1}{K^3} = \frac{1}{E^{3/2}} . \tag{3}$$

Finally, we want to point out that in their paper [4], by assuming the limiting behavior of the two-body amplitudes, the absolute square of the three-particle scattering amplitude is proportional to $(E)^{-2}$ in contrast to our result $(E)^{-3/2}$, eq.3. For an application of low energy three-body scattering in statistical mechanics, see Adhikari et al.[5].

Table A

$k\,(\text{mev})^{1/2}$	$E\,(\text{MeV})$	δ (radians)	$\sigma_T\,(\text{MeV})^{-5/2}$
0.02	$0.2\text{x}10^{-3}$	2π+0.04	$6.35\text{x}10^{8}$
0.04	$0.8\text{x}10^{-3}$	2π+0.07	$6.61\text{x}10^{7}$
0.06	$0.18\text{x}10^{-2}$	2π+0.12	$2.27\text{x}10^{7}$
0.08	$0.32\text{x}10^{-2}$	2π+0.17	$1.07\text{x}10^{7}$
0.10	$0.50\text{x}10^{-2}$	2π+0.21	$5.28\text{x}10^{6}$

k is defined by k^2=2E. σ_T is given by eq.2. Note that $\delta = 2\pi$ when k = 0.0 (see ref.[3]).

Figure A

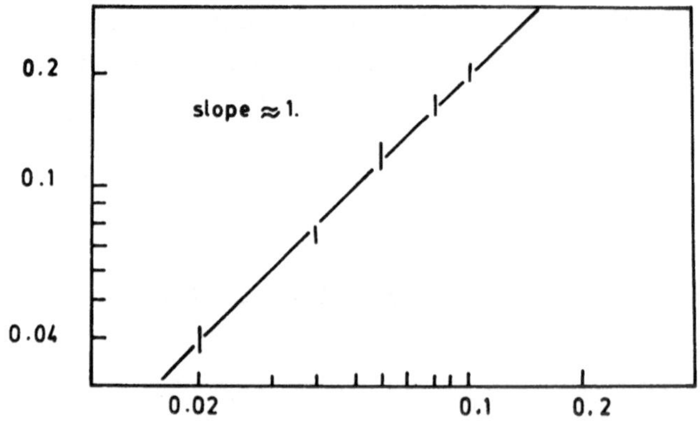

The ordinate is $\log(\delta-2\pi)$ and the abscissa is $\log(k)$.

References

[1] K.K. Fang, Phys. Rev. C16, 2117 (1977).
[2] K.K. Fang, Phys. Rev. C15, 1204 (1977).
[3] K.K. Fang, Phys. Rev. C (submitted).
[4] R.D. Amado et al., Phys.Rev.Lett. 25, 194 (1970).
[5] S.K. Adhikari et al., Phys.Rev. Lett. 27, 485 (1971).

$^3He + {}^6Li$ Elastic Scattering at 27 MeV

R.H. Bassel and M.I. Haftel
Naval Research Laboratory
Washington, D.C. 20375

In this paper we examine the elastic scattering of ^3He from ^6Li. We calculate cross-sections by two different methods: First, we calculate the cross-sections in a three-body model[1] where trinucleons are considered "elementary" and ^6Li is a ^3H-^3He bound state. Cross-sections here are calculated with the three-body Haftel-Ebenhoh code which employs the Amado model. We compare these cross-sections with experiment[2] along with those calculated from an optical model[3]. We do not consider α-d or α-n-p clustering in ^6Li in our treatment.

Heretofore three-body models have worked well in describing α+d scattering (below 30 MeV)[4], ^6Li and ^6He bound state properties[5], and the $^3He + {}^6Li \rightarrow {}^3He + {}^3He + {}^3H$ reaction[1] at 45 MeV. Success of a three-body model in the α+n+p system could be expected, but it is surprising that it worked so well in the ^6Li breakup case. The calculations of this paper provide a further test of the three-body cluster approach for the $^3He + {}^6Li$ system.

The three-body cluster model we employ is that of reference 1 and employs real separable S-wave trinucleon-trinucleon interactions. This is admittedly an oversimplication. However, the S-wave assumption should be more valid at 27 MeV than at 45 MeV. The optical model treats the system and forces as essentially two-body, explicitly ignoring anti-symmetrization although the model to some extent allows for this.

Figure 1 illustrates both the theoretical and experimental elastic scattering cross-sections. The three-body results include curves both where the Coulomb force is totally neglected and approximately included (by adding the pure Coulomb amplitude) whereas, of course, the optical model contains the Coulomb potential. Of particular note is that the optical model amplitudes are significant for $L \lesssim 8$ while the three-body model has significant contributions only for $L \lesssim 2$. This difference undoubtedly accounts for the marked structure of the optical model fit and the lack of such in the three-body cluster model. The ^6Li wave function calculated in our cluster model has a too small r.m.s. radius, which may account for some of the discrepancy. Also inclusion of α-d clustering would introduce more partial waves (due to the small binding energy) and perhaps introduce more structure.

The experimental data[2] extends only to $\Theta_{cm} \approx 115°$. The prominent feature of the $^3He-^3H$ cluster model is the backward peaking due to heavy particle stripping. This peaking does not occur in the optical model nor would it be expected in an α-d cluster model of ^6Li. The need for more extensive back angle measurements is obvious to assess the importance of $^3H-^3He$ clustering in ^6Li. The above results indicate a deficiency in the $^3H+^3He$ cluster model in ^6Li. In a separate contribution[6] we assess the cluster model in the inelastic channels.

We thank M.A. Melkanoff, J.A. Raynal and T. Sawada for the use of their optical model code SEEK.

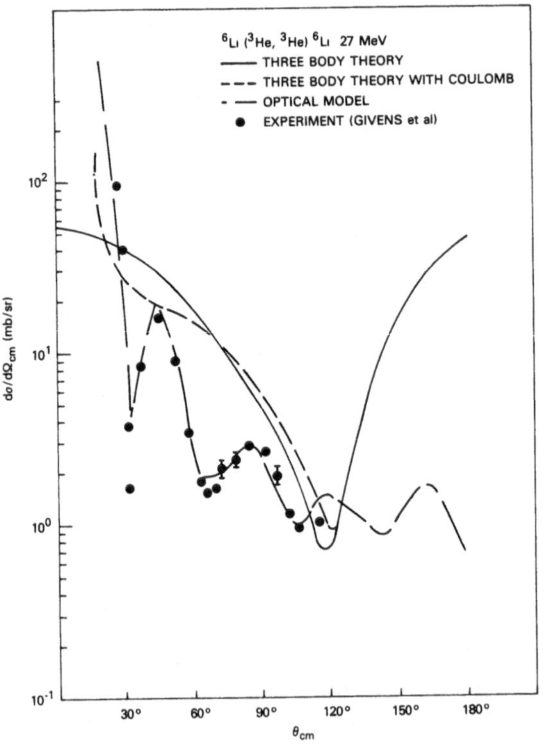

Fig. 1
Theoretical and experimental elastic scattering cross-sections at 27 MeV. The experimental data is from ref. 2.

References

[1]M.I. Haftel et al., Phys. Rev. C16, 42 (1977)
[2]R.W. Givens, M.K. Brussel and A.I. Yavin, Nucl. Phys. A187, 490 (1972)
[3]E.F. Gibson et al., Phys. Rev. 155, 1194 (1967)
[4]B. Charnmordic, C. Fayard and G.H. Lamot, Phys. Rev. C15, 864 (1977)
[5]A. Ghovanlou and D.R. Lehman, Phys. Rev. C9, 1730 (1974); D.R. Lehman, Matma Rai and A. Ghovanlou, Phys. Rev. C17, 744 (1978)
[6]R.H. Bassel and M.I. Haftel, contribution to this Conference.

The ^{6}Li(^{3}He,^{3}He)^{6}Li* and ^{6}Li(^{3}He,^{3}H)^{6}Be Reactions at 27 MeV

R.H. Bassel and M.I. Haftel
Naval Research Laboratory
Washington, D.C. 20375

This paper continues the investigation of the ^{3}He+^{6}Li system at 27 MeV begun in a previous contribution[1]. Here we compare the cross-sections of the ^{6}Li(^{3}He,^{3}He)^{6}Li* (3.56 MeV J=0,T=1 level) and the ^{6}Li(^{3}He,^{3}H)^{6}Be reactions as calculated by the ^{3}He-^{3}H cluster model[2] and by the generalized distorted-wave method (DWM) of Satchler and Love[3] with the experimental data of Givens et al.[4] One goal is to assess the three-trinucleon model of these reactions and to guide future few-body treatments of such reactions.

Figures 1 and 2 show the comparison of the three-body cluster model and the DWM to the data for transitions to the 3.56 MeV state in ^{6}Li and the ground state of ^{6}Be. The DWM is distinctly better in both cases, but requires a strength of only 60% of the strength suggested by Love and Satchler[5]. This discrepancy may occur because of the fact that there are only a few active nucleons. Alternatively, this strength could be improved by allowing ^{6}Li* to be a more-extended structure than the ground state or by allowing the optical model to be energy-dependent.

From isospin considerations the ratio $R=\sigma(^6\text{Be})/\sigma(^6\text{Li}^*)$ should be 2.0 if all forces were charge-independent. The experimental ratio is $R \approx 1.6$ while our three-body model, since it is charge-independent, predicts R=2.0. Using the same optical parameters for both ^{3}He + ^{6}Li as for ^{3}H + ^{6}Be, as in Figs. 1 and 2, gives $R \approx 2.28$. By assuming the ^{3}H + ^{6}Be potential to be isospin-dependent we get $R \approx 2.04$.[6] On the other hand, assuming that ^{6}Be is bigger (while it lives) than ^{6}Li, but with the optical potential isospin-independent, the DWM calculations then give $R \approx 1.54$.

As with elastic scattering[1], the three-body-cluster model cross-sections have the characteristic backward angle peaking, which is not predicted by DWM. One difference here from the elastic-scattering case is that one cannot appeal directly to an α-d cluster model to resolve the three-body theory discrepancy with experiment since a pure α-d cluster model would give zero cross-section. As in our previous calculation[1], our cluster model potentials do give too small r.m.s. radii in ^{6}Li* and ^{6}Be, so this may be one possible source of error. Again measurements at backward angles are crucial in assessing the importance of ^{3}He + ^{3}H clustering in ^{6}Li.

Finally our calculations indicate the need for improved few-body approaches to reactions involving light nuclei. Work has already begun in this direction by Redish and co-workers[7]. The reactions we have considered provide an excellent test for such theories. Work along these lines will help relate scattering and reaction results to the underlying inter-particle forces.

We thank P.D. Kunz for use of his code DWUCK.

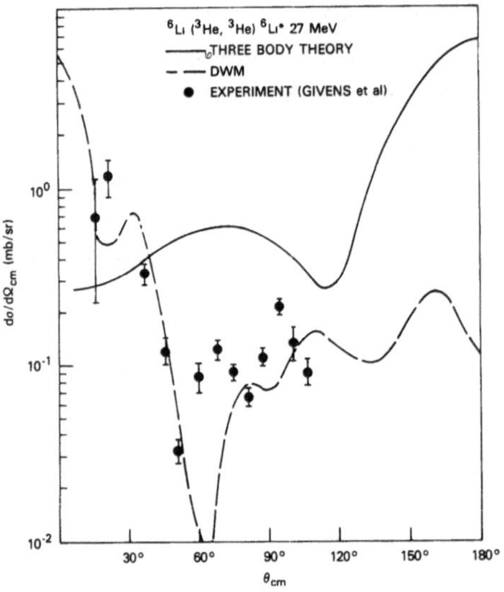

Fig. 1
Theoretical and experimental inelastic scattering cross-sections for 27 MeV incident ^{3}He. The experimental data is from ref. 4

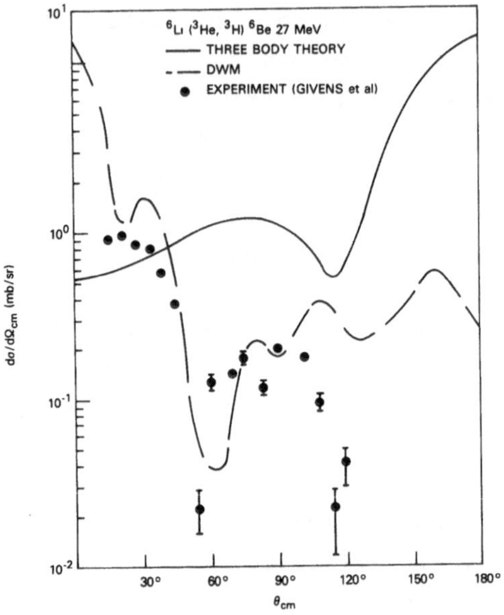

Fig. 2. Theoretical and experimental charge exchange cross-sections for 27 MeV incident ^{3}He. The experimental data is from ref. 4.

References

[1] R.H. Bassel and M.I. Haftel, contribution to this Conference.
[2] M.I. Haftel et al., Phys. Rev. C16, 42 (1977).
[3] G.R. Satchler and W.G. Love, Nucl. Phys. A172, 449 (1971).
[4] R.W. Givens, M.K. Brussel and A.I. Yavin, Nucl. Phys. A187, 490 (1972).
[5] W.G. Love and G.R. Satchler, Nucl. Phys. A159, 1 (1970).
[6] See, for example, p. 433 of G.R. Satchler's article in Isospin in Nuclear Physics ed. by D.H. Wilkinson (North Holland, Amsterdam, 1969) p.388.
[7] Gy. Bencze, Nucl. Phys. A210, 568 (1973); E.F. Redish, Nucl. Phys. A225, 16 (1974); W.N. Polyzou and E.F. Redish, University of Maryland Technical Report 78-088 (1978) (unpublished).

THREE TRION BREAKUP OF $^6Li + ^3He$ AT 132 MeV

E. Karaoglan,[a] L.T. Myers,[a] J.M. Lambert,[a] P.A. Treado,[a] M.I. Haftel,[b]
I. Slaus,[a,c] P.G. Roos,[d] A. Nadason,[d] T.A. Carey,[d] and N.S. Chant[d]

a. Georgetown University, Washington, D.C. 20057 U.S.A.; b. Naval Research Laboratory, Washington, D.C. 20375 U.S.A.; c. Rudjer Boskovic Institute, 41001 Zagreb, Yugoslavia; d. University of Maryland, College Park, MD 20742 U.S.A.

A simple model which treats the h-h and h-t quasifree scattering (QFS) in the ^3He + ^6Li interaction as a 3-body problem proved to be better than the plane wave impulse approximation (PWIA). The S-wave interaction 3-body model (3BM) predicted the shape and the absolute cross section for the h-h and h-t QFS with E_i = 45 MeV,[1)] while the PWIA did not reproduce the shape of the spectra, predicted the cross section larger by a factor of 15-30 and did not predict the ratio of the h-h vs. h-t QFS. We realize that the simple 3BM is inadequate (trion-trion interactions cannot be represented by S-wave forces, inelastic channels are always open for the trion-trion system and ^6Li is not predominantly a h + t structure) but we suggest that it is worthwhile to investigate: a) the trion-trion QFS at higher energies to determine whether the PWIA will provide a reasonable fit to the data and whether the S-wave 3BM will be less applicable and b) the ^3He + ^6Li breakup in the region of phase space where short range and effective 3-body forces play a significant role. Since the trions are bound by 15.8 MeV in ^6Li, features due to the effective 3-body force should be more pronounced than in weakly bound systems.

We have studied the reaction $^6Li(h,hh)t$ with E_i = 132 MeV and at the QFS angles 30° - 50.5^o, see Figure 1. The QFS cross section is on tenth that at E_i = 45 MeV. The h-h 2-body cross sections for corresponding CM energies also decrease by 10, yet, the sharp decrease in the QFS cross section is unexpected because the PWIA should be more applicable at higher energies. However, for other QFS reactions similar decreases vs. energy have been observed in the energy range E_{cm} = 20 to 40 MeV.[2)] The PWIA fits the shape of the QFS enhancement in the spectrum quite well, see Figure 1, but it requires a N = 0.006; compared to N = 0.025 at 45 MeV. The PWIA requires h-h elastic scattering data of E_{cm} = 45-50 MeV, while data exists only at 22,[3)] 60,[4)] 70,[5)] and 99[6)] MeV CM energy. The available 2-body data[3-6)] vary by three orders of magnitude, see Figure 2. By omitting the 70 MeV data from the PWIA analysis we obtain a N = 0.004. Also, we did not use the 60 MeV cross section data, because the numerical data are not available to us, but these data would yield a PWIA QFS cross section about 3 times smaller, i.e., N = 0.02. The variation in the h-h cross section does not affect the shape of the QFS enhancement; while the momentum transfer varies from 0.7 to 0.02 fm^{-1} for E_1 = 60 to 81.5 MeV, the h-h relative energy and angle remain about 49.5 MeV and 66°, respectively. The 2-body cross section variation yields a 1 MeV shift between the Fourier transform and the PWIA QFS peak positions.

The 3BM calculation is performed with the HS+C S-wave interaction,[1] which predicts the h-h, 90°, cross section as too small at low energies and too large at high CM energies. However, at $\theta=70^o$, $E_{cm}=50$ MeV, it predicts 2.43 ub/sr vs. an interpolated experimental value of 1.3 ub/sr. From the p-d breakup data, we expect the QFS to be very sensitive to the 2-body cross section; thus, we chose $\theta_3=30^o$, $\theta_4=50.5^o$ with $E_i=132$ MeV giving $E_{cm}=50$ MeV and $\theta_{cm}=66^o$. The large discrepancy between the 3BM QFS cross section (solid curve, Fig. 1, N=0.07) and the experimental data suggests that for tightly bound systems the QFS cross section (magnitude) has more complex dependences than for deuteron breakup. At collinearity conditions,[7] $\theta_3=\theta_4=57^o$ and $E_{cm}=50$ MeV, the 3BM vs. PWIA is 2.2 ub/sr^{2}MeV to 0.1 ub/sr^{2}MeV.

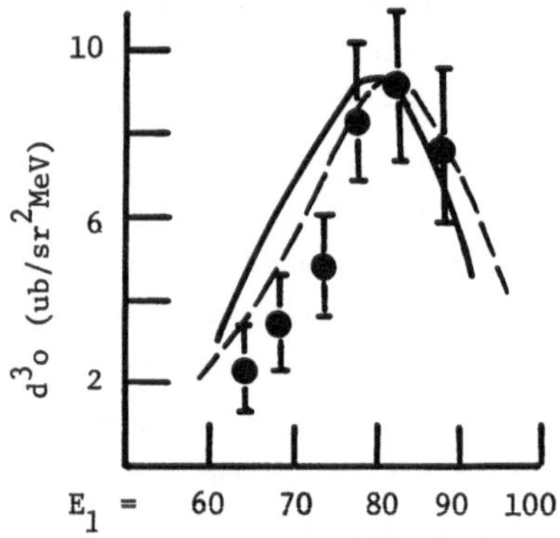

Fig. 1: Data and predictions for E_i = 132 MeV, 30° - 50.5^o. Solid curve from 3BM and dashed from PWIA.

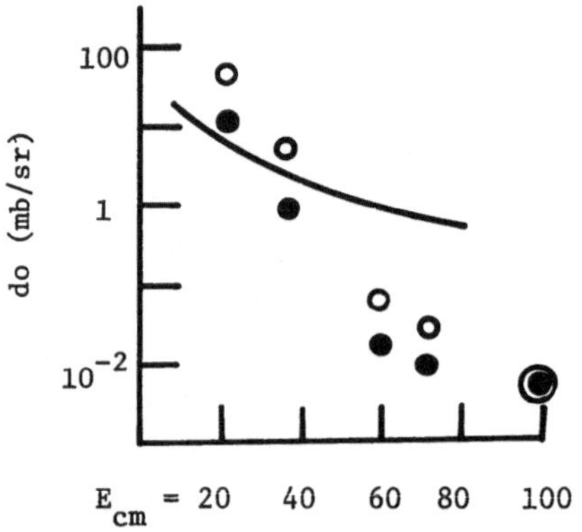

Fig. 2: h-h elastic scattering differential cross sections: 70^o(open), 90° (dots). Solid curve is the HS predictions; the difference of 70° and 90° is not visible.

1. M. Haftel, et al., Phys. Rev. C16 (1977) 42.
2. I. Slaus, et al., Nucl. Phys. A286 (1977) 67.
3. D. Thompson, et al., Nucl. Phys. A201 (1973) 301, and references therein.
4. S. Tanaka, et al., Procs. Inter. Conf. on Nuclear Structure, Tokyo (1977) 21.
5. T. Frisbee, et al., private communication.
6. P. Roos, et al., private communication.
7. J. Lambert, et al., Phys. Rev. C13 (1976) 43.

This work has been partially supported by PL480 grant no. F6F005 and SRH-SIZ-I grant 2.5.10 (I.S.) and the National Science Foundation (E.K., J.M.L., P.A.T., P.G.R, A.N., T.A.C. and N.S.C.).

THREE-ALPHA MODEL CALCULATIONS AND THE 0_2^+ STATE OF ^{12}C

O. PORTILHO† AND S. A. COON
Department of Physics, University of Arizona
Tucson, Arizona U.S.A. 85721

We consider ^{12}C as the bound state of three finite structureless alpha particles interacting with the ℓ-dependent pairwise α-α potential of Ali and Bodmer (AB) designated as (d_0', d_2, d_4).[1)] We perform a variational calculation in coordinate space with a trial wavefunction expanded in terms of harmonic oscillator (HO) states which are translationally invariant, completely symmetric, and have a well defined total angular momentum.[2)] The coefficients of the expansion, as well as the oscillator frequency, are treated as variational parameters.

The results of the calculation are in reasonable agreement with experimental features of ^{12}C except the energy of the ground state is much less than the -7.27 MeV energy of ^{12}C relative to breakup into 3α's. The variational energy of the ground state 0_1^+ is only -1.96 MeV in a basis of 31 HO states. The calculated (experimental) charge radius of the 0_1^+ state is 2.82 fm (2.46 fm) and the spatial arrangement of the α's is an equilateral triangle of side 3.8 fm. The calculated (experimental) gap between the second 0^+ state 0_2^+ and the ground state 0_1^+ is 5.15 MeV (7.655 MeV) and the reduced transition width $0_2^+ \rightarrow 0_1^+$ is 96.9 e^2 fm^4 (27.5 ± 1.8 e^2 fm^4).

We show in Figure 1 the inelastic charge form factor (labeled AB) from the wavefunctions with a basis of 16 HO states (increasing the basis set to 31 increases the height of the first maximum by only 30%). The first peak appears shifted by 0.2 fm^{-1} to the left of the data and the second peak is higher than and displaced about 0.4 fm^{-1} to the left of the experimental second maximum.[3)] A simple effective 3α force added to the AB force (labeled AB + 3B) can be paramatrized to fit the breakup energy of ^{12}C and the gaps of low-lying excited 0^+ and 2^+ states.[4)] It has little effect on the inelastic form factor.

The wave function of the 0_2^+ state depends upon $|\underset{\sim}{r}|$, $|\underset{\sim}{\rho}|$, and θ, the angle between $\underset{\sim}{r}$ connecting two of the particles and $\underset{\sim}{\rho}$ connecting their center of mass with the third particle. Figure 2 is a polar plot (the z axis is along $\underset{\sim}{r}$) in $\underset{\sim}{\rho}$ and $\underset{\sim}{\theta}$ of equiprobability curves of the third particle's position. The cross shows its most probable position and the arrows the expectation value of $\underset{\sim}{r}$. The most probable

---

†CNPq (Brazil) Fellow. Present Address: Departmento de Fisica, Universidade de Brasilia, Brasilia-DF-Brazil.

configuration of the 3α's is shown in the inset but we note that 50% of the maximum probability is at a linear chain deformation of ∿40°, and there is even a small probability for a triangular configuration. The large α-decay width of this state cannot be explained by a rigid linear chain, but seems to demand that the α's move freely in a weak coupling scheme,[5] as allowed by this wavefunction.

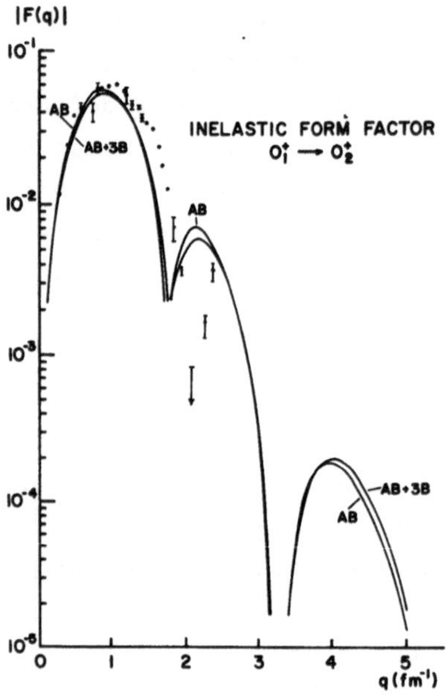

Figure 1

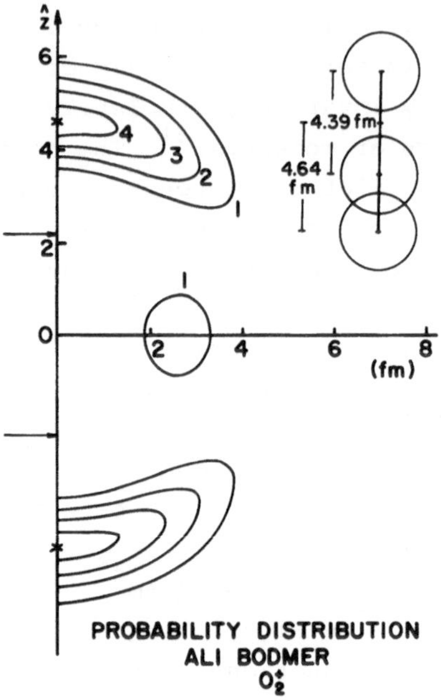

Figure 2

1. S. Ali and A. R. Bodmer, Nucl. Phys. 80 (1966) 99.
2. V. C. Aguilera-Navarro and O. Portilho, Ann. Physics 107 (1977) 127.
3. H. Crannel, et al., BAPS 23 (1978) 583, and private communication.
4. O. Portilho and S. A. Coon, University of Arizona preprint.
5. H. Horiuchi, et al., Prog. Theo. Phys. Suppl. 52 (1972) 89.

THREE-BODY CALCULATIONS FOR LIGHT NUCLEI USING SEPARABLE POTENTIALS

A. Osman
International Centre for Theoretical Physics
Trieste, Italy

and

Physics Department, Faculty of Science,
Cairo University, Egypt

The three-body problem has been proved to be one of the most interesting in the study of nuclear static properties of nuclei. Faddeev [1] introduced successfully an exact solution for the three-body problem. The Faddeev formalism is a well-behaved set of three-body equations which involves the two-body T matrix rather than the potential. This approach leads to a well-behaved set of three-body integral equations, and thus the Faddeev equations remain a well-defined system whatever the potential form is.

Separable potentials are found to be very useful in studying the three-nucleon system because it entails great simplicity in the analysis of the three-body problem. This simplicity appears clearly since the Faddeev equations are reduced to a set of coupled integral equations in one continuous variable by using separable potentials.

Let the separable two-body potential to be represented as

$$V_{ij} = \lambda f(p)\; f(q) \tag{1}$$

where the potential V_{ij} should be strong enough to give a bound state. The off-shell T matrix can easily be written in a closed form as

$$T_{ij}(p,q;z) = \frac{f(p)f(q)}{D(z)} \tag{2}$$

where

$$D(z) = \frac{1}{\lambda} + 4\pi \int_0^\infty dk\, \frac{k^2 |f(k)|^2}{Z-k^2}\ . \tag{3}$$

The T matrices will be expressed by T_{ij} with i and j taken as 1,2 and 3 in cyclic permutation. Then, we have

$$<\vec{p}_k,\ \vec{q}_k|T_{ij}(Z)|\vec{p}_k',\ \vec{q}_k'> = \delta(\vec{q}_k-\vec{q}_k')<\vec{p}_k|T_{ij}(Z-q_k^2)|\vec{p}_k'>\ . \tag{4}$$

Equation (4) shows that if $T_{ij}(Z)$ has a bound state pole at $Z = -\varepsilon_k$, then $T_{ij}(Z)$ will have a branch point there, with a cut going from $-\varepsilon_k$ to $+\infty$, being the right-hand cut.

In the present work, we used the Faddeev formalism in calculating the binding energies of the nuclei ^3H, ^3He, ^6Li, ^9Be and ^{12}C. The structure of these nuclei is described using the alpha cluster model. The three-body binding energies are obtained by a numerical solution of the resulting three-body integral equations. The two-body interactions are represented as a short-range repulsive potential surrounded by a long-range attractive potential. The used potentials are separable with potential functions of the Yamaguchi, Gaussian, Tabakin, Mongan and also of the Reid forms. The different values of the parameters of the different two-body interactions are taken to fit the corresponding phase shifts [2]. The three-body ground-state energies of the above-mentioned nuclei, with the above-mentioned potentials, are calculated with very high accuracy, using only a very limited number of summation terms. The well-behaved Schmidt-Hilbert theory of integral equations is applied, where a 36-point Gaussian integration is used in the present numerical calculations of the three-body integral equations. The eigenvalues are given as a function of the energy Z. The values of the energy Z for which a matrix eigenvalue takes the value one, are the three-body bound-state energies. In these calculations, we get the nuclear ground-state energies. Coulomb energies resulting from Coulomb repulsion should be added to the calculated nuclear energies. The Coulomb forces could be treated accurately by treating the pure Coulomb T matrix in the integral equation. Explicit treatment of the pure Coulomb T matrix is found [3] to improve more accurate values for the binding energies. The actual three-body ground-state energies are obtained by adding the calculated Coulomb energies to the calculated nuclear energies. The results obtained for the three-body binding energies for the nuclei ^3H, ^3He, ^6Li, ^9Be and ^{12}C, using the different two-body interactions are listed in Table I. The experimental values are introduced in Table I for the purpose of comparison with the theoretically calculated values. From Table I we see that our theoretically calculated values for the ground-state energies for the ^3H, ^3He, ^6Li, ^9Be and ^{12}C nuclei are quite reasonable and in good agreement with the experimentally observed values. Comparing the present results of the binding energies using separable potentials with values obtained by other methods, our method introduces accuracy for binding energies in the order of 0.8 - 1.2%. This ensures the accuracy of using a separable approximation in the three-body Faddeev equations.

Acknowledgements

I am very grateful to Professor Abdus Salam, Professor Paolo Budini, the International Atomic Energy Agency and UNESCO for hospitality at the International Centre for Theoretical Physics, Trieste, Italy. I am deeply indebted to Professor Luciano Fonda for his kind hospitality at the ICTP, Trieste, Italy, where most of this work is done. Thanks are also due to the Centro di Calcolo dell' Università di Trieste, Italy for the use of the facilities.

References

[1] L.D. Faddeev, Sov. Phys. JETP 12, 1014 (1961).
[2] A. Osman, Phys. Rev. C17, 341 (1978).
[3] A. Osman, Phys. Rev. C4, 302 (1971).

Table I Calculated binding energies in (MeV)

Nucleus	Yamaguchi	Gaussian	Tabakin	Mongan	Reid	Experimental
^{3}H	8.615	8.189	8.563	8.446	8.512	8.48
^{3}He	8.194	7.618	8.089	7.816	7.972	7.72
^{6}Li	5.614	4.989	4.824	4.587	4.683	4.53
^{9}Be	1.798	1.419	1.674	1.583	1.625	1.57
^{12}C	7.416	6.329	7.388	7.299	7.316	7.28

A MODEL THREE-PARTICLE SYSTEM FOR STUDYING LIMITATIONS OF THE SHELL-MODEL APPROACH TO NUCLEAR REACTIONS

W. Plessas
Institut für Theoretische Physik der Universität Graz
Universitätsplatz 5, A-8010 Graz

The shell-model reaction theory (or continuum shell model) [1] suffers from two basic restrictions:

(i) the test-function space is truncated to configurations with at most two clusters one being an elementary particle

(ii) the test functions themselves are not translational invariant.

As a result predictions of the theory are impaired by spurious solutions of various kinds [2]. A number of methods have been proposed (cf.e.g.[3]) to cure the situation in either of the above mentioned respects. But in most cases it is not clear how successfully these methods can be applied in practical calculations of nuclear reactions. A model three-particle system similar to one described by Beregi et al.[4] can serve as a testing-ground: Suppose two light particles having equal masses $m_1 = m_2 = m_N$ (nucleon mass) and a heavy one of mass $M = (A-2)m_N$ representing the core; let them interact via mutual forces leading to all desired properties in the behaviour of the scattering system (compound resonances,"deuteron"-induced reactions, three-particle break-up, stripping and knock-out reactions,...). Thus the model bears much (enough) resemblance to a nuclear A-particle system. In particular all effects connected with limitations (i) and (ii) show up; in that respect there is no need e.g. for antisymmetrization. On the other hand the model can be solved exactly via Faddeev equations, say. In comparing with the results of the shell-model reaction theory we are therefore given a rigorous measure on its quality and the effectiveness of modifications made with regard to (i) and (ii).

We have used the model to study a new method for the suppression of spurious centre-of-mass resonances. They occur because of the lack of translational invariance of test functions from an individual-particle theory. Our method consists in an extension of the least-squares minimization principle for the solution of the A-particle Schrödinger equation [2]. According to the following expansion of the scattering wave function into shell-model wave functions of classes 1 and 2 (we adopt the notation of ref. [1])

$$\Psi_E^{c,(+)} = \sum_{i=1}^{M} d_E^c(i)\Phi_i + \sum_{c'=1}^{\Lambda} \int_{\varepsilon_{c'}}^{\infty} dE' \, d_E^c(E',c')\chi_{E'}^{c'} = \psi_1 + \psi_2 \qquad (1)$$

we impose the condition

$$<\delta_1|\delta_1> + <\delta_2|\delta_2> + <\gamma|\gamma> \;\rightarrow\; \text{min.} \qquad (2)$$

Here $|\delta_i>$ and $|\gamma>$ represent defect vectors in the truncated Hilbert space

$$|\delta_1> = P_1 \, (H-E)|\psi_1> + P_1(H-E)|\psi_2> \qquad (3.a)$$

$$|\delta_2> = P_2 \, (H-E)|\psi_1> + P_2(H-E)|\psi_2> \qquad (3.b)$$

$$|\gamma> \; = P_1 \, (T_{cm} - \Gamma)|\psi_1> + P_1 \, (T_{cm} - \Gamma)|\psi_2> \qquad (3.c)$$

where P_1 and P_2 are the respective projection operators on subspaces spanned by test functions ψ_1 and ψ_2; H is the Hamiltonian of the system, T_{cm} the operator of the centre-of-mass energy and Γ its eigenvalue in the entrance channel. Eqs. (2) and (3) mean that we require $\Psi_E^{c(+)}$ to be a solution of the Schrödinger equation containing only a (non-spurious) centre-of-mass motion pinned down to be close to the one in the entrance channel. Solutions with spurious centre-of-mass excitations are suppressed. By our model calculation we could demonstrate the method to work excellently [5]. The problem in connection with restriction (i) (cf. e.g. ref.[6]) is still under investigation.

References

[1] C.Mahaux and H.A. Weidenmüller, Shell-Model Approach to Nuclear Reactions, North-Holland Publishing Co., Amsterdam, 1969.

[2] E.W. Schmid, Nuovo Cim., 18A (1973) 771.

[3] W. Plessas, Invited Lecture to be published in the Proceedings of the Workshop on Few-Body Problems in Nuclear Physics, Trieste, 1978.

[4] P. Beregi, I.Lovas, and J. Revai, Ann. Phys., 61 (1970) 57.

[5] W.Plessas, to be published.

[6] D. Eppel and A. Lindner, Nucl. Phys., A 240 (1975) 437.

EFFECT OF HIGHER ORDERED TERMS IN THE CLUSTER EXPANSION OF CORRELATION FUNCTION ON THE STRUCTURE OF LIGHT NUCLEI

M.A.K. Lodhi
Department of Physics
Texas Tech University
Lubbock, Texas 79409/USA

The most common approach of invoking the short-range correlations (CRS) in the structure problem is to multiply the single particle density by some type of SRC function satisfying the requisite properties of N-N interaction. This is somewhat equivalent to the approach of keeping the first ordered terms only in the cluster expansion of the matrix element of some operator with respect to the single particle wave function modified by the SRC function. In this work we take the later approach and observe that the neglect of higher ordered terms is significantly important particularly for small values of the correlation parameter and also at small distances from the center of the system. We invoke the correlation function of the type

$$1 - j_o\,(kr_{ij})$$

in the harmonic oscillator type single particle wave functions and study the two-four-and six-body systems.

For the two-body system the scalar product, the potential energy for a Gausian potential,kinetic energy and total energy differ substantially if small values of k are used. For example, for k=1 fm^{-1} the scalar product is twice as large when higher ordered terms are included as compared to the value when the first ordered terms are mainted only.

In case of the four-body the uncorrelated density at small values of r is the largest. With first ordered term it is minimum with $k = 2.8 fm^{-1}$ at the corresponding values of r. As higher ordered terms are included in the Monte Carlo method the density oscillates between these two extremes. With odd ordered terms the density moves towards the lower extreme and a vice versa trend is observed when even ordered terms are included.

An interesting feature in the six-body system viz. ^6Li is observed. A second minimum in the charge form factor is predicted at $q^2 \approx 26 fm^{-2}$ when only the first ordered terms are kept in the electron scattering calculations (see Fig. 1).

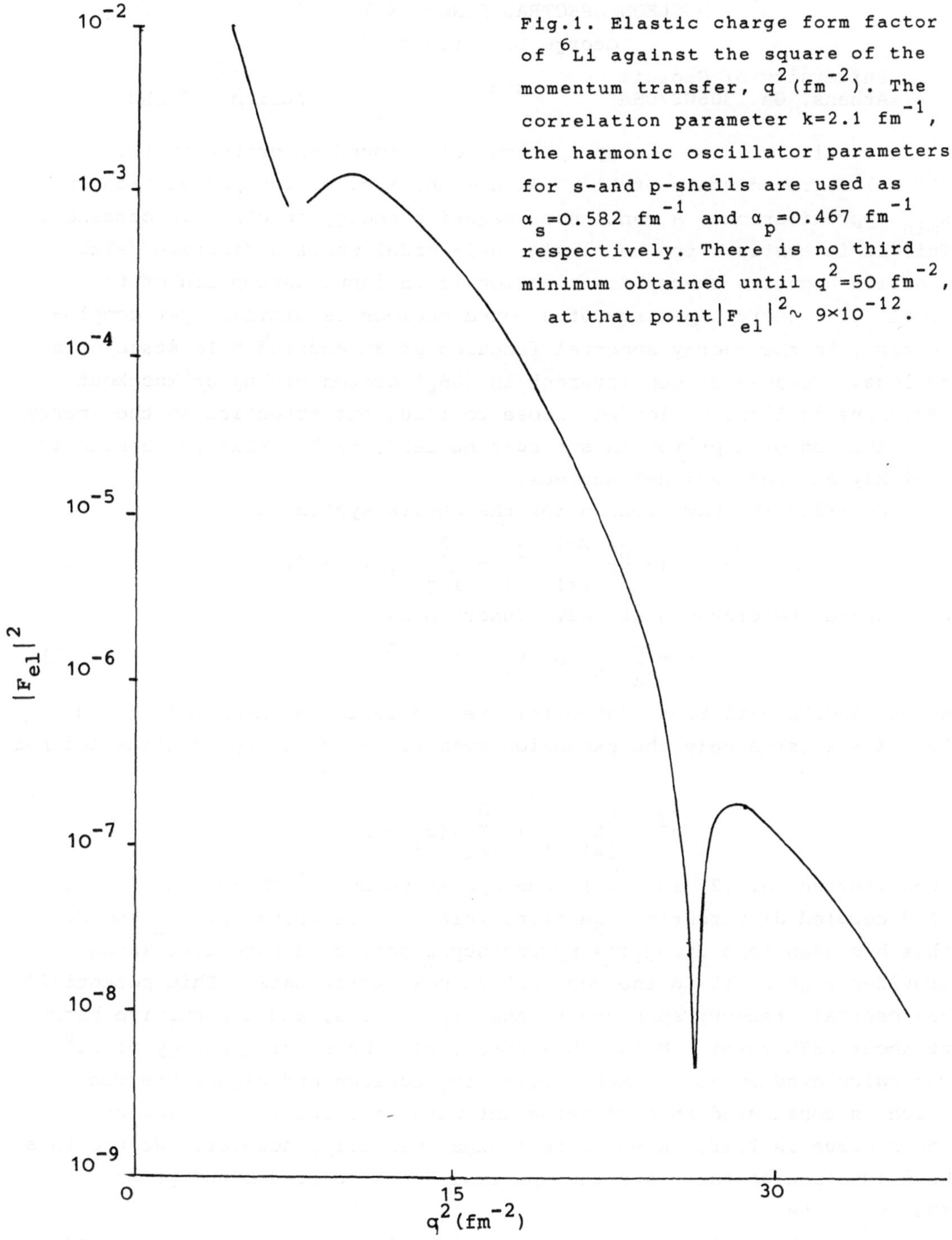

Fig.1. Elastic charge form factor of ^6Li against the square of the momentum transfer, q^2 (fm^{-2}). The correlation parameter $k=2.1\ fm^{-1}$, the harmonic oscillator parameters for s-and p-shells are used as $\alpha_s=0.582\ fm^{-1}$ and $\alpha_p=0.467\ fm^{-1}$ respectively. There is no third minimum obtained until $q^2=50\ fm^{-2}$, at that point $|F_{el}|^2 \sim 9\times10^{-12}$.

NUCLEON SPECTRAL FUNCTION IN Li^6

George L. Strobel

University of Georgia Athens, GA 30602/USA and IKP-KFA Julich 517 BRD

We calculate the energy spectrum of a bound $S_{1/2}$ proton in the $J^PT = 1^+0$ ground state of Li^6. We use the K harmonics formulation in K_{min} approximation. A continuous negative energy spectrum is obtained. This is in contrast to the simple shell model where a discrete delta function negative energy distribution is an input assumption of the model. The energy spectrum of a bound nucleon is similar, yet complementary, to the energy spectral function of an excited hole state of a nucleus. Because of our interest in $(0S_{1/2})$ proton pickup or knockout reactions in light nuclei we choose to focus our attention on the energy distribution of a proton in a target nucleus, rather than for a hole in a highly excited residual nucleus.

We write the hamiltonian for the entire system as

$$H\psi = [-\frac{\hbar^2}{2m}\sum_{i=1}^{A-1}\nabla^2_{\xi_i} + \sum_{i<j}^{A} V_{ij}]\psi = E\psi \tag{1}$$

and expand the ground state wave function as

$$\psi = \sum_a \chi_{ka}(\rho)\; Y_{ka}(\Omega)/\rho^{(3A-4)/2} \tag{2}$$

We use Jacobi variables that eliminate the center of mass motion and keep K = 2 terms only the expansion over K. ρ is the hyperradius defined as

$$\rho^2 = \sum_{i=1}^{A-1}\xi_i^2 = \sum_{i=1}^{A}(\vec{r}_i - \vec{R})^2 \tag{3}$$

Substituting eq. (2) into (1), and operating by $\int Y^*_{ka}d\Omega$ results in a set of 3 coupled differential equations which can be solved for χ_a and E. This has been done using the Sprung super soft core potential which provides a good fit to the two nucleon scattering data. This potential[1] has central, tensor, spin orbit, and other terms, and a repulsive barrier at about .833 fermi. Using this potential, the binding energy of Li^6 was calculated as 31.775 MeV, neglecting coulomb and higher K effects, which is considered in good agreement with experiment. The rms size of the nucleus is 2.2f, versus 2.56 f experimentally, however. We use this wave function to calculate the spectral function of a bound proton in the $0S_{1/2}$ state.

Since the nucleons are indistinguishable, using an isospin formation, one uses the hamiltonian of eq. (1) to calculate the energy of a nucleon. The contributions involving the nucleon with quantum numbers

$\omega_o = \{n\ell j j_z t_z\}$ must be individually picked out. Now Y_{ka} is a linear combination of m scheme Slater determinants P that have been summed to yield a state of definite J, M, T, and T_z. These Slater determinants must each be expanded into a sum of cofactors as

$$P = \sum_{j=1}^{A} \phi_{\omega_o}(\vec{r}_j)\, C_{\omega_o \vec{r}_j} \quad . \tag{4}$$

ϕ_{ω_o} is the usual monomial of the K harmonics

$$\phi_{\omega_o} = d_{n\ell}\, r^{2n+\ell} \sum_{m,\sigma} Y_{\ell m}(\hat{r})\, \chi^{\frac{1}{2}}_{t_z}\, \chi^{\frac{1}{2}}_{\sigma}\, (\ell m, \tfrac{1}{2}\sigma | j j_z) \quad . \tag{5}$$

$C_{\omega_o r}$ in eq. (4) is the cofactor of the monomial ϕ_{ω_o} where the cofactor is itself a determinant of rank A-1. The cofactor has the state ω_o missing, and the coordinate $\vec{r}_j$ also is missing. The cofactor is written in cyclic order to eliminate any minus signs due to permutations. Expanding the wave function via eq. (4) allows one to pick out the terms involving the kinetic emergy of the nucleon with the quantum numbers ω_o. The potential energy is a two body operator, and for it each cofactor $C_{\omega_o r_j}$ must be again expanded into second cofactors. This expansion allows the potential energy interaction terms involving the nucleon with quantum numbers ω_o to be identified. The potential energy of the nucleon ω_o is

$$W = \sum_a \int \chi_a^2\,(\rho)\,d\rho \sum_{i=1}^{A} \langle \omega_o \omega_i | V_{12} | \omega_o \omega_i - \omega_i \omega_o \rangle \quad , \tag{6}$$

while the kinetic energy contribution can be written as

$$T = -\frac{\hbar^2}{2m} \sum_a \int \chi_a(\rho)\,d\rho\, \{F_0 \chi_a + F_1\, \rho\, \frac{d\chi_a}{d\rho} + F_2\, \rho^2\, \frac{d^2\chi_a}{d\rho^2}\} \quad , \tag{7}$$

where the $F_{0,1,2}$ are complicated coefficients (not given) independent of ρ.

The energy distribution for a $0S_{\frac{1}{2}}$ proton calculated in this way is a broad peak. The distribution has a full width at half maximum of 20 MeV and a centroid energy of -40 MeV. This is significantly different from shell model estimates of zero width and about -8 MeV for the centroid energy. This calculated distribution depends on the two nucleon potential assumed and is much narrower for $0P_{3/2}$ nucleons (not shown). The calculated width agrees well with the width of the summed energy in (p, 2p) experiments at high initial energy. This can be understood in a one step model of the reaction if the initially bound proton has a width as calculated here.

1. R. de Tourreil, et al., Nucl. Phys. A242, 445 (1975).

NEW-RESULTS IN THE SEARCH OF HIGH ENERGY DEUTERONS FROM THE $^3He + {}^3He$ REACTION

R. Pigeon, C. Rioux, S.S. das Gupta and R.J. Slobodrian
Laboratoire de Physique Nucléaire, Département de Physique, Université Laval
Québec, G1K 7P4, Canada

The $p+p \rightarrow d+e^+ + \nu$ process initiating the nucleosynthesis in stars has had its reaction rate revised many times and, according to standards solar models, the calculated pp reaction rate is too small to be detected with presently known experimental techniques[1]. The reaction can also be studied through the $^3He + {}^3He$ reaction leading to an intermediate stage $^4He + {}^2He$, where ^2He is a p-p pair interacting strongly at small relative momentum in the final state [1]. The advantage of this method of generating the 2p system is that it is formed in the reaction,and Coulomb barrier penetrabilities do not play a significant and poorly known role in the preparation of the p-p system at very low relative momenta, as is the case in a direct proton-proton collision.

The present experiment was carried out at 11.5 MeV incident energy at the center of the target and the detection system consists of a four semiconductor detector telescope, subtending an angular width allowing the detection of statistically significant numbers with very low yields. The first three detectors measure the total energy as the sum of three pulses, ΔE, ΔE_2 and E_3 which were accumulated on-line by a PDP-9 computer and provided three independent particle identification functions. The fourth detector was operated in anticoincidence in order to reject penetrating protons that could pile-up low energy protons and might simulate deuterons. Events are displayed as three bidimensional spectra and figure 1 shows an example. The reaction $^7Li(^3He,d)^8Be$ was used as calibration in order to determine the locus of the deuteron band of each bidimensional spectrum, and thus permitting the identification of deuterons from events of the other reaction. Crosses on the deuteron band in figure 1 are events that fall on the deuteron band of the other two bidimensional plots. Air and tritium impurities, as well as metallic pieces of the set-up cannot be the source of the observed continuous spectrum. The observed events are also inconsistent with a spill-over of proton pile-ups, because they are well isolated from the proton band. The spectra are kinematically consistent with the $^3He + {}^3He \rightarrow {}^4He + {}^2H + e^+ + \nu$ process. However, a measurement using ^4He as target showed that elastic recoil ^3He's, stopped in Ni foils placed in front of the detectors could produce a small number of deuterons, which were subtracted in order to obtain cross section values which integrated over energy are 1.3 ± 0.3 nb.sr^{-1} at 20° lab. and 0.7 ± 0.2 nb.sr^{-1} at 25° lab. These values are consistent with the upper limits found by Davies et al [2] at 15 MeV and are in agreement with the results of Slobodrian et al[3] at 13.5 MeV. The cross section values cannot be explained in the context of the present picture of weak or strong interactions, if the observed events are indeed deuterons of the detected energy originating in the $^3He + ^3He$ reaction, as established by the criteria given above.

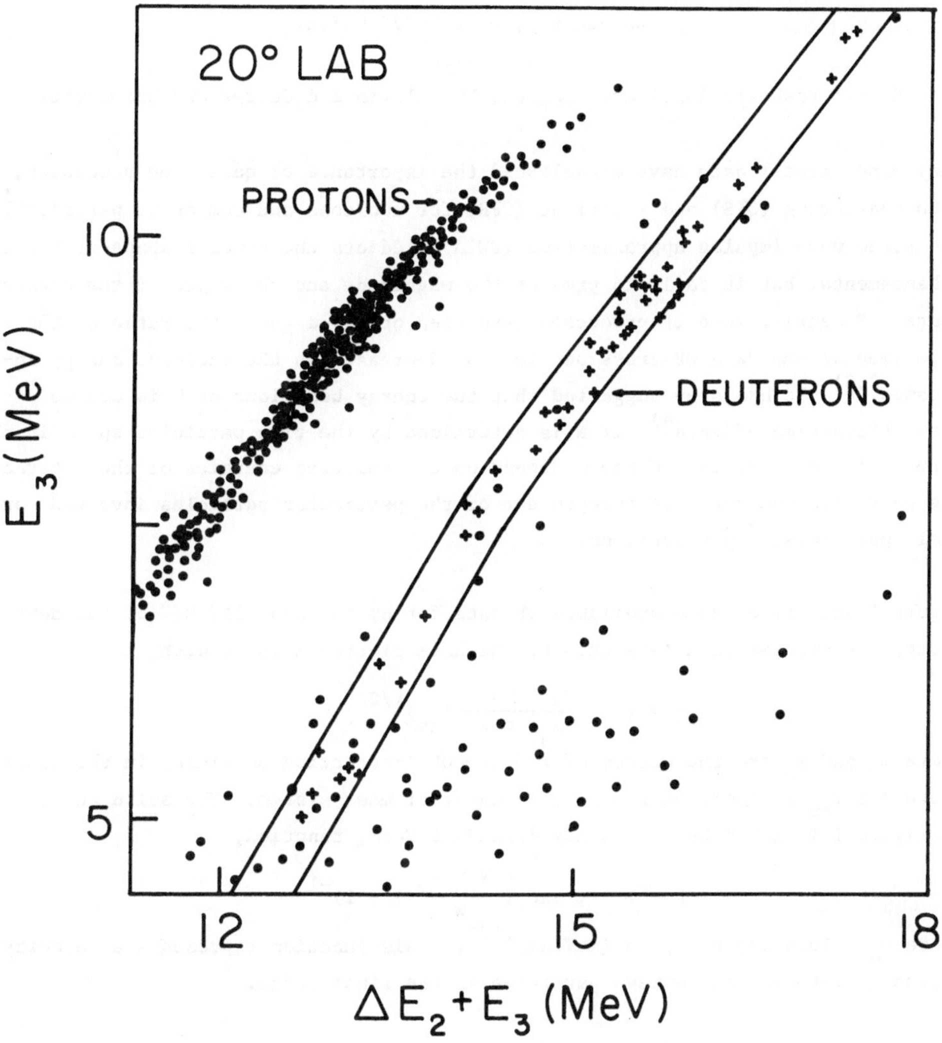

FIG. 1. A two-dimensional plot of one of the particle identification diagrams. The deuteron band locus has been determined by the ^7Li (^3He,d) ^8Be reaction.

1. R.J. Slobodrian, R. Pigeon and M. Irshad, Phys. Rev. Lett. 35, 19 (1975).
2. W.G. Davies et al., Phys. Rev. Lett. 38, 1119 (1977).
3. R.J. Slobodrian, R. Pigeon, R. Roy and M. Irshad, Proceedings of the Intl. Conf. on Nuclear Structure, page 573, Tokyo (1977).

INTERACTION-TIME EFFECTS OF QUASIFREE PROCESSES

J.M. Lambert,* P.A. Treado,* R.A. Moyle, L.T. Myers
Georgetown University and Naval Research Laboratory
Washington, D.C. 20057 U.S.A.
Ivo Slaus*
Rudjer Boskovic Institute, Zagreb, Yugoslavia and Georgetown University

Many experimental data have established the importance of quasifree processes, both scattering (QFS) and reactions (QFR) for nucleons and composite particles.[1,2] The plane wave impulse approximation (PWIA) predicts the general shape of the QF enhancements, but it fails to predict the magnitude and the width of the enhancements. Recently, some QF processes have been observed where the ratio of the experimental and PWIA QF cross section, N, decreases as the incident energy increases.[2,3] It has been suggested that the energy behaviour of N is due to interaction-time effects.[4] If N is determined by the time particles spend in the interaction region, then N would depend on the relative energies of the interacting particles and cound be independent of the particular particles involved or their particular cross sections.

Figure 1 summarizes some available QF data for up to about 150 MeV of incident energy for systems of A less than 6. We have plotted N vs. v with

$$v = \left(2\,\frac{m_a + m_b}{m_a \cdot m_b}\,T_{cm}\right)^{1/2}$$

where m_a and m_b are the masses of the two QF interacting particles in the final state and T_{cm} is their energy in the center of mass system. The solid curve in Figure 1 is given by a velocity dependent Fermi function,

$$F = 1 - \left(\exp\left(\frac{v - v_o}{w}\right) + 1\right)^{-1}$$

where $v_o = 10.5$ and $w = 3$ in $(\text{MeV/amu})^{1/2}$. This function represents a velocity dependent interaction between particles in the final state.

Any momentum tranfer occuring in a QF process can be modified in a similar manner to produce an increased velocity in the interaction region and a resulting narrowing of the Fourier transform. Such a narrowing is shown in Figure 2.

*This work has been partially supported by PL480 grant no. F6F005 (I.S.) and the National Science Foundation (J.M.L and P.A.T.).

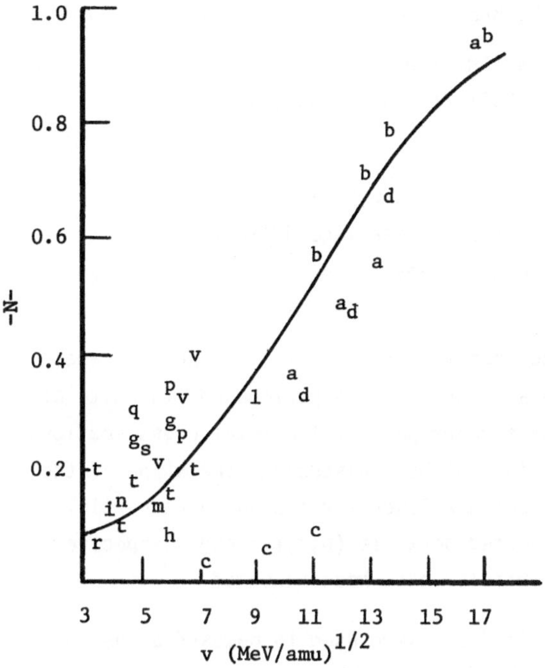

Figure 1. N vs v for the following QF reactions: a: d(p,pp)n, b: h(p,pp)d, c: α(p,pp)t, d: h(p,pd)p, f: h(h,dh)p, g: h(h,ph)d, h: h(h,pt)pp, i: h(h,dd)pp, l: h(h,pα)p, m: t(h,dh)n, p: t(h,tp)d, q: h(d,dd)p - with h(n,d)d, r: h(d,dd)p, s: t(h,dd)d, n: h(d,pd)d, t: h(d,ph)n, u: h(d,pt)p - with h(n,p)t, and v: h(d,pt)p - with d(d,p)t.

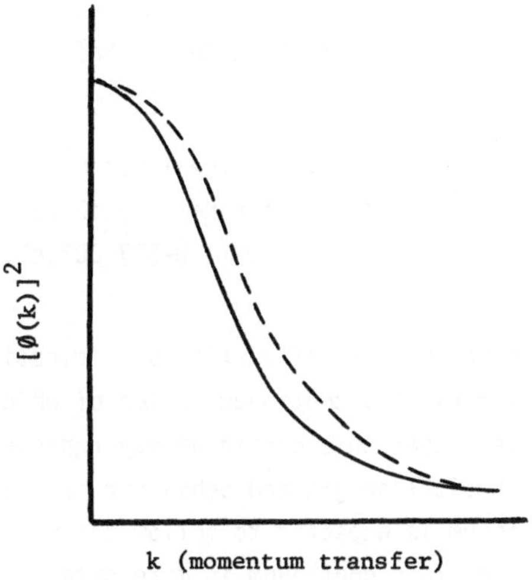

Figure 2. Example of the squared Fourier transform with reduced width produced by a velocity transformation discussed in the text.

1. Few Particle Dynamics (eds. A. Mitra, et al., North Holland, 1976) and and references therein.
2. I. Slaus, et al., Nucl. Phys. A286 (1977) 67.
3. R. Moyle, et al., to be published.
4. J. Lambert, et al., to be published.

CONTINUOUS PARTICLE SPECTRA FROM α-INDUCED BREAKUP OF α-PARTICLES AT E_α = 130 MeV

G. Paić[+] and B. Antolković[+]
Institute Rudjer Bošković, 40001 Zagreb, Yugoslavia

and

A. Djaloeis, C. Alderliesten[++], J. Bojowald and C. Mayer-Böricke
Institut für Kernphysik der Kernforschungsanlage Jülich,
D-517 Jülich, West Germany

Energy spectra of particles originating from the breakup of light nuclei generally exhibit a continuum on top of which a structure is superimposed. In spite of the fact that this continuum may represent a major part of the total cross section, its physical origin and behaviour have so far not been discussed. The shape of the continuum is expected to follow the phase space allowed for the outgoing particles. To study this continuum in more detail, charged particle (p,d,t,τ and α) spectra from the α-breakup induced by 130 MeV α-particles have been measured.

The α-particles from the Jülich cyclotron JULIC were used to bombard a gas-cell containing ^{4}He-gas at a pressure of about 250 Torr. The charged reaction products were detected by ΔE-E telescopes and identified by particle identifier units; the corresponding events were accumulated in an 8x2K multichannel analyzer. Measurements of p,d,t,τ and α-spectra at 9,11,13,15,20,25,30 and 40 degrees have been performed.

Phase space analysis of the p and d-spectra has been carried out. A computer program to calculate least-squares fits to the experimental spectra, based on any specified combination of n-body phase space distributions has been written. At each angle the relative magnitude of phase spaces pertaining to contributing n-body final state components was extracted. Regions of the experimental spectra where two body and two-step processes are present, were excluded from the fit.

Fig. 1 shows a proton and a deuteron spectrum at $\theta_p=11^0$ and $\theta_d=15^0$, each with the corresponding fit. Angular distributions of the relative magnitudes of the individual "phase space" contributions are displayed in fig. 2. From the measurement and analysis performed so far, the following conclusions can be drawn:

a) In the d and p-spectra no highly excited states of ^{6}Li and ^{7}Li have been observed.

b) The continuous part of the spectra can be well fitted with a combination of phase spaces of which the relative magnitudes vary with angle.

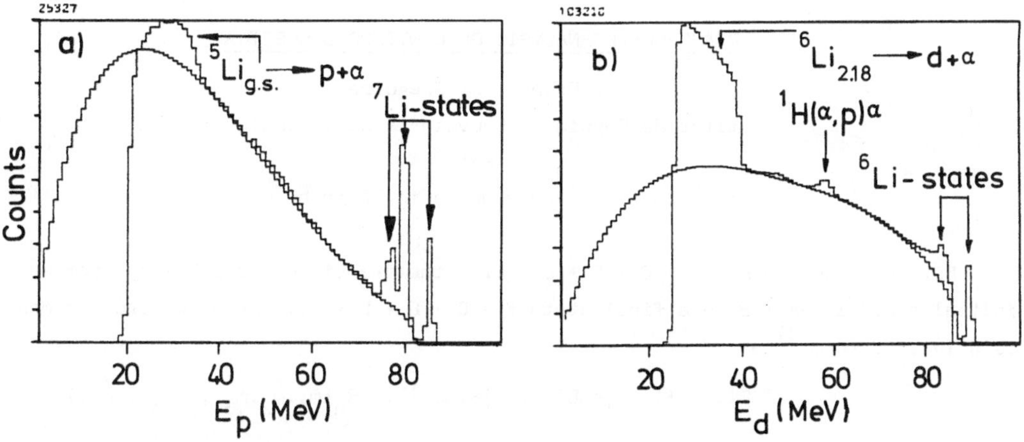

Figure 1: (a) proton spectrum at $\theta_p=11^o$, and
(b) deuteron spectrum at $\theta_d=15^o$, with phase space fits.

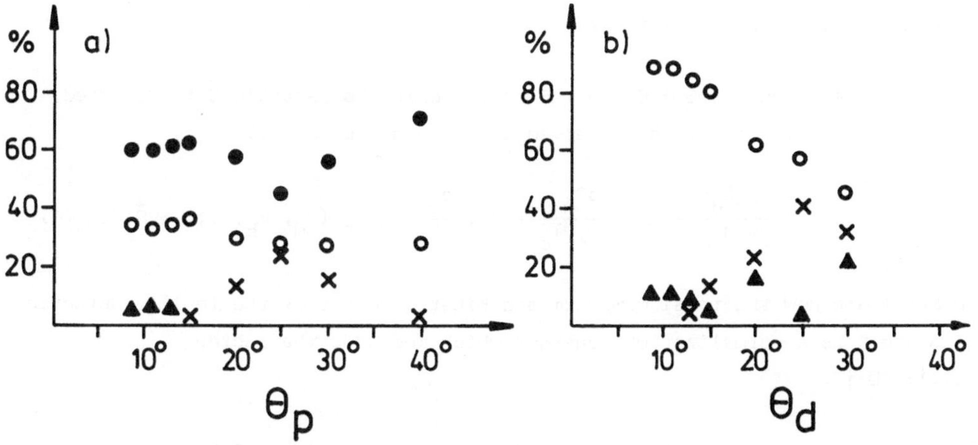

Figure 2: Angular dependence of relative contributions of different final state n-body configurations. (a) p+t+α(o), p+n+$^6Li_{2.18}$ (▲), p+n+d+α (●), p+p+n+n+α (x), (b) d+d+α (o) d+d+d+d (▲) d+p+n+α (x).

+ Guest scientist in the Institut für Kernphysik Jülich and partially supported by PL 480 Grant No. F6F005y and SIZ-I Grant 1.3.7.3.

++ Present address: Fysisch Laboratorium, Rijksuniversiteit, Utrecht, The Netherlands

PHASE-SPACE ANALYSIS OF CONTINUOUS SPECTRA

T. Delbar, G. Grégoire
Institut de Physique, Louvain-la-Neuve (Belgium)
G. Paic
Institut Ruder Boskovic, Zagreb

For the reaction A + B → C + D + E + ..., the transition probability from an initial state i = A + B to a final state f = C + D + E + ... can be written in non relativistic form [1] as

$$P(i \to f) = |\langle CDE \ldots |S|AB\rangle|^2 . \; R_n(m_1, m_2, m_3, \ldots m_n \; ; \; \vec{P}, T)$$

where $\langle CDE \ldots |S|AB\rangle$ represents the interaction of particles A,B,C,D, ... and R_n is the "phase space factor" of the n reaction products which share the total kinetic energy T and momentum $\overline{P}$ of the system.

This separation of the transition probability into 2 terms is based on the hypothesis that the interaction is independent of momentum or at least varies little in the kinematic range of interest.

If in the reaction A + B → C + D + E + ..., only the particle C is detected, then for a given angle for C, the observed spectrum can be described by

$$\frac{d^3\sigma}{d\Omega_c dT_c} = K \cdot \frac{dR_n}{d\Omega_c dT_c} = K \cdot \frac{p_c}{2} \cdot R_{n-1}\,(m_D,\, m_E,\, \ldots \; ; \; \vec{P}, T) \quad (1)$$

where $\overline{P}$ and T are now the total momentum and kinetic energy of the (n-1) undetected particles and K is a normalization constant which includes the factor $|\langle CDE \ldots |S|AB\rangle|^2$.

If a resonance appears between some of the (n-1) particles, D,E, ..., the corresponding spectrum for particle C will have the following form assuming a Breit-Wigner resonance

$$\frac{d^3\sigma}{d\Omega_c dT_c} = K \cdot \frac{p_c}{2} \cdot \frac{\Gamma_o^2/4}{(E-E_o)^2 + \Gamma_o^2/4} \cdot R_{n-1}\,(m_D, m_E, \ldots \; ; \; \vec{0},\, E') \; (2)$$

where E' is the kinetic energy available to the (n-1) breakup fragments in their centre of mass and E is the corresponding excitation energy in the composite nucleus.

The actual experimental spectrum will be represented by a sum of terms of types (1) and (2).

In order to test the ability of the above phase space model to represent the continuous part of various spectra, the model was fitted to spectra obtained for the reactions $^{9}Be(p,\alpha)^{6}Li$ at 30 MeV and $^{9}Be(p,\ )^{6}Li$, $^{9}Be(p,^{3}He)^{7}Li$ and $^{9}Be(p,t)^{7}Be$ at 75 MeV. It has proven possible to reproduce all spectra studied, in particular the results of the reaction $^{9}Be(p,\alpha)^{6}Li$ at 75 MeV has been fitted over an energy range corresponding to ^{6}Li excitation energies from 10 to 45 MeV (experimental limit)

The agreement between experiment and the theory are excellent at all angles. Parameters of known states are found to be in agreement with tabulated values.

The variation of the ratio $C_{\alpha d}/C_{\alpha np}$ as a function of angle at a given energy reflects the angular dependence of n-body breakup processes. Whereas a purely statistical model predicts a ratio $C_{\alpha d}/C_{\alpha np}$ that does not vary with angle, significant variations are seen experimentally. The lack of an adequate theory unfortunately precludes a quantitative interpretation of this result. The most that can presently be said is that $^{6}Li \rightarrow \alpha + n + p$ breakup becomes more important than $^{6}Li \rightarrow \alpha + d$ for large momentum transfers.

Investigation of an hypothetical level at approximately 8.2 MeV excitation in ^{6}Li [2] is being pursued using other reactions leading to final states with the same total number of nucleons.

The interest in the study of continuous spectra, in fact, is principally in giving a physical basis to the problem of background substraction in nuclear spectroscopy.

1) Hagedorn, "Relativistic Kinematics" (W.A. Benjamin, 1964)
2) T. Delbar, G. Grégoire, J. Lega, G. Paić, P. Wastyn, Phys. Rev. C14 (1976) 1659

FOUR-BODY APPROACH TO $\alpha(^3He,{}^3He)\alpha$ and $^6Li(p,p)^6Li$ Reactions

A. C. Fonseca†
Department of Physics and Astronomy
University of Maryland, College Park, Maryland 20742 U.S.A.

In a previous work[1] the field theoretic method of Amado[2] was used to formulate unitary four-body equations for two pairs of identical particles. Two- and three-body interactions proceed through intermediate quasiparticles and the resulting t-matrices are separable in momentum space and satisfy the constraints of unitarity. In the present work we take the first steps in building four-body reaction models that can be solved in a computer with considerable less numerical effort than a more exact formalism would allow. Since the resulting four-body equations reduce to single variable integral equations after partial wave decomposition, and the singularity structure of the Born terms and kernels is similar to that encountered in the three-body problem, the numerical techniques developed in the three-body problem can be readily applied to the solution of these equations. As an example we will formulate a very simple and crude four-body model for $^3He+\alpha$ and $p+{}^6Li$ reactions and show the results of the calculation. Since the energy required to break up an α-particle is large compared to energy of dissociation of ^6Li into n+p+α and ^3He into p+d or p+n+p, it is our hope that the internal degrees of freedom of the nucleons in the α-particle may be frozen and that $^3He+\alpha$ or $p+{}^6Li$ reactions may be treated as a four-body problem.

In the model we assume that all particles are spinless bosons and that the long range Coulomb force is neglected. In the two-body sector two quasiparticles, D (deuteron) and E, are introduced and are coupled in s-wave to $D\leftrightarrow N+N$ and $E\leftrightarrow n+\alpha$. NN scattering proceeds through the D and Nα scattering through the E and the resulting amplitudes are separable in momentum spaces. The parameters of the N-N interaction are chosen to fit the low energy triplet nucleon-nucleon observables. The parameters of the N-α interaction are those that fit neutron-alpha $p_{\frac{3}{2}}$ phase shifts. Therefore we adopt a p-wave form factor for the s-wave N-α interaction. In the three-body sector, instead of using the exact three-body amplitudes that results from the chosen two-body interactions, we introduce new quasiparticles and force three-body scattering to proceed through each one of them. The H (Helium) and L (Lithium) quasiparticles are introduced with the s-wave coupling $H\leftrightarrow D+N$, $L\leftrightarrow D+\alpha$ and $L\leftrightarrow N+E$. ND scattering proceed through the H and Dα scattering through the L. Both Dα and NE channels are coupled and the intermediate L propagator contains sums of Dα and NE bubbles where both D and E quasiparticles are fully dressed. The resulting three-body amplitudes are separable in momentum space and satisfy three-body unitarity. The parameters of the interaction are fitted to the $\ell=0$ low energy observables of nd and dα scattering. The percentage of D+α in the wave function of the L is approximately 60%. Proceeding to the four-body sector with no further approximation we obtain a set of one vector variable integral equations for $H\alpha\rightarrow H\alpha$, $H\alpha\rightarrow NL$, $NL\rightarrow NL$ and $NL\rightarrow H\alpha$. The equations for the first two processes are shown in Fig. 1 where the driving terms are sums of particle-exchange Born terms and box amplitudes. The box amplitudes contain in intermediate states the DE 2+2 channel that is treated exactly by the convolution method. Due to the absence of spin and angular momentum coupling the results obtained from the numerical solution of these equations are crude but not discouraging. In particular, we have used the model to test the importance of α-exchange to the elastic and transfer cross section. This is shown in Figs. 2-3 where curve A results from the solution of the Eq. (1) as shown and curve B from supressing all four-body diagrams that involve α or E exchange. We thank the Computer Science Center of the University of Maryland for their generous collaboration.

†Supported by the U. S. Department of Energy.

1) A. C. Fonseca and P. E. Shanley, Phys. Rev. D13, 2255 (1976).
2) R. D. Amado, Phys. Rev. 132, 485 (1963).

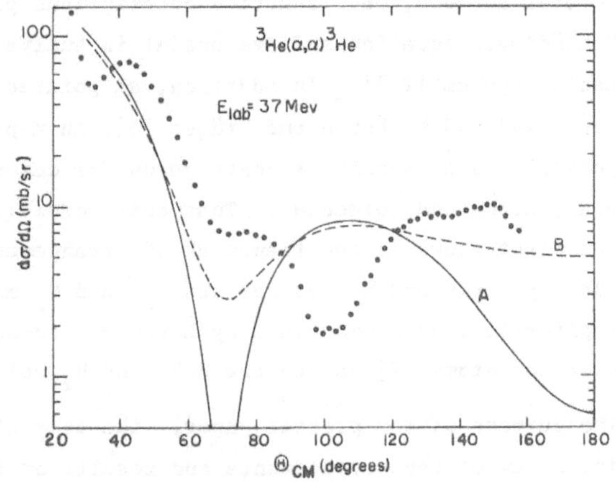

Fig. 2. Differential cross section for Hα → Hα. The black dots are experimental points.

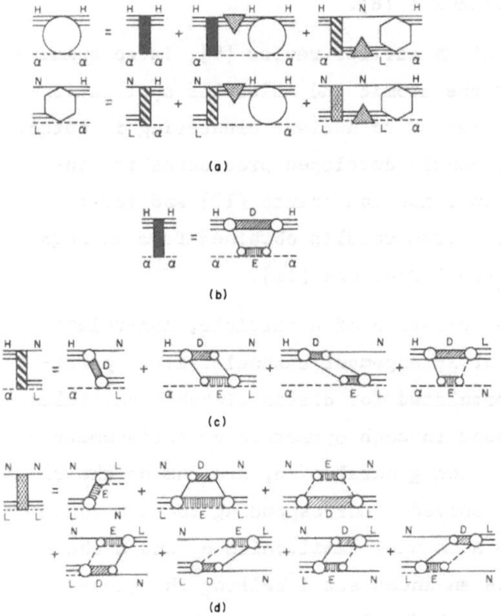

Fig. 1. αH → αH amplitudes (circle) and αH → NL amplitudes (hexagon).

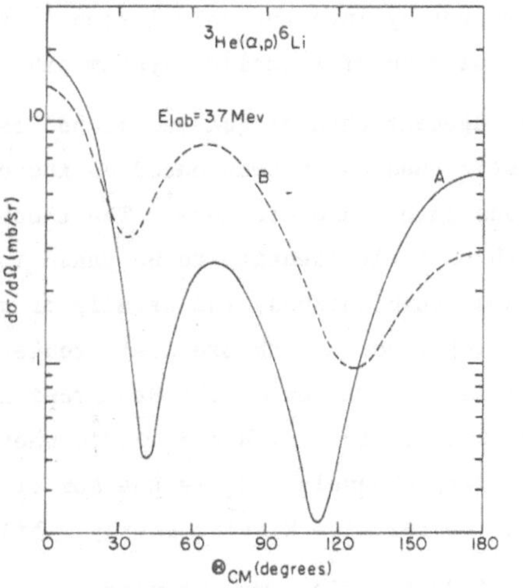

Fig. 3. Differential cross section for Hα → NL.

APPLICATION OF FEW-BODY METHODS TO ATOMIC AND MOLECULAR STRUCTURE

F. S. Levin*
Physics Department, Brown University
Providence, Rhode Island 02912/USA

A considerable effort has been devoted to the development and elucidation of various many-body scattering theories [1]. One aspect of this is very recent work establishing that some but not all n-particle transition operator formalisms have associated with them a wave-function-type formalism as well [2]. As discussed by several authors, wave function descriptions provide a framework equivalent to the Schrödinger equation and are useful in analyzing the asymptotic nature of a scattering system [2,3]. In addition, as pointed out independently by Goldflam and Kouri [4] and by Levin and Krüger [5], an n-particle wave function formalism might possibly be a useful alternate means for describing n-particle bound state systems, e.g., atoms and molecules. That this possibility could be successfully realized was first shown in the framework of arrangement channel quantum mechanics (ACQM) [3a] by Levin and Krüger for the H_2^+ and H_2 molecules [5,6]; further successful applications have been made by Levin and co-workers at Brown University to two-electron atoms [7] and to the HeH^+ and H_2 molecules [8].

The purpose of the present paper, like that of an earlier review [9], is to summarize a few of the developments and results of the atomic and molecular applications; in addition, the possible application of the method to nuclear clustering is noted. This is particularly pertinent in light of recently developed procedures for interpreting the resulting (approximate) wave function components [10] and for incorporating into the calculation of a larger system results obtained from an ACQM calculation of a smaller system embedded in the larger one [11].

Arrangement channel quantum mechanics is a description of n-particle, non-relativistic quantum systems based on the concept of arrangement channels, i.e., partitions into bound clusters. The theory is formulated for distinguishable particles, with particle identity to be taken into account through symmetric or antisymmetric linear combinations, the details of which, in the general case, are one of the remaining important theoretical problems to be solved. Corresponding to each channel is a partition of the Hamiltonian H into a channel Hamiltonian H_j and a channel interaction V^j: $H = H_j + V^j$, where $\{j\}$ is an index set labelling the (two-cluster) channels. H_j is the sum of the internal Hamiltonians for the two-clusters and the relative kinetic energy, while V^j contains all inter-cluster interactions.

The ACQM equations may be written as [3a]

$$\mathcal{H}|\psi) = E|\psi) , \tag{1}$$

where $\mathcal{H}$ is a non-hermitian, matrix "Hamiltonian" operator with elements $\mathcal{H}_{jk}$ =

$H_j\delta_{jk} + V^k\delta_{k,j+1}$ and $|\psi)$ is a column vector of channel component states $|\psi_j>$ such that $|\psi)_j = |\psi_j>$. Properties of the $|\psi_j>$ are discussed in Ref.[3a]; they are related to $|\Psi>$, the solution of the Schrödinger equation $(E-H)|\Psi> = 0$, by

$$|\Psi> = \sum_j |\psi_j> . \qquad (2)$$

In applying (1) to bound state problems, not only is it regarded as an eigenvalue problem, but of necessity it must be solved approximately. Two types of approximations have been used. In the first, the sum on j in (1) has been truncated to include only a few two-body channels, according to simple energy arguments. In the second, the allowed expansion of each $|\psi_j>$ in a complete set of states was limited to only a finite number of such states. Approximated quantities are indicated by a tilde over the symbol, so that (1), in approximate form, is

$$\tilde{\mathcal{H}}|\tilde{\psi}) = \tilde{E}|\tilde{\psi}) \quad . \qquad (3)$$

Expansion in a truncated basis, leading to a diagonalization of $\tilde{\mathcal{H}}$, is equivalent to solving (3) by using the dual eigenvectors $(\tilde{\phi}|$, as noted in [10]:

$$(\tilde{\phi}|\tilde{\mathcal{H}}|\tilde{\psi}) = \tilde{E}(\tilde{\phi}|\tilde{\psi}) = \tilde{E} , \qquad (4)$$

where a bi-orthonormality assumption has been used. This implies that the probability density to use is not $|\tilde{\Psi}|^2 = |\Sigma_j\tilde{\psi}_j|^2$, but $\tilde{\rho} = \Sigma_j\tilde{\phi}_j{}^{\dagger}\tilde{\psi}_j$, with a corresponding definition of charge density [10]. Since $\tilde{\mathcal{H}}^{\dagger} \neq \tilde{\mathcal{H}}$, $\tilde{E}$ may be complex; there is thus no Hylleraas-Undheim (H-U) variational theorem for the eigenvalues $\tilde{E}$. Complex eigenvalues can normally be ignored [5], but lack of an H-U theorem can mean a non-monotonic approach to E as the basis is enlarged.

Let us now examine a few representative results: for He, H^- and H_2, considered as two-channel systems, and for HeH^+, taken to be a three-channel system.

Since partitions of labelled particles into bound clusters define channels, there are as many two-cluster channels for an atom as there are electrons in the atom, all of the form (electron) + (bound state of n-1 electrons). There are thus two channels for He and H^-. As noted in [5] and [9], the 2 x 2 equation (1) reduces in this case to an uncoupled equation involving an exchange interaction. The expansion states to use in this equation (see [9] for details) are the pure hydrogenic states of charge Z = 2 for He and Z = 1 for H^-, one such state for each electron. To date, the most accurate He ground state result is obtained by keeping one electron in the 1s state, and allowing the other to be in any of the ns states, $1 \leq n \leq 4$. The energy, obtained by diagonalizing a 4 x 4 secular determinant is -2.89488 a.u., to be compared with the value -2.903724 a.u. determined variationally by Pekeris, who used a trial function with over 1000 variational parameters and diagonalized a 440 x 440 matrix. In the present case, energy decreases monotonically towards the Pekeris value with increasing value of n (in ns). However, when both electrons are allowed to occupy the 1s and 2s hydrogenic orbitals, the resulting energy is much less accurate, becoming -2.83099 a.u., which lies much higher than even the Hartree-Fock value of -2.8617 a.u. This is a typical example of increase in a basis set leading to a worsening of the ACQM energy, quite unlike the situation in a variational calculation.

Similar results hold for H^-. A 1s, 1s-2s basis calculation yields a ground state energy of -.57781 a.u., slightly deeper than the Pekeris value of -.57778 a.u.; increasing the second basis set from 1s-2s to 1s··4s lowers the energy slightly. However, when each electron can be in the 1s-2s linear combination, the calculated energy again rises, to -.51087 a.u.

There are six two-cluster channels for the H_2 molecule, although only the two (labelled particle) H+H channels have been used in calculations [5,9]. Using only the ground state of hydrogen to describe the ψ_j's, diagonalization of (3) yields for the ground state the values 1.42 a_o for the equilibrium separation R_{eq} and 4.435 ev for the dissociation energy D_e, accurate, respectively, to 1.5% and 6.5%. While the resulting approximate wave function is the Heitler-London valence-bond wave function Ψ^{H-L} (whose observable values are much less accurate than the preceding), the resulting ACQM probability density $\tilde{\rho}$ is just the sum of two H-atom densities, one centered on each atom. Such a density conforms to the simplest chemical structure for H_2, that of two ground state H atoms. This is in contrast to the density arising from Ψ^{H-L}, which contains an interference term not present in $\tilde{\rho}$, and is even in greater contrast with densities derived from more complicated trial functions. The implications for nuclear cluster structure are thus of some interest.

This latter remark is supported by the HeH^+ calculations [9]. Reasonable values for the ground state parameters are obtained only when the He channel is included, and use of even a crudely accurate He ground state wave function yields for R_{eq} and D_e values accurate to 2% and 35% [9]; the latter is expected to improve dramatically on use of a better He wave function. Evidently the "cluster" nature of He in HeH^+ is preserved to a good approximation, as is the much simpler cluster nature of H in H_2.

ACQM is seen in this brief survey to be an interesting alternate procedure for many-body bound state problems. In particular, since it leads to an accurate cluster structure for simple molecules, it may well do so for nuclei as well. The question of the non-monotonicity of the approximate eigenvalues is an important and interesting question, especially since the possibility that the spectrum of a non-hermitian operator is spread over more than one Riemann surface is not unlikely [12]. If this is so, then what naively may appear to be a non-monotonicity may actually be a change to other approximate eigenvalues located on a completely different Riemann surface. Investigations concerning this possibility are in progress.

REFERENCES

*Work supported in part by the United States Department of Energy

[1] See, e.g., W. Sandhas, in Few-Body Dynamics, ed. by A. N. Mitra et al., (North Holland, Amsterdam, 1976).

[2] F. S. Levin and J. M. Greben, submitted for publication.

[3a] D. J. Kouri, H. Krüger and F. S. Levin, Phys. Rev. D15, 1156 (1977).

[3b] Gy. Bencze and P. C. Tandy, Phys. Rev. C16, 564 (1977); M. L'Huillier, E. F. Redish and P. C. Tandy, Phys. Rev. C (in press).

[4] R. Goldflam and D. J. Kouri, Chem. Phys. Letts. 34, 594 (1975).

[5] F. S. Levin and H. Krüger, Phys. Rev. A15, 2147 (1977).

[6] F. S. Levin and H. Krüger, Phys. Rev. A16, 836 (1977).

[7] F. S. Levin and S. Sun, to be submitted for publication; see also [9].

[8] F. S. Levin and W. Ford, to be submitted for publication; see also [9].

[9] F. S. Levin, invited paper presented at the 18th Sanibel Symposia, to be published in Int. Journ. Quant. Chem.

[10] F. S. Levin, Quantal Rules for Non-Hermitian Bound State Formalisms Based on N-Body Scattering Theory, contributed paper, this Conference.

[11] F. S. Levin, to be submitted for publication.

[12] T. P. Živković and H. J. Monkhorst, J. Math Phys. (in press).

TWO HILBERT SPACE SCATTERING THEORY: RECENT PROGRESS

C. Chandler+)

Physikalisches Institut der Universität Bonn,
Endenicher Allee 11-13, D-5300 Bonn 1, Germany

A. G. Gibson

Department of Mathematics and Statistics
University of New Mexico, Albuquerque, NM, 87131, USA

Previously[1] we have derived a basic equation determining the scattering operator in an abstract two Hilbert space theory that is applicable to nonrelativistic multichannel quantum scattering systems. In this note we describe a method of obtaining approximate solutions to that equation.

To fix ideas and notation consider a system of N spinless distinguishable particles that interact by means of square integrable pair potentials. Denote by $\mathcal{H}_N$ the Hilbert space of N-particle wave functions, and by H_N the total N-particle Hamiltonian. Denote by $\mathcal{H}_\alpha$ the various channel subspaces, and by H_α the corresponding channel Hamiltonians. Functions in $\mathcal{H}_\alpha$ have the momentum-space form $\phi_\alpha(\xi_\alpha)f_\alpha(P_\alpha)$, where ϕ_α is the product of the bound state wave functions of the channel and is a function of the internal momentum coordinates ξ_α of the bound subsystem, and where f_α is a square integrable function of the relative momentum coordinates P_α of the bound subsystems. It is assumed that the total center of mass motion has been removed from the problem. On $\mathcal{H}_\alpha$ the operator H_α has the representation $H_\alpha \phi_\alpha f_\alpha = (\lambda_\alpha + T_\alpha(P_\alpha))\, \phi_\alpha f_\alpha$, where λ_α is the threshold energy of the channel and $T_\alpha(P_\alpha)$ is the momentum-space representation of the relative kinetic energy of the clusters of channel α.

The approximation method proceeds as follows. Define the Dirac bra $\langle \alpha j, \lambda |$ by the equation

$$\langle \alpha j, \lambda | \psi \rangle = \int d\xi_\alpha \, d P_\alpha \, \delta(\lambda - \lambda_\alpha - T_\alpha(P_\alpha))\, \phi^*_\alpha(\xi_\alpha)\, \chi^*_{\alpha j}(P_\alpha)\, \psi(\xi_\alpha, P_\alpha) \quad , \qquad (1)$$

where $\psi(\xi_\alpha, P_\alpha)$ is any vector in $\mathcal{H}_N$. The functions $\chi_{\alpha j}$, $j = 1,2,\ldots,$ are as-

+) Permanent address: Department of Physics and Astronomy, University of New Mexico, Albuquerque, NM, 87131, USA

sumed to be sufficiently differentiable and to satisfy the orthogonality condition

$$<\alpha j, \lambda \mid \alpha k, \mu> = \delta_{jk}\delta(\lambda-\mu)\,\Theta(\lambda-\lambda_\alpha) \quad , \tag{2}$$

where $\Theta(\lambda-\lambda_\alpha)$ is 1 for $\lambda \geqslant \lambda_\alpha$ and zero for $\lambda < \lambda_\alpha$. For two body channels, for examle, the functions $\chi_{\alpha j}$ could be proportional to the spherical harmonics $Y_\ell^m(\vec{q}_\alpha / q_\alpha)$. Define further $B(\lambda , \mu)$ and $C(\lambda , \mu)$ to be the matrices with the matrix elements

$$B_{\alpha j,\beta k}(\lambda , \mu) = <\alpha j, \lambda \mid H_N - H_\alpha \mid \beta k, \mu> \quad , \tag{3}$$

$$C_{\alpha j,\beta k}(\lambda , \mu) = (1-\delta_{\alpha\beta}) <\alpha j, \lambda \mid \beta k, \mu> \; . \tag{4}$$

Denote by $M(\lambda, \mu ;z)$, $\operatorname{Im} z \neq 0$, the solution to the equation

$$M(\lambda , \mu; z) = C(\lambda , \mu)(z-\mu) + K(\lambda , \mu) + \int_{-\infty}^{\infty} d\eta\, K(\lambda , \eta)(z-\eta)^{-1} M(\eta, \mu ; z) \; , \tag{5}$$

where $K(\lambda , \mu)$ is the particular solution of the equation

$$K(\lambda ,\mu) = B(\lambda , \mu) - \int_{-\infty}^{\infty} d\eta\, K(\lambda , \eta)\, C(\eta , \mu) \tag{6}$$

that is orthogonal to all solutions of the homogeneous equation. The approximate transition operators are then given by

$$\hat{T}_{\alpha\beta}(z) = \sum_{j,k} \int\int d\lambda\, d\mu \mid \alpha j, \lambda > M_{\alpha j,\beta k}(\lambda , \mu ; z) < \beta k, \mu \mid . \tag{7}$$

This approximate time-independent theory corresponds to an approximate time-dependent theory. Let $\Pi_\alpha : \mathcal{H}_N \to \mathcal{H}_\alpha$ be the orthogonal projection operators

$$\Pi_\alpha \equiv \sum_j \int d\lambda \mid \alpha j, \lambda > < \alpha j, \lambda \mid \; , \tag{8}$$

and let E be the orthogonal projection of $\mathcal{H}_N$ onto the range of $\Sigma\,\Pi_\alpha$. Define an approximate Hamiltonian $\hat{H}_N \equiv E\, H_N\, E$. Then the wave operators

$$\hat{\Omega}_\alpha^\pm = \underset{t\to\pm\infty}{\text{s - lim}}\; e^{it\hat{H}_N}\, \Pi_\alpha\, e^{-itH_\alpha} \tag{9}$$

exist, and one can define the approximate scattering operator $\hat{S}_{\beta\alpha} = \hat{\Omega}_\beta^{+*}\, \hat{\Omega}_\alpha^-$. The on-shell limit of $\hat{T}_{\beta\alpha}$ is precisely $\hat{S}_{\beta\alpha} - \delta_{\beta\alpha}\,\Pi_\alpha$.

These operators $\hat{S}_{\beta\alpha}$ are approximations for the exact scattering operators in the following sense. Suppose that equation (8) is used to define a sequence of operators $\Pi_{\alpha}(n)$, each with only a finite number of terms in the sum. And suppose that for each channel α the sequence $\{\Pi_{\alpha}(n)\}$ has as strong limit the channel projection operator $P_{\alpha}, \mathcal{H}_{\alpha} \equiv P_{\alpha}\mathcal{H}_{N}$. Let $\{\hat{S}_{\beta\alpha}(n)\}$ be the corresponding sequence of approximate scattering operators. Then this sequence has as strong limit the exact scattering operators $S_{\beta\alpha} = \delta_{\beta\alpha} - 2\pi i\,\delta(E_{out} - E_{in})\,P_{\beta}\,U_{\beta\alpha}\,P_{\alpha}$, where the $U_{\beta\alpha}$ are the transition operators defined by Alt, Grassberger, and Sandhas[3,4].

It is understood that the matrices in equations (5) and (6) are finite dimensional. Not all channels need be included for the equations to be well defined, only the physically important ones. Moreover, for any arbitrary selection of channels and functions $\chi_{\alpha j}$, the solution to the equations is unique if $\mathrm{Im}\, z \neq 0$. The problem of spurious solutions, which seems to trouble many N-particle formulations[5], is here well under control.

It is to be emphasized that the functions $\chi_{\alpha j}$ should be considered as terms in a separable expansion of the on-shell transition operators. They should hence have the known symmetries and exhibit the important known correlations of the reaction under consideration. The philosophy thus resembles that of the resonating group method[6], with equations (5) and (6) replacing a variational principle.

In this regard it is clear that the convergence of the sequence $\{\hat{T}_{\beta\alpha}(n)\}$ of approximate transition operators to the true operator $T_{\beta\alpha}$ will be very poor if $T_{\beta\alpha}$ has disconnected parts. This cannot occur if the initial channels α, which may be considered as fixed in equation (5), are two-body channels. For these initial channels, however, the breakup amplitudes are reasonably smooth and convergence of the method is expected to be good.

Finally, the on-shell limit of the lowest order approximation to the solution M of equations (5) and (6) is $M_{B}(\lambda\,,\lambda;\lambda + i\,0) = B(\lambda\,,\lambda)$. Examination of equation (3) shows that this is a sort of distorted wave Born approximation. Speculation is thus encouraged that the approximation method described here is of physical, as well as mathematical interest.

An extended account of these matters will be published elsewhere[7].

One of us (C. C.) happily acknowledges the generous support of the U.S.-German Fulbright Commission, the Minna-James-Heinemann-Stiftung in collaboration with the NATO Senior Scientists Programme, the German Academic Exchange Service (DAAD), the University of Bonn, and the University of New Mexico. It is also a pleasure to acknowledge informative conversations with W. Sandhas, E.O. Alt, and R. Perne.

References:

1.) C. Chandler and A.G. Gibson, J. Math. Phys. 14, 1328 (1973),
2.) C. Chandler and A.G. Gibson, J. Math. Phys. 18, 2336 (1977),
3.) E.O. Alt, P. Grassberger, and W. Sandhas, Nucl. Phys. B2, 167 (1967),
4.) P. Grassberger and W. Sandhas, Nucl. Phys. B2, 181 (1967),
5.) C. Chandler, Spurious solutions to N-particle scattering equations, Nucl. Phys. A (in press), and papers cited therein,
6.) H.H. Hackenbroich, W. Schütte and H. Stöwe, in Few body dynamics, ed. A.N. Mitra et al. (North Holland, Amsterdam, 1976),
7.) C. Chandler and A.G. Gibson, N-body quantum scattering theory in two Hilbert spaces, IV. An approximation method, in preparation.

RELATIONSHIPS BETWEEN DIFFERENT FORMULATIONS OF N-BODY SCATTERING AND ORIGIN OF SPURIOUS SOLUTIONS

Vittorio Vanzani

Istituto di Fisica, Università di Padova, 35100 Padova, Italy
Istituto Nazionale di Fisica Nucleare, Laboratori di Legnaro, Legnaro (Padova)

We present a general formalism for the N-body scattering problem which allows us to establish various relationships between different formulations existing in the literature. Several of the proposed approaches differ considerably in the way of treating the underlying N-body dynamics and in the structure of the resulting integral equations. Interesting results have been already obtained within the context of the so-called channel coupling scheme [1,2]. Here we shall examine Faddeev-Yakubovskiĭ—Alt-Grassberger-Sandhas—Karlsson-Zeiger—Vanzani-like (FY-AGS-KZ-V) equations, the Sloan—Bencze-Redish (S-BR) equations, the Rosenberg—Mitra-Gillespie-Sugar-Panchapakesan—Takahashi-Mishima—Alessandrini—Sasakawa (R-MGSP-TM-A-S I) equations, the Weinberg-van Winter (WW) equation and the Sasakawa—L'Huillier-Redish-Tandy (S II-LRT) equations for two-cluster wave-function components [3]. We may reformulate all these equations within a unified formalism we refer to as linked-cluster formalism. Factorization properties for the corresponding homogeneous equations (with exception of the FY-AGS-KZ-V-type) will be presented. From these properties the general possibility of spurious homogeneous solutions is inferred. Thus a simple origin of spurious solutions is exhibited. Here we present a new type of factorization relations different from the ones previously obtained in [4]. For other relevant interesting investigations and further progress see also [5,6].

Let us introduce the following components, labelled by chains β_2 of partitions, of the full N-body scattering wave function $\psi^+_{a_2}$, initiated by an incoming wave of two bound clusters Φ_{a_2},

$$\psi^+_{\beta_2;a_2} = G_o V_{b_{N-1}} G_{b_{N-1}} V^{b_{N-1}}_{b_{N-2}} \cdots G_{b_3} V^{b_3}_{b_2} \psi^+_{a_2} = F_{\beta_2} \psi^+_{a_2} . \tag{1}$$

Here, as usual, a_i denotes a partition of the N bodies into i clusters; $\alpha_i = (a_i, a_{i+1}, \ldots, a_{N-1})$ a chain of partitions satisfying the conditions $a_i \supset a_{i+1} \supset \ldots \ldots \supset a_{N-1}$; V_{a_i} $[V^{a_i}]$ is the sum of the interactions internal [external] to a_i; $V^{b_j}_{a_i}$ the sum of the interactions internal to a_i and external to b_j; $V^{a_{i+1}}_{a_i} = V_{a_i} - V_{a_{i+1}}$, $G_o = G_{a_N}$ and G are the free and full resolvents, respectively. All the interactions which bind the clusters of b_{N-1}, b_{N-2}, ..., b_2 are present in the graphs representing F_{β_2} (see (1)), because

$$V_{b_{N-1}} + V^{b_{N-1}}_{b_{N-2}} + \ldots + V^{b_{j+1}}_{b_j} = V_{b_j} \qquad (j = 2, \ldots, N-1) . \tag{2}$$

F_{β_2} has connectivity β_2 and possesses a characteristic linked-cluster structure.

Besides the sequential fragmentation components (1), other essential ingredients of our linked-cluster formalism are: the distribution property [7]

$$V^{b_{j+1}} = \sum_{b_j(\supset b_{j+1})} V_{b_j}^{b_{j+1}} , \tag{3}$$

the Lippmann-Schwinger-Glöckle-Tobocman (LS-GT) set of $2^{N-1}-1$ simultaneous equations [8]

$$\psi^+_{a_2} = \Phi_{a_2} \delta_{b_2 a_2} + G_{b_2} V^{b_2} \psi^+_{a_2} \tag{4}$$

and the homogeneous LS equations with kernels labelled by partitions into more than two clusters

$$\psi^+_{a_2} = G_{b_{2+k}} V^{b_{2+k}} \psi^+_{a_2} \qquad (k=1, \ldots, N-2). \tag{5}$$

One can verify that the <u>sequential fragmentation components</u> $\psi^+_{\beta_2;a_2}$ coincide with the FY components introduced in [9,10]. Other types of components introduced in the literature can be immediately obtained starting from the FY ones. We define the <u>m-cluster components</u> (m=1,2,...,N-1) as sums of the $\psi^+_{\beta_2;a_2}$'s over all the partitions of the chains β_2 having the same b_m (namely there is no sum over b_m):

$$\psi^+_{b_m;a_2} = \sum_{b_{N-1}\ldots b_{m+1}\; b_{m-1}\ldots b_2} \psi^+_{\beta_2;a_2} = F_{b_m} \psi^+_{a_2} \tag{6}$$

with $b_{N-1}\subset\ldots\subset b_2$. For m=2 (the sum is over β_3) we reobtain the two cluster S II-LRT components $\psi^+_{b_2;a_2}$. For m=N-1 we have the components $\psi^+_{b_{N-1};a_2}$ which are strictly related to the Rosenberg operators $^{(b_{N-1})}T$ by $\psi^+_{b_{N-1};a_2} = i\varepsilon G_o {}^{(b_{N-1})}T\, G_o \Phi_{a_2}$; in particular for N=4 we reproduce the three-cluster SI formalism. The special case m=1 (a single component, the sum is over β_2) corresponds to the full wave-function $\psi^+_{b_1;a_2} \equiv \psi^+_{a_2}$, which is the unknown of the single WW wave-function equation. An essential (and necessary) feature of all the types of components is that they sum up to give the full $\psi^+_{a_2}$

$$\sum_{\beta_2} \psi^+_{\beta_2;a_2} = \sum_{b_m} \psi^+_{b_m;a_2} = \psi^+_{a_2} \qquad (m=1, \ldots, N-1) \tag{7}$$

This follows from repeated application of (3) and (5). Furthermore appropriate use of (3) and (5) leads to a progressive exact simplification of (6) as we go from m=3 to m=N-1:

$$\psi^+_{b_3;a_2} = \sum_{\beta_4} F_{\beta_3} \psi^+_{a_2}, \;\ldots,\; \psi^+_{b_{N-2};a_2} = \sum_{b_{N-1}} F_{\beta_{N-2}} \psi^+_{a_2}, \quad \psi^+_{b_{N-1};a_2} = G_o V_{b_{N-1}} \psi^+_{a_2} \tag{8}$$

In this simplified form the (N-1)-cluster components can be more easily related to the ones introduced in the literature. Relationships between the different types of components immediately follow from their very definition. In particular the FY and the S II-LRT components are related by

$$\psi^{+}_{b_2;a_2} = \sum_{\beta_3} \psi^{+}_{\beta_2;a_2} . \tag{9}$$

We realize that other types of wave-function formalisms, different from the ones considered up to now, could be introduced (e.g. r-cluster component formalisms $(3<r<N-1)$).

Integral equations for several types of components can be rederived in a new and simple way by starting from the LS-GT set (4), multiplying it by the F-operators and using appropriate distribution properties. Multiplication of (4) by F_{β_2} leads to the FY-like equations for $\psi^{+}_{\beta_2;a_2}$ [3]. Multiplication of (4) by $F_{b_{N-1}}$ leads to the $\binom{N}{2}$ equations for the (N-1)-cluster components

$$\psi^{+}_{b_{N-1};a_2} = G_o V_{b_{N-1}} \Phi_{a_2} \delta_{a_2 \supset b_{N-1}} + \sum_{b_{N-2} \ldots b_2} F_{\beta_2} G_{b_2} G_o^{-1} \sum_{c_{N-1} \not\subset b_2} \psi^{+}_{c_{N-1};a_2} . \tag{10}$$

Multiplying (10) by G_o^{-1} we find equations of the Rosenberg type (in particular for N = 4 the four-body MGSP-TM-A-1ike equations and the S I equations). Eqs. (10) can be also obtained by summing the FY equations for $\psi^{+}_{\beta_2;a_2}$ over $b_2, \ldots, b_{N-2} (\supset b_{N-1})$. The $2^{N-1}-1$ S-BR equations (in the half-on-shell form) immediately follow from eqs. (10) after multiplying them by G_o^{-1} and summing over $b_{N-1} (\not\subset c_2)$

$$U^{c_2 a_2} \Phi_{a_2} = V^{c_2}_{a_2} \Phi_{a_2} + \sum_{b_2} \Big[\sum_{\substack{\beta_3 \\ (b_{N-1} \not\subset c_2)}} G_o^{-1} F_{\beta_2} \Big] G_{b_2} U^{b_2 a_2} \Phi_{a_2} . \tag{11}$$

Thus the connections between the FY-KZ equations and the R-MGSP-TM-A-SI equations and between the latter ones and the S-BR equations are established.

Integral equations for two-cluster components can be immediately obtained from their definition, after taking account of the fact that Φ_{a_2} remains unaltered after application of F_{a_2} $(F_{a_2} \Phi_{a_2} = \Phi_{a_2})$; one gets

$$\psi^{+}_{b_2;a_2} = \Phi_{a_2} \delta_{b_2 a_2} + (1 - F_{b_2})^{-1} F_{b_2} \sum_{c_2 \neq b_2} \psi^{+}_{c_2;a_2} . \tag{12}$$

These are the LRT equations. It is easy to see that they coincide with the S II equations derived by Sasakawa for N = 4,5 (see [3]). This proves that the LRT and the S II formalisms are substantially identical. Finally the WW equation

$$\psi^{+}_{a_2} = \Phi_{a_2} + \sum_{\beta_2} F_{\beta_2} G_{b_2} V^{b_2} \psi^{+}_{a_2} \tag{13}$$

immediately follows from the LS-GT equations multiplied by F_{β_2} and summed over β_2 or from the R-SI-like eqs. (10) summed over b_{N-1}. The relationship of the WW equation to the R-SI equations (and consequently to the S-BR equation) is thus clarified. So, a variety of systems of integral equations can be derived within the context of the linked-cluster formalism; the extreme cases are the $N!(N-1)!\,2^{-N+1}$ FY equations with two-body (or (N-1)-cluster) connectivity of the non-iterated kernel (the largest number of equations and the lowest connectivity) and the single WW equation with full connected kernel.

In the above equations the detailed structure of the kernels in terms of linked-cluster strings of resolvents and interactions can be immediately exhibited by writing down the explicit expressions for F_{β_2} and F_{b_m} (see (1) and (6) with (1) inserted in it). This facilitates the search for factorization properties for the homogeneous form of highly connected kernel equations. To factorize the homogeneous form of the R-SI-like set (10) we express the terms occurring in the linked-cluster expansion for G as follows [11]

$$F_{\beta_m} G_{b_m} = G_o V_{b_{N-1}} G_{b_{N-1}} V_{b_{N-2}}^{b_{N-1}} \cdots V_{b_m}^{b_{m+1}} G_{b_m} = \sum_{d_{N-1}} \sum_{\Gamma_{m+1}\bar{\Delta}_{m+1}} G_o \prod_{i=m+1}^{N-1} (MY)^i G_o V_{d_{N-1}} G_{d_{N-1}} \tag{14}$$

where $\Gamma_{m+1} = (\gamma_{m+1}^{(o)}, \gamma_{m+2}^{(1)}, \ldots, c_{N-1})$, $\bar{\Delta}_{m+1} = (\delta_{m+1}^{(o)}, \delta_{m+2}^{(1)}, \ldots, \delta_{N-2}^{(N-m-3)})$, $G_o(MY)^i = (MX)^i$ and the Yakubowskiĭ operators $(MX)^i$ are defined similarly to $(MX)^{k+j}$ in the Appendix A of [11] (with $a_i(o)$ replaced by d_{k+j-1}^{j-2}). Since $G_oV_dG_dG^{-1}\psi = \sum_e F_{de}\psi_e$, the homogeneous set associated with (10) takes the form

$$\sum_d S_{bd} \left(\sum_e F_{de}\, \psi_e\right) = 0 \qquad \text{or} \quad \mathbf{S}\,\mathbf{F}\,\boldsymbol{\psi} = 0, \tag{15}$$

where $S_{bd} = \delta_{bd} + \sum_{m=2}^{N-2} \sum_{\bar{\beta}_m} \sum_{\Gamma_{m+1}\bar{\Delta}_{m+1}} G_o \prod_{i=m+1}^{N-1} (MY)^i$, $\quad F_{de} = \delta_{de} - G_o T_d \bar{\delta}_{de}$;

$b = b_{N-1}$, $d = d_{N-1}$ and so on; $\bar{\beta}_m = (b_m, \ldots, b_{N-2})$. In particular for N = 4 our factorization property (15) reduces to the one found by Karlsson and Zeiger by means of a different procedure [12]. In addition to the physical solutions generated by the Faddeev-like operator $\mathbf{F}$ (namely the solutions of $\mathbf{F}\boldsymbol{\psi} = 0$), the homogeneous R-SI-like set of $\binom{N}{2}$ equations (15) can admit spurious solutions, in correspondence to vectors annihilated by the spurious multiplier $\mathbf{S}$. The homogeneous S-BR equations can have spurious solutions too. In fact, suppose that the homogeneous R-SI equations have a spurious solution $\psi_{b_{N-1}}^{(s)}$ and construct $\chi_{c_2}^{(s)} = G_o^{-1} \sum_{b_{N-1} \not\supset c_2} \psi_{b_{N-1}}^{(s)}$; one verifies that the $\chi_{c_2}^{(s)}$'s form a spurious solution of the homogeneous S-BR equations. Viceversa if the S-BR equations have spurious solutions, the R-SI equations have spurious solutions too.

We proved that starting from the three-body Weinberg equation and constructing its four-body analogue, according to the rules of the AGS N-body matrix technique, we exactly obtain the six R-MGSP-TM-A-SI-like eqs. (10) for N=4 [4]; furthermore, the factorization property (15) for N=4 is the exact four-body analogue of the factorization property found by Newton with reference to the three-body Weinberg equation.

Let us now consider the general factorization property

$$\left\{\sum_{m=k}^{N} \sum_{\beta_m} F_{\beta_m} G_{b_m}\right\} G^{-1} \psi = 0 \qquad (16)$$

for the equation $\psi = \sum_{\beta_k} F_{\beta_k} G_{b_k} V^{b_k} \psi$ $(k=2, \ldots, N-1;\ F_o=1)$. Spurious solutions to this equation may occur for any k. In correspondence to k=2,3 we have two interesting cases. For k=2 the eq. (16) is the factorized form of the homogeneous WW N-body equation.

Let us now consider the case k=3 and suppose that the above equation has a spurious solution $\psi^{(s)}$. We define $\psi^{(s)}_{b_2} = F_{b_2} \psi^{(s)}$. Since $\psi^{(s)} = \sum_{b_2} \psi^{(s)}_{b_2}$, because of (3) with j=2, one immediately sees that the $\psi^{(s)}_{b_2}$'s form a solution of the homogeneous SII-LRT equations, precisely a spurious solution, being $G^{-1} \psi^{(s)} \neq 0$ by definition of $\psi^{(s)}$. This establishes the possibility of spurious solutions for the homogeneous SII-LRT equations too. The homogeneous Avishai equation or, equivalently, the homogeneous Kouri-Levin-Tobocman (KLT) equations (with the channel permuting array choice) inserted into each other, factorize too and they can have spurious solutions (see also [4-6]).

References

1. K.L. Kowalski, Phys. Rev. C 16, 7 (1977).
2. Gy. Bencze and P.C. Tandy, Phys. Rev. C 16, 564 (1977).
3. Most of the relevant references are quoted in: V. Vanzani, The N-Body problem, in Few-Body Nuclear Physics (IAEA, Vienna 1978), in press.
4. V. Vanzani, contribution to the Summer Meeting on Few-Body Nuclear Physics (Uppsala, 1977), and to be published.
5. C. Chandler, Budapest preprint (1977), Bonn preprint (1977).
6. K.L. Kowalski, Case Western Reserve University preprint (1978).
7. V. Vanzani, Lett. Nuovo Cimento 16, 1 (1976).
8. W. Glöcke, Nucl. Phys. A141, 620 (1970).
9. B.R. Karlsson and E.M. Zeiger, Phys. Rev. D9, 1761 (1974); D10, 1291 (1974).
10. W. Sandhas, in Few-Body Dynamics, ed. A.N. Mitra et al. (North-Holland, Amsterdam 1976), p. 540.
11. V. Vanzani, Nuovo Cimento 2A, 525 (1971).
12. B.R. Karlsson and E.M. Zeiger, private communication.

Spurious solutions in few-body equations

S.K. Adhikari, W. Glöckle[+]
Departamento de Fisica, Universidade
Federal de Pernambuco
50000 Recife, Pe, Brazil

Few body formulations may suffer from the disease to be not equivalent to the underlying Schrödinger equation. This has been discovered by Federbush and Newton in the Weinberg formulation. We investigate[1] the possibility and consequence of the existence of spurious solutions in some of the few body formulations. Contrary to proofs[2] the channel coupling array scheme (CCAS) of Kouri, Levin and Tobocman[3] admits spurious solutions. We demonstrate this for three particles. The basic set of three Lippmann Schwinger equations[4] can be rewritten into coupled sets of equations with kernels which get connected after one or two iterations. This is one way to formulate the CCAS[5]. We can relate these sets to a system of three coupled equations, called the Faddeev choice[5], which follows naturally[4] from the basic set of Lippmann Schwinger equations and which is equivalent to the Schrödinger equation. The link is a matrix multiplier, which can be shown to admit spurious solutions. The CCAS is also formulated in terms of the channel components of the wavefunction $\chi_\mu, \mu = 1,2,3$. The various possible sets for χ_μ can again be related by a matrix multiplier to a set which is equivalent to the Schrödinger equation. In that case it is the set of Faddeev equations for the wavefunction components. As an example we state the multiplier and the corresponding spurious eigenvalue problem for the choice of the channel coupling matrix $W = \begin{pmatrix} 0 & 1 & 0 \\ 0 & 0 & 1 \\ 1 & 0 & 0 \end{pmatrix}$ ($V_1 = V_{23}$ etc.)

$$\begin{pmatrix} 1 & G_0 V_1 & -G_0 V_2 \\ -G_0 V_3 & 1 & G_0 V_2 \\ G_0 V_3 & -G_0 V_1 & 1 \end{pmatrix} \begin{pmatrix} \chi_1 \\ \chi_2 \\ \chi_3 \end{pmatrix} = 0$$

Clearly all the solutions have the property $\sum_\mu \chi_\mu = 0$, a possibility which has been overlooked in [2]. That eigenvalue problem can be related to the one studied by Federbush.

In a similar manner we investigate the six coupled four body equations proposed by Mitra, Gillespie ...[6] and the seven coupled equations proposed by Sloan[6] and generalised by Bencze[6] and Redish[6]. Again they can be related by matrix multipliers to sets of dimensions six or seven, which are equivalent to the Schrödinger equation. These multipliers are likely to give spurious solutions to these equations.

In all these cases spuriosities are shown to have no hazardous consequence if one is interested in studying the scattering problem.

+ Permanent adress: Inst. für theoret. Physik, Ruhr-Universität Bochum, D-4630 Bochum, Germany

1) S.K. Adhikari, W. Glöckle, to be published
2) Y. Hahn, D.J. Kouri, F.S. Levin, Phys. Rev. C10 (1974) 1620
   B. Bencze, P.C. Tandy, Phys. Rev. C16 (1977) 564
3) D.J, Kouri, F.S. Levin, Phys. Lett. 50B (1974) 421
   W. Tobocman, Phys. Rev. C9 (1974) 2466
4) W. Glöckle, Nucl. Phys. A141 (1970) 620
5) W. Sandhas, in Few-Body Dynamics, Delhi, 1976
6) for references see 5)

TIME DELAY IN N-BODY SCATTERING

T.A. Osborn
Department of Physics, University of Manitoba,
Winnipeg, Manitoba, Canada R3T 2N2

D. Bollé
Instituut voor Theoretische Fysica, Universiteit Leuven,
B-3030 Leuven, Belgium

Time delay and its relation to the S-matrix has been studied in both the two- and three-body problem. We outline a derivation that extends the theory of time delay to the N-body scattering problem. For all scattering processes initiated by the collision of two clusters we find a simple proof establishing the connection of time delay to the on-shell S-matrix. The physical interpretation of time delay is the total spatial retardation of a scattering state induced by the collision process. This idea is mathematically implemented by placing the center of a 3(N-1) dimensional sphere of radius r about the N-body c.m. position. The transit time of the asymptotic wave packet (evolving without the inclusion of the inter-cluster interactions that cause multichannel scattering) is subtracted from the corresponding transit time for the fully interacting wave. Then the radius r is taken to ∞. The resulting difference is the time delay. It is a function of the initial scattering channel, the incident asymptotic wave packet, and all the interactions in the N-body system.

Our proof is carried out in the two-Hilbert space formalism valid for non-relativistic N-body scattering[1]. The basic results of this formalism are easy to outline. The c.m. motion is of no importance and so is systematically removed. Each distinct asymptotic channel will be labelled by α. The symbol α denotes both a partition, A, of the N particles into clusters and the specification of the internal eigenfunction of each cluster. Let $\Phi_\alpha(t)$ denote the incident wave packet in channel α and $\Psi_\alpha(t)$ the corresponding exact solution to the Schroedinger equation. $\Phi_\alpha(t) = J_\alpha \exp(-iH_\alpha t) f_\alpha$, where f_α is the wave packet giving the incident relative motion of the clusters in α, J_α is multiplication by the product of all internal cluster eigenfunctions and H_α is the Hamiltonian including the relative-motion kinetic energy minus the total binding energy of all clusters. The wave operator is defined by the limit $\exp(iHt) J_\alpha \exp(-iH_\alpha t) \to \Omega_\alpha^{(\pm)}$, where $t \to \mp\infty$ and the convergence is in the strong sense. Then $\Psi(0) = \Omega_\alpha^{(-)} f_\alpha$. Define $R_\alpha^{(\pm)}$ to be the projection operator onto the range of $\Omega_\alpha^{(\pm)}$. Channel orthogonality, energy conservation, and completeness are

$$\Omega_\alpha^{(\pm)\dagger}\,\Omega_\beta^{(\pm)} = \delta_{\alpha\beta} I_\alpha\,, \quad H\Omega_\alpha^{(\pm)} = \Omega_\alpha^{(\pm)} H_\alpha\,, \quad \sum_\alpha R_\alpha^{(+)} = \sum_\alpha R_\alpha^{(-)}\,. \tag{1}$$

where I_α is the identity in the Hilbert space containing functions f_α. One also constructs the S-matrix from $\Omega_\alpha^{(\pm)}$, viz, $S_{\alpha\beta} = \Omega_\alpha^{(-)\dagger}\,\Omega_\beta^{(+)}$. The S-matrix energy conservation and unitarity are the statements that

$$S_{\alpha\beta}H_\beta = H_\alpha S_{\alpha\beta} \ , \quad \sum_\gamma S_{\alpha\gamma}\, S^\dagger_{\beta\gamma} = \sum_\gamma S^\dagger_{\gamma\alpha}\, S_{\gamma\beta} = \delta_{\alpha\beta} I_\alpha \ . \tag{2}$$

The last element of scattering theory we need is the time integral estimate for the convergence of the wave operator. Let $f'_\beta = S_{\beta\alpha} f_\alpha$. The set $\{f'_\beta\}$ gives all the outgoing wave packets. Then

$$\int_{-\infty}^{0} dt||\Psi(t) - \Phi_\alpha(t)|| < \infty \ , \quad \int_{0}^{\infty} dt||R^{(-)}_\beta \Psi(t) - \Phi'_\beta(t)|| < \infty \ , \tag{3}$$

For N-particle systems whose interactions are local pairwise potentials that are also L^2 functions, all the statements above have rigorous proof with the one exception of the completeness property[2]. For the duration of the paper we take it as an hypothesis that N-particle systems are complete.

Our first objective is to give a universally valid definition of N-body time delay for an arbitrary two-cluster incident channel. Let $r^2 = (\Sigma m_i r_i^2)/(\Sigma m_i)$, where m_i and $\vec{r}_i$ are the mass and position of particle i. Denote by P_r the projection operator onto functions with support inside a 3(N-1) dimensional sphere of radius r. For times prior to the collision $\Psi(t)$ is approximated by $\Phi_\alpha(t)$. After the collision $\Psi(t)$ behaves like $\sum_\beta \Phi'_\beta(t)$, where $\Phi'_\beta(t) = J_\beta \exp(-iH_\beta t) f'_\beta$. For sphere r we define time delay to be

$$T(f_\alpha;r) = \frac{1}{2}\int_{-\infty}^{\infty} dt\ [2||P_r\Psi(t)||^2 - ||P_r\Phi_\alpha(t)||^2 - \sum_\beta ||P_r\Phi'_\beta(t)||^2] \ . \tag{4}$$

At this point two interrelated problems arise. For a definition of time delay to be physically reasonable the limit as $r \to \infty$ must be finite. Secondly, a method of extracting the relation to the on-shell S-matrix needs to be developed. We indicate briefly the solution to these problems.

First we extend a method developed by Ph. Martin to treat the two-body problem[3]. The method allows us to transform the definition (4) into an equivalent form that contains only the asymptotic channel operators J_β, H_β, and $S_{\beta\alpha}$. We write $T(f_\alpha;r) = \tilde{T}(f_\alpha;r) + \Delta(f_\alpha;r)$, where

$$\tilde{T}(f_\alpha;r) = \frac{1}{2}\int_{-\infty}^{\infty} dt(\mathrm{sgn}t)[\sum_\beta ||P_r\Phi'_\beta(t)||^2 - ||P_r\Phi_\alpha(t)||^2] \ , \tag{5}$$

$$\Delta(f_\alpha;r) = \int_{-\infty}^{0} dt[||P_r\Psi(t)||^2 - ||P_r\Phi_\alpha(t)||^2] + \int_{0}^{\infty} dt[||P_r\Psi(t)||^2 - \sum_\beta ||P_r\Phi'_\beta(t)||^2] \ . \tag{6}$$

It turns out that $\Delta(f_\alpha;r) \to 0$ as $r \to \infty$, so we need only compute $\tilde{T}(f_\alpha;r)$. To understand behavior of $\Delta_\alpha(f_\alpha;r)$ consider the left integral term of (6). Add and subtract $(\Psi(t),\ P_r\Phi_\alpha(t))$ to the integrand. Then use the standard Hilbert space inequalities,

$|(f,g)| \leq ||f||\ ||g||$. It is easy to see that $2||f_\alpha||\ ||\Psi(t) - \Phi_\alpha(t)||$ is an r-independent upper bound for the integrand, which by (3) is integrable in t. Thus the Lebesgue dominated convergence theorem allows us to interchange the limit $r \to \infty$ and the t integration. But the $r \to \infty$ limit of the integrand is $||\Psi(t)||^2 - ||\Phi_\alpha(t)||^2 = ||\Omega_\alpha^{(-)} f_\alpha||^2 - ||f_\alpha||^2 = 0$, by (1). A similar but more elaborate analysis shows that the integrand of the remaining integral has an r-independent integrable bound. Again the dominated convergence theorem applies. In this case the integrand is $\lim_{r\to\infty} [||P_r \Psi(t)||^2 - \sum_\beta ||P_r \Phi'_\beta(t)||^2] = ||f_\alpha||^2 - \sum_\beta ||f'_\beta||^2 = 0$. The last equality is a consequence of the unitarity relation in (2).

The computation of the $r \to \infty$ limit of $\tilde{T}(f_\alpha;r)$ is too long to report in detail, so we shall indicate just a few of the major steps. If we do the t-integration in $\tilde{T}(f_\alpha;r)$ one is lead to,

$$\tilde{T}(f_\alpha;r) = i\sum_\beta \int d\vec{Q}'_\alpha d\vec{Q}_\alpha (\tilde{Q}'^2_\alpha - \tilde{Q}^2_\alpha)^{-1} f^*_\alpha(\vec{Q}'_\alpha) < \vec{Q}'_\alpha | K^\beta_\alpha(r) | \vec{Q}_\alpha > f_\alpha(\vec{Q}_\alpha), \quad (7)$$

where $K^\beta_\alpha(r) = S^\dagger_{\beta\alpha} J^\dagger_\beta P_r J_\beta S_{\beta\alpha} - \delta_{\alpha\beta} J^\dagger_\alpha P_r J_\alpha$. Here $\vec{Q}_\alpha$ is the relative momentum of the two colliding clusters incident in channel α, and $\tilde{Q}^2_\alpha$ is the relative motion kinetic energy. The singular integral in (7) is a principal-value integral. By using the fourier transform properites of the sphere P_r we can explicitly compute the $r \to \infty$ limit of (7). It is finite and consists of a momentum space wave packet average of the S-matrix times energy derivatives of the S-matrix. Let us indicate the mechanism for introducing the energy derivative. As an example consider a typical term from $K^\beta_\alpha(r)$, say $\delta_{\alpha\beta} J^\dagger_\alpha P_r J_\alpha$. The momentum space value of the projection operator P_r is a Bessel function. If the singular factor $(\tilde{Q}^2_\alpha - \tilde{Q}'^2_\alpha)^{-1}$ where absent from (7) then the $r \to \infty$ limit would lead to a δ-function $\delta(\vec{Q}_\alpha - \vec{Q}'_\alpha)$. However, it is an exercise in integration by parts to show that the result with singular factor is altered to

$$\lim_{r\to\infty} \int d\vec{Q}'_\alpha d\vec{Q}_\alpha f^*_\alpha(\vec{Q}'_\alpha) \frac{<\vec{Q}'_\alpha | J^\dagger_\alpha P_r J_\alpha | \vec{Q}_\alpha>}{Q'^2_\alpha - Q^2_\alpha} f_\alpha(\vec{Q}_\alpha) = - \int d\vec{Q}'_\alpha f^*_\alpha(\vec{Q}'_\alpha) \frac{d}{dQ_\alpha} \left| \left(\frac{Q_\alpha}{Q'_\alpha}\right) \frac{f_\alpha(Q_\alpha \hat{Q}_\alpha)}{Q_\alpha + Q'_\alpha} \right|_{Q_\alpha = Q'_\alpha}$$

The remaining factor in $K^\beta_\alpha(r)$, namely $S^\dagger_{\beta\alpha} J^\dagger_\beta P_r J_\beta S_{\beta\alpha}$, leads energy derivatives that act on the S-matrix as well as the wave packet f_α. When all the terms in $K^\beta_\alpha(r)$ are summed over β the terms with a energy derivative acting on a wave packet cancell. All surviving terms have the energy derivative restricted to the S-matrix elements.

We summarize the final result in the next three equations. Refering to (4) we find that the t-integration leads to a δ-function in energy, so we may write it in terms of an on-shell kernel $q_\alpha(E;r)$

$$T_\alpha(f_\alpha;r) = \int d\vec{Q}'_\alpha d\vec{Q}_\alpha (m_\alpha Q_\alpha)^{-1} \delta(E_\alpha - E'_\alpha) f^*_\alpha(\vec{Q}'_\alpha) < \hat{Q}'_\alpha | q_\alpha(E_\alpha;r) | \hat{Q}_\alpha > f_\alpha(\vec{Q}_\alpha) . \quad (9)$$

Here m_α is the reduced mass that relates momentum to kinetic energy $\tilde{Q}^2_\alpha = \vec{Q}^2_\alpha / 2m_\alpha$.

Let us also remove the energy conserving δ-function from the S-matrix, viz.

$$< \vec{Q}_\beta | S_{\beta\alpha} | \vec{Q}_\alpha > \; = \; \delta(E_\beta - E_\alpha)(m_\beta Q_\beta^{n-2} m_\alpha Q_\alpha)^{-\frac{1}{2}} \; < \hat{Q}_\beta | s_{\beta\alpha}(E_\alpha) | \hat{Q}_\alpha > \quad , \tag{10}$$

where n is the dimension of the vector $\vec{Q}_\beta$ describing channel β. In these formula E_α represent the total energy available in channel α, i.e. $\tilde{Q}_\alpha^2 - \varepsilon^\alpha$ where ε^α is the total binding energy of all clusters in channel α. Equation (10) defines an on-shell energy dependent S-matrix $s_{\alpha\beta}(E)$. Our principal result may be stated

$$q_\alpha(E) \equiv \text{w-}\lim_{r\to\infty} q_\alpha(E;r) = -i \sum_\beta s^\dagger_{\beta\alpha}(E) \frac{d}{dE} s_{\beta\alpha}(E) \quad . \tag{11}$$

The theory of time delay is of interest for several reasons. It represents an observable that is a residual consequence of the detailed time-dependent dynamical evolution of the N-body scattering system. For two-particle systems considered in a specific partial wave the time delay is proportional to the energy derivative of the phaseshift. Thus the general theory of time delay provides a method of defining a universal phase-shift like functional that is a characteristic of the scattering process. So far the applications of time delay have exploited this analogy with the phaseshift. The theory has been used to study the role of the collision process in statistical mechanics. A good example of what one can find with this method is the two-body case. The second virial coefficient, giving the non-ideal gas law behavior, is the Laplace transform of the trace of the time-delay operator, tr q(E). In fact one may obtain a infinite class of moment relations for this function. These relations involve $E^\ell \mathrm{tr} q(E)$ where ℓ is an integer. When ℓ = o the corresponding moment statement is equivalent to Levinson's theorem[4)]. By systematically utilizing all the moment relations one can obtain an explicit inverse temperature expansion of the second virial coefficient[5)]. In the few and N-body case, two-Hilbert space scattering theory together with the time-delay formalism has been used to give a new definition of the canonical partition function suitable for describing states that are a mixture of free particles and bound clusters[6)]. The multispecies density expansion of the equation of state again involves only Laplace transforms of the trace of the time-delay operator $q_\alpha(E)$. Thus the equation of state is sensitive only to the time delay of the collision process.

References

1. C. Chandler and A.G. Gibson, J. Math. Phys. 14, 1328 (1973).
2. W. Hunziker, in Lectures in Theoretical Physics edited by A.O. Barut and W.E. Brittin (Gordon and Breach, New York, 1968), Vol. X-A.
3. Ph.A. Martin, Commun. Math. Phys. 47, 221 (1976).
4. T.A. Osborn and D. Bollé, J. Math. Phys. 18, 432 (1976).
5. D. Bollé and H. Smeeters, Phys. Letts. 62A, 290 (1977).
6. T.A. Osborn, Phys. Rev. A16, 334 (1977).

DERIVATION OF AN APPROXIMATE THREE-BODY MODEL OF DEUTERON STRIPPING FROM N-BODY SCATTERING THEORY

G. Cattapan and V. Vanzani

Istituto di Fisica, Università di Padova, 35100 Padova - Italy
Istituto Nazionale di Fisica Nucleare - Sezione di Padova e L.N. di Legnaro

Owing to the weakly bound structure of the deuteron, one is led to suspect calculations treating the deuteron as an elementary particle. Johnson and Tandy (JT) [1] proposed an approximate three-body model of deuteron stripping, by which three-body effects due to deuteron breakup in the nuclear field can be included in a rather simple and practical way. In the first order approximation, the model can be related to the so-called adiabatic model of deuteron stripping, which has been rather successful in several phenomenological applications. Here, we shall derive the JT model, starting from general N-body scattering theory, and pointing out the relevant approximations one has to introduce. Let us recall the distribution property [2].

$$V^{b_2} = \sum_{(b_2\neq)c_2(\supset b_3)} V^{b_3}_{c_2} \tag{1}$$

where, as usual, a_i denotes a partition of the N bodies into i clusters, $b_3 \subset b_2$, V^{b_2} is the interaction external to the two-cluster partition b_2, and $V^{b_3}_{c_2}$ the interaction internal to c_2 but external to the three-cluster partition b_3. By means of Eq. (1) one can derive a set of three coupled equations for the N-body transition operators $U^{b_2a_2}$, in correspondence to each three-cluster partition b_3 [3]. Full connectivity of the (iterated) kernel is never achieved for these equations; however, if each cluster in b_3 is assumed to be tightly bound (or elementary), partially connected formalisms can be regarded as reasonable approximations, and one obtains

$$U^{b_2a_2} = G^{-1}_{b_3}\,\bar{\delta}_{b_2a_2} + \sum_{c_2\neq b_2} T^{b_3}_{c_2}\, G_{b_3}\, U^{c_2a_2} \tag{2}$$

with $b_3 \subset a_2$, b_2, c_2 (G_{a_i} is the a_i-partition resolvent operator; $V^{b_3}_{c_2}G_{c_2} = T^{b_3}_{c_2}\, G_{b_3}$). Let us now introduce the usual three-body notation. If a,b,c are the three clusters in b_3, then $G_{b_3} \equiv G_o$ can be regarded as an effective free resolvent operator, $T^{b_3}_{c_2} \equiv T_c$ is the effective two-body T-operator for the a-b scattering, and Eqs.(2) take the form of the well-known three-body Feddeev-like equations for $U^{b_2a_2} \equiv U^{b\,a}$. In our case, a is the (tightly bound) target nucleus, b the proton and c the neutron transferred from the deuteron (b+c) to the residual nucleus (a+c).

Eliminating $U^{c\,a}$ in Eqs.(2) for $U^{b\,a}$, $U^{c\,a}$, $U^{a\,a}$ one gets

$$U^{b\,a} = (G_o^{-1}+T_c)(1+G_oT_aG_oU^{a\,a}) + T_cG_oT_bG_oU^{b\,a} \tag{3}$$

Introducing $T_{c(a)} = V_c + V_c\,G^a\,V_c$ (G^a the resolvent constructed with V^a) and combining (3) with the integral equation

$$T_{c(a)} = T_c + T_{c(a)} G_o T_b G_o T_c \tag{4}$$

for $T_{c(a)}$, one obtains the exact representation

$$T^{ba} = \langle \Phi_b | (1 + T_{c(a)} G_o T_b G_o) (G_o^{-1} + T_c) (1 + G_o T_a G_o U^{aa}) | \Phi_a \rangle \tag{5}$$

for the stripping transition amplitude $T^{ba} = \langle \Phi_b | U^{ba} | \Phi_a \rangle$ ($|\Phi_a\rangle$ the a-channel state). In the usual heavy-target approximation, one can take the proton-target interaction V_c as distorting potential in the final channel b; thus, $\langle \Phi_b | (1+T_{c(a)} G_o T_b G_o)$ can be approximated by $\langle f_b^{(-)} | \langle \psi_b | V_b G_o$, with $|f_b^{(-)}\rangle$ the two-body optical scattering state in channel b, and $|\psi_b\rangle$ the bound-state of the residual nucleus (a+c). If the unitary pole expansion (UPE) for the neutron-proton scattering operator T_a is introduced [1] ($T_a = \sum_{r,s} |a\,r\rangle t_{a,rs} \langle a\,s|$; $|a\,i\rangle = V_a|\phi_i\rangle$, with $\{|\phi_i\rangle\}$ a set of Weinberg states evaluated in correspondence to the deuteron binding energy) one has

$$T^{ba} = \langle f_b^{(-)} | \langle \psi_b | V_b G_c \sum_r |a\,r\rangle | F^{(+)}_{ar,a1} \rangle \tag{6}$$

where

$$|F^{(+)}_{ar,a1}\rangle = N_a (\delta_{r1} + \sum_s t_{a,rs} u_{as,a1}) |\vec{p}_a\rangle \tag{7}$$

and $u_{ar,a1} = \langle ar | G_o U^{aa} G_o | a1 \rangle$. Apart from the unessential normalization factor N_a, the states $|F^{(+)}_{ar,a1}\rangle$ coincide with the generalized spectator states considered in Ref. [4]. Practically, the three-body state on the left of T^{ba} in (5) has been written as the product of two two-body states in virtue of the heavy-target approximation, while the three-body state on the right of T^{ba} has been expanded in products of two-body states: the form-factor state-vectors $|ar\rangle$, coming from the UPE for T_a, and the generalized spectator states $|F^{(+)}_{ar,a1}\rangle$, belonging to the a-channel relative motion space. The "effective transition potentials" $\langle \psi_b | V_b G_c | ar \rangle$ admit of a simple physical interpretation in terms of a polar diagram, describing the c-transfer, and of a triangular one, describing both the c-transfer and the a-b off-shell scattering [4]. Since in heavy-target approximation, $\langle f_b^{(-)} | \langle \psi_b | V_b$ can be replaced by $\langle f_b^{(-)} | \langle \psi_b | G_c^{-1}$, one recognizes that the polar and triangular contributions contained in (6) sum up to give the JT representation for the deuteron stripping transition amplitude.

References

1. Johnson R.C., Tandy P.C.: Nucl. Phys. 235 A, 56 (1974).
2. Vanzani V.: Lett. Nuovo Cim. 16, 1 (1976).
3. Bencze Gy., Vanzani V.: Contribution to the Intern. Symposium on Nuclear Reaction Models (Balatonfüred, 1977).
4. Vanzani V.: in Few-Body Dynamics, edited by A.N. Mitra et al. (North-Holland, Amsterdam, 1976), p. 394.

TIME DEPENDENT APPROACH TO THE COLLISION OF TWO CHARGED COMPOSITE PARTICLES

E.O. Alt

Institut für Physik, Universität Mainz, Mainz, West-Germany

W. Sandhas

Physikalisches Institut, Universität Bonn, Bonn, West-Germany

Scattering states and amplitudes are most naturally introduced by means of Møller operators defined as limits of time evolution operators. It is the purpose of the following investigation to demonstrate how this concept can be applied to collision processes of two fragments each containing an arbitrary number of charged particles. In this way we corroborate a representation of the corresponding transition amplitudes as a superposition of a pure Coulomb and a Coulomb-modified short-range term which was derived previously on the basis of integral equations[1] and resolvent identities[2].

Let us consider the Hamiltonian for N particles interacting via short-ranged pair potentials V_{ij}^{S} and screened Coulomb potentials V_{ij}^{R},

$$H^{(R)} = H_0 + \sum_{i<j}^{N} V_{ij}^{S} + \sum_{i<j}^{N} V_{ij}^{R}. \tag{1}$$

Denoting two-fragment partitions by a, b, ... the Hamiltonian is split according to $H^{(R)} = H_a^{(R)} + \overline{V}_a^{(R)}$ into a channel Hamiltonian $H_a^{(R)}$ and a channel interaction

$$\overline{V}_a^{(R)} = \overline{V}_a^{S} + \overline{V}_a^{R} = \sum_{i<j,(ij)\not\subset a} V_{ij}^{S} + \sum_{i<j,(ij)\not\subset a} V_{ij}^{R} . \tag{2}$$

For finite screening radius R the two-fragment Møller operator exists as (strong) limit

$$\Omega_{a(R)}^{(\pm)} = \lim_{t\to\mp\infty} e^{iH^{(R)}t} e^{-iH_a^{(R)}t} \tag{3}$$

We, furthermore, introduce a (screened) Coulomb potential v_a^R acting between the total charges of the two clusters conceived to be accumulated in their respective centers of mass. That is, v_a^R is obtained by replacing in $\overline{V}_a^R$ the relative coordinates $\vec{r}_{ij}$ between particles i and j by the two-cluster relative variable $\vec{\rho}_a$:

$$\sum_{i<j,(ij)\not\subset a} e_i e_j / r_{ij}\, f_{ij}^R(r_{ij}) \longrightarrow \sum_{i<j,(ij)\not\subset a} e_i e_j/\rho_a\, f_a^R(\rho_a) = v_a^R(\rho_a) \tag{4}$$

(f_{ij}^R and f_a^R are the screening functions). Herewith, the Coulomb-distorted channel Hamiltonian is defined as

$$h_a^R = H_a^{(R)} + v_a^R . \tag{5}$$

Thus, the full Møller operator (3) can be decomposed into the product

$$\Omega_{a(R)}^{(\pm)} = \lim_{t\to\mp\infty} e^{iH^{(R)}t} e^{-ih_a^R t} e^{ih_a^R t} e^{-iH_a^{(R)}t} = \Omega_{a,SR}^{(\pm)}\, \omega_{a,R}^{(\pm)}, \tag{6}$$

which evidently exists for finite R. When the full, the Coulomb-modified short-range, and the pure Coulomb S- and T-operators (with screening) are introduced as

$$S_{ba}^{(R)} = \Omega_{b(R)}^{(-)\dagger} \Omega_{a(R)}^{(+)} = \delta_{ba} - 2\pi i\, T_{ba}^{(R)} \tag{7}$$

$$S_{ba}^{SR} = \Omega_{b,SR}^{(-)\dagger} \Omega_{a,SR}^{(+)} = \delta_{ba} - 2\pi i\, T_{ba}^{SR} \tag{8}$$

$$s_a^R = \omega_{a,R}^{(-)\dagger} \omega_{a,R}^{(+)} = 1 - 2\pi i\, t_a^R , \tag{9}$$

the following relationship between the latter can be derived by means of (6)

$$T_{ba}^{(R)} = \delta_{ba} t_a^R + \omega_{b,R}^{(-)} T_{ba}^{SR} \omega_{a,R}^{(+)} \quad . \tag{10}$$

Applied onto two-fragment channel states $|\psi_a>|\vec{q}_a>$, built up by products $|\psi_a>=|\psi_a^{(1)}>\times|\psi_a^{(2)}>$ of cluster wave functions and the relative momentum states $|\vec{q}_a>$, the Møller operators $\omega_{a,R}^{(\pm)}$ map the latter onto the two-body scattering states $|\vec{q}_{a,R}^{(\pm)}>$,

$$\omega_{a,R}^{(\pm)}|\psi_a>|\vec{q}_a> = |\psi_a>|\vec{q}_{a,R}^{(\pm)}>, \tag{11}$$

which describe pure (screened) Coulomb scattering of the centers of mass of the two fragments of partition a. The corresponding amplitude be denoted by $t_a^R(\vec{q}_a',\vec{q}_a)$. Then, sandwiched between $|\psi_a>|\vec{q}_a>$ the basic relation (10) becomes

$$T_{ba}^{(R)}(\vec{q}_b',\vec{q}_a) = \delta_{ba} t_a^R(\vec{q}_a',\vec{q}_a) + <\vec{q}_{b,R}'^{(-)}|<\psi_b|T_{ba}^{SR}|\psi_a>|\vec{q}_{a,R}^{(+)}> \quad . \tag{12}$$

The transition to zero screening can now be performed in a straightforward way. Multiplying $|\vec{q}_{a,R}^{(\pm)}>$ and $t_a^R(\vec{q}_a',\vec{q}_a)$ by the usual two-body renormalization factors the unscreened Coulomb scattering states $|\vec{q}_{a,C}^{(\pm)}>$ and Coulomb amplitudes $t_a^C(\vec{q}_a',\vec{q}_a)$ are obtained for $R\to\infty$. Furthermore, the Coulomb-modified short-range Møller operator exists (as strong limit) on $|\psi_a>|\vec{q}_{a,C}^{(\pm)}>$ even after the screening is switched off,

$$\Omega_{a,SC}^{(\pm)} = \lim_{t\to\mp\infty} e^{iHt}\, e^{-ih_a^C t} \tag{13}$$

For, in $H-h_a^C=\overline{V}_a^S+\overline{V}_a^C-\upsilon_a^C$ not only the short-ranged part $\overline{V}_a^S$, but also the difference $(\overline{V}_a^C - \upsilon_a^C)$ vanishes sufficiently fast for large relative distance ρ_a between the two clusters, when applied onto the two-fragment states $|\psi_a>|\vec{q}_a>$ (compare Eq.(4) and, for more details, Refs.2). The Coulomb-modified short-range transition operators T_{ba}^{SC} are, therefore, defined by (8) with $\Omega_{a,SR}^{(\pm)}$ replaced by the operators (13). Consequently, relation (12), multiplied by two-body renormalization factors, yields in the zero screening limit the scattering amplitude for two charged fragments[1,2)]

$$T_{ba}(\vec{q}_b',\vec{q}_a) = \delta_{ba} t_a^C(\vec{q}_a',\vec{q}_a) + <\vec{q}_{b,C}'^{(-)}|<\psi_b|T_{ba}^{SC}|\psi_a>|\vec{q}_{a,C}^{(+)}> \quad . \tag{14}$$

We should emphasize that this result can also be obtained without any screening procedure. In fact, the limit (3) exists if, instead of $H^{(R)}$, the unscreened Hamiltonian H, and, instead of $H_a^{(R)}$, Dollard's modified channel Hamiltonian is used[3)]. With these replacements, and with h_a^C instead of h_a^R in (6), we immediately get the representation (14).

1. E.O. Alt, in Few Body Dynamics, ed. by A.N. Mitra et al. (North Holland, Amsterdam, 1976); E.O. Alt, W. Sandhas, H. Zankel and H. Ziegelmann, Phys. Rev. Lett. 37, 1537 (1976); E.O. Alt, W. Sandhas and H. Ziegelmann, to be published in Phys. Rev.C; E.O. Alt, Invited Talk at the Workshop on Few Body Problems in Nuclear Physics, Trieste, 1978; E.O. Alt, Invited Contribution to this Conference.

2. E.O. Alt and W. Sandhas, Contribution to this Conference, and to be published.

3. J.D. Dollard, J. Math. Phys. 5, 729 (1964)

SCATTERING AMPLITUDES FOR TWO CHARGED FRAGMENTS

E.O. Alt
Institut für Physik, Universität Mainz, Mainz, West-Germany
W. Sandhas
Physikalisches Institut, Universität Bonn, Bonn, West-Germany

In recent publications[1)] three-body scattering processes with two or three charged particles have been investigated by employing the quasiparticle concept which leads to the formulation of the three-body problem as an effective two-body one. This enabled us to rigorously define scattering amplitudes for (in-)elastic and rearrangement collisions by making use of the screening procedure well known in the genuine two-charged particle theory, and to derive manageable (one-dimensional) integral equations.

Here we present an alternative approach[2)] which is based on the resolvent equations and fundamental operator relations of the problem. In this way the above mentioned definitions of scattering amplitudes are reproduced in a direct and transparent manner. Moreover, all derivations can be performed immediately for an arbitrary number of charged particles.

Let us denote by a,b,... two-fragment partitions of N distinguishable particles interacting via short-ranged pair potentials V^S_{ij} and screened Coulomb potentials V^R_{ij}. Besides the full resolvent

$$G^{(R)}(z) = (z - H_0 - \sum_{i<j}^{N} V^S_{ij} - \sum_{i<j}^{N} V^R_{ij})^{-1} \tag{1}$$

we introduce the channel resolvent $G^{(R)}_a(z) = (z - H_0 - V^S_a - V^R_a)^{-1} = (z - H^{(R)}_a)^{-1}$ which contains only the potentials acting within the two clusters. The channel interaction consists of all potentials not occurring in $H^{(R)}_a$ and is denoted by

$$\overline{V}^S_a + \overline{V}^R_a = \sum_{i<j,(ij)\not\subset a} (V^S_{ij} + V^R_{ij}). \tag{2}$$

In addition we define a screened Coulomb potential v^R_a by replacing in all Coulomb potentials $V^R_{ij}(\vec{r}_{ij})$ occurring in $\overline{V}^R_a$ the variables $\vec{r}_{ij}$ by the relative coordinate $\vec{\rho}_a$ between the centers of mass of the two fragments of partition a. In other words,

$$v^R_a(\vec{\rho}_a) = \sum_{i<j,(ij)\not\subset a} e_i e_j/\rho_a \, f^R_a(\rho_a) \tag{3}$$

can be interpreted as the Coulomb interaction between the total charges of the two clusters concentrated in their respective centers of mass. Here $f^R_a(\rho_a)$ represents a suitably chosen screening function. With the help of v^R_a a Coulomb distorted channel resolvent is introduced, $g^R_a(z) = (z - H^{(R)}_a - v^R_a)^{-1}$. Extracting from the full resolvent $G^{(R)}$ the resolvents $G^{(R)}_a$ or g^R_a, the total transition operator $U^{(R)}_{ba}$ and the Coulomb-modified short-range transition operator U^{SR}_{ba}, respectively, are defined similarly to Ref. 3 as

$$G^{(R)} = \delta_{ba} G^{(R)}_a + G^{(R)}_b U^{(R)}_{ba} G^{(R)}_a \tag{4}$$

and

$$G^{(R)} = \delta_{ba}\, g_a^R + g_b^R\, U_{ba}^{SR}\, g_a^R. \qquad (5)$$

They are related by means of a Gell-Mann-Goldberger-type formula

$$U_{ba}^{(R)} = \delta_{ba}\, t_a^R + (1 + t_b^R\, G_b^{(R)})\, U_{ba}^{SR}\, (1 + G_a^{(R)} t_a^R), \qquad (6)$$

with the operator t_a^R being determined via $g_a^R = G_a^{(R)} + G_a^{(R)} t_a^R\, G_a^{(R)}$.

In order to perform the zero screening limit the fundamental relation (6) has to be sandwiched between channel states $|\Phi_a\rangle = |\psi_a^{(1)}\rangle|\psi_a^{(2)}\rangle|\vec{q}_a\rangle$ which are products of the bound state wave functions of the two fragments of partition a and of the plane waves describing their relative motion, and which belong to the energy E_a,

$$\langle\Phi_b|U_{ba}^{(R)}(E_a+io)|\Phi_a\rangle = \delta_{ba} t_a^R(\vec{q}_a{}',\vec{q}_a) + \langle\vec{q}\,'^{(-)}_{b,R}|\langle\psi_b^{(2)}|\langle\psi_b^{(1)}|U_{ba}^{SR}(E_a+io)|\psi_a^{(1)}\rangle|\psi_a^{(2)}\rangle|\vec{q}^{\,(+)}_{a,R}\rangle. \qquad (7)$$

Here $t_a^R(\vec{q}_a{}',\vec{q}_a)$ is the two-body on-shell amplitude for screened Coulomb scattering of the centers of mass of the two clusters, and $|\vec{q}^{\,(\pm)}_{a,R}\rangle$ are the corresponding scattering states. After multiplication by the usual two-body renormalization factors they go over for $R \to \infty$ into the unscreened quantities $t_a^C(\vec{q}_a{}',\vec{q}_a)$ and $|\vec{q}^{\,(\pm)}_{a,C}\rangle$.

Thus, still the existence of the unscreened quantity

$$\langle\Phi_b|U_{ba}^{SC}(E_a+io)|\Phi_a\rangle = \ldots\ldots + \langle\Phi_b|(\overline{V}_b-v_b^C)G(E_a+io)(\overline{V}_a-v_a^C)|\Phi_a\rangle \qquad (8)$$

remains to be shown for all values of the momenta. This, however, follows from the fact that G appears here between normalizable states $(\overline{V}_a^S+\overline{V}_a^C-v_a^C)|\Phi_a\rangle$. Indeed, since $\overline{V}_a^S$ is of short range, $\|\overline{V}_a^S\Phi_a\|$ exists. Moreover, denoting collectively by $\vec{x}_a^{(i)}$ the internal variables of cluster i of fragmentation a, the part $(\overline{V}_a^C-v_a^C)|\Phi_a\rangle$ reads in position space

$$\langle\vec{\rho}_a|\langle\vec{x}_a^{(1)}|\langle\vec{x}_a^{(2)}|(\overline{V}_a^C-v_a^C)|\Phi_a\rangle = \sum_{\substack{i<j\\(ij)\notin a}} e_i e_j (r_{ij}^{-1}-\rho_a^{-1})\psi_a^{(1)}(\vec{x}_a^{(1)})\psi_a^{(2)}(\vec{x}_a^{(2)})\frac{e^{i\vec{q}_a\cdot\vec{\rho}_a}}{(2\pi)^{3/2}} . \qquad (9)$$

The strong decrease of the cluster wave functions justifies a multipole expansion of r_{ij}^{-1} in powers of ρ_a^{-1}. Consequently, the first nonvanishing term of $(r_{ij}^{-1}-\rho_a^{-1})$ is of the order ρ_a^{-2}, which guarantees the normalizability of the wave function (9).

Hence, multiplying the whole equation (7) by the familiar two-body renormalization factors the limit $R \to \infty$ exists yielding the following representation of the transition amplitude for the scattering of two charged fragments

$$T_{ba}(\vec{q}_b{}',\vec{q}_a) = \delta_{ba} t_a^C(\vec{q}_a{}',\vec{q}_a) + \langle\vec{q}\,'^{(-)}_{b,C}|\langle\psi_b^{(2)}|\langle\psi_b^{(1)}|U_{ba}^{SC}(E_a+io)|\psi_a^{(1)}\rangle|\psi_a^{(2)}\rangle|\vec{q}^{\,(+)}_{a,C}\rangle. \qquad (10)$$

1. E.O. Alt, in Few Body Dynamics, ed. by A.N. Mitra et al.(North Holland, Amsterdam, 1976); E.O. Alt, W. Sandhas, H. Zankel and H. Ziegelmann, Phys. Rev. Lett. 37,1537 (1976); E.O. Alt, W. Sandhas and H. Ziegelmann, to be published in Phys. Rev. C; E.O. Alt, Invited Talk at the Workshop on Few Body Problems in Nuclear Physics, Trieste, 1978; E.O. Alt, Invited Contribution to this Conference.

2. E.O. Alt and W. Sandhas, to be published.

3. E.O. Alt, P. Grassberger and W. Sandhas, Nucl. Phys. B2, 167 (1967).

ON OFF-SHELL TRANSFORMATIONS AND WAVE FUNCTION FORMALISMS IN MANY-BODY SCATTERING THEORIES

F. S. Levin* and J. M. Greben*
Physics Department, Brown University
Providence, Rhode Island 02912/USA

Most of the theoretical and calculational effort in the area of n-particle scattering theory has been based on various sets of transition operator equations. Nevertheless, a wave function approach is not without interest [1,2]. We have recently completed a study of the role that wave function formalisms may play in n-particle scattering [3] and report a few results here. Amongst our findings, we particularly stress the fact that wave function formalisms can have a previously unnoticed interpretive and predictive function: their non-existence predicts that an approximate numerical calculation will fail to yield unitary amplitudes.

The general form of the transition operator equations for almost all n-particle scattering theories is

$$R(z) = D(z) + C(z)R(z) , \tag{1}$$

where R is a matrix of transition operators, D is the inhomogeneous or driving term, C is the kernel, and z is a complex energy parameter. An element or a linear combination of elements of R will act on an asymptotic state Φ_k in incident channel k. Let $\underset{\sim}{\Phi}$ be a column vector of such states: $(\underset{\sim}{\Phi})_j = \delta_{jk}\Phi_k$. Then the only sensible definition of a vector of wave function components $\underset{\sim}{\psi}$ is [3]

$$R(E + i0)\ \underset{\sim}{\Phi} = D(E + i0)\ \underset{\sim}{\psi} , \tag{2}$$

from which it follows that

$$\underset{\sim}{\psi} = \underset{\sim}{\Phi} + D^{-1}(E + i0)\ C(E + i0)\ \underset{\sim}{\psi} . \tag{3}$$

When either C(z) factors, i.e., $C(z) = D(z)\ \mathcal{G}(z)$, where $\mathcal{G}(E + i0)$ is a diagonal matrix of outgoing wave Green's functions, or when R obeys both the right-handed equation (1) and a corresponding left-handed equation, then (3) defines components displaying the correct asymptotic boundary conditions leading to (1), such that the sum of the components is the Schrödinger wave function. It thus follows that not all transition operator formalisms have corresponding (proper) wave function components. On the other hand, more than one set of transition operators may correspond to the <u>same</u> set of wave function components. The following statement is verified for the n = 3 case, and we conjecture that it is valid in general: If R is a matrix of transition operators for which (proper) wave function components exist, then these components will (will not) exist for operators which are phase equivalent to R through a multiplicative (additive) transformation. Thus wave function components exist for the AGS operators [4] but not for the Lovelace operators [5], and for the CCA operators obtained using a CPA [6] but not for the Cattapan-Vanzani "symmetric" version of these operators [7]. In addition, these components do not exist for the Baer-Kouri operators [8], a fact that allows for a straightforward explanation of the non-unitary (approximate) results found by Baer and Kouri [8] and by Lewanski and Tobocman [9] using the Baer-Kouri equations. Finally we note that the AGS transition operators [4] and the CCA transition operators obtained using a Faddeev-Lovelace channel coupling array [6], which are related by a multiplicative off-shell term [10], correspond to the same set of wave function components, those originally derived by Faddeev [11]. This is a particularly interesting result, since the asymptotic form of the wave function component equations yield the CCA transition operators, not the AGS operators. Nevertheless, the latter operators are true transition operators, while the former are defined only as right-half-shell quantities [10]. Details concerning this investigation can be found in Ref. [3].

REFERENCES

*Work supported in part by the United States Department of Energy.

[1] D. J. Kouri, H. Krüger and F. S. Levin, Phys. Rev. D15, 1156 (1977).

[2] Gy. Bencze and P. C. Tandy, Phys. Rev. C16, 564 (1977); M. L. Huillier, E. F. Redish, and P. C. Tandy, Phys. Rev. C (in press).

[3] F. S. Levin and J. M. Greben, submitted for publication.

[4] E. O. Alt, P. Grassberger and W. Sandhas, Nucl. Phys. B2, 167 (1967).

[5] C. Lovelace, Phys. Rev. 135, B1225 (1964).

[6] D. J. Kouri and F. S. Levin, Nucl. Phys. A250, 127 (1975).

[7] C. Cattapan and V. Vanzani, Nuovo Cimento 41, 553 (1977).

[8] M. Baer and D. J. Kouri, J. Math. Phys. 14, 1637 (1973).

[9] A. J. Lewanski and W. Tobocman, Phys. Rev. C (in press).

[10] D. J. Kouri, F. S. Levin and W. Sandhas, Phys. Rev. C13, 1825 (1976).

[11] L. D. Faddeev, Sov. Phys. J.E.T.P. 12, 1014 (1961).

HAMILTONIAN FORMULATION OF N-BODY THEORIES†

Wayne N. Polyzou‡
Department of Physics and Astronomy
University of Maryland, College Park, Maryland 20742 U.S.A.

Edward F. Redish*
NASA/Goddard Space Flight Center, Greenbelt, Maryland 20771 U.S.A.

Since the work of Faddeev, research in few-body theory has taken on many different directions. Two interesting areas of current research involve developing generalized Faddeev equations for N>3, and constructing phenomenological three-body models for systems of more than three particles. As both experimental and computational techniques get more refined, it becomes increasingly important to understand how few-body models are imbedded in full N-particle theories. This is important for both understanding the structure of effective interactions and for generating the corrections due to many-body effects.

To carry out such a program one needs a good understanding of how the few-body physics is imbedded in the N-body operators. Our approach is to use solutions of fewer (<N)-body problems to extract as much physics from the full Hamiltonian as can be obtained without actually solving the N-body problem. To do this we utilize the general combinatoric theorem:[1,2]

$$[A]_{disc} = \sum_a{}' \, C_a A_a \tag{1}$$

(where $[A]_{disc}$ is the disconnected part of the operator A, A_a is the sum of all parts of A with connectivity internal to the n_a clúster partition a, and $C_a = (-)^{n_a}(n_a-1)!$) to express the full Hamiltonian in terms of all proper (n_a>1) partition Hamiltonians[2]

$$H = \sum_a{}' \, C_a H_a . \tag{2}$$

We let A_0 denote the set of asymptotic channels of H, and divide it into the physically important channels, A, and the remainder, A'. We call a set of asymptotic channels a reaction mechanism (RM). By making a spectral resolution of each partition Hamiltonian, identifying the parts corresponding to each RM, we can use (2) to obtain a decomposition of the full Hamiltonian by RM:

$$H = H(A) + H(A'). \tag{3}$$

If we let P(A) be the projector on the subspace spanned by those eigenstates of the full H corresponding to the channels A, one can show that H(A) is the disconnected part of P(A)H. Since all fewer body solutions are disconnected operators in N-body Hilbert space, it follows that this decomposition by RM is optimal with respect to the available fewer body physics. We use this decomposition as a starting point for discussing the physics contained in N-body operators. In particular we define the RM resolvents $G(A) = (z - H(A))^{-1}$ and $G(A') = (z - H(A'))^{-1}$. One can use the RM Hamiltonian H(A) (resp. H(A')) to construct effective interaction and partition resolvents for each RM, using only solutions of fewer body problems. Using these operators one can derive dynamic equations for G(A) and G(A') whose iterated kernels are connected for any choice of $A \cup A' = A_o$. These solutions are related to the resolvent of the full Hamiltonian by the equation:

$$G = G(A) + G(A)H(A')G(A') + G(A)H(A')G(A')H(A)G \tag{4}$$

which also has a connected kernel for any choice of $A \cup A' = A_o$.

These equations can be recast in transition operator form. The RM transition operators, $T^{ab}(A)$ (resp. A'), that arise can be shown to *satisfy a truncated optical theorem* where all scattered flux is conserved and comes out in one of the open channels of A (resp. A'). For states of A one can show:

$$T^{ab} = T^{ab}(A) + \Omega^a(A)^\dagger \hat{T}(A')\Omega^b(A) \qquad \text{(on shell)}. \tag{5}$$

The operators $\Omega^a(A)^\dagger$ and $\Omega^b(A)$ are wave operators for the A dynamics, while $\hat{T}(A')$ satisfies an equation similar to (4) and guarantees that at least one A' effective interaction appears.

The formalism outlined here provides a flexible foundation for constructing and studying few-body models of N-body systems. It gives a physical, unitary division of the dynamics based only on spectral properties and solutions of fewer body problems. It provides a multi-channel connected kernel generalization of the well-known Feshbach projection operator formalism.[3] In contrast to the Feshbach approach, it permits a correct treatment of the asymptotic behavior of the many-body wave function, including rearrangement and breakup. RM's involving overlapping clusters can be treated with no additional complications. When A contains only few cluster channels, the equations for $T^{ab}(A)$ have the numerical simplicity of a few-body problem, while corrections for excluded channels can be systematically included by perturbation theory.

† Supported in part by U. S. Department of Energy.

‡ This work is based on thesis research for the degree of Ph.D. at the University of Maryland.

* N.A.S.-N.R.C. Resident Research Associate. Permanent address: Department of Physics and Astronomy, University of Maryland, College Park, MD 20742 U.S.A.

1. P. Benoist-Gueutal, M. L'Huillier, E. F. Redish and P. C. Tandy, to be published in Phys. Rev. C (1978).
2. W. Polyzou and E. F. Redish, Univ. of Maryland technical report no. 78-088, to be published.
3. H. Feshbach, Ann. Phys. (N.Y.) 5, 357 (1958); 19, 287 (1962).

QUANTAL RULES FOR NON-HERMITIAN BOUND STATE FORMALISMS BASED ON N-BODY SCATTERING THEORY

F. S. Levin*

Physics Department, Brown University

Providence, Rhode Island 02912/USA

One of the interesting recent developments in the areas of few-body physics and many-body scattering theory has been the application of some of the scattering theory equations to the structure of simple (few-body) atoms and molecules [1,2]. The results obtained to date fall into two classes: (a) that consisting of relatively accurate values of observables obtained using relatively simple approximate wave functions, and (b) that containing all other results, consisting mainly of complex energy eigenvalues, spurious (real energy) solutions, and non-monotonically converging sets of energies. Since calculations are based on approximations to a non-hermitian operator equation, questions naturally arise concerning the meaning of these two classes, the nature of the eigenvalue spectrum, and the rules for interpreting results, e.g., how to compute probability and charge densities. Aspects of the first two questions are considered in Ref. [2]; the question of interpretation is briefly considered below [3].

Let the system of interest consist of n distinguishable particles, 1..n. Effects of particle identity are to be included by taking appropriate linear combinations [1], but do not change any basic conclusions reached below. Arrangement channels are defined by partitions of the particles into bound clusters of particles, which are labelled by an index set {j}, which here runs over the set of two-cluster channels. Corresponding to these channels are partitions of the Hamiltonian H into a channel Hamiltonian H_j and a channel interaction V^j for each channel j: $H = H_j + V^j$, all j, as noted in [1] and [2]. The formalism considered here is that of arrangement channel quantum mechanics, the basic equations of which are

$$\mathcal{H}|\psi) = E|\psi) \quad , \tag{1}$$

where $\mathcal{H}$ is a non-hermitian channel-space matrix "Hamiltonian" operator given by $\mathcal{H}_{jk} = H_j\delta_{jk} + V^k\delta_{k,j+1}$ and $|\psi)$ is a column vector of channel component states $|\psi_j>$: $|\psi)_j = |\psi_j>$. Properties of the channel component states are discussed in Ref. [4]; they are related to the solution $|\Psi>$ of the Schrödinger equation (E-H) $|\Psi> = 0$ by

$$|\Psi> = \sum_j |\psi_j> \ . \tag{2}$$

In general, approximations must be introduced in order to solve (1). Approximate quantities are indicated by a tilde and obey:

$$\tilde{\mathcal{H}}|\tilde{\psi}^{(\alpha)}) = \tilde{E}^{(\alpha)}|\tilde{\psi}^{(\alpha)}) \quad , \tag{3}$$

where α labels a particular state. Since $\tilde{\mathcal{H}}^\dagger \neq \tilde{\mathcal{H}}$, $\tilde{E}$ may be complex, while $\tilde{\mathcal{H}}$ (and also $\mathcal{H}$) can only be diagonalized by a bi-orthogonal pair, the $|\tilde{\psi}^{(\alpha)})$ and their dual eigenvectors $(\tilde{\phi}^{(\beta)}|$:

$$(\tilde{\phi}^{(\beta)}|\tilde{\mathcal{H}}|\tilde{\psi}^{(\alpha)}) = \tilde{E}^{(\alpha)}(\tilde{\phi}^{(\beta)}|\tilde{\psi}^{(\alpha)}) = E^{(\alpha)}\delta_{\alpha\beta} , \tag{4}$$

or in component form

$$\sum_{j,k}<\tilde{\phi}_j^{(\beta)}|\tilde{\mathcal{H}}_{jk}|\tilde{\psi}_k^{(\alpha)}> = \tilde{E}^{(\alpha)}\sum_j<\tilde{\phi}_j^{(\beta)}|\tilde{\psi}_j^{(\alpha)}> = \tilde{E}^{(\alpha)}\delta_{\alpha\beta} , \tag{5}$$

where $(\tilde{\phi}^{(\beta)}|$ obeys $(\tilde{\phi}^{(\beta)}|\tilde{\mathcal{H}} = \tilde{E}^{(\beta)}(\tilde{\phi}^{(\beta)}|$.

The analogy of (4) and (5) with the Schrödinger eigenvalue equation

$$<\psi^{(\beta)}|H|\psi^{(\alpha)}> = E^{(\alpha)}<\psi^{(\beta)}|\psi^{(\alpha)}> = E^{(\alpha)}\delta_{\alpha\beta} \tag{6}$$

is evident. This analogy suggests a definition of probability density $\tilde{\rho}$ that is of consistent form and goes over to the exact result $\rho = \Psi^{\dagger}\Psi$ when $|\tilde{\psi}) \to |\psi)$ and $(\tilde{\phi}| \to (\phi|$. We define $\tilde{\rho}(\xi)$, where ξ denotes all coordinates, by

$$\tilde{\rho}(\xi) \equiv \sum_j \tilde{\phi}_j^{\dagger}(\xi)\tilde{\psi}_j(\xi) . \tag{7}$$

To show that $\tilde{\rho}(\xi) \to \rho(\xi)$, when $\tilde{\phi}_j^{\dagger}(\xi)$ and $\tilde{\psi}_j(\xi)$ become exact, we use the following result [3], first established by Kouri and Levin [5]:

$$\phi_j^{\dagger}(\xi) \equiv \Psi^{\dagger}(\xi) \quad , \text{ all } j. \tag{8}$$

Substitution of (8) into (7) followed by use of (2) gives $\tilde{\rho}(\xi) \to \rho$. It then follows that the single particle probability density [e.g., the charge density] is $\tilde{\rho}_1(\xi_1) = \int d\xi_2 \cdot\cdot d\xi_m \tilde{\rho}(\xi)$, etc.

This definition clearly allows one to use the results of bound state calculations based on (3) to define a charge density, which can then be compared with those from other calculations or with experiment. The most important consequence of (7) for simple (one-state) calculations, such as those for H_2^+, H_2 and HeH^+ for example [1], is that such calculations lead to charge densities without interference terms [2]. Extensions of these results are straightforward and will be discussed elsewhere [3].

REFERENCES

*Work supported in part by the United States Department of Energy

[1] F. S. Levin and H. Krüger, Phys. Rev. A15, 2147 (1977); A16, 836 (1977); F. S. Levin, Invited paper presented at the 18th Sanibel Symposia, to be published in Int. Jour. Quant. Chem.

[2] F. S. Levin, Application of Few-Body Methods to Atomic and Molecular Structure, invited paper, this Conference.

[3] F. S. Levin, submitted for publication.

[4] D. J. Kouri, H. Krüger, and F. S. Levin, Phys. Rev. D15, 1156 (1977).

[5] D. J. Kouri and F. S. Levin, Nucl. Phys. A253, 395 (1975).

METHOD OF ORTHOGONALIZED DISTORTED WAVES IN THE MANY-BODY SCATTERING THEORY

V.N. Pomerantsev and V.I. Kukulin
Institute of Nuclear Physics, Moscow State University;
Moscow 117234, USSR.

Direct reactions are widely analyzed using the distorted wave Born approximation (DWBA) the applicability of which is not quite clear. DWBA may be obtained from the Faddeev equations (FE) for three-body system with two-body potentials $V_i = V_{jk}(\vec{r}_{jk})$ if the distorting optical potentials $W_i(\vec{\rho}_i)$, i =1,2,3 affecting the relative motion Yacobi variables are introduced. Let FE set with distorting potentials (FEDP) be presented for the function of scattering in channel 1, i.e. scattering of particle 1 by bound state φ_{23}:

$$\Psi_1^{(i)} = \varphi_{23}\chi_1^+\delta_{i1} + \mathcal{G}_i\left(V_i - \frac{W_j+W_k-W_i}{2}\right)\left(\Psi_1^{(j)}+\Psi_1^{(k)}\right) \tag{1}$$

ijk=123, 231, 312

Here $\mathcal{G}_i = (E-H_0-V_i-W_i)^{-1}$ is the distorted Green function of channel i; χ_i^+ is the distorted wave in the potential W_i. If we are interested in the rearrangement process 1 → 2, it may be set that W_3=0. In this case, symmetrized DWBA for the amplitude

$$A_{21}^{(1)} = \langle \varphi_{13}\chi_2^- | \frac{V_1+V_2}{2} | \varphi_{23}\chi_1^+ \rangle \tag{2}$$

follows from the iteration series of the system (1). Though many versions of FEDP similar to (1) exist, the iteration convergence has proved for none of them. We propose to ensure the convergence by orthogonalization to the bound states using the orthogonal projecting method /1/. In this case, we obtain, instead of (2), for a first approximation the following improved Born approximation with orthogonalized distorted waves (ODWBA):

$$\tilde{A}_{21}^{(1)} = \langle \tilde{\Phi}_2^- | \frac{V_1+V_2}{2} | \tilde{\Phi}_1^+ \rangle \tag{3}$$

where the orthogonalized distorted waves (ODW) $\tilde{\Phi}_i$ are related to the conventional DW $\Phi_i = \varphi_{jk}\chi_i$ as $\tilde{\Phi}_i = \Phi_i - \mathcal{G}_i\Gamma(\Gamma\mathcal{G}_i\Gamma)^{-1}\Gamma\Phi_i$, Γ is the projector to the bound states.

The proposed formalism of ODW makes it possible to find a new approach to determination of the optical potentials W_i which are usually determined from fitting to the cross sections of elastic scattering in channel i. This requirement is partly justified by the fact that, if the iteration series of FEDP (1) converges,then in the elastic channels the one-particle amplitudes corresponding to the potentials W_i should be actually more or less good approximation. If one proceeds from such observations basing on orthogonalized FEDP, it is necessary for the iteration series convergence that the two-body scattering amplitudes in the potentials W_i with orthogonality conditions should be a good approximation for elastic channels. Calculations show that in many cases the Saito orthogonality condition model /2/, in which the direct potential V_D of interaction between clusters is taken as W_i ,is a good approximation for composite-particle scattering. Thus, orthogonalization not only may improve DWBA by improving the FEDP iteration convergence but also makes it possible to reject the conventional fitting of optical potentials.

The proposed FEDP set /1/ is also of assistance in describing the system with Coulombic interation. It may be shown that the use of the distorting Coulombic potentials in variables ρ_i excludes the well known Coulombic singularities from FE kernels (below the three-body threshold) and is, in this respect, similar to the Veselova regularization method /3/. Thereby, the proposed ODW method is also suitable for the system with Coulombic interactions, in particular for p-d and d-d scattering in terms of the Faddeev-Yakubovsky equations, and for direct nuclear reactions.

REFERENCES:

1. V.I. Kukulin and V.N. Pomerantsev, Ann.Phys. (N.Y.)111(1978)330

2. S. Saito ,Progr. Theor. Phys. 41 (1969) 705.

3. A.N. Veselova, Theor. Mat. Phys. 3 (1970) 326.

GHOST STATES AND PAULI PRINCIPLE IN MANY-BODY SCATTERING

V.I. Kukulin and V.N. Pomerantsev
Institute of Nuclear Physics, Moscow State University;
Moscow 117234, USSR.

Solutions of the Schrödinger and Faddeev-Yakubovsky equations for a system containing identical particles are classified in accordance with the permutation group. All the rest solutions are usually called ghost solutions relative to the solutions of the given symmetry. Interest is usually taken in only the physical states of needed symmetry, namely the antisymmetrical states for fermions and symmetrical ones for bosons. However, we have showed that the information about the states which are ghosts relative to physical channel can be of help in many cases when determining physical solutions. We have analyzed the effect of the bound ghost states (ghosts) on the scattering problem using the quartet channel of the n-d system as an example. Three-body system with two interactions is a good model of the n-d system in quartet state since in the given case the neutrons fail to interact in the S-wave. Only two classes of solutions of Faddeev equations (FE) exist in this simplest case, namely the antisymmetrical Ψ_a relative to neutron transposition (i.e. physical solutions) and the symmetric Ψ_s (ghosts) which are forbidden by the Pauli principle. The spin part of the wave functions is insignificant in this case since "all three spins are parallel". The total wave function is $\Psi_{a,s} = (1\mp P)\,\Psi_{a,s}^{(i)}$ where $\Psi_{a,s}^{(i)}$ is the Faddeev component determined by the equation (FE)

$$\Psi_{a,s}^{(i)} = \Phi_i \mp G_i V_i P \Psi_{a,s}^{(i)} \qquad (1)$$

Here V_i is the interaction between proton and neutron; $G_i = (E - H_0 - V_i)^{-1}$; Φ_i is the initial state function of the scattering problem. The only bound state in the system (at the values of the parameters corresponding to real deuteron) is the unobservable "quartet triton", whose function $\varphi_G = (1+P)\,\varphi_G^{(i)}$ and energy E_G are determined by homogeneous equation

$\varphi_G^{(i)} = G_i(E_G) V_i P \varphi_G^{(i)}$. It is this ghost state that contributes to the generalized Levinson theorem and gives rise to divergence of the iteration series of FE(1) in S-wave up to energies of about 100 MeV. We have applied the orthogonality projecting method /1/ to FE(1) by including the projector to ghost state $|\varphi_G\rangle\langle\varphi_G|$ with infinite constant λ in the Hamiltonian and then solved the equation (1) for the physical channel using the iteration method. Such ghost projecting has proved to give not only a convergent iteration series at all energies but also the zero-order approximation which is very close to the accurate amplitude of the quartet scattering:

$$\tilde{F}_0 = \frac{\mu}{2\pi} \frac{\langle \Phi_i | \varphi_G \rangle \langle \varphi_G | \Phi_i \rangle}{\langle \varphi_G | G_i | \varphi_G \rangle} \tag{2}$$

Expression (1) is a three-body generalization of Saito's orthogonality condition model /2/ (with $V_D=0$) in two respects. First, the Green function G_i includes approximately the virtual disintegration of deuteron. Second, even if only the deuteron pole is taken in G_i, the resultant one-body expression for $\tilde{F}_0$ will contain the formfactor $\langle \varphi_d | \varphi_G \rangle$ of the accurate three-body function of ghost state. The latter is of special importance since the Saito model usually uses the so called forbidden states in resonating group method (RGM) which exist only in exceptional cases (for the oscillator functions of scattered clusters). As regards our approach, the ghost states are always unambiguously defined, and their absence or presence are independent of the choice of cluster functions. It is clear, therefore, that the appearance of the forbidden states in RGM reflects the fact that bound states exist in the ghost channels.

Thus, the quartet scattering is mainly determined by the condition of orthogonality to ghost state. This is another formulation of the known assertion that the quartet n-d scattering is determined by the Pauli principle. We have also shown /3/ that a similar situation also takes place for other systems in the channels with maximum spin, for example for the quintet d-d scattering.

REFERENCES:

1. V.I. Kukulin and V.N. Pomerantsev, Ann. Phys. (N.Y.) 111 (1978) 330
2. S. Saito, Progr. Theor. Phys. 41 (1969) 705.
3. V.I. Kukulin, Fizika 9, Suppl. 4 (1977) 395.

THE GREEN FUNCTION HIERARCHY AND THE THREE-BODY PROBLEM

A.G.Sitenko
Institute for Theoretical Physics
of the Ukrainian SSR Academy of Sciences,
Kiev, U.S.S.R.

The usual treatment of the many-body problems is to introduce a succession of multiparticle Green functions, like in the relativistic field theory [I,2]. Particle interaction is interpreted as annihilation of those in the initial state and creation in the final one. The main difficulty of the method is, that, to find the Green function, the solution of an infinite set of coupled equation is needed. The aim of this contribution is to describe the few body systems by means of the Green function approach. As far as such systems are nonrelativistic, one can cut off the infinite set of the Green function equations and thus obtain a closed description.

Define the n-particle Green function for N interacting particles according to

$$G_n^N(1,2,\dots,n;1',2',\dots,n') = -i\langle T\psi(1)\psi(2)\dots\psi(n)\psi^+(n')\dots\psi^+(2')\psi^+(1')\rangle, \quad (I)$$

where ψ and ψ^+ are the annihilation and creation operators in Heisenberg representation ($1\equiv r_1$, t_1 , etc.), T is the chronological operator; the averaging is performed over the ground state of the system. Suppose for simplicity, that all particles are spinless and different; only retardless pair interaction is taken into account. Using the generalized Wick theorem [3], one obtains the reccurence relations for the Green functions with different n:

$$G_n^N(1_1,\dots,1_\nu,2,\dots,(n-\nu+1);1_1',\dots,1_\nu',2',\dots,(n-\nu+1)') = i\sum_{\alpha=1}^{\nu} G_1^{(0)}(1_1,1_\alpha')G_{n-1}^N(1_2,\dots,1_\nu,2,\dots,(n-\nu+1);1_1',\dots,1_{\alpha-1}',1_{\alpha+1}',\dots,1_\nu',2',\dots,(n-\nu+1)')+$$

$$+\sum_{k\neq 1_\alpha}\int d1''\int dk\, G_1^{(0)}(1_1,1_1'')V(1_1'',k)\,G_{n+1}^N(1_1'',1_2,\dots,1_\nu,2,\dots,(n-\nu+1),k;1_1',1_2',\dots,1_\nu',2',\dots,(n-\nu+1)',k^+), \quad (2)$$

where $G_1^{(0)}(1,1')$ is the free singl-body Green function, $V(1'',k)$ - pair potential and $k^+=r_k, t_k+0$. The infinite succession of reccurence relations (2) is the total description of multiparticle Green functions. The consideration is essentially simplified in the so-called ladder approximation, when only intermediate states with fixed number of particles n are taken into account. Then the Green functions with $n > N$ may be expressed in terms of those with $n \leq N$ and so the set of equations is cut off. Besides that, the Lippman-Schwinger equation or (under more correctly formulated boundary conditions) the Faddeev-Yakubovsky one for G_N^N follow from (2). Having found the solution of the latter equation, one can determine the whole succession of Green functions G_n^N with $n < N$ by means of recurrence relations.

As an example we consider the simplest three-body system. In such a case $G_4^3(1'',2,3,3'';1',2',3',3''+) \to \int dr_2'' G_2^2(2,3,2'',3'') G_3^3(1'',2'',3'';1',2',3')$ and Lippman-Schwinger equation for $G_3^3(1,2,3;1',2',3')$ follows immediately from (2):

$$G_3^3(1,2,3;1',2',3') = G_3^{(23)}(1,2,3;1',2',3') +$$

$$+\int d1''\int d2''\int d3''\, G_3^{(23)}(1,2,3;1'',2'',3'')\{V(1'',2'')+V(1'',3'')\}\,\delta(t_2''-t_3'')\,G_3^3(1'',2'',3'';1',2',3'), \quad (3)$$

where $G_3^{(23)}(1,2,3;1',2',3') \equiv i\,G_1^{(0)}(1,1')\,G_2^2(2,3;2',3')$. The two-particle Green function in the three-body system $G_2^3(1,2;1',2')$ is governed by

$$G_2^3(1,2;1',2') = G_2^{(23)}(1,2;1',2') + \int d1''\int d2''\, G_2^{(23)}(1,2;1'',2'')\,V(1'',2'')\,G_2^3(1'',2'';1',2') +$$

$$+\int d1''\int d2''\int d3\, G_2^{(23)}(1,2;1'',2'')\,V(1'',3)\,\delta(t_2''-t_3)\,G_3^3(1'',2'',3;1',2',3^+), \quad (4)$$

where $G_2^{(23)}(1,2;1',2') \equiv i\,G_1^{(0)}(1,1')\,G_1^{(23)}(2,2')$ and $G_1^{(23)}(2,2')$ is the single-particle Green function taking into account the interaction between particles 2 and 3. Once the two-body function $G_2^3(1,2;1',2')$ is known, one can find the two-body t-matrix, which describes the interaction between particles 2 and 3 in the presence of the third one. The single-particle Green function in the three-body system $G_1^3(1,1')$ is expressed in terms of the two-body one $G_2^3(1,2;1',2')$ as follows:

$$G_1^3(1,1') = G_1^{(0)}(1,1') + \sum_{k=2,3}\int d1''\int dk\, G_1^{(0)}(1,1'')\cdot V(1'',k)\,G_2^3(1'',k;1',k^+). \quad (5)$$

If the pair potentials are taken separable [4], then (4) may be reduced to the one-dimensional integral equation, the solution of which makes it possible to analyze the third particle influence on the nature of the pair interaction.

1. P.Martin, J.Schwinger. Phys.Rev. 115, 1342, 1959.
2. A.B.Migdal. Theory of Finite Fermi Systems and Properties of Atomic Nuclei, Interscience, N.Y., 1967.
3. N.N.Bogoliubov, D.V.Shirkov. Introduction to the Theory of Quantized Fields. Interscience, N.Y., 1959
4. A.G.Sitenko, V.F.Kharchenko. Soviet Physics Uspekhi 14, 125, 1971.

EXAMINATION OF MANY-BODY RESONANCES BY MEANS OF ANALYTICAL CONTINUATION OF BOUND STATES

V.I. Kukulin and V.M.Krasnopol'sky
Institute of Nuclear Physics, Moscow State University;
Moscow 117234, USSR

In many cases it is quite difficult to study the resonance states even for the two-body system (broad resonances $\Gamma \sim E$, non-spherical potential, etc.), whereas only the Gilbert-Schmidt three-body method /1/ is used in practice to study three-body resonances. The method is based on solution of the Faddeev equation at complex values of energy. At the same time, it is clear from the examination of the motion of the S-matrix pole in dependence on coupling constant λ that resonance may be defined as such pole of the S-matrix on nonphysical sheet of energy which is an analytical continuation in λ of the bound state pole. In other words, resonance (for such definition) is the state which is obtained from the bound state in case of the corresponding weakening of interaction. Thus, if the position of the S-matrix singularity at considerable λ on physical sheet of E (i.e. the bound state energy) and the analytical properties of the function $E_B(\lambda)$ are known, the parameters of the resonance state (energy, width, wave function etc.) may be found from bound state parameters by numerical analitical continuation in the coupling constant. For this purpose, at first the Schrödinger equation

$$\{H_0 + \lambda V - k^2(\lambda)\} \Psi(\lambda) = 0 \tag{1}$$

for the bound states is solved at some values of $\{\lambda_i\}$, $\lambda_i < \lambda_0$ where $E(\lambda_0)=0$; and the corresponding sets $\{k_i \equiv k(\lambda_i)\}$ and $\{\psi_i\}$ are determined. Then, the technique of Pade approximants (PA) of second type /2/ is used to make analytical continuation to the point $\lambda = 1$ where the state questioned is already resonant. To make the continuation, the account of the analytical properties of the continued values as functions of the coupling constant is of primary importance. In the two-body case the function $k_\ell(\lambda)$ at $\lambda = \lambda_0 (k_\ell(\lambda_0)=0)$ has a root point of branching, i.e. is an analytical function of the variable $x = (\lambda - \lambda_0)^{1/2}$ in which the continuation should be made to achieve convergence of PA sequence /2/

$$K^{[N,M]}(x) = P_N(x)/Q_M(x) \,. \qquad (2)$$

The position and width of resonance are determined from PA $K^{[N,M]}(x_0)$ where $x_0 = (1-\lambda_0)^{1/2}$. The Table lists the so calculated /3/ resonances in the two- α -particle system (the ^{8}Be nucleus levels).

Table

J^π	2^+			4^+			6^+	
Order of PA	E MeV	Γ MeV	λ_0	E MeV	Γ MeV	λ_0	E MeV	Γ MeV
2.2	2.679	1.532	1.4583	10.77	2.66	1.3732	44.2	32.7
3.3	2.782	1.280	1.4577	11.42	2.69	1.3725	30.88	39.4
4.4	2.795	1.228	1.4577	11.44	2.68	1.3725	31.07	40.3
experimental values	2.99± 0.03	1.45 ± 0.06		11.5 ± 0.3	~7		~28.5	~20

It can be seen that even for very wide resonances ($\Gamma \sim E$) a proper accuracy can be achieved at already small values of N and M. It is also of importance that the analytical form of PA(2) determined the S-matrix pole trajectory in the complex k-plane with changing λ . This trajectory can be used to continue the wave function and matrix elements already directly in k. The procedure described above can be naturally generalized for the case of three and more particles. In case of three-body system, probably, it is simpler to use the Faddeev equations, where the three-body eigenvalue of $\eta(E^2 \ln(-E), k_i)$ is an analytical function of the variable $E^2 \ln(-E)$ and k_i (E is the three-body energy; k_i is the wave vector in channel i). By calculating η for $E < 0$ (i.e. for bound states region) and continuiting it to non-physical sheet of energy, the three-body resonances can be found from the condition $\eta(E^2 \ln(-E), k_i) = 1$. The approach described above may in particular prove convenient to calculate the nuclear configurations when two particles (above the core) are in continuum, which is of importance to the nuclear reaction theory.

REFERENCES:

1. V.B. Belyaev and K.Möller, JINR-Preprint,E4-9911,Dubna 1976.
2. G.A. Baker and J.L. Gammel, The Padé approximant in Theoretical Physics, Acad. Press, N.-Y., 1976.
3. V.I. Kukulin and V.M. Krasnopol'sky, J.Phys.A. Math.Gen.10(1977)33; Yad. Fiz. to be published.

CONVERGENCE OF THE DISTORTED WAVE SERIES†

Daniel S. MacMillan
Department of Physics and Astronomy
University of Maryland, College Park, Maryland 20742 U.S.A.

Edward F. Redish*
NASA/Goddard Space Flight Center, Greenbelt, Maryland 20771 U.S.A.

The basic equation of the formal theory of (potential) scattering is the Lippmann-Schwinger equation which after iteration leads to the Born series. If this series converges then perturbation theory can be used and we may consider making the Born approximation, keeping only the first term of the series. Even if the Born series fails to converge we may introduce a quasiparticle[1] into the theory, thereby reducing the interaction treated perturbatively to $V_1 = V - V|\Gamma\rangle\langle\Gamma|V$. Proper choice of quasiparticles can guarantee convergence of the remaining series. In practice they may be difficult to find. An easier alternative which is often used is distorted wave theory. The full potential is separated into two parts $V = V_1 + V_2$. We solve the V_1 part exactly and treat the remainder perturbatively. The Lippmann-Schwinger equation that we obtain is

$$|\Psi\rangle = |\Psi_1\rangle + G_1V_2|\Psi\rangle \qquad (1)$$

in which $|\Psi_1\rangle$ is the distorted wave produced when the potential V_1 distorts the incoming plane wave state. The Green's function $G_1 = G_0 + G_0V_1G_1$, where G_0 is the free resolvent $(E-H_0)^{-1}$. If the full wave function $|\Psi\rangle$ is approximated by the distorted wave $|\Psi_1\rangle$ in the expression for the transition operator, then we get the DWBA (Distorted Wave Born Approximation). Iteration of (1) leads to a multistep series. In order to understand the convergence of this series, we look at the eigenvalues of the kernel G_1V_2 for different separations of the full potential.

We have chosen a two-channel model which might describe the scattering of a projectile from an excitable target. The interaction matrix is the set of folding potentials

$$\{\langle\Psi_A^i|V|\Psi_A^j\rangle\} = |h\rangle\begin{pmatrix}\Lambda_0 & \Lambda\\ \Lambda & \Lambda_0\end{pmatrix}\langle h| \qquad (2)$$

where Ψ_A^i is a target wave function and h(k) is the S-wave Yamaguchi form factor $(k^2+1/R^2)^{-1}$. We now make a separation of the potential into $V = V_1 + V_2$ by

$$\begin{pmatrix}\Lambda_0 & \Lambda\\ \Lambda & \Lambda_0\end{pmatrix} = \begin{pmatrix}\lambda & 0\\ 0 & \lambda\end{pmatrix} + \begin{pmatrix}\Lambda_0-\lambda & \Lambda\\ \Lambda & \Lambda_0-\lambda\end{pmatrix} \qquad (3)$$

and obtain the two eigenvalues of the kernel G_1V_2

$$\eta_\pm = \{(\Lambda_0-\lambda) \pm \Lambda\}\langle h|G_1(E)|h\rangle. \qquad (4)$$

To obtain convergence of the distorted wave series, we must require that $|\eta_\pm|^2_{max} < 1$. For the folding separation we put all of the diagonal part of V into the distortion potential V_1 by setting $\lambda = \Lambda_0$. To see the regions of convergence we have plotted the contour $|\eta_\pm^{fold}|^2_{max} = 1$ in Fig. 1 for various values of Λ_0 and K (where $E = \hbar^2K^2/2\mu$) with

R and Λ fixed.

In DWBA calculations it is often argued phenomenologically that the distorted waves should be generated by an optical potential that gives the correct elastic scattering cross section. In our model, we construct that optical potential λ_{opt} analytically. Taking $\lambda = \lambda_{opt}$ in (4) gives us the curves $|\eta_{\pm}|^2_{max} = 1$ which are plotted in Fig. 2. The width of the divergent region for the optical contours of Fig. 2 is roughly twice that for the folding contours of Fig. 1. *Therefore the convergence of the distorted wave series is better when we use the folding potential for the distortion potential rather than the optical potential.*

To see what causes the difference in the two results we plot in Figs. 3, 4 the trajectories of the eigenvalues $(\eta_{\pm}^{fold}(E))^{-1}$ and $(\eta_{\pm}^{opt}(E))^{-1}$ in the complex η^{-1}-plane. It is the presence of the bound states which causes the divergence of the Born series. The positions of the bound state energies in the η^{-1}-plane are shifted for the optical distortion, thereby causing the distorted wave series to diverge at a range of energies where the folding distortion series converges. One must go to higher energies to get convergence in the optical case.

†Supported in part by U. S. Department of Energy
*N.A.S.-N.R.C. Resident Associate. Permanent address: Department of Physics and Astronomy, University of Maryland, College Park, MD 20742 U.S.A.

1) S. Weinberg, Phys. Rev. 131, 440 (1963).

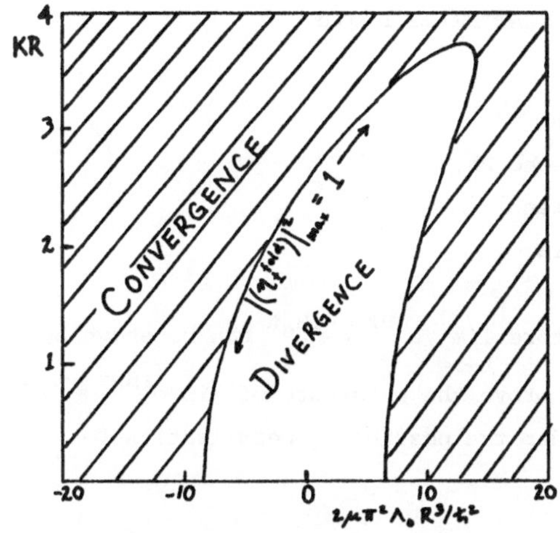

Fig. 1. Folding Distortion
Contour plots of
$|\eta_\pm(K,R,\Lambda_0,\Lambda)|^2_{max} = 1.$

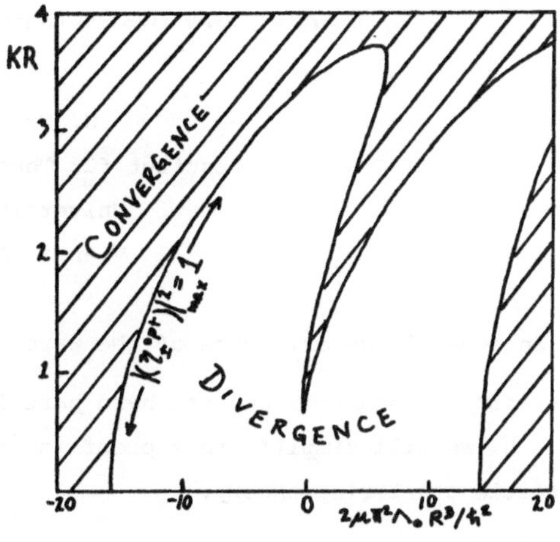

Fig. 2. Optical Distortion
(Λ = -10 MeV/fm and
R = 1.2 fm)

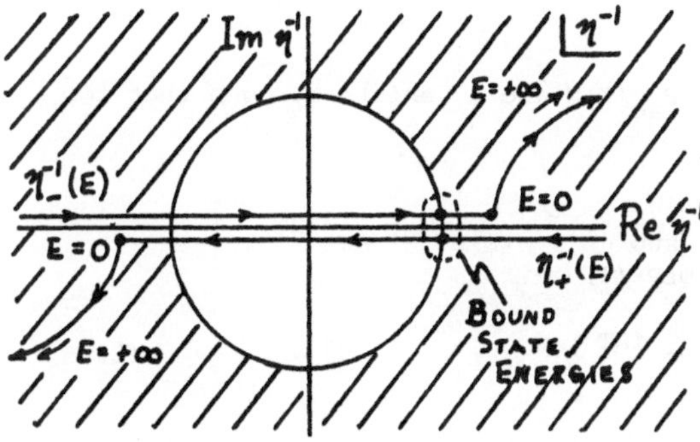

Fig. 3. Folding Distortion

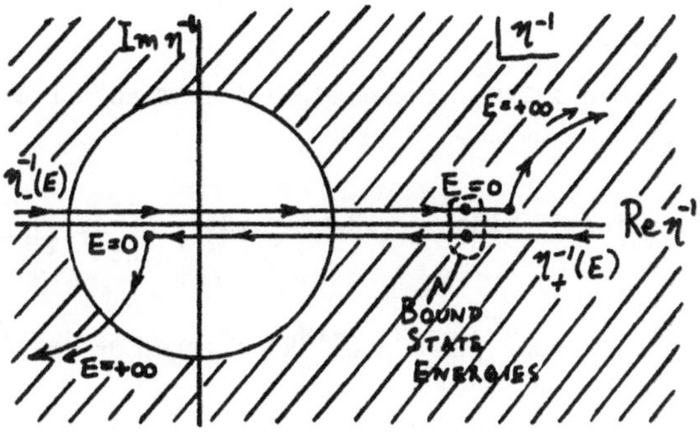

Fig. 4. Optical Distortion

The trajectories of the eigenvalues $(\eta_\pm(E))^{-1}$ of the distorted wave kernel as E increases from $-\infty$ to $+\infty$. (The trajectories have been displaced from the real axis for clarity.)

ON ASYMPTOTIC COMPLETENESS IN SCATTERING THEORY

H. Narnhofer
Institut für Theoretische Physik
Universität Wien
Vienna, Austria

In general the existence of the wave operators $\lim_{t\to\pm\infty} e^{iHt} e^{-iH_0 t}$ can be shown by rather simple arguments and the hard part is to prove the existence of $\lim e^{iH_0 t} e^{-iHt}$ on $P_{ac} \mathcal{H}$. We will simplify this proof and consider the possible generalization to n-particle scattering theory.

Theorem 1a

Let $H = H_0 + V$, V H_0-bounded, $r^{1+\varepsilon} V \varepsilon \mathcal{L}^\infty$ at infinity. Assume $P_{sc}(H) = 0$. Then

$$\text{st} \lim_{t\to\pm\infty} e^{iH_0 t} e^{-iHt} P_{ac} = \Omega_\pm^* \tag{1}$$

exists.

Proof

Choose $I \subset R_+$, I closed, $0 \notin I$. Then for $\phi \,\varepsilon\, P_{ac}(I)\mathcal{H}$ and ε, exists T, such that for all t', $t'' > T$

$$\|(e^{iH_0 t'} e^{-iHt'} - e^{iH_0 t''} e^{-iHt''}) P_{ac}(I)\phi\| = \sup_{\psi \varepsilon P^0(I)} \left|\int_{t'}^{t''} \langle\psi|e^{iH_0 t} V e^{-iHt}|\phi\rangle\right| \leq$$

$$\leq \sup_{\psi} \int_{-\infty}^{+\infty} \|r^{\varepsilon/2} V e^{-iH_0 t} P^0(I)\psi\| \, dt \, \|r^{-\varepsilon/2} e^{-iHT} \phi\| < \varepsilon \tag{2}$$

since the integral is bounded and $r^{-\varepsilon}$ relatively compact.

Theorem 1b

Let $H = H_0 + V$, $r^{\varepsilon} V \varepsilon \mathcal{L}^\infty$ at infinity. Assume that $\exists$ $f(H_0,t)$ such that

$$\text{st} \lim_{t\to\pm\infty} e^{iHt} e^{-if(H_0,t)} = \Omega_\pm \quad \exists .$$

Then

$$\text{st} \lim_{t\to\pm\infty} e^{if(H_0,t)} e^{-iHt} P_{ac} = \Omega_\pm^* \quad . \tag{3}$$

Proof

We can use the same argument and have to calculate

$$\sup_{\psi} \int_{-\infty}^{+\infty} dt \; \| r^{\varepsilon/2} (V - \dot{f}(H_o,t)) \; e^{-if(H_o,t)} \; P^o(I) \; dt \| < \infty \quad . \tag{4}$$

Example: $V = 1/r$, $f(H_o,t) = H_o t - \frac{1}{|p|} \ln t$, $\dot{f} = \frac{1}{|p|t}$,

$$\underset{t\to\pm\infty}{\text{st lim}} \; e^{if(H_o,t)} \; t^{\varepsilon+1} \left(\frac{1}{r} - \frac{1}{|p|t}\right) \; e^{-if(H_o,t)} = 0 \quad .$$

<u>Theorem 2</u>

Generalization to $H = K_1 + V_1 + K_2 + V_2 + V_{12}$ for negative energy. Asymptotic completeness requires that $P_{ac} = \sum P_{\alpha\pm}$, such that

$$\underset{t\to\pm\infty}{\text{st lim}} \; e^{iH_\alpha t} \; e^{-iHt} \; P_{\alpha\pm} = \Omega^{*}_{\alpha\pm} \tag{5}$$

with H_o, $H_1 = K_1 + V_1 + K_2$, $H_2 = K_1 + K_2 + V_2$, $H_{12} = K_1 + K_2 + V_{12}$.

Following [1] the unknown projection operator P_α can be replaced by functions in space J_α concentrated on the region where the particles in channel α move. A possible choice is

$$J_1 = \frac{r^k}{r^k + |x_1|^{k+1+\varepsilon}} \, , \qquad J_2 = \frac{r^k}{r^k + |x_2|^{k+1+\varepsilon}} \, , \qquad J_{12} = \frac{r^k}{r^k + |x_1 - x_2|^{k+1+\varepsilon}} \, ,$$

$$J_o = 1 - J_1 - J_2 - J_{12} \quad . \tag{6}$$

(5) can be replaced by the condition

$$\underset{t\to\pm\infty}{\text{st lim}} \; e^{iH_\alpha t} \; J_\alpha \; e^{-iHt} \; P_{ac} \;\; \exists \; .$$

Then $\underset{t\to\pm\infty}{\text{st lim}} \; e^{iHt} \; J_\alpha \; e^{-iHt} \; P_{ac} = P_{\alpha\pm}$ and $\sum_\alpha P_{\alpha\pm} = P_{ac}$ follows from $\sum_\alpha J_\alpha = 1$.

(2) becomes, with $I \in R_-$, $0 \notin I$,

$$\sup_{\psi} \int_{t'}^{t''} \langle \psi | P(H_1,I) \; e^{iH_1 t} \; ([H_o,J_1] - (V_2 + V_{12})J_1) \; e^{-iHt} \; \phi \rangle \leq$$

$$\leq \sup_{\psi} \int_{-\infty}^{+\infty} \{ \| r^{\varepsilon/2} (V_2 + V_{12}) J_1 \; e^{-iH_1 t} \; P(H_1,I)\psi \| + \| r^{\varepsilon/2} [H_o,J_1] \; e^{-iH_1 t} \; P(H_1,I)\psi \| \} dt \; \cdot$$

$$\cdot \; \| r^{-\varepsilon/2} \; e^{-iHT} \; \phi \|$$

with $|(V_2 + V_{12})J_1| < \frac{\gamma}{r^{1+\varepsilon}}$ for $r \to \infty$ and k sufficiently large:

$$|[H_o,J_1]| \leq \frac{1}{r^2} + ||p| \; \frac{1}{|x_1|} \; \frac{r^k \; |x_1|^{k+1+\varepsilon}}{(r^k + |x_1|^{k+1+\varepsilon})^2}|$$

which is integrable since $|x_1|$ is bounded (exponential decay of the wave functions) and $r \sim pt$, $p > 0$ by the appropriate choice of I.

Theorem 2b

For the Coulomb potential (or potentials behaving like $r^{-\varepsilon}$ at infinity) $e^{iH_1 t}$ is to be replaced by $e^{itH_1 + if(K_2,t)}$.

Conjecture

For positive energy $r^{\varepsilon}[H_o, J_1]$ is not uniformly integrable because strictly positive kinetic energy for particle 1 (or 2) is not guaranteed. The result would hold if one could show that

$$\lim_{t\to\pm\infty} \left\|\left(\frac{r^k |x_1|^{k+1+\varepsilon}}{r^k + |x_1|^{k+1+\varepsilon}}\right)^{\gamma} e^{-iHt}\,\phi\right\| = 0 \quad ?$$

For n particles the J_α can be generalized correspondingly to the possible clusters. Estimates for the sub-Hamiltonians can be reduced to estimates for H_o, according to the existence of scattering theory between H_o and H for n-k particles.

Reference

[1] P. Deift, B. Simon, A Time Dependent Approach to the Completeness of Multiparticle Quantum Systems, Preprint.

MULTIPARTICLE CORE IN NUCLEI

V.K.Lukyanov and A.I.Titov

Joint Institute for Nuclear Research, Dubna, USSR

In this paper we pay attention to the existence of specific many body repulsion correlations in nuclei at short distances of an order of the nucleon radius r_0 . Naturally, in very small correlation volumes $V_\xi = \frac{4}{3}\pi r_\xi^3$ nucleons lose their individuality so that one should consider them as objects consisting of many quarks (the"quark bags" in nuclei). Such configurations have been called "fluctuons" because of their fluctuation nature.

The idea of fluctuation of nuclear matter was first suggested by D.I.Blokhintsev [1] in 1958. Now the idea is supported by a number of experiments at large momentum transfers such as elastic[2] and deep inelastic electron-nuclear scattering, the cumulative particle production from high energy hadron nuclei collisions[3] and others. The detailed analysis of these data is carried out in a series of papers, see e.g.[4,5,6]. The specific multinucleon correlations at short distances may turn out to be useful also in solving some old problems of a few nucleon systems such as the explanation of the binding energy of the lightest nuclei ^3H , ^3He, ^4He , the behaviour of the second maximum $(q \sim 1\,GeV/c)$ of their form factors and others, which cannot be solved using the two-body NN-forces only.

One of the main problems of our consideration is the calculation of the probabilities β_k^A for the existence of fluctuons in nuclei (k is the number of nucleons correlated in a fluction). In all the previous papers[1,4] β_k^A have been estimated qualitatively by the classical gas fluctuation theory

$$\beta = \binom{A}{k} \left(V_\xi / V_0\right)^{k-1} A^{1-k} \tag{1}$$

and have been treated further like parameters. However, the analysis of the mentioned experimental data has shown that the correlation radius r_ξ is approximately constant for all fluctuons with $k=2,3,4$ and is equal to 0.75 fm the magnitude of an order of the NN-forces core radius. Thus, this seems to indicate a deep relation between the phenomenological conception of a core and the micro-structure of several nucleons at short distances, that is what we call now fluctuon or many quark bag in nuclei.

To calculate probabilities β_k^A on a more realistic basis, we use the main hypothesis that the nucleons composing the fluctuons lose their individuality and in the correlation region $r < r_\xi$ are transfered to the new phase, the many quark object, which can be treated in the framework of quantum chromodynamic models.

Let us define β_k^A as follows:

$$\beta_k^A = b_k^A D_k \tag{2}$$

where b_k^A is the probability of finding the usual noncorrelated nuclear cluster of k nucleons in a nucleus A, and D_k is the probability of finding this cluster in the fluctuon compressional state. In fact D_k is the probability of the phase transition of k nucleons to the $3k$ quark state. The calculation of b_k^A may be carried out in the framework of conventional nuclear physics. The quantity D_k is defined as an integral over the fluctuon volume:

$$D_k = \int_{\tau} |\Psi(1 \ldots k)|^2 d\tau \tag{3}$$

where Ψ is the wave function of k nucleons in their center-of-mass system. Thus, the problem is to find the Ψ -function in a small region (of an order of the core radius), what needs the solution of the basic problem, to determine the nuclear many-body repulsion potential at short distances. By its nature this potential is determined as difference between the energy of the $3k$ -quark hadronic bag (the true state of the hadronic matter in the core region) and the mass of the k -nucleon cluster (the state outside the core)

$$V_k^c = E(3k) - k\,Mc^2$$

In calculating $E(3k)$, we use the spherical hadronic "MIT-bag" model[7]. Though the model has been adapted only for the light hadronic masses the same parametrization can be used in estimating more complicated systems. We deal with the quark bags, fluctuons, in nuclei[8] and other authors even with the quark component in stars[9]. The results of our calculation are the following values of the many-body potential barriers

$V_2 = 0.27$, $V_3 = 0.8$, $V_4 = 0.99$. Thus, the many quark configurations in nuclei give rise to a strong manyparticle repulsion which cannot be reduced to the pair interactions only. Figure 1 represents calculations of the multibarionic configuration probabilities D_k with the help of the aforesaid potentials. One can see a strong decrease in D_k with increasing k. It can be seen also that the magnitude of D_k decreases sharply when k is change from $k=4$ (^4He) to $k=5$ (^5Li) because of the Pauli principle which forbids the five nucleons to be together in a small volume. The calculated magnitude for the quark configuration probability in the deuteron ($\sim 8\%$) turned out to be in good agreement with the ed-elastic scattering data /2/. The order of magnitudes of D_k $(k>2)$ are compared with the results of the cumulative particle production from high energy pA - collisions. The magnitudes of $D_k^{(exp)} = \beta_k^{A(exp)}/b_k^A$ are given by crosses in figure 1. One can conclude that the theoretical predictions of D_k are in a qualitative agreement with the corresponding experimental values.

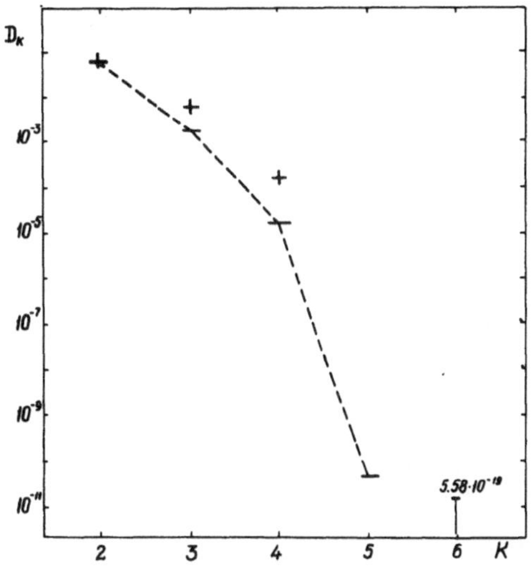

1. Blokhintsev D.I. JETP, 1957, 33, p. 988.

2. Arnold R.G. e.a. Phys.Rev.Lett., 1975, 35, p. 776.

3. Baldin A.M. In: High Energy Physics and Nuclear Structure 1975, eds. D.E.Nagle et al (American Institute of Physics New York, 1975) p. 621.

4. Burov V.V., Lukyanov V.K., Titov A.I. Phys.Lett., 1977, 67B, p.46. Burov V.V., Lykuanov V.K., Titov A.I. Proc. of Int. Conf. on Selected Topics in Nuclear Structure, VII, JINR, D9920, Dubna, 1977, p, 432.

5. Efremov A.V.Yad. Fiz. 1973, 18, 1246.
6. Burov V.V. e.a. JINR E2-11091, Dubna, 1977.
7. De Grand T. e.a. Phys.Rev., 1975, D12, p. 2060.
8. Lukyanov V.K. Titov A.I. Dorkin S.M. JINR, P2-11049, Dubna, 1977.
9. Baym G., VII International Conf. on High Energy Physics and Nuclear Structure, ed. by. M.P. Locher, 1977 (Birkhauser).

THE TWO PION EXCHANGE THREE-NUCLEON POTENTIAL AND NUCLEAR MATTER

B. R. Barrett*, D. W. E. Blatt†, S. A. Coon*, B. H. J. McKellar††, P. C. McNamee*, and M. D. Scadron*

*Department of Physics, University of Arizona, Tucson, AZ U.S.A.
†School of Mathematics, University of Newcastle, Newcastle, Australia 2308
††School of Physics, University of Melbourne, Parkville, Victoria, Australia 3052

In this paper we present a new calculation of the contribution to the energy of symmetric nuclear matter at saturation density ($k_F = 1.36\ fm^{-1}$), due to the three-nucleon (3N) potential which we displayed at the Delhi meeting.[1] The contribution is -1.9 ± 0.2 MeV per nucleon.

The potential arises from 2π exchange and is therefore based on subthreshold πN scattering amplitudes, which are extrapolated off the pion mass shell via PCAC and are subjected to current algebra constraints.[2] These amplitudes, although off-shell, are approximately model-independent, in that they depend primarily on πN data. On-shell they map out very well the experimentally determined amplitudes in the subthreshold region, agree fairly well at and above threshold with s-wave scattering lengths and effective ranges and yield excellent agreement with p-wave scattering volumes. To avoid double counting of iterated π-exchange 2N potentials, we made an explicit subtraction of the forward propagating nucleon pole term for each amplitude.

The energy contribution E_3 is estimated using a perturbative formalism[4] with a new Eular function,[5] appropriate for an effective mass approximation of single-particle energies in nuclear matter. To first order in the 3N potential W

$$E_3 = \sum_{i<j<k} \sum_{p} \varepsilon_p \langle \psi_{ijk} | W | \psi_{p(ijk)} \rangle$$

where the p are the permutations of ijk and ε_p is the parity of the permutation. We approximate the three-nucleon wave function by the simple form

$$\psi_{123} = g(r_{12}) g(r_{23}) g(r_{31})\ \Phi \quad ,$$

where $g(r_{ij})$ is a two-nucleon correlation function obtained from a Reid soft core (RSC) potential calculation and Φ is the uncorrelated plane wave function. We represent the tensor correlations by the perturbation theory term $\frac{Q}{e} V_2 \psi$, where V_2 is the RSC potential and Q/e is calculated with the modified Eular function. Introducing an effective

2N potential V_3 by summing W over one of the nucleon states and taking correlations into account, we arrive at ($E_3 = E_3^{(1)} + E_3^{(2)}$ in an obvious notation)

$$E_3 \approx \sum_{i<j} \varepsilon_p \left\{ \langle \psi_{ij} | V_3 | \psi_{p(ij)} \rangle + \langle \psi_{ij} | V_2 \frac{Q}{e} V_3 + V_3 \frac{Q}{e} V_2 | \psi_{p(ij)} \rangle \right\} .$$

With these approximations to the 3N wavefunction the spin and isospin dependent parts of the 3N potential make no contribution to symmetric nuclear matter. The resulting potential is based on the isospin-symmetric, non-spinflip πN amplitude

$$T_3 \propto K(q^2)K(q'^2)[a + b\underset{\sim}{q}\cdot\underset{\sim}{q}' + c(\underset{\sim}{q}^2 + \underset{\sim}{q}'^2) + \ldots] ,$$

where $a = +1.13 \pm .12\ m_\pi^{-1}$, $b = -2.58 \pm .33\ m_\pi^{-3}$, and $c = 1.05 \pm .10\ m_\pi^{-3}$. We choose for the πNN form factor $K(q^2)$ a parametrization known as form factor III, which roughly corresponds to a monopole form factor with a cutoff mass of about $4m_\pi$. Finally we present here results computed with the conventional effective mass spectrum obtained by a self-consistent nuclear matter calculation[6] with self-consistent hole energies, kinetic energies for particle states, and an energy gap at the Fermi momentum k_F. Our final result is $E_3 = -1.9 \pm 0.2$ MeV per nucleon. The s-wave (a & c) and p-wave (b) parts of the amplitude contribute to E_3 in the following way (contributions due to the central values of T_3 are in MeV).

	Contribution of a	Contribution of b	Contribution of c	Total
$E_3^{(1)}$	-1.15	+4.02	-3.30	-0.43
$E_3^{(2)}$	+0.25	-5.28	+3.58	-1.45
E_3	-0.90	-1.26	+0.28	-1.88

1. S. A. Coon, et al., "Few Body Dynamics," (ed A. N. Mitra et al., North Holland, Amsterdam, 1976) p. 739.
2. S. A. Coon, M. D. Scadron, P. C. McNamee, B. R. Barrett, D. W. E. Blatt, and B. H. J. McKellar, Nucl. Phys. A (to be published).
3. B. H. Wilde, S. A. Coon, and M. D. Scadron, University of Arizona preprint.
4. B. H. J. McKellar and R. Rajaraman, Phys. Rev. C3 (1971) 1877.
5. D. W. E. Blatt and B. H. J. McKellar, University of Melbourne preprint UM-P-77/42.
6. P. K. Banerjee and D. W. L. Sprung, Can. J. of Phys. 49 (1971) 1899.

CHARGE ASYMMETRY OF NUCLEAR FORCES AND BINDING ENERGIES OF HEAVY NUCLEI

P.Haensel
Institute of Theoretical Physics, Warsaw University,
Hoża 69, PL-00-681 Warszawa, Poland

Existence of a small charge symmetry breaking /CSB/ component of nuclear forces should imply existence of a "nonconventional" term $E_a(N-Z)$ in the volume part of semiempirical mass formula describing the systematics of <u>nuclear</u> binding energies of heavy nuclei.[1] The magnitude of E_a has been estimated in calculations of binding energy of a slightly asymmetric nuclear matter. The N-N interaction has been assumed to be of the form $V(1,2)= v(1,2)+ V^{CSB}(1,2)$, with $V^{CSB}=(\tau_z^1+\tau_z^2)U(1,2)$ and hence $V_{pp}-V_{nn}=4U$. Nuclear matter calculations have been performed to first order in V^{CSB} using standard lowest order Brueckner theory /LOBT/[2] and two-body cluster approximation of variational approach /LOVA/.[3] With standard approximations of LOBT[2] the formula for E_a reads[4]

$$E_a = -\frac{12\pi}{k_F^4}\rho\int_0^{k_F} dk\,(1-k)k^2\sum_{JS}\sum_{LL'L''}(2J+1)[1+(-1)^{L+S}]\times$$
$$\times\int_0^\infty dr\,r^2\,u^{JS}_{L'L}(k,r)\,U^{JS}_{L'L''}(r)\,u^{JS}_{L''L}(k,r),$$

where $\rho=2k_F^3/3\pi^2$, k is momentum in the centre of mass system in units of $\hbar$, and $u^{JS}_{LL'}$ is correlated two-body radial wave function generated by /dominating/ charge symmetric v. The corresponding formula of LOVA, with suitable approximations, reads[4]

$$E_a = -\frac{1}{4}\pi\rho\sum_{JLS}(2J+1)[1+(-1)^{L+S}]\int_0^\infty dr\,r^2\,[\tfrac{1}{3}k_F r\,I_L'(k_F r)+$$
$$+\;2\,I_L(k_F r)]\,f^2_{JLS}(r)\,U^{JS}_{LL}(r)\;,$$

where f_{JLS} is correlation function in JLS channel generated by v, function $I_L(x)$ is defined as

$$I_L(x)\equiv 48\int_0^1 dz\,z^2\,(1-\tfrac{3}{2}z+\tfrac{1}{2}z^3)\,j_L^2(zx)\,,$$

and small tensor component of the correlation operator f in the $^3P_2$-$^3F_2$ channel has been neglected. The LOBT and LOVA calculations have been performed at normal nuclear density ρ=0.17 nucleons/fm^3. The soft core potential of Reid[5] has been used as a charge symmetric potential v. In LOVA realistic state dependent correlation operator of Pandharipande and Wiringa[7] has been used. For all V^{CSB}'s considered numerical values of E_a obtained in LOBT are very similar to those obtained in LOVA. For phenomenological V^{CSB}'s [4,6] which are consistent with Coulomb displacement energies of mirror nuclei $E_a \cong -0.7$ MeV. When introducing the term $E_a(N-Z)$ into semiphenomenological nuclear mass formula one replaces, in fact, the familiar term $(M_n + M_p)c^2$ by $(\overline{M}_n + \overline{M}_p)c^2$, where $\overline{M}_n = M_n + E_a/c^2$, $\overline{M}_p = M_p - E_a/c^2$. The present estimate of E_a corresponds to $\overline{M}_n \cong \overline{M}_p$.

This result appears to be in accordance with an unpublished finding of Myers and Swiątecki,[8] who found, that putting the n-p mass difference <u>much</u> <u>smaller</u> than the observed one helped "to a nontrivial extent"[8] to improve the Droplet Model fit to nuclear masses in heavy element region.

References

1) P.Haensel, J.Phys.G <u>3</u>, 383(1977); ibid <u>3</u>, L237(1977).
2) D.W.L.Sprung, Advances in Nuclear Physics, <u>5</u>, 225(1972).
3) J.C.Owen, R.F.Bishop and J.M.Irvine, Nucl.Phys. A<u>277</u>, 45 (1977); see also Ref.7.
4) P.Haensel, preprint and to be submitted for publication.
5) R.V.Reid, Ann.Phys. <u>50</u>, 411(1968).
6) S.Shlomo and D.O.Riska, Nucl.Phys.A<u>254</u>, 81(1975) and references therein.
7) V.R.Pandharipande and R.B.Wiringa, Nucl.Phys. A<u>266</u>, 269(1976).
8) W.Swiątecki, private communication.

SOLUTION OF THE INVERSE SCATTERING PROBLEM IN THE FINITE-DIFFERENCE APPROXIMATION

V.N.Melnikov and B.N.Zakhariev
Laboratory of Theoretical Physics
Joint Institute for Nuclear Research
Head Post Office P.O.Box 79,
Moscow U.S.S.R.

Recently[1] a very simple and exactly solvable finite-difference (f-d) model for reconstruction of a potential from scattering data has been suggested. It is impossible to apply immediately the corresponding method for approximate solution of the inverse scattering problem. This is because <u>only low energy part</u> of the scattering data is approximately the same for differential and f-d Schroedinger equations. But for inverse problem we need the <u>whole</u> spectrum as the initial data (the completeness relation of eigensolutions of Schroedinger equation is the key property for the potential reconstruction).

We can use the fact that for high energy the spectral parameters converge to the corresponding free values for Schroedinger equation without potential (perturbative corrections depending on average potential value only can be taken into account). Unfortunately, the spectrum of f-d equation has analogous asymptotic properties only in its middle part. But in spite of this difficulty the high energy half of f-d scattering data can be calculated due to their symmetry relative to the middle point of the spectrum, if we transform the f-d Schroedinger equation to the form with two-diagonal Schroedinger operator matrix (with zero diagonal matrix elements, as for example in[2]).

So, the whole set of initial f-d spectral data can be calculated from the given corresponding values for differential Schroedinger equation. Then the inverse problem is solved exactly as within f-d model[1].

References

1. B.N.Zakhariev, V.N.Melnikov, B.V.Rudjak, A.A.Suzko. Review article in Particles and Nucleus <u>8</u>, part 2, 290 (1977).
2. K.M.Case, M.Kac. J.Math.Phys. <u>14</u>, 594 (1973).

THE REACTION $\pi^+ d \to pp$ AND THE ABSORPTIVE P-WAVE PION-NUCLEUS OPTICAL POTENTIAL

C. M. Ko and D. O. Riska
Department of Physics
Michigan State University
East Lansing, Michigan 48824

As pions cannot be absorbed on free nucleons, nuclear pion absorption involves at least a two-nucleon mechanism, which may either be viewed as absorption on a single but bound nucleon or as an explicit two-nucleon mechanism. The evidence from theoretical investigations of the reaction $\pi^+ d \to pp$ is that one-body absorption is of minor importance [1,2]. To emphasize the importance of the two-nucleon aspect of nuclear pion absorption the absorptive P-wave pion-nucleus optical potential is usually parametrized as

$$2\omega \; \mathrm{Im}\, U_{opt} = 4\pi \, \vec{\nabla} \; \mathrm{Im}\, C_o \; \rho^2 \vec{\nabla} \; , \tag{1}$$

where ω is the pion energy, ρ the nuclear density and Im C_o an energy dependent parameter.

Using the second order term in the multiple scattering series expansion of U_{opt} one may relate the absorptive part of the potential to the total nuclear pion absorption rate. This leads to the following expression for Im C_o:

$$\mathrm{Im}\, C_o = \frac{\omega}{2k^2\rho^2\Omega} \sum_f \delta(E_f - E_i - \omega) \; |T_{fi}|^2 \; . \tag{2}$$

Here k is the pion momentum, Ω the nuclear volume and T_{fi} the matrix element of the pion absorption operator between the initial (i) and final (f) nuclear states and the sum is taken over all final states allowed by energy conservation.

The P-wave part of the fundamental reaction $\pi^+ d \to pp$ can be well explained in terms of pion and ρ-meson rescattering through the Δ_{33} resonance [1,2]. Considering the same mechanisms for two-body pion absorption in nuclei we have calculated Im C_o as a function of pion momentum using Eq.(2). We used the pair wavefunctions given by the Fermi gas model. The calculational method is similar to that used recently by Bertsch and Riska [3] and Riska and Chai [4] to derive the S-wave part of the absorptive potential.

The results for Im C_o are shown in Fig. 1. The curve A is obtained with π and ρ rescattering, with the parameters of Ref. 1 without hadronic form factors but setting the energy of the rescattered meson to be half of that of the initial pion. Setting the energy of the rescattered meson to 0 as was done in Ref. 1 leads to curve B which thus corresponds to the model used in Ref. 1 for $\pi^+ d \to pp$. Introducing a hadronic form factor of monopole type with a mass scale of 1.2 GeV c^2 at the vertices of the rescattered mesons leads to curve C. The values for Im C_o obtained by a phenomenological analysis of pionic atom levels near threshold are of the order $1\mu^{-6}$. The curve C is in reasonable agreement with those values [5]. We therefore conclude that the same two-body absorption mechanisms that explain the total cross section for $\pi^+ d \to pp$ satisfactorily seem to be able to account for the two-body contribution to the absorptive P-wave pion nucleus optical potential.

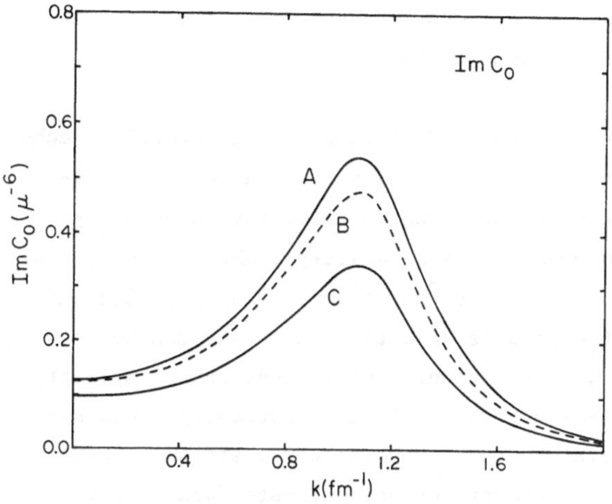

Fig. 1. Pion and ρ-rescattering contributions to Im C_o.

1 D.O. Riska, M. Brack and W. Weise, Phys. Lett. 61B, 41 (1976)
2 M. Brack, D.O. Riska and W. Weise, Nucl. Phys. A287, 425 (1977)
3 G.F. Bertsch and D.O. Riska: "Threshold Pion Absorption in Nuclei", MSU Preprint 1978, to appear in Phys. Rev. C
4 J. Chai and D.O. Riska: "The Energy Dependence of the Absorptive S-wave Pion-Nucleus Optical Potential", MSU Preprint 1978
5 M. Krell and T.E.O. Ericson, Nucl. Phys. B11, 521 (1969).

Elastic Scattering of $\pi^{\pm}$ on ^3H and ^3He at Forward Angles

R. Minehart, J. McCarthy, D. Roeder and E.A. Wadlinger, Univ. of Va.,
Charlottesville, Va. 22901/USA
H. Brandle, P. Glodis, R. Haddock, I. Kostoulas, N. Matz, B. Nefkens,
W. Plumlee, and O. Sander, Univ. of Cal., Los Angeles, Cal. 90024/USA
J. Novak, F. Shively, J. Pratt, R. Sherman, J. Spencer, LASL,
Los Alamos, New Mex. 87545/USA

Pion interactions with light nuclei can be used to study the interaction mechanisms for pions with bound nucleons, to test the validity of a variety of many-body techniques, to learn about nuclear structure, and to check fundamental physical principles. At the Clinton P. Anderson Meson Physics Facility (LAMPF) we have measured the elastic scattering of $\pi^{\pm}$ mesons on ^3H and ^3He at momenta of 232, 252 and 295 MeV/c. Besides checking pion reaction models, elastic scattering can be used to study differences in the ^3H and ^3He wave functions, and to test charge symmetry.

In this paper we report on some results for the elastic scattering cross sections at small angles (30° - 65°), obtained with a spectrometer designed to measure the scattered pion momentum and angle. A separate detection system was used for larger angles where the recoiling target particle had enough energy to be detected and identified. A common target and beam monitoring system was used for both systems which operated simultaneously.

The small angle spectrometer consisted of a uniform field bending magnet (76.2 x 101.6 x 20.3 cm) with a set of three horizontal and three vertical multi-wire proportional planes on the entrance side and a similar set on the exit side. The scattering plane was horizontal and the momentum analysis plane was vertical. A nominal bend angle of 43.5^o was used. The proportional chambers were used both for position information and for ΔE. Using ΔE enabled us to reject more than 99.5% of the protons and heavier particles passing through the spectrometer.

The target for the ^3H and some of the ^3He runs was a vertical cylinder with 63.5 μm aluminum walls, 15.2 cm in diameter, cooled to 38^oK, and filled with gas at 580 mm Hg pressure. A horizontal cylinder with 127 μm Kapton walls was used for some of the ^3He data.

The pion beam was monitored by an ionization chamber and by two sets of two-element scintillation telescopes looking into the beam at an angle of 5°, placed symmetrically on each side of the beam upstream from the target. These telescopes were sensitive to decay muons as well as to particles scattered from pole tips and apertures in the beam line.

Beam monitors were calibrated against the number of counts in a scintillation counter in the beam during low intensity pulses. Beam composition was measured with a time-of-flight technique using the r-f signal from the accelerator as a stop pulse and the output of a scintillation counter near the target position as the start signal. Measurements of the π-p cross section were made to determine the spectrometer acceptance by comparing them to calculations based on phase shifts fitted to the data of Bussey et al.[1]

The overall relative stability of the monitor system was ~1%. The absolute pion flux was measured to about 5% for π^+, and to somewhat less accuracy for π^-. The target density was uncertain to about 3% because of a 1°K uncertainty in its temperature. The spectrometer detection efficiency was monitored continuously and was typically 99.0 ± 0.5%.

The coordinates from the 12 wire planes were fitted by a χ^2 method to a momentum and scattering angle. A loose cut in χ^2 was made to reject badly fitting events (less than 10%), and the remainder were then used to calculate an effective recoil mass, M_r, for the reaction $\pi + M_t \rightarrow \pi + M_r$. Data taken with the target empty were subtracted from data taken with the target full to obtain recoil mass spectra, an example of which is shown in fig. 1. The recoil mass spectra were fitted to a broad background attributed to pion decay in the spectrometer and to the phase space for break-up reactions to remove background events under the elastic peak.

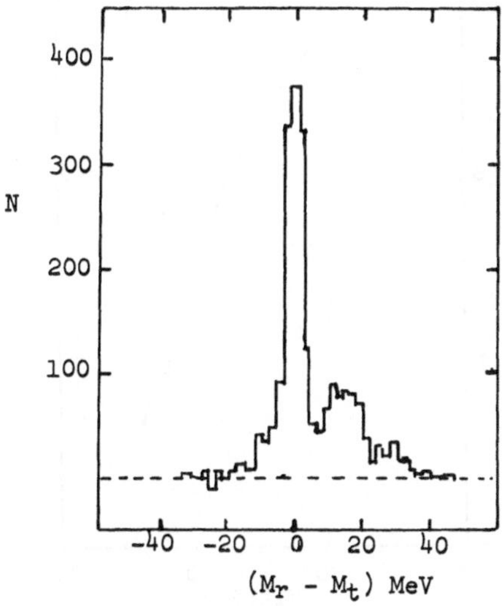

Fig. 1. Recoil mass spectrum for $\pi^- + {}^3H$ at 252 MeV/c and 45° in the laboratory system.

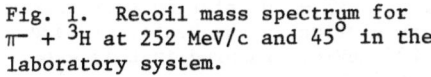

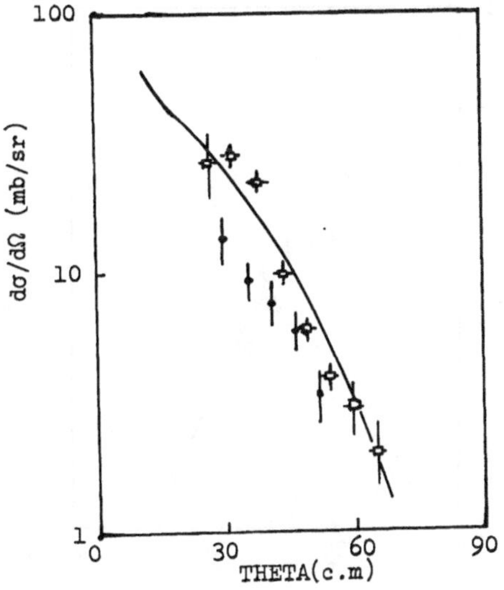

Fig. 2. Pion elastic scattering cross section in the center of mass. Open squares are $\pi^+ + {}^3H$ at 252 MeV/c. Solid circles are the data of Shcherbakov et al. for $\pi^- + {}^3He$ at 248 MeV/c. The solid line represents the calculation of Landau at 252 MeV/c.

The remaining events in the channels spanned by the elastic peak were then summed to obtain the cross section for elastic scattering.

Some of our results with ^3H are displayed in Figs. 2 & 3 along with data of Shcherbakov et al.[2)] for pion scattering on ^3He at similar energies. The curves show the calculations of Landau[3)] for pion scattering on ^3He at the same energies as our data. Landau's calculations are based on an optical model potential, and include uncertainties due to poor knowledge of the magnetic form factor of ^3He. This effect can be as large as 30–40% at 60°, but decreases with decreasing angle.

Our data are exhibited with angle error bars that denote the full range in angle accepted by the spectrometer. In contrast, the data of Shcherbakov et al. are integrals over the angular range between successive points.

The trend of our results is roughly in agreement with the theoretical model and with the data of Shcherbakov et al., but there are disagreements in magnitude as well as in detailed shape. Individual points disagree by as much as 40% with the model. At 232 MeV/c our data agree better with Landau than do the measurements of Shcherbakov et al. Although not shown in the figures, we have a few measurements

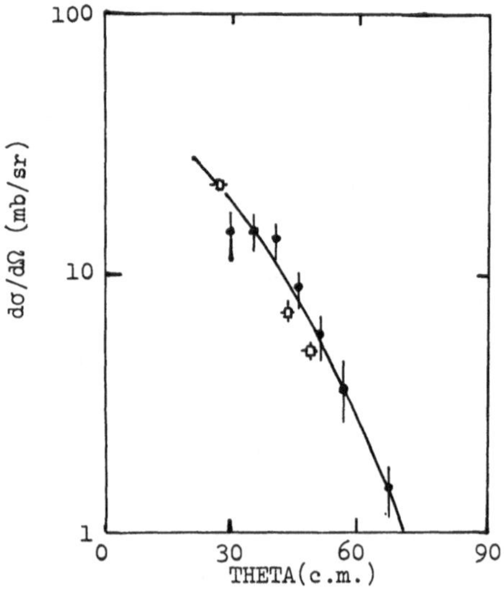

Fig. 3. Pion elastic scattering cross section in the center of mass. Open squares are $\pi^+ + {}^3H$ at 232 MeV/c. Solid circles are the data of Shcherbakov et al. for $\pi^- + {}^3He$ at 236 MeV/c. The solid line represents the calculation of Landau at 233 MeV/c.

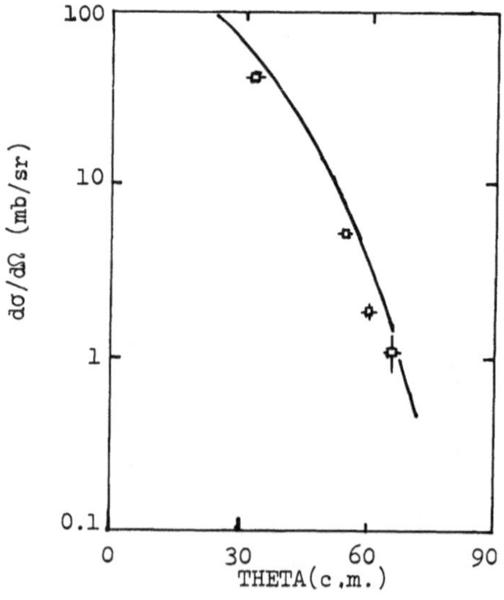

Fig. 4. $\pi^+ + {}^3He$ elastic scattering cross section in the center of mass at 295 MeV/c, measured in this experiment. The solid line is the calculation of Landau for this momentum.

on π^+ scattering from ^3H at 295 MeV/c , which are in fair agreement with the data of Shcherbakov et al.

The ratio, $R = (\pi^- + {}^3He)/(\pi^+ + {}^3H)$, which is expected to be unity, was measured at 45° lab angle at 232 and 252 MeV/c to be 0.92±0.14 and 1.14±0.16 respectively. These provide a first check, although weak, on charge symmetry for pion interactions with the three nucleon systems.

Measurements made with the Kapton flask on ^3He are shown in Fig. 4. The data follow the shape of the Landau calculations but ten to be 30-40% lower. The normalization of these data, which were taken much earlier in the experiment than were the data from the aluminum flask, is somewhat uncertain as it depends on calibrations made during the latter period. The uncertainty, however, is not expected to exceed 15%.

This work was supported in part by the U. S. Dept. of Energy. We would like to thank the LAMPF staff for their dedicated support of this experiment.

References

1. P. J. Bussey et al., Nucl. Phys. B58 (1973) 363
2. Yu. A. Shcherbakov et al., Il Nuovo Cim. 31A (1976) 262
3. Rubin H. Landau, Ann. of Phys. 92 (1975) 205, and private communication.

NEW APPROACH TO LOW ENERGY π-He^3-SCATTERING

V.B.Belyaev, J.Wrzecionko*, M.I.Sakvarelidze **

Joint Institute for Nuclear Research
Head Post Office, P.O.B. 79 Moscow, U.S.S.R.

At present there exist the measurements of the level shift and width in $\pi^- He^3$ mesoatom/1/. For the S-state the corresponding numbers are following $\Delta E=(44 \pm 5)$ eV, $\Gamma=(42 \pm 14)$ eV. This experiment bring to the following value of the πHe^3 scattering amplitude at zero energy $a_{\pi^- He^3} = [0.050 \pm 0.005 + i(0.034 \pm 0.012)] m_\pi^{-1}$. Moreover there exist the interpolation formule/2/ on A and Z, from which one can obtain the value of $\pi^- H^3$ level shift (using the corresponding shifts measured on neighbourhood light nuclei). Extracted in such a way level shift for $\pi^- H^3$ leads to value -0,226 fm of the $\pi^- H^3$ scattering length. Due to the charge symmetry this number must be equal to $\pi^+ He^3$ scattering length.

In these contribution we present the results of our calculation of the $\pi^\pm He^3$ scattering lengths, performed on the base of an approximate potential 4-body equations/3,4/. Comparying these calculations with the above experimental data one can estimate, the contribution of an inelastic processes, to the considered effects.

The essence of the approximation presented in paper is as follows. Let us write the Hamiltonian pion - 3N system in the form

$$H = h_o + H_c + V_{\pi N} \qquad (1)$$

where h_o is the kinetic energy of the relative motion pions, and the center of mass of the 3N system.

H_c is the nuclear Hamiltonian. $V_{\pi N} = \sum_{i=1}^{3} V_{\pi N}^{i}$ is the sum of the elementary pion nucleon potential.

In expression (1) we replace the nuclear Hamiltonian by the approximate one.

---

* Institute for Nuclear Research, Warsaw.
** Tbilisi State University.

$$\tilde{H}_c \equiv \mathcal{E} |\chi\rangle\langle\chi| \tag{2}$$

where $\mathcal{E}$ is the binding energy of the nucleus, and $|\chi\rangle$ is the corresponding eigenstate. The problem on the elastic scattering of pions on the three nucleon target can be solved by using the approximation (2) practically exactly. Indeed, for the elastic scattering amplitude $\langle \vec{K} | \mathcal{T} | \vec{K} \rangle$ one can easily obtain the following equation

$$\langle \vec{K} | \mathcal{T} | \vec{K}' \rangle = \langle \vec{K} | \mathcal{T}_0 | \vec{K}' \rangle + \int d\vec{p} \left[\frac{\langle \vec{K} | \mathcal{T}_0 | \vec{p} \rangle}{\frac{p^2}{2\mu} - E} - \frac{\langle \vec{K} | \mathcal{T}_0 | \vec{p} \rangle}{\frac{p^2}{2\mu} - (E - \mathcal{E})} \right] \times \langle \vec{p} | \mathcal{T} | \vec{K}' \rangle \qquad E = p_0^2/2\mu + \mathcal{E} \tag{3}$$

The function $\langle \vec{K} | \mathcal{T}_0 | \vec{p} \rangle$ is real at $E < 0$, at $E > 0$ it becomes complex, and for $p_0^2/2\mu \gg |\mathcal{E}|$ coincides with the scattering matrix of pions by the three fixed scatterers, avareged over the ground state wave function He^3.

Equation (3) has been solved for the S-wave component at the negative energy of the system. The scattering lengths and the phase shifts in the states with total isospin T=1/2 and T=3/2 have been computed. In these calculations the one term separable potentials, which reproduced the pion-nucleon S-phase shifts (at energies $0 \leq E \leq 80$ MeV) and the scattering lengths /5/ in the states with $t_{\pi N}=1/2$ and 3/2 have been used. The ground state wave function $\chi(\vec{\tau}_{12}, \vec{\tau}_3)$ of the three-nucleon system has been choosen in the form proposed by Irving /6/.

Our results are summarised in the Table. As one can see from it for the $\pi^- He^3$ scattering the strong cancellation of the pion nuclear amplitude take place though the amplitudes with the given total isospin are essentially different from zero.

To compare our results with experiment one has to take in mind that calculated in this paper scattering lengths, are elastic one. Let us assume the Bruckner /7/ mechanism for the formation of the level shifts, then for the measurable shifts we get the following expression

$$\Delta E_{exp} = \Delta E_{el} + 1.05\, \Gamma_{exp}$$

Comparying the last expression with the experiment /1/ we see that the "elastic" part of the $\pi^- He^3$ scattering lengths comprise only few percent of the experimental value. Within the experimental errors it agrees with our calculation. Calculated by us $\pi^+ He^3$ scattering length (see Table) agrees very well with the value -0.226 fm, extracted from the interpolation procedure for $\pi^- H^3$ level shift /2/.

From the aforesaid we conclude that at very low energy the $\pi^- He^3$ amplitude is formed mainly due to the inelastic processes while $\pi^+ He^3$ is mainly elastic.

Table

$a^1_{\pi He^3}$ (fm)	$a^3_{\pi He^3}$ (fm)	$a_{\pi^- He}$ (fm)	$a_{ch.exch.}$ (fm)
0.1053	-0.2101	0.00016	-0.1487

Experiment: $a^3_{\pi He^3} = -0.226\ fm$ [2];

$$a_{\pi^- He^3} = [0.050 \pm 0.005 + i(0.034 \pm 0.012)]\,m_\pi^{-1} \quad [1]$$

References

1. R.Abela et al. Phys.Lett. B68, 429 (1977).
2. L.T.Cheon and T.von Egidy. N.P. A234, 401 (1974).
3. L.Rosenberg. Phys.Rev. C13, 1406 (1976).
4. V.B.Belyaev, J.Wrzecionko, Preprint JINR, E2-10668 (1972), Sov.Nucl.Phys. 7, 8 (1978).
5. I.R.Afnan and A.W.Thomas. Phys.Rev. C10, 109 (1974).
6. I.Irving. Phil.Mag. 42, 338 (1952).
7. K.A.Brueckner. Phys.Rev. 98, 769 (1955).

A MEASUREMENT OF THE PANOFSKY RATIO IN ^{3}He*

M.D. Hasinoff, F. Corriveau, D.F. Measday, M. Salomon and J-M. Poutissou
Universities of British Columbia and Montréal, Canada

The nucleus ^{3}He is one of the simplest nuclear systems; hence it is often used to study the complications introduced by the presence of additional nucleons on basic processes such as pion absorption on a free nucleon. It is also the only nucleus (other than hydrogen) for which both the pion charge exchange and radiative capture processes can occur at rest with reasonable probability.

The ratio of these two processes is the well-known Panofsky ratio, $P_3 = \omega(\pi^- {}^3He \rightarrow \pi^\circ T)/\omega(\pi^- + {}^3He \rightarrow \gamma T)$. This ratio can be accurately measured experimentally since it does not entail the measurement of two absolute rates but merely the ratio of two relative rates which can be measured simultaneously using the same experimental equipment. The Panofsky ratio in ^{3}He has been used by several authors[1-4] to check the validity of the impulse approximation (IA) for π^- absorption in light nuclei. Other authors[5,6] have used it to test the elementary particle model for the 3N system using the partial conservation of the axial vector current (PCAC). Some authors[1,2] have also chosen to limit the amount of the D state contribution to the 3N wavefunction by comparing their calculated rates with the measured ratio.

Experimentally there have been two previous measurements of P_3: 2.28 + 0.18[7] and 2.68 ± 0.13[8] which differ by considerably more than their quoted errors. The former number is in reasonable agreement with the PCAC elementary particle approach which yields[6] a value of $\simeq$2.0 ± 0.1 after corrections for ρ meson exchange, NN correlations and nucleon intermediate states have been applied. The latter value agrees well with the IA calculations[1-4] which yield $P_3 \simeq 2.8$.

We have used the stopping π^- beam from the stopped π/μ channel at TRIUMF to obtain a third value for P_3. A 96 MeV/c π^- beam was degraded by 2.7 cm CH_2 and stopped in a 1.9 cm thick liquid ^{3}He target. The high energy γ-rays from both the γT and $\pi^\circ T$ final states were detected in a large NaI detector (46 cmϕ x 51 cm) placed 2.8 m from the target. This large detector-to-target distance allowed an excellent separation of γ and neutron events using the time-of-flight technique (Δt = 2 nsec). The solid angle for the NaI was limited to 2.34 msr ($\theta_{\frac{1}{2}} = 1.53^\circ$) in order to produce an energy resolution of 4.5% (FWHM) at 135 MeV. A typical energy spectrum of γ-rays is shown in fig. 1. The clean separation of the γT events from those in the radiative break-up channels (γnd + γnnp) is clearly evident in these data.

Before extracting the γT and $\pi^\circ T$ yields it was necessary to subtract the two small backgrounds shown at the bottom of the figure. The higher energy background events arise from the (π^-,γ) reaction in the target frame, mylar windows and superfluid ^{4}He used to cool the ^{3}He target. This contribution was determined from a run in which the ^{3}He cell was empty. The "T2 neutron" background was produced by neutrons from the pion production target leaking through the shielding wall and arriving within the γ window in the time-of-flight spectrum.

The corrected data were then fitted with 4 lineshapes corresponding to the 3 possible final states (γT, $\gamma nd + \gamma nnp$ and $\pi^\circ T$) plus a low energy bremsstrahlung background. The radiative break-up lineshape was obtained by folding the NaI response function into the theoretical curves of Phillips and Roig[1] (who used the Amado model) and then extrapolating the results down to zero energy.

The results of the least-squares fit to several spectra similar to fig. 1 yield a value for $P_3 = 2.83 \pm 0.07$. The error is half statistical and half systematic. Our value agrees reasonably well with Truöl et al[8] and disagrees strongly with Zaimidoroga[7]. It also agrees very well with the value predicted by Gibbs et al[3] when the enhancement in the (π^-,γ) rate produced by the N* anomalous magnetic moment is included. Our measured value is also in agreement with the calculations of Phillips and Roig[1] if the D state of the 3N wavefunctions is 5%.

The ratio of the radiative break-up rate to the radiative capture rate, $B_3 = \omega(\pi^- He^3 \rightarrow \gamma nd + \gamma nnp)/\omega(\pi^- {}^3He \rightarrow \gamma T)$ was found to be 1.35 ± 0.11. This value is slightly higher than that obtained previously[8], 1.12 ± 0.18 but also within the range predicted by Phillips and Roig[1].

* Work supported in part by National Research Council of Canada.

1. A.C. Phillips and F. Roig, Nucl. Phys. A234, 378 (1974).
2. M. Mizuta, Y. Kohyama and A. Fujii, VI Int. Conf. High-Energy Physics and Nuclear Structure, Santa Fe (1975) 175.
3. W.R. Gibbs, B.F. Gibson and C.J. Stephenson Jr., LASL preprint LA-UR-78-863 (1978).
4. B. Goulard, A. Laverne and J.D. Vergados, preprint, U. Montreal (1977).
5. M. Ericson and A. Figureau, Nucl. Phys. B3, 609 (1967); B11, 621 (1967).
6. M. Ericson and M. Rho, Phys. Rep. 5C, 59 (1972).
7. O.A. Zaimidoroga et al., Soviet Phys. JETP 21, 848 (1965).
8. P. Truöl et al., Phys. Rev. Lett. 32, 1268 (1974).

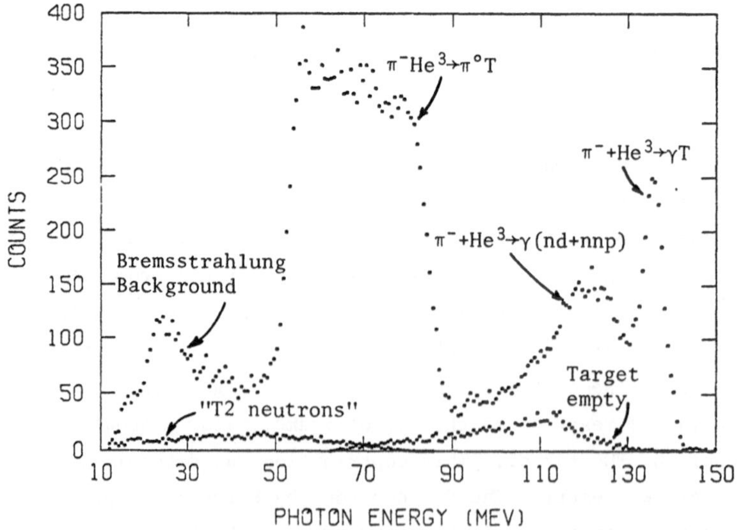

ABSORPTION OF NEGATIVE PIONS AT REST IN ^{3}HE

G. Backenstoss*, P. Blüm**, M. Dörr**, W. Fetscher**, D. Gotta**, W. Klotz**, H. Koch**, W. Kowald*, U. Raich**, G. Schmidt**, I.Schwanner*, L. Simons**, L. Tauscher*, H. Ullrich**, H.J. Weyer*

*University Basel, Switzerland

**Kernforschungszentrum and University Karlsruhe, Germany

In nuclei absorption of π^- proceeds on at least two nucleons. Therefore, the study of the π^{-3}He-system is of particular interest being the simplest system where more than 2 nucleons are offered to the pion. There are two possible nonradiative reactions:

(1) $\pi^- + {}^3He \rightarrow p + n + n$ (56%)   (2) $\pi^- + {}^3He \rightarrow d + n$ (16%)

In our experiment reaction (1) is measured for the first time kinematically complete by observing the momenta of two outgoing particles. An initial s state is selected by demanding coincidences with pionic (np→1s) X-ray transitions. Because of the spatial symmetry (pions at rest) the fivefold differential absorption probability depends only on two variables enabling us to map a large part of phase space with one counter setting. The second reaction (2) is registered simultanuously and serves for calibration and normalization purposes.

π^- from the πE1-channel at SIN are stopped in a gaseous ^{3}He target cooled by liquid ^{4}He to 5°K. We measure the energy (FWHM: 4 % at 50 MeV) and time of flight (300-400ps) of the charged particles with a large area plastic scintillator hodoscope (Ω=0.5sr) thus also identifying the particles. Their trajectories are determined with two proportional multiwire chambers. The momenta of the emitted neutrons are measured by time of flight by two large position sensitive neutron counters (0.5x2 m^2 each, time resolution 1ns). A schematic drawing is shown in Fig. 1.

Since data taking is still in progress we present here preliminary results from our on-line data. Fig. 2 shows a two dimensional spectrum of neutron TOF vs. charged-particle energy, where p and d are displayed simultaneously. The pronounced peak at lower energies is due to the monoenergetic n-d-pairs from reaction (2). At the upper right on the dashed kinematical curve a second peak is seen which corresponds to a region where the two neutrons have a small relative momentum. Background events seen above the kinematical curve due to pion absorption on surrounding material are expected to be substantially reduced by the off-line analysis. The single X-ray spectrum of pionic ^{3}He has been measured yielding results on the total pion absorption[1]. Spectra as shown in Fig. 2 coincident with X-rays detected by an array of thin NaI detectors are being accumulated.

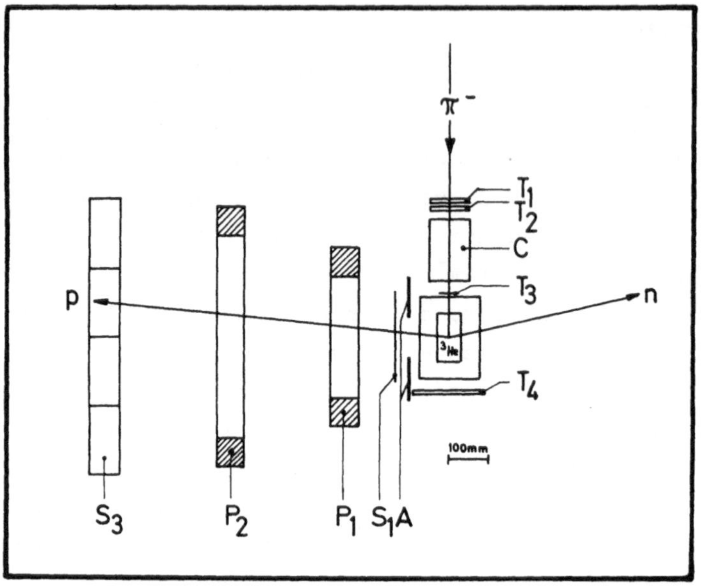

Fig.1

π-stop-telescope, Target, charged particle counter: T_1-T_4: telescope counters; C: graphite moderator; A: anti-counters; S_1: Coincidence counters; P_1 and P_2: proportional chambers; S_3: energy counter (plastic scintillator)

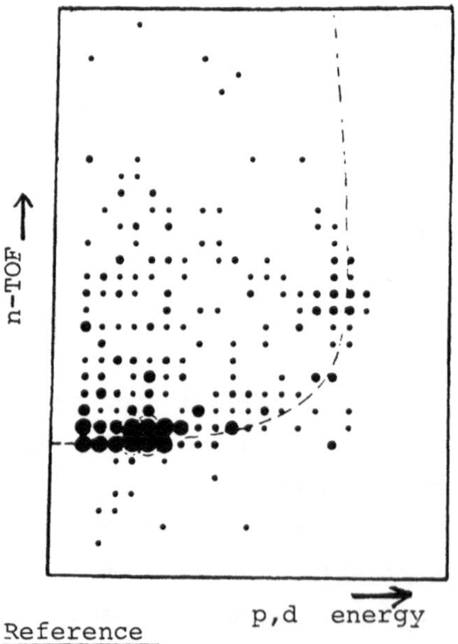

Fig.2

Two dimensional spectrum of the (np) and (n,d) pairs emitted around 180° after the absorption of π^- in ^{3}He. The area of the circles is proportional to the counting rate. The dashed line shows the kinematical boundary for the reaction π^{-3}He→pnn at ϑ_{np}=180°.

Reference

1) R. Abela, G. Backenstoss, A. Brandao d'Oliveira, M. Izycky, H.O. Meyer, J. Schwanner, L. Tauscher, P. Blüm, W. Fetscher, D. Gotta, H. Koch, H. Poth, L.M. Simons, Phys. Lett. 68B (1977) 429.

Absorption Effects in π-He-Scattering in the Resonance Region

H. M. Hofmann
Institute for Theoretical Physics, Erlangen, Germany

The pion-nucleon interaction is dominated by the Δ(33) resonance for 100 MeV $\leq T_\pi \leq$ 300 MeV. The corresponding multiple scattering expansion, however, fails to reproduce the pion-nucleus-scattering-data quantitatively [1,2].

Once the total cross section is fitted phenomenologically, good agreement is achieved also for the differential cross sections [1,2,3]. This additional input reflects the strong coupling to other excitation-modes, which do not exist in pion nucleon interactions, such as true pion absorption.

In a Δ-dominance model the interaction of a pion with a nucleus is assumed to occur always via the excitation of ($\Delta\bar{N}$) doorway-states. In the particle-hole propagator we include the terms shown in Fig. 1a - 1f. To evaluate those diagrams we use effective Lagrangians [4] .

We find that the absorption terms are very important to reproduce quantitatively the experimental data of both, total and differential cross section (see Fig. 2,3): Without true pion absorption the calculated elastic cross section exceeds the data [5] by far.

Furthermore it turns out that (i) only few ($\Delta\bar{N}$) states carry all elastic transition strength, (ii) the exchange diagram 1d contributes about the same as diagram 1c, (iii) the vertex corrections 1e and 1f are relatively small compared to 1c for pion elastic channels, but reach about 50 % for other channels.

References

1) M. Hirata, F. Lenz and K. Yazaki, Ann. Phys. (N.Y.) 108 (1977) 116
2) L. S. Kisslinger and W. L. Wang, Ann. Phys. (N.Y.) 99 (1976) 374
3) N. Auerbach, Phys. Rev. Lett. 38 (1977) 804
4) M. Brack, D. O. Riska and W. Weise, Nucl. Phys. A 287 (1977) 425
5) F. Binon et al., Nucl. Phys. A 298 (1978) 499

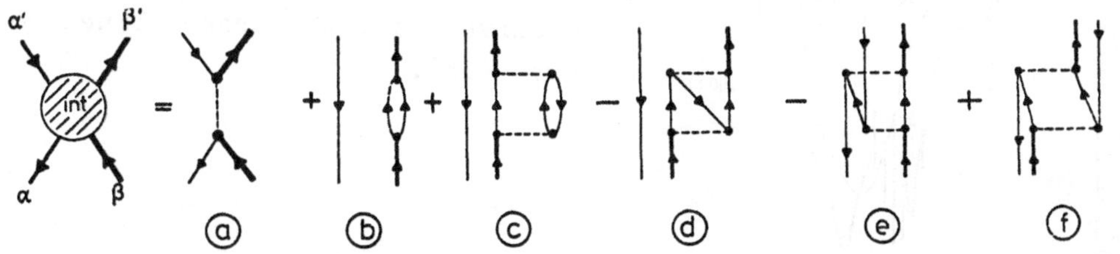

Fig. 1: Channels contributing to the ($\Delta\bar{N}$)-propagator.

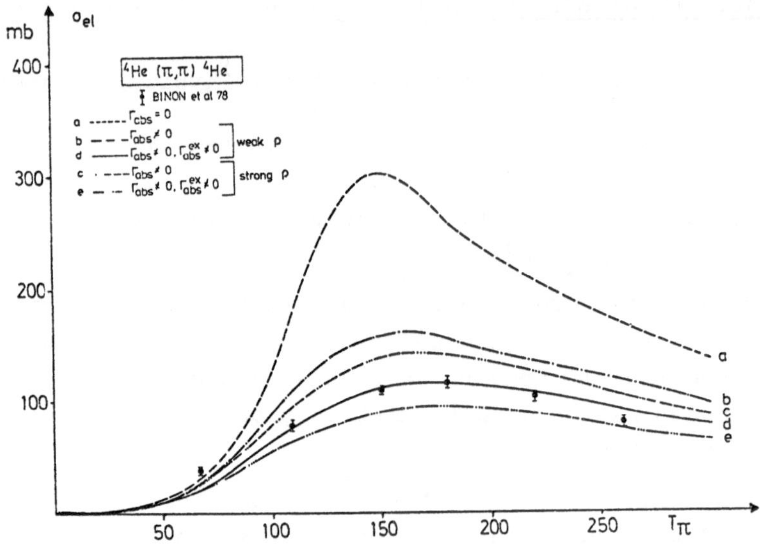

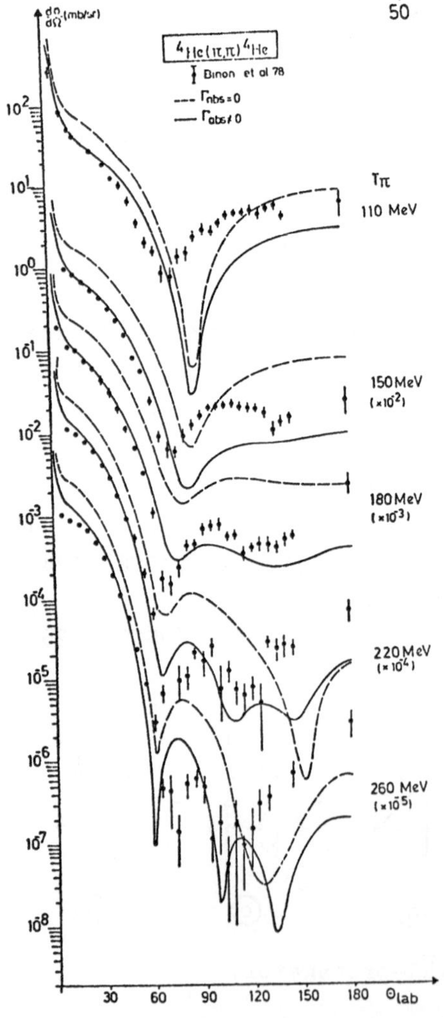

Fig. 2:
Comparison of the integrated elastic cross section for different assumptions concerning the pion absorption: (a) without absorption, (b) and (c) only absorption diagram 1c taken into account, (d) and (e) all absorption diagrams 1c to 1f considered. The curves (b) and (d) resp. (c) and (e) are calculated using weak resp. strong ρ coupling.

Fig. 3:
Comparison of differential cross sections. The full lines correspond to curve (2d), the broken lines to curve (2a).

A DERIVATIVE SUM RULE AND INCONSISTENCY IN THE LOW-ENERGY π^4He SCATTERING PARAMETRIZATION

Olgierd Dumbrajs
Research Institute for Theoretical Physics
University of Helsinki
Finland

We apply a sum rule, which we call a derivative sum rule in analogy with similar sum rules known in particle physics [1], to the symmetric elastic π^4He forward scattering amplitude $f(\omega)$. This sum rule follows from the condition that the subtraction constant defined by the dispersion relation should be independent of the energy at which this relation is evaluated:

$$\frac{d}{d\omega}\mathrm{Re}f(\omega) = \frac{d}{d\omega}\left[\frac{2\omega^2}{\pi} P \int_{\omega_0}^{\infty} \frac{\mathrm{Im}f(\omega')}{\omega'(\omega'^2-\omega^2)} d\omega'\right] .$$

Here ω is the pion lab energy and ω_0 is the beginning of the unphysical region.

Although this sum rule utilizes the same input data as the conventional dispersion relation it is entirely independent of the value of the subtraction constant and provides a strong consistency requirement. Numerical tests of the sum rule show that the behaviour of $\mathrm{Im}f(\omega)$ around the threshold with a bump at $\sim$ 20 MeV above it, as deduced from phase shift analysis [2], is inconsistent with the remainder of our knowledge of π^4He interactions. The same conclusion had been reached already earlier [3] on the basis of data from mesic atoms.

References

1. N. M. Queen, S. Leeman and F. E. Yeomans, Nucl. Phys. B11, 115 (1969).
2. Yu. A. Shcherbakov et al., Nuovo Cimento 31A, 249 (1976).
3. H. Pilkuhn, N. Zovko and H. G. Schlaile, Z. Physik A279, 283 (1976).

Effect of P-Wave Pion Absorption on Two Nucleons and Low-Energy Pion-Nucleus Scattering

R. Rockmore
Physics Dept., Rutgers University
New Brunswick, New Jersey 08903 USA

and

E. Kanter
Physics Division, Argonne National Laboratory
Argonne, Illinois 60439 USA

Theoretical efforts in π-nucleus scattering have recently been directed at incorporating the effect of the pion absorption, which becomes important in the low-energy domain, into the usual optical model description. We report here the results of a calculation of the effect of P-wave pionic absorption on low-energy elastic and inelastic pion-nucleus scattering in the energy interval $60 \text{ MeV} \leq T \leq 120 \text{ MeV}$.

The P-wave pion-nucleus optical potential $\sim \rho^2(R)$ which derives from P-wave pion absorption on two nucleons is constructed using a πNN absorption model with π- and ρ- rescattering through the 33-resonance. Both the lowest-order optical potential[1] and that due to P-wave pionic absorption are cast into local Laplacian form in our study of $\pi^- - {}^{12}C$ scattering. The logarithmic compatibility of our lowest-order local Laplacian calculation with best fits to experimental data within the energy domain of interest are shown in Fig. 1. (For ${}^{12}C$ a modified Gaussian density with a = 1.47 F is used.)

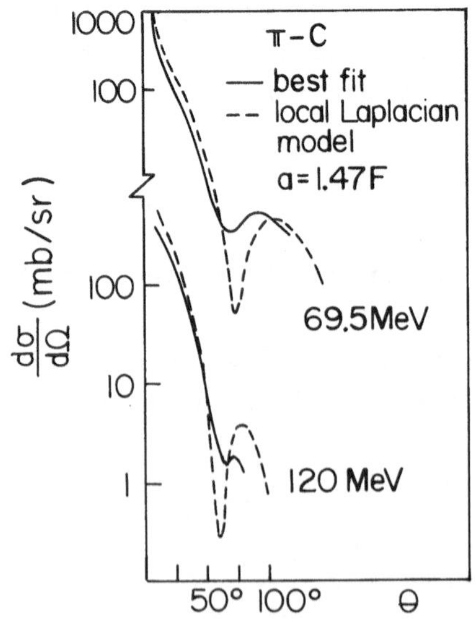

Fig. 1. Lowest-order optical model fits ($\pi^- - {}^{12}C$).

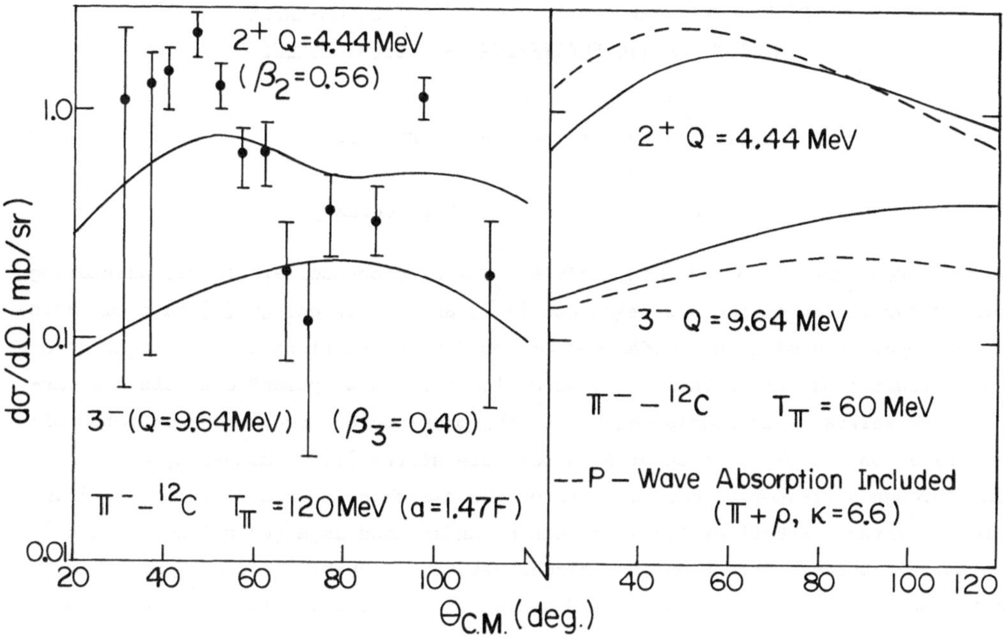

Fig. 2. Lowest-order optical model predictions (T_π = 120 MeV) with 2^+ data of Binon et al. [Nucl. Phys. B17, (1970) 168] for comparison.

Fig. 3. Inelastic scattering optical model predictions (T_π = 60 MeV) with and without the ρ^2-contribution.

In Figs. 2 and 3 are seen the predictions of the same model for the scattering of 120 MeV and 60 MeV π^- from ^{12}C leading to excitation of 2^+ (4.44 MeV) and 3^- (9.64 MeV) states.

In the ρ^2 - potential monopole hadronic form factors[2] with Λ_π = 1200 MeV, Λ_ρ = 1500 MeV are introduced; we also take $g_\rho^2/4\pi = 0.52$ and $\kappa = 6.6$ in the determination of f_ρ.[2] Short-range nucleon-nucleon correlations are taken into account in the calculation through the schematic hard-core correlation function[3] $C(r) = -\theta(r_c-r)$, with $r_c = 0.7$ F.

One sees in Fig. 3 that the ρ^2-potential from P-wave pionic absorption produces at 60 MeV sizeable but parameter-sensitive effects on the lowest order prediction for both 2^+ and 3^- states, although it should be mentioned that the use of the local Laplacian version at the lower end of the T_π domain is not so realistic.

References:

1. E. H. Auerbach, D. M. Fleming, and M. M. Sternheim, Phys. Rev. 162 (1967) 1683.
2. M. Brack, D. O. Riska, and W. Weise, Nucl. Phys. A287 (1977) 425.
3. G. Baym and G. E. Brown, Nucl. Phys. A247 (1975) 441.

MICROSCOPIC CALCULATION OF ABSORPTION CHANNELS IN PION ELASTIC SCATTERING FROM LIGHT NUCLEI

E. Oset +) and W. Weise

Institute of Theoretical Physics

University of Regensburg

D-8400 Regensburg, Fed. Rep. Germany

Recent investigations of the many-body aspects of pion-nuclear elastic scattering both in the region of the 3.3 resonance [1,2] and at low energy [3] have emphasised the importance of higher order medium corrections to first order optical potential calculations. Among these, the so-called "true absorption" channels are particularly relevant. In models which describe pion-nucleus scattering in terms of an excitation and propagation of Δ-isobar-hole states [1,2], many-body effects appear through self-energy modifications of the Δ-isobar Green's function [1,2] and through vertex corrections [2] which can be understood as a generalised form of the Lorentz-Lorenz correction [5]. Altogether, the many-body corrections contribute to the energy dependent shifts and widths of the relevant isobar-hole doorway states [4].

Unlike the purely phenomenological treatment [1] of absorption and other contributions (e.g. reflection corrections to quasi-elastic scattering channels), our aim is to incorporate these, together with Pauli and binding effects, into a microscopic many-body calculation [6] of the π-nuclear scattering amplitude. The starting point are effective meson-baryon Lagrangians, in particular πNΔ and ρNΔ vertex operators, the off-shell properties of which are adjusted to yield a good description of the πd → pp reaction above threshold [7]. Absorption channels are treated in terms of the coupling of Δ-hole states to the two-nucleon-two-hole continuum. At least for ^{4}He, proper incorporation of such two-body mechanisms accounts for the major part of p-wave absorption in the 3.3 resonance region, inasmuch as this takes into account the contribution of the (π, 2N) reaction to the elastic channel. Assuming Δ-isobar dominance, the pion-nuclear T matrix is shown schematically in Fig. 1.

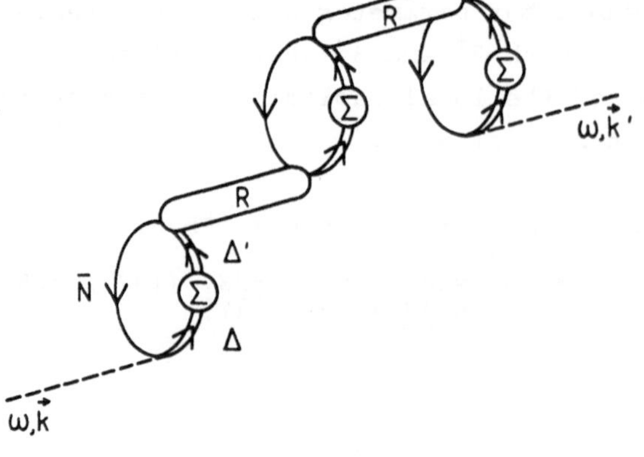

Figure 1:

Schematic representation of the pion-nuclear T matrix in the isobar-hole RPA approach. The isobar-hole interaction is denoted by R, self-energy corrections of the isobar propagator by Σ. Not shown here are binding corrections on the nucleon lines.

The incoming pion excites Δ-hole states, $|s\rangle = |(\Delta\bar{N})\, J^\pi, T=1\rangle$ coupled to angular momentum $J^\pi = 0^-, 1^+, 2^-, \ldots.$ and isospin one. We choose a harmonic oscillator basis for these states. The T matrix for scattering of a pion with frequency ω and initial/final momentum $\vec{k}$ and $\vec{k}'$, resp., for a N=Z spin saturated nucleus, becomes

$$\langle\vec{k}'|T(\omega)|\vec{k}\rangle = \frac{1}{2\omega} \sum_{ss'} \langle\vec{k}'|\delta H|s'\rangle\, G_{s's}(\omega)\langle s|\delta H|\vec{k}\rangle \tag{1}$$

where δH is the πNΔ vertex operator, chosen in accordance with ref. [7]. Included in δH are relativistic and recoil corrections and a πNΔ formfactor of monopole type, with cutoff mass Λ = 1.2 GeV. [6,7]. Furthermore, $G_{s's}$ is the full isobar-hole Green's function, written in the RPA approximation:

$$G_{s's}(\omega) = \left\{ \left[\left(\omega\delta_{s's} - M_{s's}(\omega)\right)^{-1} - \left(\omega\delta_{s's} + \tilde{M}_{s's}(\omega)\right)^{-1} \right]^{-1} - R_{s's}(\omega) \right\}^{-1} \tag{2}$$

where the complex mass matrix $M_{s's}$ contains single particle energies ε_s of the isobar-hole states and additional self-energy corrections $\Sigma_{s's}$:

$$M_{s's}(\omega) = \varepsilon_s(\omega)\ \delta_{s's} + \Sigma_{s's}(\omega). \tag{3}$$

Included in ε_s are the free isobar width, a possible shift due to an attractive (energy independent) Hartree potential, and the single particle energy of the nucleon hole. Energy dependent Fock terms (Pauli corrections) and absorptive self-energies are summarized in $\Sigma_{s's}$. The quantity $\tilde{M}_{s's}(\omega) = M_{ss'}(-\omega)$ takes into account crossed terms. The complex matrix $R_{s's}$ contains matrix elements of the isobar-hole interaction. The leading piece of this is one-pion exchange, which generates multiple scattering and elastic broadening of the Δ-hole states. Also included in $R_{s's}$ are ρ meson exchange and additional short range-correlations, the main sources of the Lorentz-Lorenz correction in the interpretation of ref. [5].

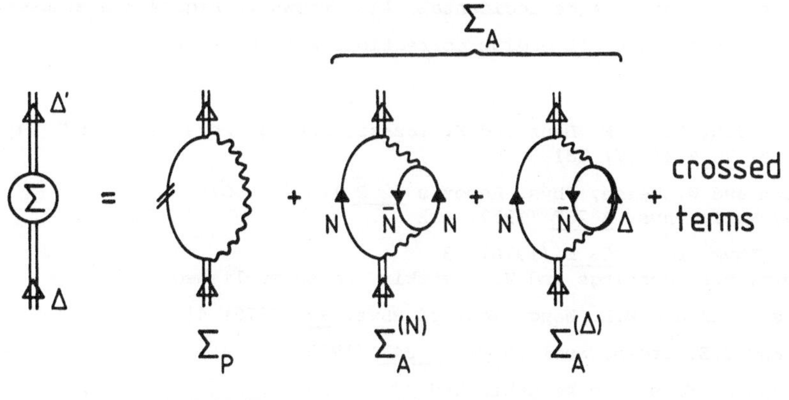

Figure 2: Isobar self-energies discussed in this work; Σ_P: Pauli corrections (Fock term); Σ_A: coupling of isobar to two-nucleon-one-hole or ΔN-hole configurations. The wavy lines include π and ρ exchange plus additional baryon-baryon correlations [2].

The calculation of $\Sigma_{s's}$ includes the diagrams shown in Fig. 2. Σ_P describes Pauli effects and additional binding corrections for the intermediate nucleon which moves in a single particle potential. The partial blocking of the $\Delta \to \pi N$ decay due to the Pauli principle gives a reduction of the free isobar decay width by about 30 MeV at resonance for ^4He. The real part of Σ_P is attractive (once ρ exchange is included) and energy dependent due to retardation in the one-pion exchange. The term Σ_A involves the coupling of the isobar to the 2-nucleon-1-hole continuum ($\Sigma_A^{(N)}$) and resonant reflection contributions to the quasielastic channel ($\Sigma_A^{(\Delta)}$). These terms involve the full complexity of box diagrams and are calculated accordingly. Both the imaginary parts of $\Sigma_A^{(N)}$ and $\Sigma_A^{(\Delta)}$ contribute to a broadening of the isobar-hole states. While the energy dependence of Im $\Sigma_A^{(N)}$ is smooth, Im $\Sigma_A^{(\Delta)}$ shows a strong variation with energy due to the resonant behaviour of the rescattering mechanism. At threshold ($\omega = m_\pi$) only Im $\Sigma_A^{(N)}$ survives. The real parts of Σ_A are altogether attractive. At threshold, most of this attraction comes from Re $\Sigma_A^{(\Delta)}$. This combines with the attraction from a (not well determined) isobar Hartree potential.

The results for the self-energies Σ are sensitive to the ρNΔ coupling strength and the off-shell properties of the meson-baryon Lagrangians. It is therefore crucial to minimize this off-shell freedom by using effective πNΔ and ρNΔ Lagrangians consistent with $\pi d \rightarrow pp$ [7], from where we have taken coupling constants and formfactors.

Results for the effects of various medium corrections on the $\pi\,^4He$ total cross section are shown in Fig. 3. The T matrix has been obtained by combined inversion of the mass (self-energy) and isobar-hole interaction matrices according to eq. (1). Included are all π-nuclear partial waves $J^\pi = 0^-, 1^+, \ldots$ up to 6^-, the most important ones being 1^+ and 2^- for ^4He. There is partial cancellation between Pauli (plus binding) effects versus absorption and reflection terms, although these effects are individually large. The surprisingly good agreement sometimes obtained with simple fixed scatterer treatments may thus be accidental. Also shown in Fig. 4 are examples of angular distributions, once all medium corrections are incorporated.

References:

[1] M. Hirata, J.H. Koch, F. Lenz and K. Yazaki; Ann. of Phys. 108 (1977) 16; Phys. Lett. 70 B (1977) 281

[2] G.E. Brown and W. Weise, Phys. Reports 22 C (1975) 280; W. Weise, Nucl. Phys. A 287 (1977) 402

[3] M. Thies, Phys. Lett. 63 B (1976) 43; G.E. Brown, B.K. Jennings and V. Rostokin, to be published

[4] L.S. Kisslinger and W.L. Wang, Ann. of Phys. 99 (1976) 41

[5] G. Baym and G.E. Brown, Nucl. Phys. A 247 (1975) 345

[6] E. Oset and W. Weise, to be published

[7] M. Brack, D.O. Riska and W. Weise, Nucl. Phys. A 287 (1977) 425

+) On leave from University of Barcelona

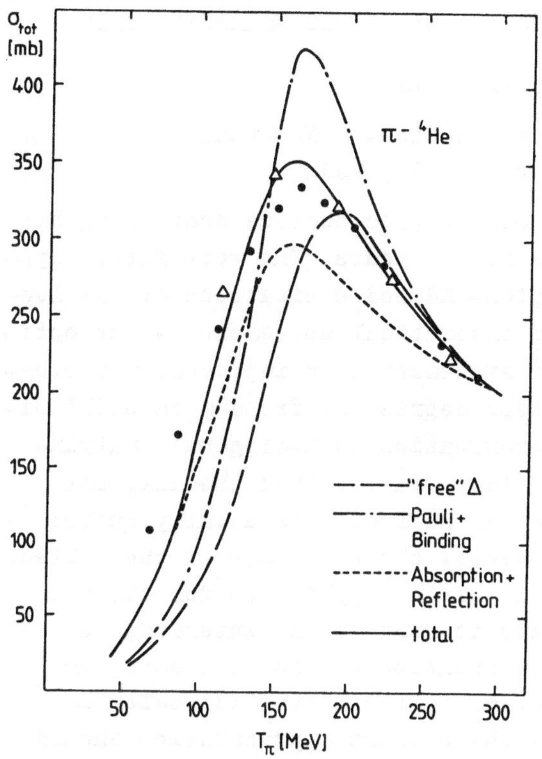

Figure 3:

Effects of various medium corrections on total π^4He cross section:

"free Δ": Free Δ-isobar (no self-energy interactions), but binding of nucleon-hole and full isobar-hole interaction (π and ρ exchange plus additional short range correlations) included (Omission of ρ exchange and nucleon binding shifts peak downward by 50 MeV and raises σ_{tot} beyond 350 mb)

"Pauli + Binding":
Fock terms (Σ_P) and isobar Hartree potential (with depth V_O = - 50 MeV) added to "free Δ"

"Absorption + Reflection:
Combined effect of $\Sigma_A^{(N)}$ and $\Sigma_A^{(\Delta)}$ added to "free Δ".

"Total": Summed effects of "Pauli + Binding" and "Absorption + Reflection" added to "free Δ".

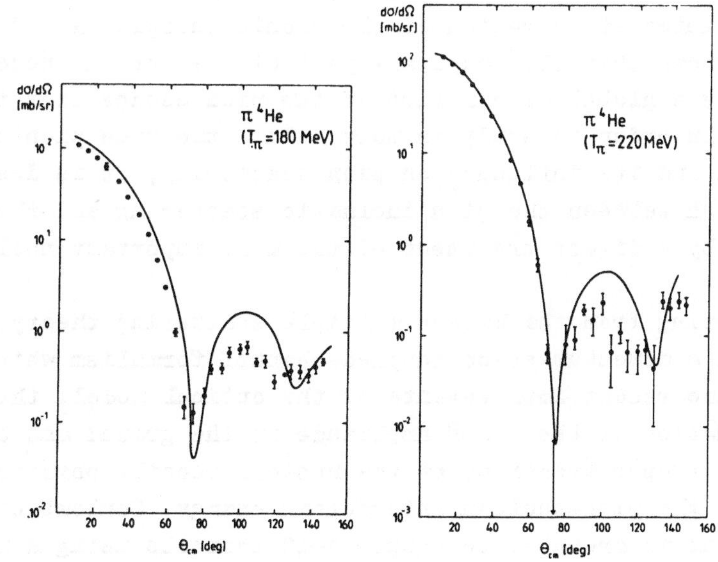

Figure 4: Examples of differential cross sections for π^4He, including all medium corrections

LOW ENERGY PION-NUCLEUS SCATTERING IN THE COUPLED-CHANNEL MODEL

M.Gmitro and R.Mach

Laboratory of Theoretical Physics, JINR Dubna
101 000 Moscow, PO Box 79, USSR

The experimental information on the pion-nucleus scattering has been substantially enriched in the recent years. The very first experimental data also appear on the pion-induced excitations of the low-lying nuclear levels. The standard theoretical approaches - the optical model and DWIA - are indeed [1] systematically improved, nevertheless a serious account of the nuclear degrees of freedom is still missing. The coherent scattering approximation (non-diagonal matrix-elements $\langle m \mid t_{\pi N} | n \rangle$ being A times smaller than the diagonal ones are neglected, A is the number of nucleons) is usually quoted in this respect. It disregards, however, the existence of the collective nuclear excitations and also is hardly applicable for the very light nuclei. A possible approximate account of the intermediate nuclear states is the "closure approximation". This is common for the Glauber model and the second order optical potential calculation [2]. If applicable at all [3], the closure approximation should be valid in the domain of high energies only.

There is another important reason why at least some of the nuclear excited states should be explicitly taken into account in the very interesting region of the pionic energies E = 0-50 MeV. It is well known that the imaginary part of the optical potential provides only a global description of the pion escape from the elastic channel. In order to study in more detail the true pion absorption by nuclei and its influence on pion scattering, it is desirable to distinguish between the pion inelastic scattering and the pion absorption by a direct treatment of the most important nuclear excited states.

Starting from the Watson multiple scattering theory we have developed a momentum-space coupled-channel formalism which comprises most of the recent improvements of the optical model: the angular transformation of the π-N amplitude in the ground and excited states, a proper treatment of the nuclear recoil, physically sound choice of the pion-nucleon interaction energy, Coulomb effects, etc. At present we are able to couple 5-10 channels using a medium size computer. The formalism is meant for the description of the low-energy $\pi - {}^4He$ scattering. The spectrum of the ^4He can be approximately described by a few 1p-1h shell-model configurations plus one collective 0^+ state (breathing mode) if the region of 0 - 30 MeV excitation energies is considered. Via coupling of these states we expect to reach a fairly realistic description of the pion elastic and inelastic scattering in the energy interval 0-50 MeV. Let us recall also the results of the energy-dependent phase-shift analysis [4] of the π-^{4}He scattering, total cross sections and mesoatomic data which exhibit a remarkable structure in the region of 20-30 MeV, where the most pronounced nuclear states are located.

Finally we would like to stress that this formalism is also fully applicable for the studies of the pion interactions with more complex nuclei, especially:

(i) low-energy scattering of pions by the p-shell nuclei including the induced tensor pion-nuclear forces. Due to the large quadrupole momenta exhibited by these nuclei, the role of the tensor forces is expected to be important;

(ii)in the interactions of pions with the p- and (sd)-shell nuclei we expect strong effects from the few collective ("giant multipole resonances") excited states in the region of 20-30 MeV excitation energy;

(iii) a more precise treatment of the Coulomb effects is required by the recent high-quality scattering data on heavier nuclei. The multiple-scattering series can be re-arranged to yield a coupled system of two integral equations which take into account the Coulomb distortion effects in the <u>individual scattering acts</u> in variance with the usual optical model.

(iv) the coupled channel formalism represents a natural ground for the description of the multistep processes as are e.g. the charge-exchange reactions.

References:

1) A.W.Thomas, 7th Int.Conf. on High Energy Physics and Nuclear Structure, Zurich, 1977.
2) T.-S. H. Lee and S.Chakravarti, Phys.Rev. C16 (1977) 273.
3) T.de Forest and J.L. Friar, Phys.Lett. 58B (1975) 397.
4) Yu.A.Shcherbakov et al. Nuovo Cim. 31A (1976) 249.

NON ELASTIC INTERACTION OF π^+ MESONS ON ^4He AT 120, 145 AND 165 MeV

F.Balestra,M.P.Bussa,L.Busso,R.Garfagnini,G.Piragino and A.Zanini
Istituto di Fisica dell' Università, Torino, Italy
I.N.F.N. - Sezione di Torino, Italy

C.Guaraldo,A.Maggiora and R.Scrimaglio
Laboratori Nazionali dell'I.N.F.N. di Frascati, Italy

I.V.Falomkin,G.B.Pontecorvo and Yu.A.Shcherbakov
Joint Institute for Nuclear Research, Dubna, U.S.S.R.

The (π^+, ^4He) non elastic reactions at 120, 145 and 165 MeV have been investigated by means of a diffusion cloud chamber in magnetic field. The quasi elastic scattering and charge exchange of π^+ on bound nucleon have been investigated and the results compared with those obtained in the (π^-, ^4He) interactions and with the ($\pi^\pm$, p) reactions.

The analysis of the data shows that the pion inelastic reactions, involving a single nucleon, seem to occur through the isobaric resonance excitation and the π^+ absorption in ^4He to occur on two nucleons. In the (π^+ 2p2n) and the absorption reactions the isobaric resonance seems to give negligible contribution.

Angular Distribution of the Reaction $np \to d\pi^o$ between 350 MeV $< T_n <$ 580 MeV +)

W. Hürster, Th. Fischer, G. Hammel, K. Kern, R. Kettle, M. Kleinschmidt, L. Lehmann, E. Rössle, H. Schmitt

Fakultät für Physik der Universität Freiburg
D - 7800 Freiburg i.Br.

The reaction $np \to d\pi^o$ is one of the fundamental coherent pion production processes. Recently these reactions have attracted much interest, particularly in the region where the Δ_{33} resonance dominates the reaction mechanism through a resonant intermediate state. Several models have been proposed, but so far no decision has been possible. On the other hand there are still inconsistencies between the different experimental results.

We have measured, at the SIN neutron beam, the angular distribution of the deuterons within the unambiguous region of the forward center of mass emission. The momentum analysis has been performed by a drift chamber magnet spectrometer with a resolution of about 3 % [1]. The energy range of the primary neutrons covered simultaneously was from about 350 MeV to 580 MeV. Typical angular distributions are shown in Fig. 1. The solid lines are fits to the experimental data by the relation

$$d\sigma/d\Omega \propto A + \cos^2\theta + B \cos^4\theta \qquad (1)$$

The energy dependence of the angular distribution coefficients A and B is given in Fig. 2 together with the results of other experiments [2 - 5]. For the parameter A, there is no significant discrepancy; the difference between our results and those of Bartlett et al. [2] is due to the fact that they completely neglect the $\cos^4\theta$ term. In our data this term varies from small positive values around 400 MeV neutron energy to B = -.3 at 580 MeV, with $B \approx 0$ at about 480 MeV, in striking contrast to the data of Wilson et al. [3], whose B values reach large negative values below 450 MeV. This behaviour is hard to understand, since B represents d-wave pion production which should decrease in magnitude by approaching threshold. The small positive value we find below 480 MeV is in agreement with the finding of Preedom et al. [4] for B, on the $\pi^+d \to pp$ reaction, which is related to $np \to d\pi^o$ by isospin invariance and time reversal.

+) Work supported by the Bundesministerium für Forschung und Technologie

[1] SIN Jahresbericht 1977, E 25
[2] D. F. Bartlett et al., Phys. Rev. D1 (1970), 1984
[3] S. S. Wilson et al., Nucl. Phys. B33 (1971), 253
[4] B. M. Preedom et al. Physics Letters 65B (1976), 31
[5] L. C. Northcliffe, Int.Conf. on the Interactions of neutrons with nuclei, Lowell, Mass. (1976)

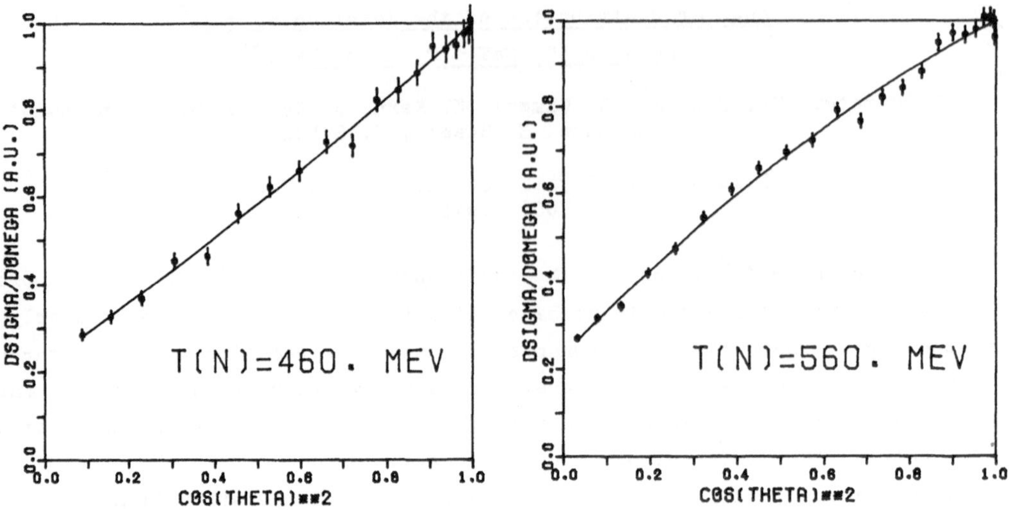

Fig. 1. Angular distribution at 460 MeV and 560 MeV, respectively, as function of $\cos^2\theta$. The curves are fits of eq. (1) to the data, normalized to 1.0 at $\cos^2\theta$ = 1.

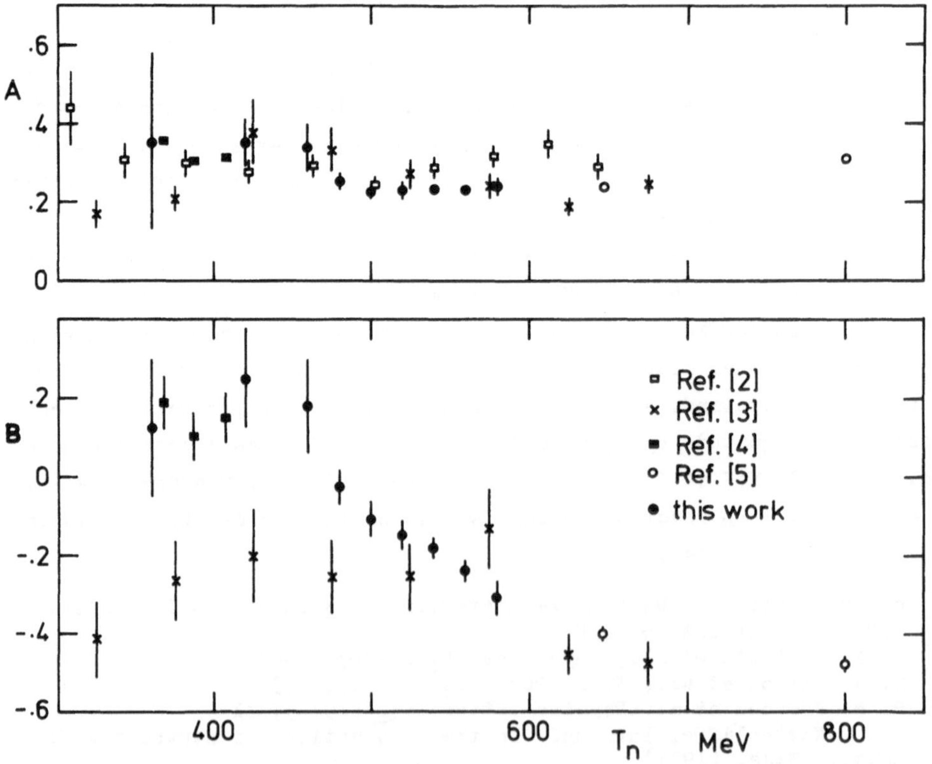

Fig. 2. Energy dependence of A and B of eq. (1).

Positive Pion Production from Neutron-Proton Collisions between 460 and 580 MeV [+)]

M. Kleinschmidt, Th. Fischer, G. Hammel, W. Hürster, K. Kern, R. Kettle, L. Lehmann, E. Rössle, H. Schmitt

Fakultät für Physik der Universität Freiburg
D - 7800 Freiburg i.Br.

Experimental data on pion production by np-collisions are still very scarce in the energy region from threshold to 600 MeV because of the lack of suitable neutron beams. On the other hand the theoretical situation is also unsatisfactory particularly for the pion production in the initial isoscalar state (σ_{01}) for which the Mandelstam model [1] cannot be applied because it describes the production only via a resonant (Δ_{33}) intermediate state. Therefore the pion production cross section σ_{01} is of particular interest.

We have measured the energy spectra of positive pions and their angular distribution within the lab. angular range from 0° to 20° using the magnetic spectrometer at the SIN neutron beam [2]. The data have been converted event by event into the c.m. system. The resulting energy spectrum is given in fig. 1. As can be seen, the phase space distribution fails to reproduce the measured energy spectrum.

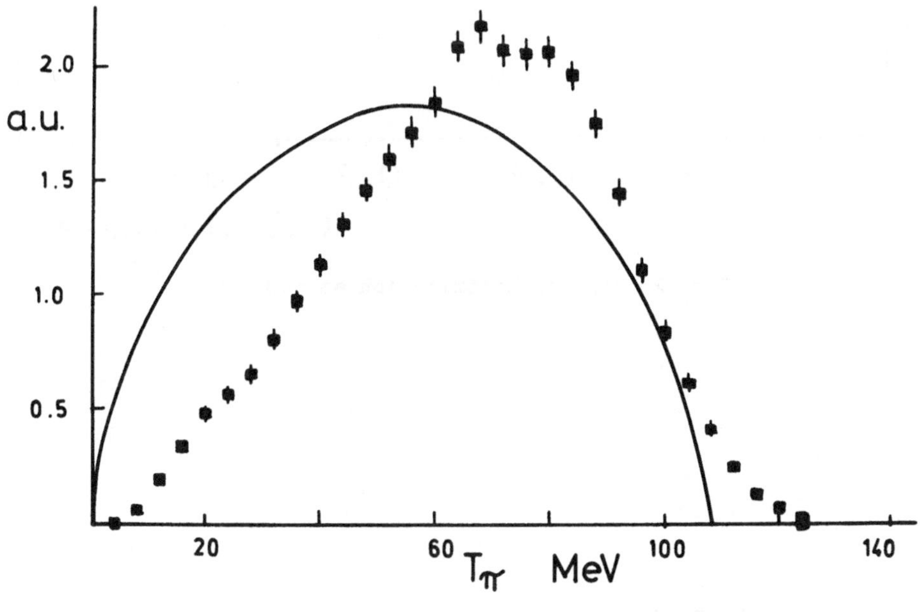

Fig. 1. Pion Energy Distribution for E_n = 540 MeV and ϑ_{cm} = 11.5°.
Solid line: phase space.

The angular distribution (fig. 2.) has been fitted to an expression

$$\frac{d\sigma}{d\Omega} \propto A + \cos^2\theta ,$$

thus neglecting a linear term as well as higher order terms. From the fit we obtain A = 0.86 ± 0.32 at 540 MeV. If one assumes an isotropic distribution for the isovector σ_{11}-part as has been measured earlier [3] the value of the parameter A requires a non-vanishing σ_{01}-contribution. We conclude that at 540 MeV the isoscalar and the isovector contributions are of about the same magnitude.

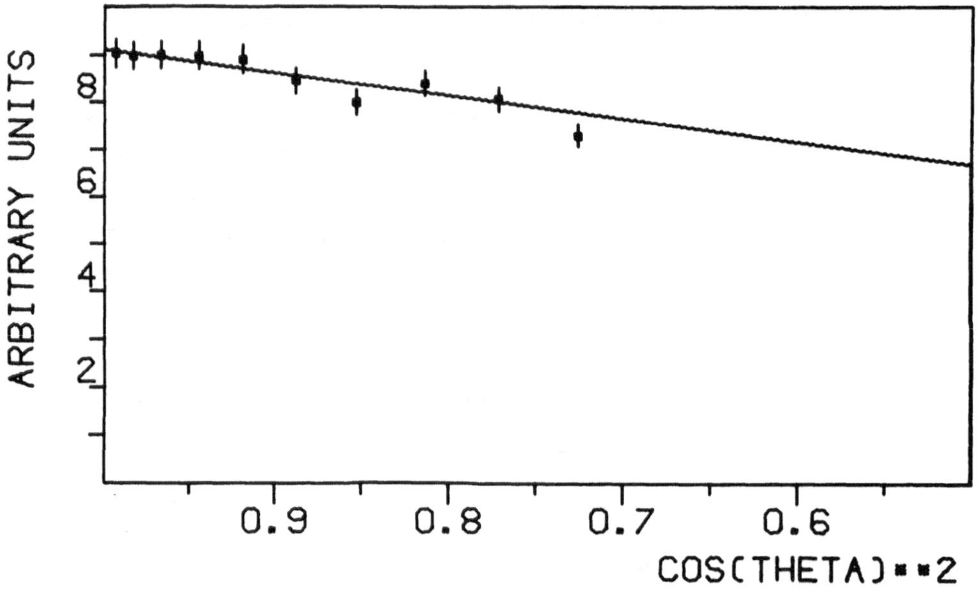

Fig. 2. Angular Distribution at 540 MeV.

+) Work supported by Bundesministerium für Forschung und Technologie

[1] S. Mandelstam, Proc.Roy.Soc. A244 (1958) 491
[2] SIN Jahresbericht 1976, E25
[3] A. F. Dunaitsev, Yu.D.Prokoshkin JETP 36 (1959) 1179

A TEST OF ISOSPIN INVARIANCE IN THE np → dπ° REACTION AT 795 MeV*

C. L. Hollas, C. R. Newsom, and P. J. Riley
University of Texas, Austin, Texas 78712

B. E. Bonner
Los Alamos Scientific Laboratory, University of California
Los Alamos, New Mexico 87545

G. Glass
Texas A & M University, College Station, Texas 77843

We have carried out differential cross section measurements for the np → dπ° reaction at 795 MeV. Yang suggested in 1952 that a comparison of the two reactions $n + p \rightarrow d + \pi^0$ and $p + p \rightarrow d + \pi^+$ should provide a severe test of the hypothesis of charge independence in strong interactions. This test requires that the differential cross sections at the same center-of-mass energy and angle have the ratio 1:2 for the two reactions, neglecting $\pi^+ - \pi^0$ and n-p mass differences.

Three possible experimental observations are as follows: first, the np → dπ° angular distribution in the center-of-mass system must be symmetric about 90°. This follows from the identity of the particles in the initial state in the $pp \rightarrow d\pi^+$ reaction. This consequence allows a test for the np → dπ° reaction which is independent of any other measurement. Second, the angular distributions observed for the two reactions at the same c.m. energy must have the same shape. Finally, the magnitude of the cross sections for the two reactions at the same c.m. energy must have the ratio of 1:2. This observation requires accurate absolute cross section measurements for both reactions.

The development at LAMPF of an intense, nearly monoenergetic neutron beam made possible a new measurement of np → dπ° with greater precision than had previously been obtained.[1] The np → dπ° measurements were carried out using the LAMPF 800-MeV d(p,n)2p 0° neutron beam and the MWPC spectrometer in area B of LAMPF. The accelerator was operated in a chopped beam mode, providing 40 ns between micropulses. Using neutron time-of-flight measurements, we were able to select events that were initiated only by neutrons within the sharp high-energy peak of the neutron momentum spectrum. Charged particle mass identification was provided by simultaneous measurement of momentum and flight time through the spectrometer, allowing a direct rest mass calculation for each event.

Using five angular positions of the spectrometer, we have obtained a complete angular distribution in the center-of-mass system. The data are presented in the figure, the errors shown being statistical only. Corrections to the data to accommodate deuteron breakup in the target and spectrometer system, and deuteron contamination from the np → dγ process are not yet included, but will not alter the data significantly. Absolute np → dπ° cross-section data were not measured in the experiment; the data have therefore been normalized to the $\pi^+ d \rightarrow pp$ work of Richard-Serre et al.[2] for a pion energy corresponding to 810-MeV proton energy, and the solid curve is their fit to their data. The solid curve closely approximates our observed angular distribution, although the data would indicate a more pronounced leveling in shape at the extreme angles, in qualitative agreement with the behavior observed at slightly higher incident energies,[3] where the cross section actually shows a pronounced dip near 0° at 1.3 GeV. The $\pi^+ d \to pp$ data set extends from 16° to 90° c.m. only, and is thus not inconsistent with our data at the extremes. In addition, differences in the detailed shapes of the $np \to d\pi^\circ$ and $pp \to d\pi^+$ angular distributions might arise solely from the difference in c.m. energies of the two reactions.

To summarize, at the present level of analysis, no evidence of violation of isospin conservation in the strong interaction is indicated. The shape of the angular distribution of the $np \to d\pi^\circ$ data appears to be symmetric about 90°. A detailed statistical analysis will be made to search for any odd powers in cos θ required to describe the data. A precision measurement of the $pp \to d\pi^+$ reaction at the c.m. energy of our work, and extending to angles near 0°, is necessary for a more detailed comparison with the shape of our $np \to d\pi^\circ$ data.

REFERENCES

*Work supported by the U. S. Department of Energy.

1. S. S. Wilson et al., Nucl. Phys. B33 (1971) 253; D. F. Bartlett et al., Phys. Rev. D 1 (1970) 1984.
2. C. Richard-Serre et al., Nucl. Phys. B20 (1970) 413.
3. R. M. Heinz et al., Phys. Rev. 167 (1968) 1232; D. Dekkers et al., Phys. Letters 11 (1964) 161.

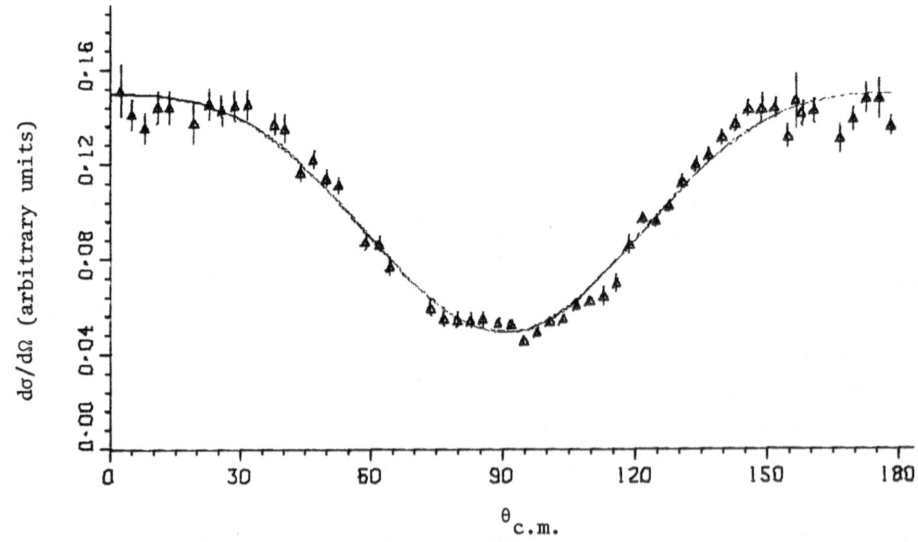

Angular distribution for the $np \to d\pi^\circ$ reaction at 795 MeV. The solid curve is from a fit to the $\pi^+ d \to pp$ data of Ref. 2.

THE πNN VERTEX AND THE RELATIVISTIC EVALUATION OF $\pi^+ d\to pp$ AMPLITUDE*

Shin-ichi Morioka, & I.R. Afnan

School of Physical Sciences, The Flinders University of South Australia,

Bedford Park, S.A., 5042, AUSTRALIA.

Recently [1], it has been shown that (p,π^+) reactions in nuclei are sensitive to the choice of πNN vertex. In particular, it was found, using pionic stripping and Dirac-shell-model wave function for the captured neutron, that the pseudoscalar(PS) coupling gives a larger cross section than the pseudovector(PV) coupling. To further investigate the importance of the πNN vertex in pion production (absorption) we present a field theoretic formulation of the reaction $\pi^+ d\to pp$, with the hope clarifying the above difference between the two choices for the πNN vertex.

The amplitude for $\pi^+ d\to pp$ is given by (see Fig. 1)

$$M_{12} = \bar{u}(p_2)\Lambda\frac{\gamma k + M}{k^2 - M^2}\varepsilon\Gamma u^c(p_1), \qquad (1)$$

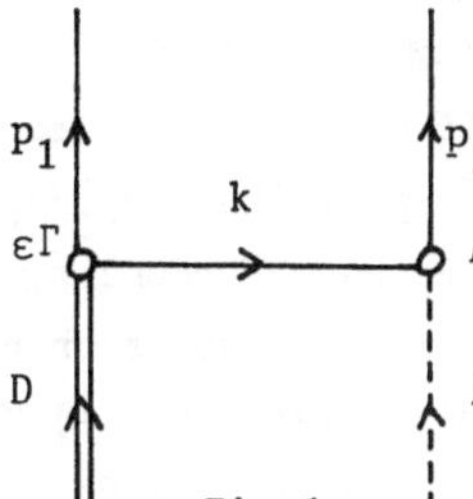

Fig.1

where u^c, ε, and Γ are the charge-conjugate spinor, deuteron polarization vector, and the Blankenbeckler-Cook vertex function[2]. In Eq.(1) M is the nucleon mass and Λ is the πNN vertex given by;

$$\begin{aligned}\Lambda &= g\gamma_5 && \text{for PS coupling,}\\ &= G\gamma_5\gamma_\mu\ell^\mu && \text{for PV coupling,}\end{aligned} \qquad (2)$$

with g and G the coupling constans. The four momenta p_1, p_2, k, ℓ and D are defined in Fig. 1. Decomposing $\gamma k + M$ in terms of the positive and negative energy spinors[3], allows us to write Eq. (1) as

$$M_{12}\propto \bar{u}(p_2)\Lambda u(\bar{k})\psi_+ + \bar{u}(p_2)\Lambda v(p_1)\psi_-, \qquad (3)$$

with $\bar{k} = k + \delta(p_1 + k)$ and $\delta = (M^2 - k^2)/M_D^2$. Here M_D is the deuteron mass, while ψ_+, and ψ_- are the relativistic deuteron wave functions, defined in the deuteron rest frame, and given by

$$\psi_+ = \phi_0 P_0 + \phi_2 P_2 \;\; ; \;\; \psi_- = \phi_{10}P_{10} + \phi_{11}P_{11} \qquad (4)$$

where ϕ_0, ϕ_2, ϕ_{10}, and ϕ_{11} denote the $^3S_1$, $^3D_1$, $^1P_1$ and $^3P_1$ deuteron wave function respectively[3], while the P's are the corresponding projection operator. The first term on the r.h.s. of Eq.(3) is the contribution from the positive energy spinor, and reduces to the usual non-relativistic result, while the second term is purely relativistic. We now examine these two contributions to the amplitude separately: (i) the positive energy part: In this case the πNN vertex is given by;

$$\begin{aligned}\bar{u}_{\nu_2}(p_2)\Lambda u_{\nu_1}(p_1) &= \chi^\dagger_{\nu_2} g\vec{\sigma}\cdot(\tilde{p}_2 + \tilde{p}_1)\chi_{\nu_1}, \quad \text{for PS}\\ &= \chi^\dagger_{\nu_2} 2MG\vec{\sigma}\cdot\{(\tilde{P}_2 + \tilde{P}_1) + ((\ell_0 + E_{P_1} - E_{P_2})/2M)(\tilde{P}_1 - \tilde{P}_2)\}\chi_{\nu_1}, \quad \text{for PV}\end{aligned} \qquad (5)$$

where $\tilde{p} = \vec{p}\,/\,(E_p + M)$. In the above we have dropped the normalization factors. The first term on the r.h.s. of Eq.(5) is the same for the two couplings provided $G = g/2M$. Using this equivelence we get in the static limit $\Lambda_{PS}\rightarrow(g/2M)\vec{\sigma}\cdot\vec{\nabla}_{\pi}$ and $\Lambda_{PV}\rightarrow(g/2M)\vec{\sigma}\cdot[\vec{\nabla}_{\pi} + (\ell_0/2M)(\vec{\nabla}_{p_1} - \vec{\nabla}_{p_2})]$. The latter being the well known Galilean invariant vertex.

(ii) The negative energy part: In this case the πNN vertex is

$$\bar{u}_{\nu_2}(P_2)\ \Lambda\ u_{\nu_1}(p_1) = \chi^{\dagger}_{\nu_2} g\{ - 1 + \tilde{p}_2\cdot\tilde{p}_1 + i\vec{\sigma}\cdot(\tilde{p}_2 \times \tilde{p}_1)\}\chi_{\nu_1} \quad \text{for PS,}$$

$$=\chi^{\dagger}_{\nu_2} G\{- E_B + [\tilde{p}_2\ \tilde{p}_1 + i\vec{\sigma}\cdot(\tilde{p}_2 \times \tilde{p}_1)](4M - E_B)\}\chi_{\nu_1} \quad \text{for PV,} \qquad (6)$$

where E_B is the binding energy of the deuteron. If we are to set an equivelence between PS and PV in lowest order, we have to take $G = g/E_B$ which is different that $G = g/2M$ used in (i).

If we employ the equivelence at positive energy i.e. $G = g/2M$ then at threshold for $\pi^+d\rightarrow pp$ we have $\sigma_{PS}/\sigma_{PV} \sim 10^3$, where σ_{PS} and σ_{PV} are the total cross section using PS and PV coupling for the πNN vertex respectively. This result is consistant with earlier results on pion production in nuclei. This large value for σ_{PS}/σ_{PV} is due to two factors. a) The negative energy contribution to the πNN vertex in the PS case goes as g (see Eq.(6)). This is much larger than the PV vertex. b) For low energy pions the value of the momentum $q = |(\vec{P}_1 - \vec{P}_2 + \vec{\ell})/2| \sim 1.85\ \text{fm}^{-1}$ at which we need the deuteron wave function is close to where the S-wave deuteron wave function has its node in momentum space. Thus ϕ_0 is of the same order of magnitude as the D- and P- wave component, i.e. ψ_+ and ψ_- are comparable in magnitude. This allows the negative energy (i.e. the second term in Eq.(3) to be comparable to the first term). This effect may be partially reduced by introducing distortion in the p-p channel.

We thus may conclude that low energy $\pi^+d\rightarrow pp$ is sensitive to relativistic components of the deuteron wave function, and may be used to distinguish between PS and PV coupling for the πNN vertex. Finally, we note that a similar dominance of the relativistic components of the deuteron wave function is present in the evaluation of the circular polarization of γ-rays in np→γd[4].

* work supported by ARGC, and Flinders University research Budget.

REFERENCES:

1. L.D. Miller and H.J. Weber, Phys. Rev. C17, 219 (1978).
2. R. Blankenbecler and L.F. Cook, Phys. Rev. 119, 1745 (1960).
3. E.A. Remler, Nucl. Phy. B42, 56 (1972)
4. S. Morioka and T. Ueda, Contribution to this Conference.

CONTRIBUTOR INDEX

Selected Issues from

Lecture Notes in Mathematics

Lecture Notes in Physics

Vol. 44: R. A. Breuer, Gravitational Perturbation Theory and Synchrotron Radiation. VI, 196 pages. 1975.

Vol. 45: Dynamical Concepts on Scaling Violation and the New Resonances in e^+e^- Annihilation. Edited by B. Humpert. VII, 248 pages. 1976.

Vol. 46: E. J. Flaherty, Hermitian and Kählerian Geometry in Relativity. VIII, 365 pages. 1976.

Vol. 47: Padé Approximants Method and Its Applications to Mechanics. Edited by H. Cabannes. XV, 267 pages. 1976.

Vol. 48: Interplanetary Dust and Zodiacal Light. Proceedings 1975. Edited by H. Elsässer and H. Fechtig. XII, 496 pages. 1976.

Vol. 49: W. G. Harter and C. W. Patterson, A Unitary Calculus for Electronic Orbitals. XII, 144 pages. 1976.

Vol. 50: Group Theoretical Methods in Physics. 4th International Colloquium. Nijmegen 1975. Edited by A. Janner, T. Janssen, and M. Boon. XIII, 629 pages. 1976.

Vol. 51: W. Nörenberg und H. A. Weidenmüller. Introduction to the Theory of Heavy-Ion Collisions. IX, 273 pages. 1976.

Vol. 52: M. Mladjenović, Development of Magnetic β-Ray Spectroscopy. X, 282 pages. 1976.

Vol. 53: D. J. Simms and N. M. J. Woodhouse, Lectures on Geometric Quantization. V, 166 pages. 1976.

Vol. 54: Critical Phenomena. Sitges International School on Statistical Mechanics, June 1976. Edited by J. Brey and R. B. Jones. XI, 383 pages. 1976.

Vol. 55: Nuclear Optical Model Potential. Proceedings 1976. Edited by S. Boffi and G. Passatore. VI, 221 pages. 1976.

Vol. 56: Current Induced Reactions. International Summer Institute, Hamburg 1975. Edited by J. G. Körner, G. Kramer, and D. Schildknecht. V, 553 pages. 1976.

Vol. 57: Physics of Highly Excited States in Solids. Proceedings 1975. Edited by M. Ueta and Y. Nishina. IX, 391 pages. 1976.

Vol. 58: Computing Methods in Applied Sciences. Proceedings 1975. Edited by R. Glowinski and J. L. Lions. VIII, 593 pages. 1976.

Vol. 59: Proceedings of the Fifth International Conference on Numerical Methods in Fluid Dynamics. 1976. Edited by A. I. van de Vooren and P. J. Zandbergen. VII, 459 pages. 1976.

Vol. 60: C. Gruber, A. Hintermann, and D. Merlini, Group Analysis of Classical Lattice Systems. XIV, 326 pages. 1977.

Vol. 61: International School on Electro and Photonuclear Reactions I. Edited by C. Schaerf. VIII, 650 pages. 1977.

Vol. 62: International School on Electro and Photonuclear Reactions II. Edited by C. Schaerf. VIII, 301 pages. 1977.

Vol. 63: V. K. Dobrev et al., Harmonic Analysis on the n-Dimensional Lorentz Group and Its Application to Conformal Quantum Field Theory. X, 280 pages. 1977.

Vol. 64: Waves on Water of Variable Depth. Edited by D. G. Provis and R. Radok. 231 pages. 1977.

Vol. 65: Organic Conductors and Semiconductors. Proceedings 1976. Edited by L. Pál, G. Grüner, A. Jánossy and J. Sólyom. 654 pages. 1977.

Vol. 66: A. H. Völkel, Fields, Particles and Currents. VI, 354 pages. 1977.

Vol. 67: W. Drechsler and M. E. Mayer, Fiber Bundle Techniques in Gauge Theories. X, 248 pages. 1977.

Vol. 68: Y. V. Venkatesh, Energy Methods in Time-Varying System Stability and Instability Analyses. XII, 256 pages. 1977.

Vol. 69: K. Rohlfs, Lectures on Density Wave Theory. VI, 184 pages. 1977.

Vol. 70: Wave Propagation and Underwater Acoustics. Edited by J. Keller and J. Papadakis. VIII. 287 pages. 1977.

Vol. 71: Problems of Stellar Convection. Proceedings 1976. Edited by E. A. Spiegel and J. P. Zahn. VIII, 363 pages. 1977.

Vol. 72: Les instabilités hydrodynamiques en convection libre forcée et mixte. Edité par J. C. Legros et J. K. Platten. X, 202 pages. 1978.

Vol. 73: Invariant Wave Equations. Proceedings 1977. Edited by G. Velo and A. S. Wightman. VI, 416 pages. 1978.

Vol. 74: P. Collet and J.-P. Eckmann, A Renormalization Group Analysis of the Hierarchical Model in Statistical Mechanics. IV, 199 pages. 1978.

Vol. 75: Structure and Mechanisms of Turbulence I. Proceedings 1977. Edited by H. Fiedler. XX, 295 pages. 1978.

Vol. 76: Structure and Mechanisms of Turbulence II. Proceedings 1977. Edited by H. Fiedler. XX, 406 pages. 1978.

Vol. 77: Topics in Quantum Field Theory and Gauge Theories. Proceedings, Salamanca 1977. Edited by J. A. de Azcárraga. X, 378 pages 1978.

Vol. 78: Böhm, The Rigged Hilbert Space and Quantum Mechanics. IX, 70 pages. 1978.

Vol. 79: Group Theoretical Methods in Physics. Proceedings, 1977. Edited by P. Kramer and A. Rieckers. XVIII, 546 pages. 1978.

Vol. 80: Mathematical Problems in Theoretical Physics. Proceedings, 1977. Edited by G. Dell'Antonio, S. Doplicher and G. Jona-Lasinio. VI, 438 pages. 1978.

Vol. 81: MacGregor, The Nature of the Elementary Particle. XXII, 482 pages. 1978.

Vol. 82: Few Body Systems and Nuclear Forces I. Proceedings, 1978. Edited by H. Zingl, M. Haftel and H. Zankel. XIX, 442 pages. 1978.